Roots, Tubers, and Bulb Crop Wastes: Management by Biorefinery Approaches

Ramesh C. Ray
Editor

Roots, Tubers, and Bulb Crop Wastes: Management by Biorefinery Approaches

Springer

Editor
Ramesh C. Ray
Microbiology
ICAR- Central Tuber Crops Research
Institute, Regional Center
Bhubaneswar, Odisha, India

ISBN 978-981-99-8265-3 ISBN 978-981-99-8266-0 (eBook)
https://doi.org/10.1007/978-981-99-8266-0

This Springer imprint is published by the registered company Springer Nature Singapore Pte Ltd.
The registered company address is: 152 Beach Road, #21-01/04 Gateway East, Singapore 189721, Singapore

Paper in this product is recyclable.

Preface

The non-edible portion of roots, tubers, and bulb crops, after processing, such as peels, skins, pulps, rejects, and unmarketable tubers, account for about 20–40% of the total weight of the fresh produce. Tropical tuber crops such as cassava, yams and aroids, and sweet potato generate bagasse, the by-product of the starch processing industry, which contains about 30–50% starch and fibrous materials on a dry weight basis. Similarly, the husk is the leading waste from onion and garlic processing. The amount of waste and by-products of the potato industry is estimated to be around 12–20% of their total production. These wastes pose increasing disposal and potentially severe pollution problems and represent a loss of valuable biomass and nutrients. Most of these wastes are dumped in landfills, used as animal feeds, or burned as an alternative for elimination toots, tubers, and bulb crop wastes (RTBCWs) usually have a composition of sugar, starch, proteins, phenolic phytochemicals, fibers, and minerals. Therefore, they should not be considered as "wastes" but raw materials for other industrial processes in the biorefinery approach. RTBCWs are thus a particular group of solid waste (biomass) that must be characterized to understand the nature of applications as raw materials and propose appropriate methodologies for bioprocessing into value-added commodities. RTBCWs provide conditions amenable for the growth of microorganisms and open up great opportunities for their reuse in fermentation processes to develop value-added bioproducts such as enzymes, single-cell proteins, bio-absorbents, phenolic bioactive compounds, aroma and flavor compounds, organic acids, and biofuels. The book deals with all these aspects of the valorization of RTBCWs into different biocommodities and platform chemicals using primarily fermentation processes.

There are 16 chapters in this book. Chapter 1 is the introductory chapter discussing the industrial applications of wastes, in general with emphasis on RTBCWs. Chapters 2–14 discuss the valorization prospects of the individual crop wastes. Chapter 15 focuses on the life cycle assessment of waste valorization processes of roots and tuber crops taking cassava as a case study. The last chapter (Chap. 16) discusses the perspectives for biorefinery development from RTBCWs and by-products—value addition and circular economy and critically analyzed the techno-economic constraints in scaling up the valorization processes.

The book provides up-to-date information on research and development in the field of valorization of RTBCWs. The editor is immensely thankful to the authors for their full cooperation and timely submission of the manuscripts.

Bhubaneswar, Odisha, India Ramesh C. Ray

Contents

1 **Roots, Tubers, and Bulb Crops Wastes: Residue Utilization for Industrial Biotechnology** . . . 1
Preshanthan Moodley and Cristina Trois

2 **Biovalorization of Potato Peel Waste: An Overview** . . . 19
Mansi Singh, Vibhuti Sharma, and Reena Gupta

3 **Potato Peel Enrichment in Functional Food and Feed** . . . 43
Harsh Kumar, Shivani Guleria, Rajni Dhalaria, Richard Cimler, Vandana Guleria, Ruchi Sharma, Rajni Choudhary, Sivakumar Manickam, and Kamil Kuča

4 **Management of Potato Peel Waste Through Biorefinery Approaches** . . . 65
Omojola Awogbemi and Daramy VandiVon Kallon

5 **Bioprocessing Cassava Bagasse: Part I—Bioproducts and Biochemicals** . . . 85
Nizzy Albert Mariathankam and Kannan Suruli

6 **Bioprocessing of Cassava Bagasse: Part II—Potential for Renewable Biofuels** . . . 113
Yeshona Sewsynker-Sukai and Edson L. Meyer

7 **Bio-Valorization of Sweet Potato Bagasse into Food Additives, Feeds, and Fuels** . . . 133
Ivi Chakraborty, Arup Chattopadhyay, Pinky Maity, Aditi Pahari, Subhradeep Pramanik, Hirak Banerjee, Rajib Nath, and Ramesh C. Ray

8 **Yams and Aroid Crop Waste: Bio Valorization into Bioproducts and Platform Chemicals** . . . 149
Chintha Pradeepika, Deep J. Kalita, C. Visalakshi Chandra, Senthil and Sankar, and K. Hanume Gowda

9 Valorization of Carrot and Turnip Processing Wastes and By-Products 183
Akashdeep Kaur and Gargi Ghoshal

10 Sugar Beet Waste as Substrate for Microbial Production of Food Ingredients 215
Negar Rostami and Kianoush Khosravi-Darani

11 Valorization of Beetroot Waste for Extraction of Natural Dye for Textile and Food Applications 237
Balwinder Singh Sooch, Navdeep Sandhu, Manpreet Kaur Mann, and Ramesh C. Ray

12 Valorization of Jerusalem Artichoke and Its Crop Residues Using Green Technologies 261
Kurappalli Rohil Kumar and Kiran Babu Uppuluri

13 Onion Solid Waste as a Potential Source of Functional Food Ingredients 281
Meenakshi, Yogesh Kumar, Suvarna Bhoj, C. Nickhil, Narashans Alok Sagar, Deepak Kumar, Ranjna Sirohi, and Ayon Tarafdar

14 Biovalorization of Garlic Waste to Produce High Value-Added Products 309
Fatma Kallel and Semia Ellouz Chaabouni

15 Life Cycle Assessment of Valorization of Root and Tuber Crop Wastes for Bio-commodities and Biofuels: Cassava as a Case Study 333
Noé Aguilar-Rivera

16 Prospective for Biorefineries Development from Roots, Tubers, and Bulb Crop Wastes and By-Products: Value Addition and Circular Economy 351
Ramesh C. Ray, Noé Aguilar-Rivera, and B. B. Sooch

About the Editor

Ramesh C. Ray is an agriculture and food microbiologist, author, and editor. He is a former Principal Scientist (Microbiology) and Head of the ICAR-Central Tuber Crops Research Institute (Regional Centre), Bhubaneswar, India. He has more than 40 years of research experience in agriculture, food, and industrial microbiology; published 140 research and review papers in national and international journals, 70 book chapters; edited/co-edited 24 books published by leading Publishers like Springer, CRC, and Elsevier; co-authored three books and received more than 8000 citations. He was a Visiting Professor/Scientist at several universities and research institutes in the USA, Brazil, France, Bangladesh, Indonesia, Vietnam, and Consultant to International Potato Center (Lima, Peru) on sweet potatoes. He is a distinguished fellow of the prestigious National Academy of Agricultural Sciences and National Academy of Biological Sciences, India.

Roots, Tubers, and Bulb Crops Wastes: Residue Utilization for Industrial Biotechnology

1

Preshanthan Moodley and Cristina Trois

Abstract

The escalating global demand for both sustenance and energy has led to intensified agricultural activities, yielding substantial amounts of waste from vegetables, including root, tuber, and bulb crops. This chapter details the utilization of these agricultural residues for industrial biotechnology applications. Since disposal of such waste becomes a mounting concern, an urgent need arises to convert these underutilized resources into valuable commodities. Root, tuber, and bulb crops, spanning a diverse range, including potatoes, cassava, and onions, offer an enticing prospect for biorefineries. This study explores their potential as a sustainable feedstock for biofuel production, particularly in the face of dwindling fossil fuel reserves. As these crops are widely cultivated across the globe, their waste materials present an abundant and accessible resource for bioconversion processes. Additionally, this chapter presents the viability of transforming these residues into bio-plastics, and other high-value chemicals. Through innovative approaches such as enzymatic hydrolysis, fermentation, and thermal conversion, these crop wastes can be harnessed to generate bioethanol, biogas, and other bioproducts.

Keywords

Biofuel · Bioplastic · Biorefinery · Cassava waste · Potato waste

P. Moodley (✉) · C. Trois
SARChI Chair Waste and Climate Change, School of Engineering, University of KwaZulu-Natal, Durban, South Africa

R. C. Ray (ed.), *Roots, Tubers, and Bulb Crop Wastes: Management by Biorefinery Approaches*, https://doi.org/10.1007/978-981-99-8266-0_1

Abbreviations

COD	Chemical oxygen demand
CPP	Cassava pulp and peel
CWP	Cassava waste pulp
LA	Lactic acid
OWs	Onion wastes
PHAs	Polyhydroxy alkanoate
PHB	Polyhydroxy butyrate
PLA	Polylactic acid
PPWs	Potato peel waste
SBRs	Sequencing batch reactors
SSCF	Simultaneous saccharification and co-fermentation
SSF	Simultaneous saccharification and fermentation
SSTF	Simultaneous saccharification and two-step fermentation
VFAs	Volatile fatty acids
VS	Volatile solids

1.1 Introduction

A significant part of human nourishment comes from vegetables, roots, and tubers, which are also plentiful and relatively inexpensive sources of vitamins and minerals. They serve as a vital source for maintaining good health and offer protection from several degenerative disorders. Some significant roots, tubers, and bulb crops are the potato, sweet potato, radish, garlic, leeks, carrot, beet, and yam among many others, as indicated in Fig. 1.1. After harvest, the product is kept in locally made containers, in open warehouses, or bamboo baskets for selling and consumption all year long (Arutselvan et al. 2023). Like any other vegetables, roots, tubers, and bulbs are perishable but some crops such as yams, taro, and elephant foot yams can be stored for a longer period after adequate curing and suberization (Ray and Ravi 2005; Ray and Nedunchezhiyan 2012; Arutselvan et al. 2023).

Crops with starchy roots, tubers, rhizomes, corms, or stems are known as root and tuber crops. They are mostly used to produce starch, alcohol, and fermented drinks like beer, as well as human food (either raw or processed), animal feed, and starch. Root and tuber crops provide a significant source of food and nutrition for many of the poorest farmers and households in developing countries (Scott et al. 2000). Further, 737 million tons of root and tuber crops, including potatoes, sweet potatoes, cassava, and yams, are produced yearly on over 54 million ha of land globally. In terms of yearly amount of production, the potato, sweet potato, cassava, and yam are some of the most significant food crops globally. Cassava, sweet potatoes, and potatoes are three of the top ten food crops grown in developing nations.

Fig. 1.1 Common crops in the tuber, bulb, and root crop categories

Due to poor agricultural storage techniques for protecting food against microbes-induced spoilages, many of these crops incur significant post-harvest losses (Kana et al. 2012). Because they can withstand drought and have a large harvest window, these crops can serve as famine and pandemic reserves (Arutselvan et al. 2023). The most significant source of output loss, however, and a major factor in the failure of long-term root tuber storage is post-harvest deterioration brought on by microbial invasion of the tubers. The area where the tubers are separated from the stems at harvest, the broken root tips, and other natural openings and cracks on the tuber's surface sustained during harvesting, transit, or storage are likely points of entry for pathogenic organisms, including fungi (Ray and Ravi 2005; Okigbo 2009). When food is handled improperly during harvest, transportation, or storage, bruises are left on the product that might subsequently become contaminated by adhering dirt or airborne propagules that eventually colonize such abrasions to induce rot (Okigbo and Nmeka 2005). Because of the warm and humid environment in the tropics, postharvest degradation of these crops during storage causes substantial issues and results in significant losses. Various sources claim that rotting typically begins in the field and advances in storage (Okigbo and Ikediugwu 2000). Because of flaws in the storage facility, unsanitary store conditions cause produce to absorb moisture, which promotes the growth of mold and bacteria.

1.2 Residue Composition

The residue compositions of these crops are discussed in the following sections.

1.2.1 Root Crops

Root crops include crops such as sugar beets, carrots, and radishes, among many others. Sugar beets account for 16% of global sugar production (FAO 2019). The global production of sugar beet in 2017 exceeded 300 million tons. Currently, around 70% of the world's output is attributed to Europe. In 2017, France and Germany collectively produced more than 68 Mt of sugar beet, or more than 22% of global output (FAO 2018). The European rule for sugar quotations and minimum prices was removed in 2017, which significantly enhanced competition and resulted in a decrease in pricing (Hunt 2019). In light of this, it is anticipated that sugar producers will search for fresh business ideas to diversify the sugar beet industry's revenue (Tomaszewska et al. 2018). Using hot water, sucrose is traditionally produced from sugar beets. After that, the juice is distilled, condensed, and crystallized. The residual pulp is cleaned, condensed, and crystallized before being used as animal feed or biogas. The residual pulp is used to make biogas or animal feed. Typically, molasses is used to produce ethanol, as animal feed, or as a production medium. The molasses is often used to produce ethanol, animal feed, or as a substrate for the growth of yeast biomass (Tomaszewska et al. 2018).

1.2.2 Tuber Crops

Starch, protein, fat, and lignocelluloses make up the majority of tuber waste, such as potato peel waste (PPWs). Furthermore, this waste includes significant levels of phenolic chemicals and glycoalkaloids (Soltaninejad et al. 2022). A possible source for the creation of bio-based goods is the lignocellulosic fraction, which makes up most tuber waste (44–65%) and contains cellulose, hemicellulose, and lignin. However, lignocelluloses are resistant to biological degradation due to their stubborn and crystalline structure. To destroy this structural barrier, it is essential to perform a pretreatment phase. The starch fraction, which makes up 16–35%, has a significant potential for being turned into products with additional value. The production of fermentable sugars from starch requires starch hydrolysis since a wide range of microbes cannot directly ferment starch. For biological transformations, lipids and proteins are less effective than carbohydrates. However, they can be used for anaerobic digestion to produce biogas. According to Rodríguez-Martínez et al. (2023), PPWs have a range of 0.6–3.6 mg/g of glycoalkaloids, depending on cultivation conditions, harvest, various types, and storage.

1.2.3 Bulb Crops

Bulb crops include vegetables such as onion and garlic. In many regions of the world, including Asia, North Africa, South Europe, the Middle East, and South and Central America, garlic (*Allium sativum* L.), a member of the onion genus *Allium*, is used extensively for both culinary and medicinal uses. Around 20 million tons of garlic are produced worldwide each year (Lee et al. 2017), and further growth is anticipated due to the therapeutic properties of garlic. In the fiscal year 2018–2019, India produced over 1,100,000 tons of garlic, with Madhya Pradesh, Rajasthan, and Uttar Pradesh coming in first, second, and third positions sharing 62.9%, 16.9%, and 6.6%, respectively. Other significant garlic producing states in India are Gujarat, Punjab, Assam, Maharashtra, Himachal Pradesh, Karnataka, Tamil Nadu, and Chhattisgarh. After onions, garlic is the second-most significant bulb crop grown in India. According to Lee et al. (2017), 25–30% of the solid waste produced by garlic is made up mostly of garlic husk and straw. As a result, over 6 million tons of garlic-related agricultural waste are generated annually, making optimal use of this agricultural waste necessary. The importance of garlic wastes in the bioenergy sector for the manufacture of biofuels or chemicals was underlined in a recently published in-depth review study on the subject (Kallel and Chaabouni 2017). It is projected that garlic consumption will rise significantly in the next few years, producing greater amounts of garlic residues. It should be noted that a significant amount of garlic waste decomposes in landfills, causing damage to the environment and global warming. Onion, a significant agricultural commodity, experienced a global production surge to 97.8 million tons in 2017 due to their recognized medicinal and nutritional benefits. Despite their substantial nutritional worth, onions and their by-products contain abundant fiber and bioactive compounds. Remarkably, around 500,000 tons of discarded onion wastes (OWs) are generated annually in Europe, posing environmental concerns, particularly in Spain, the Netherlands, and the United Kingdom. This trend has elevated the significance of researching innovative upcycling approaches for transforming onion waste into environmentally friendly chemicals, an urgency accentuated by Asia's prominent role, contributing to approximately 63.7% of the world's onion production (Kim et al. 2019).

1.2.4 Residue Disposal

Disposal of root, bulb, and tuber vegetable waste through unsustainable methods, such as burning and landfill disposal, poses significant negative impacts on the environment. Burning vegetable waste releases harmful air pollutants, including particulate matter, volatile organic compounds, and greenhouse gases like carbon dioxide and methane. These emissions contribute to air pollution, smog formation, and climate change, exacerbating respiratory issues and compromising overall air quality. Furthermore, burning releases black carbon, a potent climate forcer that significantly contributes to climate change (Bond et al. 2013).

Landfill disposal of vegetable waste is equally problematic. As this waste decomposes anaerobically in landfills, it produces methane, a greenhouse gas with a global warming potential significantly higher than carbon dioxide. Methane emissions from landfills contribute significantly to climate change, intensifying the greenhouse effect and its associated consequences (IPCC 2014). Additionally, decomposing vegetable waste in landfills generates leachate, a toxic liquid that can seep into groundwater and surface water, contaminating water bodies, and posing health risks to humans and wildlife.

Improper disposal of root, bulb, and tuber waste can also lead to the spread of pests and diseases. When waste is inadequately managed, it can attract insects, rodents, and other vectors, promoting the proliferation of disease-causing pathogens. The spread of pests and diseases can harm agricultural crops, affecting food production and potentially leading to economic losses for farmers and food shortages in affected regions.

Moreover, landfilling and burning vegetable waste result in the wastage of valuable organic matter and nutrients that could otherwise be recycled back into the soil (Ray 2022). The loss of these nutrients from the waste stream leads to a depletion of soil fertility, necessitating higher dependence on synthetic fertilizers to maintain agricultural productivity. The excessive use of chemical fertilizers can lead to nutrient imbalances in the soil, water pollution through nutrient runoff, and negative impacts on aquatic ecosystems.

1.3 Biological Conversion of Crop Residue

The pretreatment processes are briefly discussed.

1.3.1 Pretreatment Processes

Tuber crops, bulb crops, and root crops are essential components of the human diet, serving as major sources of carbohydrates and other nutrients. These plant-based foods share common characteristics in their composition, being lignocellulosic and starchy in nature. Lignocellulosic materials, consisting of cellulose, hemicellulose, and lignin, form the structural framework of these crops, providing strength and rigidity to the plant tissues. However, the presence of lignocellulosic components presents a challenge in utilizing these crops for biochemical processes, such as biofuel production or biorefining, as they are inherently resistant to enzymatic degradation (Ragauskas et al. 2006; Panda et al. 2018). In addition to lignocellulosic components, tuber crops, bulb crops, and root crops contain significant amounts of starch, a polysaccharide that serves as a primary energy reserve in plants. Starch consists of glucose units and can be broken down into fermentable sugars, making it a valuable substrate for bioconversion processes (El Sheikha and Ray 2022). However, the dense structure of starch granules in these crops can also hinder enzymatic access, necessitating pretreatment to improve enzymatic digestibility. Table 1.1 outlines some common crops and their residue composition.

Pretreatment plays a crucial role in maximizing the bioconversion of crop residues, particularly tuber, root, and bulb crop residues, which exhibit a lignocellulosic and starchy composition, as evidenced by the data in Table 1.1. The lignin content of carrot and garlic residues, at 32% and 29% respectively, creates formidable barriers to enzymatic accessibility, limiting efficient hydrolysis of cellulose and

Table 1.1 Lignocellulosic composition of common crop residues

Crop residue	Lignin	Cellulose	Hemicellulose	Starch	Reference
Sugar beet	11	45	25	–	Li et al. (2014)
Carrot	32	52	12	–	Varanasi et al. (2018)
Potato peels	20	30 (polysaccharide)		25	Pathak et al. (2018)
Cassava peels	24	24	9	28	Ona et al. (2019)
Onion	26	45	25	–	Reddy and Rhim (2018)
Garlic	29	50	18	–	Reddy and Rhim (2018)

hemicellulose. To address this challenge, pretreatment methods have been explored, such as acid or alkali treatments, which effectively degrade lignin, thereby facilitating enzymatic breakdown and improving overall bioconversion efficiency (Alvira et al. 2010; Ragauskas et al. 2006). Furthermore, crop residues with significant starch content, such as cassava peels (28%), possess great potential for bioconversion. However, the dense structure of starch granules restricts enzymatic attack, hindering optimal sugar release. Considering this, pretreatment techniques have been explored to disrupt starch granules and enhance enzymatic accessibility, leading to improved starch hydrolysis and enhanced sugar yields (Alvira et al. 2010). The utilization of potato peels, onion residues, and sweet potato peels as feedstocks for bioconversion also warrants consideration. Their lignocellulosic nature highlights the significance of pretreatment in improving bioconversion efficiency. Pretreatment strategies have been extensively investigated to break down lignocellulosic barriers, enhancing the susceptibility of these residues to enzymatic degradation, and ultimately leading to increased bioconversion yields (Alvira et al. 2010). Table 1.2 presents the main findings of recent studies employing pretreatment to enhance sugar yield from crop residues.

Fermentable sugar plays a pivotal role in biorefinery and biofuel production processes due to its ability to serve as a versatile and sustainable feedstock. The conversion of fermentable sugars into biofuels and other value-added products is primarily accomplished through microbial fermentation. During this process, microorganisms, such as yeast, bacteria, and fungi, metabolize these sugars to produce a variety of bio-based products, including ethanol, butanol, organic acids, and other valuable biochemicals. Fermentable sugars are obtained from various biomass sources, such as lignocellulosic materials. The biorefinery concept aims to convert these raw materials into a range of biofuels and bioproducts, thereby promoting sustainable and eco-friendly alternatives to conventional fossil fuels. In biofuel production, fermentable sugars are first extracted from the biomass feedstock through pretreatment and hydrolysis processes, as outlined in Table 1.1. Subsequently, these sugars are fed to microorganisms capable of fermenting them into specific products. For instance, *Saccharomyces cerevisiae* is used to convert sugars into ethanol, while *Clostridium* species can produce butanol and acetone through acetone-butanol-ethanol fermentation. Moreover, advancements in metabolic engineering have enabled the manipulation of microorganisms to produce a wider array of valuable chemicals, such as succinic acid and lactic acid (LA), from fermentable

Table 1.2 Some important findings related to pretreatments of roots, tubers, and bulb crop residues

Crop residue	Pretreatment conditions	Main finding	Reference
Potato peel waste (PPWs)	Organosolv 50% (v/v) ethanol at 120 °C	High hydrolysate yield, 539.8 g glucose/kg PPWs	Soltaninejad et al. (2022)
sPPWs	1% H_2SO_4, 56 h at 68 °C	Highest yield of 46.17 g/L glucose obtained	Malakar et al. (2020)
Sugar beet residue	Aqueous ammonia treatment at 80 °C for 6 h	Maximum yield of 487.8 mg/g reducing sugar was obtained, 3 times higher than the control	Li et al. (2018)
Sugar beet residue	Steam at 8 bar for 30 min	60%, 11%, and 8% arabinose, galacturonic and glucose releases respectively	Hamley-Bennett et al. (2016)
Cassava residue	Combination of organic acids (0.33 wt% acetic acid + 0.45 wt % L lactic acid+ 0.20 wt% formic acid) at 180 °C for 60 min	Yield of C5 sugars after pretreatment and C6 sugar after enzymatic hydrolysis were 89.06% and 78.56%, respectively	Lyu et al. (2019)
Cassava residue	1% NaOH at 120 °C for 30 min	Maximum 670.58 ± 10.13 mg/g reducing sugar yield with 81.25% ± 3.20% saccharification efficiency	Papathoti et al. (2021)
Carrot residue	121 °C for 15 min followed by a combination enzyme treatment with 0.15 g/g CellicCTec 2 and 0.3 g/g viscozyme	87.2% total sugar recovery (glucose, arabinose, and galactose)	López-Linares et al. (2023)
Carrot residue	Electric field pretreatment, with intensity of 1250 V/cm, 10 pulses 0	75.32% sugar yield	Dastangoo et al. (2020)
Onion waste	Popping technique at a pressure of 3 kg/cm^2	74.7% total sugar recovery	Cho et al. (2021)

sugars (Panda et al. 2016). These bio-based chemicals can replace petroleum-derived counterparts in various industrial applications, reducing greenhouse gas emissions and promoting sustainable development.

The results from Table 1.2 indicate the vast potential application of root, tuber, and bulb crop residues as indicated by the sugar yields. Potato peel residue has long been considered a potential reservoir of fermentable sugar owing to its high cellulose and starch content. Malakar et al. (2020) examined dilute acid pretreatment to enhance sugar yield from PPWs, employing 1% sulfuric acid at 68 °C for 56 h resulting in a high glucose yield of 46.17 g/L. More recently, Soltaninejad et al. (2022) explored organosolv pretreatment on the same substrate. These authors employed a 50% ethanol solution at 120 °C and reported a yield of 539.8 g glucose/kg PPWs. Cassava residue, another starchy crop, was pretreated with 1% NaOH at 120 °C for 30 min, and 70.58 mg/g reducing sugar yield was obtained with

81.25% saccharification efficiency (Papathoti et al. 2021). This high saccharification efficiency is a highly desirable outcome since it indicates the pretreated biomass is amenable to either enzymatic hydrolysis or direct microbial conversion. Upon successful pretreatment, the pretreated hydrolysate or biomass can then be directed toward fermentative processes.

1.4 Application of Roots, Tubers, and Bulb Crops Waste for Industrial Biotechnology

The various industrial applications of roots, tubers, and bulb crop wastes such as biofuels, biopolymers, etc. are discussed in the following sections.

1.4.1 Biofuels

1.4.1.1 Potato Waste

Several studies have reported the production of different biofuels from crop waste. For instance, Chohan et al. (2020) reported on the valorization of PPWs for bioethanol production in a simultaneous saccharification and fermentation (SSF) system. These authors reported a maximum bioethanol concentration and yield of 22.54 g/L and 0.32 g/g, respectively under optimal process conditions of 40 °C (temperature), 5.78 (pH), and 12.25% w/v (solid loading). An SSF system has the advantage of using a single vessel for both pretreatment and fermentation which can potentially reduce process costs. Additionally, conditions in an SSF vessel have been shown to enhance biomass digestibility. A similar study exploring SSF production of bioethanol from PPW reported a maximum ethanol concentration of 116.5 g/L and a yield of 80.4% after 72 h of fermentation (Rodríguez-Martínez et al. 2023). Biohydrogen production has also been extensively investigated. A recent study (Lee et al. 2023) explored PPWs as a substrate for hydrogen (H_2) production. Two strains of bacteria, *Thermococcus onnurineus* NA1 parent strain and WG-100T were examined. The potato peels were dried and ground before being added to serum vials for the experiment. Both strains were able to produce hydrogen from the PPWs, but WG-100T outperformed the parent strain significantly. WG-100T showed higher rates of H_2 production (2.83 vs. 0.97 mmol/L/h) and higher H_2 productivity (1.29 vs. 0.37 mmol/L/h). The results indicate that WG-100T, which had been acclimated to glucose, exhibited improved abilities to consume carbohydrate-containing food waste, leading to enhanced H_2 production from potato peels. This suggests the potential of PPWs as a suitable candidate for biohydrogen production.

1.4.1.2 Cassava Waste

A recent study explored the use of cassava waste pulps (CWP) to produce biofuel, specifically biohydrogen, through dark fermentation (Balbin et al. 2022). Acid hydrolysis was utilized to prepare the CWP, and heat-shocked cow dung was selected as the inoculum. Results exhibited that 99.89% of the total biogas produced through dark fermentation was biohydrogen. At 6.09 pH and 83.24 h hydraulic retention time, a kilogram of fresh CWP produced 12.39 ± 0.8 L of biohydrogen. This study demonstrated that underutilized CWP from starch-producing industries can be channeled to generate a sustainable biofuel, providing a potential solution to the improper disposal of organic wastes and reducing fossil-fuel-related emissions. Another study evaluated the potential of cassava stems as an immobilizing agent for inoculum in biohydrogen production (Bolonhesi et al. 2023). Hydrogen production was assessed in an anaerobic sequencing batch biofilm reactor with maximum volumetric hydrogen productivity and yield of 1.48 LH_2/L/day and 1.98 $molH_2$/kg carbohydrates, respectively. The results demonstrated that cassava stem can be utilized as an alternative support material and co-substrate for efficient biohydrogen production through dark fermentation. Bioethanol production has also been considered. Efeovbokhan et al. (2019) investigated the production of bioethanol from three varieties of cassava pulp and peel (CPP) using acid and biological hydrolysis. Acid hydrolysis yielded 110.7 g/L, 100.4 g/L, and 96.7 g/L of reducing sugar from three cassava varieties, while 98.9 g/L was obtained from cassava peel. The highest ethanol yields were 54.8% and 33.1%, respectively, obtained from a heat-pretreated variety and cassava peel, indicating a more promising approach to managing cassava wastes through bioethanol production. Another study explored the enzymatic saccharification of soaking-assisted and thermal pretreated cassava peel waste, focusing on optimizing fermentable sugar yield for bioethanol production using response surface methodology (Aruwajoye et al. 2020). This study investigated the effect of substrate loading, α-amylase concentration, amyloglucosidase concentration, and cellulase concentration on sugar yield. The optimized conditions resulted in a fermentable sugar yield of 0.58 g/g and saccharification efficiency of 78.66%. This sugar was channeled to fermentation and gave an ethanol yield of 0.53 g/g, demonstrating the potential of cassava peel waste for bioethanol production with *S. cerevisiae* fermentation.

1.4.1.3 Sugar Beet Waste

Sugar beet waste is rich in fermentable sugars that can be directed toward biofuel processes. Vučurović et al. (2019) compared sugar beet molasses and thick juice as raw materials for bioethanol production. Immobilized *S. cerevisiae* in calcium alginate beads showed better fermentative performance, resulting in maximum ethanol concentrations of 84.6 g/L and 109.5 g/L for molasses and thick juice, respectively, at initial sugar concentrations of 200 g/L and 250 g/L. The highest ethanol yields of 31.7 L/100 kg of molasses and 37.6 L/100 kg of thick juice were obtained at an initial sugar concentration of 175 g/L, indicating the advantage of immobilized yeast for efficient high-gravity bioethanol production from sugar beet. In another study, Cieciura-Włoch et al. (2020b) demonstrated the potential of sugar beet pulp for dark fermentative hydrogen production. The hydrogen yield achieved ranged from 17 to 37 cm^3/g VS (volatile solids), showcasing sugar beets' capability as a feedstock for biohydrogen. Another study by the same authors explored the impact of different iron compounds on hydrogen production via mesophilic dark fermentation of hydrolyzed sugar beet pulp (Cieciura-Włoch et al. 2020a). The batch tests revealed that the addition of 0.1 g Fe_2O_3/dm^3 resulted in a maximum hydrogen yield of over 200 dm^3 H_2/kg VS, twice greater than the control. Additionally, a methane yield of 19.4 dm^3 CH_4/kg VS was achieved indicating the versatility of this feedstock to produce multiple biofuels in a single process.

1.4.1.4 Carrot Waste

Khoshkho et al. (2022) explored the utilization of dried carrot pulp as a raw material for bioethanol production, employing *S. cerevisiae* as the inoculum. The results indicated that the highest bioethanol yield (10.3 mL or 40.63 g/L) was obtained using a combination of 50 mL inoculum, 150 mL of water, and 10 g of dried feedstock. This highlights the potential of dried carrot pulp as a valuable source for producing bioethanol. Alternatively, de Vrije et al. (2010) explored biohydrogen production from carrot pulp hydrolysate and untreated carrot pulp using extreme thermophiles *Caldicellulosiruptor saccharolyticus* and *Thermotoga neapolitana*. The hydrogen yields of both thermophiles ranged from 2.7 to 2.8 mol H_2/mol hexose with 10 g/L sugars from carrot pulp hydrolysate. Enzymatic hydrolysis of the polysaccharide fraction prior to fermentation increased the hydrogen yield by almost 10% to 2.3 g/kg of hydrolyzed carrot pulp, further suggesting the need for pretreatment of biomass prior to fermentation.

1.4.2 Bioproducts and Biomaterials

1.4.2.1 Bioplastic

Bio-based plastics sourced from renewable materials have garnered significant attention as a promising solution to address disposal issues associated with conventional and petroleum-based polymers, offering reduced environmental impact. Notable bioplastics include thermoplastic starch, cellulose esters and ethers, polylactic acid (PLA), polyhydroxyalkanoates (PHAs), and bio-derived polyethylene. Among PHAs, polyhydroxybutyrate (PHB) is the most prevalent and extensively studied due to its biodegradability and biocompatibility, making it suitable for various applications in the agriculture, marine, and medical sectors. Although PHB is currently more expensive than conventional plastics, efforts are being made to lower production costs by employing blending strategies with other biodegradable polymers and incorporating fillers such as PLA, bast fiber, and agricultural residues to enhance its properties and biodegradability (Wei et al. 2015). To this end, several studies have explored the use of crop residue for the production of bioplastics. For instance, Liang et al. (2015) explored an alternative method for producing LA, a crucial feedstock for bioplastic production from PPWs through anaerobic fermentation. The researchers employed a statistical design of experiments using sequencing batch reactors (SBRs) with gelatinized PPWs at different solids content and retention times to optimize LA yield and productivity. The results showed that the maximum LA production yield of 0.25 g/g PPWs and highest productivity of 125 mg/g/day were achieved in the 0.8 L SBRs, while the scale-up SBR trial at 3 L scale resulted in a yield of 0.14 g/g PPWs and productivity of 138 mg/g/day with a 1-day solids retention time. This innovative approach demonstrates the potential of utilizing potato waste as a sustainable resource for bioplastic production. A different study investigated the application of a biorefinery approach to convert cassava pulp, an abundant agricultural residue, into valuable products (Prasertsilp et al. 2023). Microwave-assisted hydrolysis of cassava pulp was investigated, and the optimized conditions yielded a high fermentable sugar content of 88.1 g/L, mainly composed of glucose. The resulting crude hydrolysate was utilized for PHB production using *Cupriavidus necator* strain A-04, achieving a PHB content of 66.8% and a yield (YP/S) of 0.35 g/g. The study demonstrates a promising method for converting cassava waste into valuable PHB, providing a sustainable management solution for agricultural residues. Another study focused on the biosynthesis of PHB in the extremely halophilic archaeon *Haloarcula* sp. TG1 (Kurt-Kızıldoğan et al. 2023). These researchers optimized PHB production in mineral salt medium cultures with 5 M NaCl and 4% glucose, resulting in a PHB content of 34.6%. Thereafter, the study explored the use of sugar beet pulp treated with recombinant endoglucanase, as low-cost carbon sources, leading to the highest PHB content of 45.6%.

1.4.2.2 Organic Acids

Kim et al. (2019) explored the potential of producing acetic acid, a green chemical, from carbohydrate-rich OWs as an environmentally friendly alternative to natural gas-based feedstocks. The simultaneous saccharification and two-step fermentation (SSTF) process demonstrated greater efficiency in acetic acid production from OWs compared to the simultaneous saccharification and co-fermentation (SSCF) process. SSTF achieved a notable acetic acid concentration of 19.3 g/L and an impressive conversion yield of 90.5% from OWs, highlighting the possibility of sustainable and efficient acetic acid production through this biorefining approach. PPWs were also examined as a substrate for anaerobic digestion to produce volatile fatty acids (VFAs) (Lu et al. 2020). The study aimed to assess the impact of various pH levels (pH 5.0, pH 7.0, pH 11.0, and uncontrolled pH) on VFA concentration, composition, intermediate products, and metabolic state. The results revealed that the highest total VFA production was observed at pH 7.0 [41.9 g COD (chemical oxygen demand)/L and 632.2 mg COD/g VS fed], with uncontrolled pH following closely behind. The acidic pH favored butyric acid as the primary product, while alkaline pH led to the dominance of acetic acid, indicating a mixed-acid type acidogenic fermentation at pH 7.0.

1.4.2.3 Other Compounds

Manimaran et al. (2023) explored the extraction of inulin, a polysaccharide rich in dietary fiber, from garlic using water as a solvent. The study optimized the extraction process by varying temperature, time, and material ratio and employed response surface methodology with the central composite design. The results showed that the optimal conditions for maximum yield (44.73 g/L) were achieved at 77.35 °C, 3.5 h, and a material ratio of 5 mL/g. The study highlights an effective method for obtaining inulin from garlic, offering potential alternatives for industries seeking inulin sources for nutrition and pharmaceutical applications. Another study investigated the sequential production of levulinic acid from cassava roots through an integrated biorefinery process (Phachwisoot et al. 2021). Initially, the cassava roots underwent pretreatment in KOH solution at 120 °C for 1 h. Subsequently, the valorization of cassava roots to levulinic acid was achieved via a catalytic hydrothermal process in 0.1–0.7 M HCl solution at temperatures ranging from 150 to 200 °C for 1–3 h. The experimental conditions that yielded the maximum levulinic acid content, 19.62 wt%, were found to be at 200 °C for 1 h in 0.4 M HCl. The successful conversion of cassava roots to levulinic acid showcases the potential of cassava rhizome as a valuable feedstock to produce this valuable chemical.

1.5 Future Prospects

The utilization of various agricultural and food waste feedstocks, including root, tuber, and bulb crops, for biotechnological applications in the industrial sector, holds great promise in the quest for sustainable and environmentally friendly practices. Over the years, extensive research has been conducted to explore the potential of these feedstocks in producing valuable products such as biofuels, bioplastics, and biochemicals, thus contributing to the promotion of a circular economy.

One of the key feedstocks that has shown tremendous potential is cassava waste, a root crop. With its abundance as a byproduct of cassava processing, cassava waste, particularly cassava peels, has emerged as a suitable candidate for bioethanol production. Through enzymatic saccharification and fermentation processes, high yields of fermentable sugars have been achieved, paving the way for efficient bioethanol production. Additionally, cassava pulp has been harnessed to produce biohydrogen, a clean and sustainable energy carrier. The successful application of cassava waste in bioethanol and biohydrogen production highlights its valuable contribution to the renewable energy sector. Another promising feedstock is sugar beet pulp, a tuber crop, which has been extensively investigated for bioplastic production. With the growing demand for eco-friendly alternatives to conventional plastics, sugar beet pulp provides an attractive solution due to its high carbohydrate content. By employing enzymatic hydrolysis and fermentation processes, bioplastics like PHA have been synthesized from sugar beet pulp, offering a sustainable alternative to petroleum-based plastics.

Inulin-rich waste, such as garlic, classified as a bulb crop, also presents an exciting opportunity in the biorefining arena. Inulin extraction from garlic waste has been achieved through solvent-phase extraction methods, providing a renewable source of this valuable polysaccharide with various applications in the nutrition and pharmaceutical industries. This exemplifies the potential for utilizing underutilized agricultural residues to obtain high-value products. Additionally, the sequential production of levulinic acid from cassava roots, classified as a root crop, through an integrated biorefinery process showcases the versatility of these feedstocks. The production of levulinic acid, an important platform chemical, from cassava roots not only demonstrates the efficient valorization of this feedstock but also opens up possibilities for diverse downstream applications in the chemical industry.

However, as research moves forward, several areas warrant further exploration and improvement to enhance the integration of these feedstocks into industrial biotechnology. Optimization of process conditions, such as temperature, pH, and reaction time, is crucial to maximize product yields and improve overall efficiency. Furthermore, the development of robust and cost-effective pretreatment methods will be essential in enhancing the accessibility of carbohydrates and other valuable components in these feedstocks. Moreover, research efforts should be directed toward the development of innovative and sustainable technologies to harness the full potential of these feedstocks. Biotechnological advancements, such as metabolic engineering and synthetic biology, hold potential in tailoring microbial strains to efficiently utilize these feedstocks and produce desired bio-based products. Additionally, the integration of biorefinery processes, where multiple products are simultaneously derived from a single feedstock, could enhance the overall economic viability and sustainability of these ventures.

The utilization of agricultural and food waste feedstocks, specifically root, tuber, and bulb crops, in industrial biotechnology represents a promising pathway toward a more sustainable and resource-efficient future. The successful applications of these crops in bioprocesses showcase their immense potential as valuable resources. By capitalizing on these feedstocks and implementing innovative technologies, the industrial sector can transition towards greener practices, reducing environmental impacts and promoting a circular economy. Continued research and development efforts in optimizing processes and exploring new avenues for product diversification will undoubtedly accelerate the integration of these feedstocks in industrial biotechnology, fostering a more sustainable and prosperous future.

References

Alvira P, Tomás-Pejó E, Ballesteros M, Negro MJ (2010) Pretreatment technologies for an efficient bioethanol production process based on enzymatic hydrolysis: a review. Bioresour Technol 101(13):4851–4861

Arutselvan R, Raja K, Pati K et al (2023) Tuber crops and their potential in food and nutritional security. In: Anal AK (ed) Pandemics and innovative food systems. CRC Press, pp 122–136

Aruwajoye GS, Faloye FD, Kana EG (2020) Process optimisation of enzymatic saccharification of soaking assisted and thermal pretreated cassava peels waste for bioethanol production. Waste Biomass Valor 11:2409–2420

Balbin AJFT, Gonzales HA, Bornillo LAM, Tagupa JHA, Jaro RH, Ido AL, Arazo RO (2022) High biohydrogen yield from fresh cassava waste pulps via gas dissolution in a dark fermentative environment. Cleaner Waste Syst 3:100039

Bolonhesi IBDTM, Andreani CL, de Melo MR, Gomes SD, Lopes DD (2023) Biomass immobilization in hydrolyzed lignocellulosic material can enhance biohydrogen production from cassava residues. Biochem Eng J 190:108725

Bond TC, Doherty SJ, Fahey DW et al (2013) Bounding the role of black carbon in the climate system: a scientific assessment. J Geophys Res Atmos 118(11):5380–5552

Cho EJ, Choi Y-S, Bae H-J (2021) Bioconversion of onion waste to valuable biosugar as an alternative feed source for honey bees. Waste Biomass Valor 12(6229):4503. https://doi.org/10.1007/s12649-021-01344-x

Chohan NA, Aruwajoye GS, Sewsynker-Sukai Y, Kana EG (2020) Valorisation of potato peel wastes for bioethanol production using simultaneous saccharification and fermentation: process optimization and kinetic assessment. Renew Energy 146:1031–1040

Cieciura-Włoch W, Borowski S, Domański J (2020a) Dark fermentative hydrogen production from hydrolyzed sugar beet pulp improved by iron addition. Bioresour Technol 314:123713

Cieciura-Włoch W, Borowski S, Otlewska A (2020b) Biohydrogen production from fruit and vegetable waste, sugar beet pulp and corn silage *via* dark fermentation. Renew Energy 153: 1226–1237

Dastangoo S, Hamed Mosavian MT, Yeganehzad S (2020) Optimization of pulsed electric field conditions for sugar extraction from carrots. Food Sci Nutr 8(4):2025–2034

De Vrije T, Budde MA, Lips SJ, Bakker RR, Mars AE, Claassen PA (2010) Hydrogen production from carrot pulp by the extreme thermophiles *Caldicellulosiruptor saccharolyticus* and *Thermotogane apolitana*. Int J Hydrog Energy 35(24):13206–13213

Efeovbokhan VE, Egwari L, Alagbe EE, Adeyemi JT, Taiwo OS (2019) Production of bioethanol from hybrid cassava pulp and peel using microbial and acid hydrolysis. Bioresources 14(2): 2596–2609

El Sheikha AF, Ray RC (2022) Bioprocessing of horticultural wastes by solid-state fermentation into value-added/innovative bioproducts: a review. Food Rev Int 39:3009. https://doi.org/10.1080/87559129.2021.2004161

FAO (Food and Agriculture Organization) of the United Nations Sugarbeet (2019). http://www.fao.org/land-water/databases-and-software/crop-information/sugarbeet/en/

FAO (Food and Agriculture Organization) of the United Nations: Crops (2018). http://www.fao.org/faostat/en/#data/QC/visualize

Hamley-Bennett C, Lye GJ, Leak DJ (2016) Selective fractionation of sugar beet pulp for release of fermentation and chemical feedstocks; optimisation of thermo-chemical pre-treatment. Bioresour Technol 209:259–264

Hunt N (2019) EU Sugar companies struggle to survive as prices plunge post-quotas. https://www.reuters.com/article/us-europe-sugar/eu-sugar-companies-struggle-to-survive-as-prices-plunge-post-quotas-idUSKCN1HA1K5

IPCC (Inter-governmental Panel on Climate Change) (2014) Climate change 2014: synthesis report. Contribution of working groups I, II, and III to the fifth assessment report of the Intergovernmental Panel on Climate Change. IPCC

Kallel F, Chaabouni SE (2017) Perspective of garlic processing wastes as low-cost substrates for production of high-added value products: a review. Environ Prog Sustain Energy 36:1765–1777

Kana HA, Aliyu IA, Chammang HB (2012) Review on neglected and underutilized root and tuber crops as food security in achieving the millennium development goals in Nigeria. J Agric Vet Sci 4:27–33

Khoshkho SM, Mahdavian M, Karimi F, Karimi-Maleh H, Razaghi P (2022) Production of bioethanol from carrot pulp in the presence of *Saccharomyces cerevisiae* and beet molasses inoculum; a biomass based investigation. Chemosphere 286:131688

Kim HM, Choi IS, Lee S, Yang JE, Jeong SG, Park JH, Ko SH, Hwang IM, Chun HH, Wi SG, Kim JC (2019) Biorefining process of carbohydrate feedstock (agricultural onion waste) to acetic acid. ACS Omega 4(27):22438–22444

Kurt-Kızıldoğan A, Türe E, Okay S, Otur Ç (2023) Improved production of poly (3-hydroxybutyrate) by extremely halophilic archaeon *Haloarcula* sp. TG1 by utilization of rCKT3eng-treated sugar beet pulp. Biomass Convers Biorefinery 13:10911–10921

Lee YH, Kim YI, Oh YK, Ahmadi F, Kwak WS (2017) Yield survey and nutritional evaluation of garlic stalk for ruminant feed. J Anim Sci Technol 59:22

Lee SH, Lee S, Lee SM et al (2023) Biohydrogen production from food waste using glucose-adapted hyperthermophilic archaeon. Waste Biomass Valor 14:2923. https://doi.org/10.1007/s12649-023-02049-z

Li M, Wang LJ, Li D, Cheng YL, Adhikari B (2014) Preparation and characterization of cellulose nanofibers from de-pectinated sugar beet pulp. Carbohydr Polym 102:136–143

Li G, Sun Y, Guo W, Yuan L (2018) Comparison of various pretreatment strategies and their effect on chemistry and structure of sugar beet pulp. J Clean Prod 181:217–223

Liang S, McDonald AG, Coats ER (2015) Lactic acid production from potato peel waste by anaerobic sequencing batch fermentation using undefined mixed culture. Waste Manag 45: 51–56

López-Linares JC, Coca M, Plaza PE, Lucas S, García-Cubero MT (2023) Waste-to-fuel technologies for the bioconversion of carrot discards into biobutanol. Renew Energy 202: 362–369

Lu Y, Zhang Q, Wang X, Zhou X, Zhu J (2020) Effect of pH on volatile fatty acid production from anaerobic digestion of potato peel waste. Bioresour Technol 316:123851

Lyu H, Zhang J, Zhou J, Lv C, Geng Z (2019) The byproduct-organic acids strengthened pretreatment of cassava straw: optimization and kinetic study. Bioresour Technol 290:121756

Malakar B, Das D, Mohanty K (2020) Optimization of glucose yield from potato and sweet lime peel waste through different pre-treatment techniques along with enzyme assisted hydrolysis towards liquid biofuel. Renew Energy 145:2723–2732

Manimaran DR, Sugapriya G, Mahima S, Sivamani S, Sivarajasekar N, Prabhu SV (2023) Response surface analysis and optimization of inulin extraction from garlic. Biomass Convers Biorefinery 13:10845–10851. https://doi.org/10.1007/s13399-021-01953-5

Okigbo RN (2009) Variation in phytochemical properties of selected fungicidal aqueous extract of some plant leaves in Kogi State, Nigeria. Am Eur J Sustain Agric 3(3):407–409

Okigbo RN, Ikediugwu FEO (2000) Studies on biological control of post-harvest rot in yams (*Dioscorea* spp.) using *Trichoderma viride*. J Phytopathol 148:351–335

Okigbo RN, Nmeka IA (2005) Control of yam tuber rots with leaf extract of *Xylopia aethiopica* and *Zingiber officinale*. Afr J Biotechnol 4(8):804–807

Ona JI, Halling PJ, Ballesteros M (2019) Enzyme hydrolysis of cassava peels: treatment by amylolytic and cellulolytic enzymes. Biocatal Biotransformation 37(2):77–85

Panda SK, Mishra SS, Kayitesi E, Ray RC (2016) Microbial-processing of fruit and vegetable wastes for production of vital enzymes and organic acids: biotechnology and scopes. Environ Res 146:161–172

Panda SK, Ray RC, Mshra SS, Kayitesi K (2018) Microbial processing of fruit and vegetable wastes into potential biocommodities: a review. Crit Rev Biotechnol 38:1–16

Papathoti NK, Laemchiab K, Megavath VS, Keshav PK, Numparditsub P, Le Thanh T, Buensanteai N (2021) Augmented ethanol production from alkali-assisted hydrothermal pretreated cassava peel waste. Energ Source A Recov Util Environ Eff. https://doi.org/10.1080/15567036.2021.1928338

Pathak PD, Mandavgane SA, Puranik NM, Jambhulkar SJ, Kulkarni BD (2018) Valorization of potato peel: a biorefinery approach. Crit Rev Biotechnol 38(2):218–230

Phachwisoot G, Nakason K, Chanthad C, Khemthong P, Kraithong W, Youngjan S, Panyapinyopol B (2021) Sequential production of levulinic acid and supercapacitor electrode materials from cassava rhizome through an integrated biorefinery process. ACS Sustain Chem Eng 9(23): 7824–7836

Prasertsilp P, Pattaragulwanit K, Kim BS, Napathorn SC (2023) Microwave-assisted cassava pulp hydrolysis as food waste biorefinery for biodegradable polyhydroxybutyrate production. Front Bioeng Biotechnol 11:1131053

Ragauskas AJ, Williams CK, Davison BH, Britovsek G, Cairney J, Eckert CA, Frederick WJ Jr, Hallett JP, Leak DJ, Liotta CL, Mielenz JR (2006) The path forward for biofuels and biomaterials. Science 311(5760):484–489

Ray RC (2022) Fruits and vegetable wastes: valorization into bioproducts and platform chemicals. Springer, Singapore. 448p

Ray RC, Nedunchezhiyan M (2012) Post-harvest rots of sweet potato and control measures (Invited Mini-review). Fruits Veg Cereal Sci Biotechnol 6(Spl. Issue 1):134–138

Ray RC, Ravi V (2005) Spoilage of sweet potato in tropics and control measures (Review). Crit Rev Food Sci Nutr 45:623–644

Reddy JP, Rhim JW (2018) Extraction and characterization of cellulose microfibers from agricultural wastes of onion and garlic. J Nat Fibers 15(4):465–473
Rodríguez-Martínez B, Coelho E, Gullón B, Yáñez R, Domingues L (2023) Potato peels waste as a sustainable source for biotechnological production of biofuels: process optimization. Waste Manag 155:320–328
Scott GJ, Rosegrant M, Ringler C (2000) Global projections for root and tuber crops to the year 2020. Food Policy 25:561–597
Soltaninejad A, Jazini M, Karimi K (2022) Sustainable bioconversion of potato peel wastes into ethanol and biogas using organosolv pretreatment. Chemosphere 291:133003
Tomaszewska J, Bieliński D, Binczarski M, Berlowska J, Dziugan P, Piotrowski J, Stanishevsky A, Witońska IA (2018) Products of sugar beet processing as raw materials for chemicals and biodegradable polymers. RSC Adv 8(6):3161–3177
Varanasi S, Henzel L, Sharman S, Batchelor W, Garnier G (2018) Producing nanofibres from carrots with a chemical-free process. Carbohydr Polym 184:307–314
Vučurović VM, Puškaš VS, Miljić UD (2019) Bioethanol production from sugar beet molasses and thick juice by free and immobilised *Saccharomyces cerevisiae*. J Inst Brew 125(1):134–142
Wei L, Liang S, McDonald AG (2015) Thermophysical properties and biodegradation behavior of green composites made from polyhydroxybutyrate and potato peel waste fermentation residue. Ind Crop Prod 69:91–103

Biovalorization of Potato Peel Waste: An Overview

2

Mansi Singh, Vibhuti Sharma, and Reena Gupta

Abstract

The potato is one of the world's most widely grown and consumed food crops and its production has significantly increased in recent years. Its industrial processing produces several waste products that negatively impact the environment, potato peel stands out among them, accounting for up to 10% of all the potato waste obtained after processing. Potato peel is a valuable source of bioactive substances that can be processed to be converted into value-added bioproducts. It is one of the essential substrates for industrial aspects such as cosmetic, medicinal, bioenergy, etc., to make economically profitable and environmentally safe industries. The integrated biorefinery model for potato peel can increase the value of this waste by producing several value-added products like biofuels, dietary fiber, biofertilizers, biogas, biocomposite, antioxidants, organic acids, livestock feed, etc., and reducing extensive waste generation. Researchers are looking forward to producing bio-renewable energy using potato peel. This chapter provides an overview of the biorefinery approaches used in the valorization of potato peel waste which includes the formation of various valuable products such as biohydrogen, bioethanol, etc.

Keywords

Potato peel · Biofuel · Biorefinery model · Antioxidants · Bioactive substances

M. Singh · V. Sharma · R. Gupta (✉)
Department of Biotechnology, Himachal Pradesh University, Shimla, Himachal Pradesh, India

R. C. Ray (ed.), *Roots, Tubers, and Bulb Crop Wastes: Management by Biorefinery Approaches*, https://doi.org/10.1007/978-981-99-8266-0_2

Abbreviations

ABTS	2,2′-Azino-bis(3-ethylbenzothiazoline-6-sulfonic acid)
DPPH	2,2-Diphenyl-1-picrylhydrazyl
PP	Potato peels
PPF	Potato peel fibers
PPW	Potato peel waste
PW	Potato waste
RBC	Red blood cells
SSF	Solid state fermentation

2.1 Introduction

The agriculture and agri-food sectors significantly contribute to organic waste, with up to 90 million tons per year of production in EU28 countries in 2014. Almost half of the organic waste (around 46 million tons) is lignocellulosic waste (Gil et al. 2015). Additionally, it is estimated that the agri-food sector contributes to approximately 30% of global food losses and waste, highlighting the need for proper waste management to ensure sustainability (Bancal and Ray 2022). If organic waste is left to decompose without adequate control, it can cause significant contamination of soil, water, and air (Pandey et al. 2011). The decomposition of organic waste can also release carbon dioxide and methane into the atmosphere, with 1 metric ton of organic solid waste potentially releasing 50–110 m^3 of carbon dioxide and 90–140 m^3 of methane (Macias-Corral et al. 2008). The potato is ranked as the third most important food crop globally. Over 100 countries are involved in potato production. China and India are the primary producers, accounting for approximately 40% of the total (FAO et al. 2020). The Food and Agriculture Organization of the United Nations (FAO) reported that its production has increased steadily and reached nearly 390 million tons in 2021 (Barampouti et al. 2023). Of over 4000 known potato varieties, *Solanum tuberosum* L. is the most commonly cultivated variety. Due to the growing demand for fast food, the industrial processing of potatoes has increased in recent years, leading to significant waste generation (Atitallah et al. 2019; Chohan et al. 2020; Ude et al. 2020).

The use of potatoes in the food industry, including canning, starch, flour, chips, and french fries, generates a significant amount of potato peel waste (PPW) during the peeling, cutting, and trimming processes, which accounts for 15–40% of the initial potato weight (Liang and McDonald 2014; Pathak et al. 2018; Barampouti et al. 2023). The PPW is usually discarded in landfills, negatively impacting the environment by releasing toxic leachates and greenhouse gases (Chohan et al. 2020; Ebrahimian and Karimi 2020). However, PPW is a valuable source of starch and lignocelluloses that can produce bio-based products (Taher et al. 2017). The biovalorization concept can help extract multiple bioproducts from PPW and

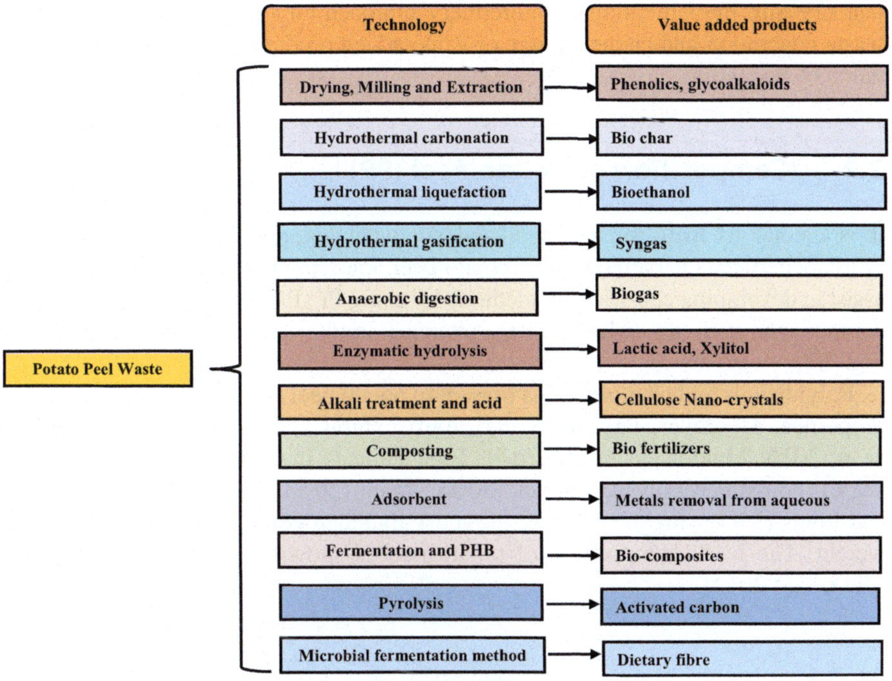

Fig. 2.1 The valorization options of potato peel wastes

minimize the amount of waste residue generation, thereby supporting the development of a circular bio-economy (Cristóbal et al. 2018).

The processing of PPW can be carried out through biological or thermochemical technologies. Biological processing involves using microorganisms to convert PPW into bioproducts through fermentation or anaerobic digestion (Abedini et al. 2020; Chohan et al. 2020; Adeleye et al. 2022; Patel et al. 2021). Thermochemical processing involves the thermal decomposition of PPW into desired bioproducts through pyrolysis, hydrothermal liquefaction, and hydrothermal carbonization (Chen et al. 2018, 2020; Daimary et al. 2022). The valorization options of potato peel wastes are illustrated in Fig. 2.1.

Several high-value products, such as organic acids, biofuels, biopolymers, enzymes, biochemicals, biochar, and biosorbents have been the subject of extensive research for their production using PPW as a low-cost raw material. This chapter focuses on the physicochemical composition of potato peel (PP) and biorefinery methods for managing PPW and extracting its potential for valorization. Potato peel contains bioactive compounds that can be transformed into high-value products. The products obtained by fermenting PP can be utilized for various purposes, such as adsorbents, biocomposites, and packaging materials. In addition to being used as dietary fiber or medicine, these products can also be utilized for energy generation, developing biopolymer films, inhibiting corrosion, and synthesizing cellulose nanocrystals. Implementing the biorefinery approach for PP can significantly reduce waste generation and create a range of value-added products, thereby increasing the value of this waste.

2.2 Composition of Potato Peel (PP)

It is crucial to focus on physical and chemical composition to understand the physicochemical characteristics of potato peel, knowing that these properties could assist in developing eco-friendly methods for using PP. PP contains several bioactive components such as carbohydrates, phenols, lipids, and alkaloids as shown in Fig. 2.2.

Polyphenols and phenolic acids present in potato peel contribute to its antioxidant properties. Moreover, fatty acids and lipids exhibit antibacterial activity (Jeddou et al. 2016; Maldonado et al. 2014). Biochemicals in potato peel include starch (25%), non-starch polysaccharides (30%), lignin (20%), protein (18%), ash (6%), and lipids (1%), with starch being the most abundant component based on the dry weight. The basic composition of potato peel is shown in Table 2.1 (Liang and McDonald 2014).

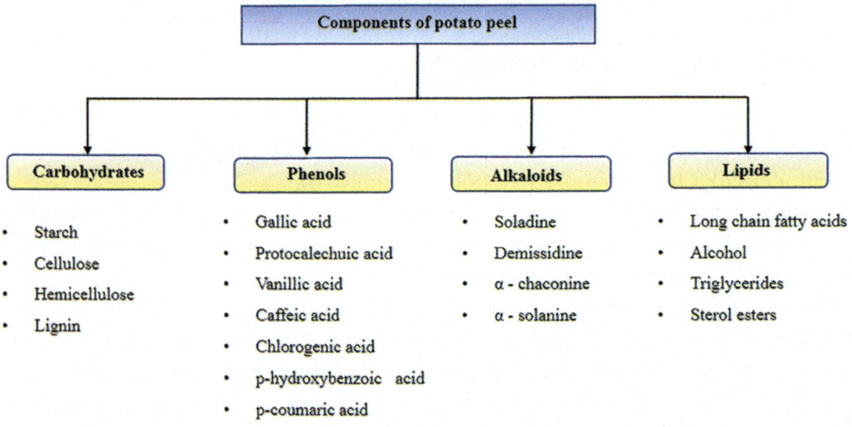

Fig. 2.2 Various bioactive components present in potato peel (Rasheed et al. 2022)

Table 2.1 The basic composition of potato peel (% dry weight)

Component	Amount (% dry weight)
Non-starch polysaccharides	30
Starch	25
Lignin	20
Protein	18
Ash	6
Lipids	1

Source: Liang and McDonald (2014)

The lipid fraction includes long-chain fatty acids, alcohols, triglycerides, and sterol esters. Lignin units are also present in the cell walls of potatoes (Liang et al. 2014). However, fermentable reducing sugars in potato peel are limited (0.6% dry weight), making fermentation impractical without initial hydrolysis of carbohydrates through enzymatic or acidic means. Elemental analysis shows that potato peel contains carbon, hydrogen, nitrogen, and oxygen. It has a C/N ratio equivalent to 10.7, pH is 6.5 (Liang et al. 2015a), and the calorific value is 17.37 ± 0.38 MJ/kg (Liang et al. 2014). Potato peel extract contains various phenolic acids such as gallic acid, protocatechuic acid, vanillic acid, caffeic acid, chlorogenic acid, *p*-hydroxybenzoic acid, and *p*-coumaric acid. The lipid content in potato peel is higher than peeled potatoes, with higher levels of polyunsaturated fatty acids (PUFAs) (Arapoglou et al. 2010). The relative quantities of individual fatty acids seem to be influenced by the color of potato peels (Onyeneho and Hettiarachchy 1993).

2.3 Utilisation of Potato Peel in Different Areas/Valorization

Potato peels can be utilized in different areas depending on their properties. Valorization of PP into valuable bioproducts such as enzyme production, bioethanol, biogas, and organic acid for fermentation and extraction of chemicals for uses are elaborated in further sections. An overview of the valorization process of potato peel is shown in Fig. 2.3.

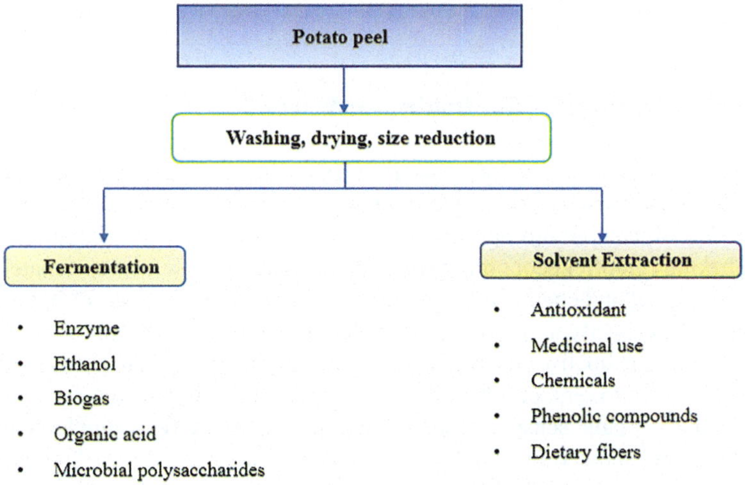

Fig. 2.3 Overview of the valorization process of potato peel (Pathak et al. 2018)

2.3.1 Pharmaceutical Application

Due to the high starch content of potato peel, it can be utilized as a burn-healing agent in the form of medicinal bandages. PP bandages offer advantages over gauze bandages by minimizing or eliminating desiccation, which enhances the viability of the outermost skin cells and accelerates epithelial tissue regeneration. PP bandages also help in reducing intrinsic antibacterial activity. Antioxidant characteristics of extracts from PP have been observed to counteract liver damage caused by CCl_4 and provide protection against oxidative stress and diabetes in rats (Singh et al. 2005a). Due to these properties, PP can be utilized effectively as an element in functional foods and health products (Singh et al. 2005b, 2008). Secondary metabolites found in PP are steroidal alkaloids that provide protection against insect attacks and possess anti-inflammatory, antifungal, and antibacterial properties. The steroidal alkaloids found in PP include two aglycones (solanidine and demissidine) and two glycoalkaloids (α-solanine and α-chaconine). These compounds can be extracted from PP using ultrasound-assisted extraction (UAE), solid-liquid extraction (SLE), or a combination of both (Hossain et al. 2014). Due to their straightforward application and cost-effective production, PP bandages are an excellent option for treating burn wounds (Singh et al. 2008).

They are cleaned with sterile saline water before being applied to burns and are placed to face the wound being covered with two layers of saline gauze and additional bandages for protection. The benefits of using PP bandages include faster restoration of the skin's color and texture, rapid epithelial regeneration, increased comfort, and less painful bandage removal. Additionally, they do not shrink in size during their application period, which is advantageous. However, placing them edge-to-edge on a large wound can be exhausting and time-consuming for nursing staff. As an alternative, simplified roller bandages are available for burn treatment, which are easy to apply and remove and reduce treatment time (Keswani and Patil 1985).

2.3.2 Application in the Baking Industry

The baking industry often uses dietary fibers extracted from various sources in combination with wheat flour. One potential source of dietary fiber is potato peel fiber (PPF), which contains approximately 50% dietary fiber by weight. The use of PPF in baking can offer economic benefits by providing a low-cost raw material with superior physical and chemical properties compared to wheat bran. PPF has a better water-holding capacity and lower starch content hence, using extruded potato peels can help eliminate quality defects such as musty odors. According to research, using 5% and 10% PPF to replace wheat flour in biscuit processing results in smaller stack weight and acceptable sensory scores. In another study, muffins supplemented with 25% potato peel showed fine quality and resistance to compression (Abd-El-Magied 1991; Al-Weshahy and Rao 2012).

2.3.3 Biotechnological Applications

There are various applications of PP in the field of biotechnology, some of which are described below:

2.3.3.1 Organic Acid Production

Organic acids have significant industrial importance because of their extensive range of uses (Kumari et al. 2022). Some of the organic acids produced using potato peels are as follows

Acetic acid. It is utilized in pharmaceutical, food, and intermediate chemical industries and is vital in manufacturing various industrial products (Uslu et al. 2010; Panda et al. 2020).

Lactic acid. It is commonly used for synthesizing polylactic acid. During the fermentation of PP, lactic acid is the main product produced, with smaller amounts of acetic acid and ethanol also being produced as by-products. Hydrolytic enzymes such as mixed α-amylase, glucoamylase, and cellulase have been utilized to convert PP into lactic acid and other substances (Liang et al. 2014). Another technique for producing lactic acid from PP is using anaerobic fermentation in a sequencing batch reactor. This method uses an unspecified mixture of microorganisms obtained from a municipal wastewater treatment plant. It has been shown to yield a high overall production of lactic acid, with a rate of 138 mg/g of substrate per day (Liang et al. 2015b).

Citric acid. It is the most commonly used organic acid in various industries such as food, beverage, metal processing, pharmaceutical, cosmetics, metallurgical chemical, and detergent industries (Swain et al. 2012). *Aspergillus niger* CA 16 was employed to generate citric acid from PP hydrolysate (Faruk et al. 2014; Pathak and Mandavgane 2015).

2.3.3.2 Glucose Production

Hydrolyzing starch present in PP is utilized to generate glucose for fermentation. The dried PPs are subjected to extrusion at a temperature of 135 °C with 22% feed moisture and then treated with amyloglucosidase and α-amylase to facilitate digestion. The conversion of starch to glucose is generally observed within 1–2 h, indicating that PP is a suitable source for enzymatic fermentation in glucose production (Camire and Camire 1994).

2.3.3.3 Dietary Fibers and Nutrition

Dietary fibers are a type of chemical compound that can provide health benefits to humans. They are derived from plant cells and are not easily broken down by the digestive enzymes in the human body. Some of the dietary fibers include cellulose, waxes, lignin, pectin, hemicellulose, pectin, oligosaccharides, and gums. These fibers can be found in various amounts and forms in cereals, vegetables, fruits, and nuts. To maintain good health, it is recommended that a healthy adult should consume between 20 and 35 g of dietary fiber per day. PP, which contains about

76.40% dietary fiber (dry weight basis), is a good source of nutrients (Arora and Camire 1994; Dhingra et al. 2012). Dietary fibers can reduce serum glucose levels in rats (Singh et al. 2005a, b). However, the composition of PP's dietary fiber can be altered under certain cooking conditions, such as extrusion cooking, which can also affect other macromolecules (Camire et al. 1997).

In terms of overall dietary fibers, certain minerals, and water-holding capacity, PP outperforms wheat bran. PP contains fewer starchy components and no phytate than total dietary fiber. These nutritional benefits remain as such even after baking PP bread. An enzymatic-chemical method is used to extract dietary fibers from PP. These fibers have functional properties such as oil-holding capacity, water-holding capacity, swelling capacity, and water solubility index. In a study, it was observed that PP, when mixed with wheat flour, resulted in acceptable-quality biscuits, even when up to 5% of wheat flour was replaced with PP fiber (Dhingra et al. 2012).

To make cinnamon muffins, some wheat flour is replaced with both extruded and non-extruded PP dietary fiber. Muffins made with 25% PP and 75% wheat flour are shorter, darker, and more compression-resistant. There are no significant differences in shape, odor, color, and flavor between the muffins made up of extruded and non-extruded PP (Arora and Camire 1994).

2.3.3.4 Application of Potato Peel Waste for Energy Production

Biogas Production

Reducing carbon dioxide emissions by 70% till 2050 can help stabilize the weather and combat global warming. However, a single energy system based on fossil fuels can cause environmental damage. Diversifying energy sources and their locations can be a good energy production and distribution strategy. Biogas production through anaerobic digestion can be a suitable replacement for traditional energy sources (Moodley et al. 2023). This process involves multiple stages, including hydrolysis, acidification, enzyme digestion, and methanogenesis, all in a dynamic equilibrium as shown in Fig. 2.4.

Global biogas production has increased considerably from 0.28 Exa Joules (EJ) in 2000 to 1.37 EJ in 2018 (Awogbemi et al. 2022). Its production from 2010 to 2018 is shown in Fig. 2.5. Waste residue from potato starch processing is now being used for biogas production. For example, a potato residue biogas fermentation system with an average biogas production ratio of 0.55 L/g volatile solid (VS) content, much higher than other fermentation methods, is available (Fu et al. 2013). This process has an energy efficiency of 92%, indicating optimal control of the process (Sanaei-Moghadam et al. 2014). As potato peel waste becomes a more common replacement for traditional starch extraction, it is essential to research recycling methods for this waste. Although the processing technique may be more complex, the cost will remain relatively high. This technology has great potential for future development because it generates energy from waste residue in potato starch processing with high efficiency, standard technical requirements, and low cost.

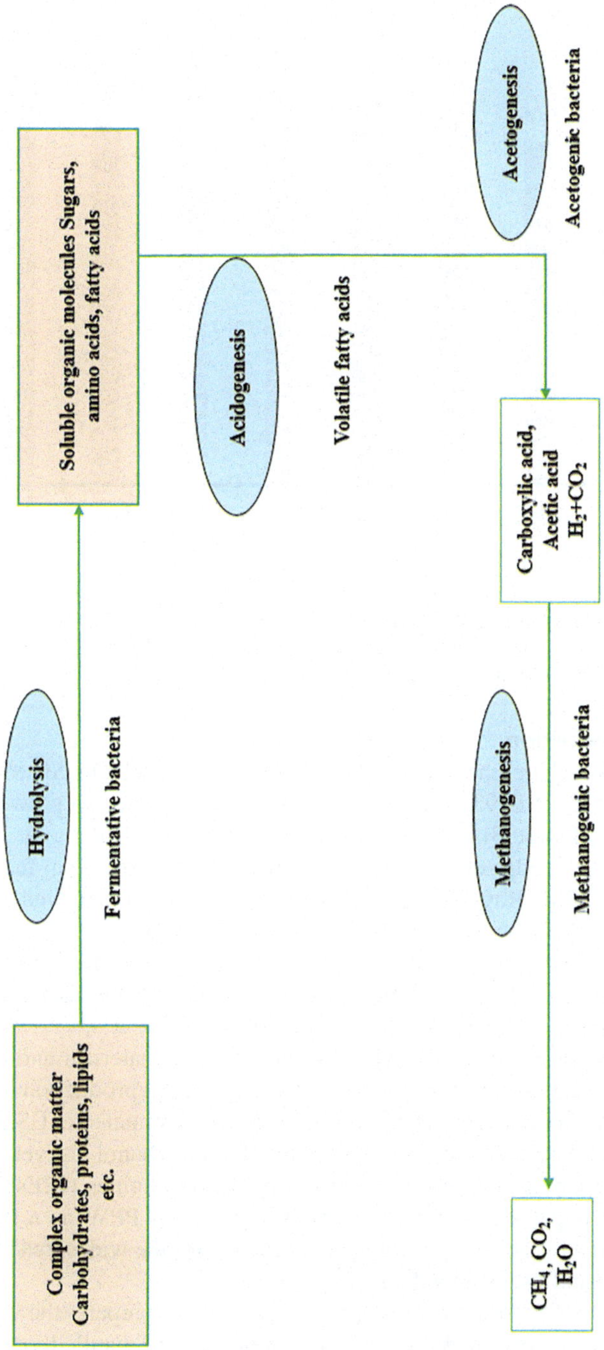

Fig. 2.4 Different stages of the biogas production process

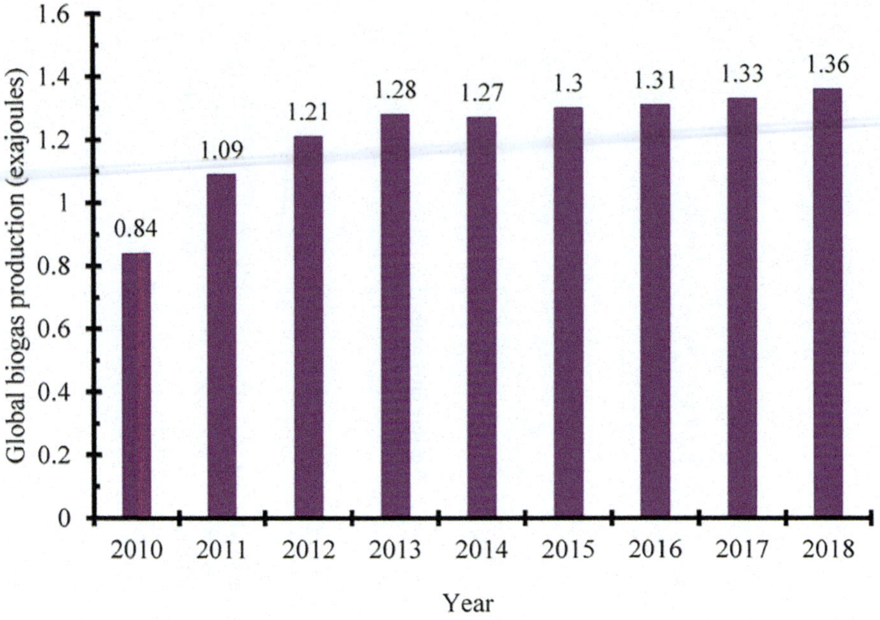

Fig. 2.5 Global biogas production from 2010 to 2018 (Awogbemi et al. 2022). (Derived from open-access article distributed under the terms and conditions of the Creative Commons Attribution (CC BY) license (https://creativecommons.org/licenses/by/4.0/))

Bioethanol Production

Regarding market share, the worldwide ethanol market reached USD 89.1 billion in 2019, increased to USD 93.7 billion in 2020, and is expected to grow by 5.2% by 2030 (Ethanol Market Size Worth Around USD 155.6 Billion by 2030.). The projected growth is driven by the rising use of ethanol as a fuel (Ray and Ramachandran 2018; Ray 2020). Global ethanol production moved from 13.123 Million Gallons (MG) in 2007 to 29.026 MG in 2019. Its global production from 2007 to 2020 is shown in Fig. 2.6 (Awogbemi et al. 2022). At least 73% of globally produced ethanol is estimated to be utilized as fuel, with the remainder used for beverages and industrial purposes (Arapoglou et al. 2010).

Like other liquid biofuels, the high expense of raw materials and the negative impact on the food industry have hindered the commercial production of bioethanol (Ray and Ramachandran 2018). With production costs estimated at USD 0.40/L and the use of edible feedstocks, it is not surprising that bioethanol has yet to be widely adopted as an alternative fuel for internal combustion engines (ICEs) (Arapoglou et al. 2010; Mohapatra et al. 2018). However, utilizing PPW as a feedstock for producing bioethanol offers a promising opportunity for the widespread and sustainable production and use of bioethanol.

Several physical pretreatment steps are needed to prepare the potato waste (PW) for further processing. Initially, PW is washed with distilled water and dried under sunlight for 2–3 days. Subsequently, it is dried in an oven at a temperature of

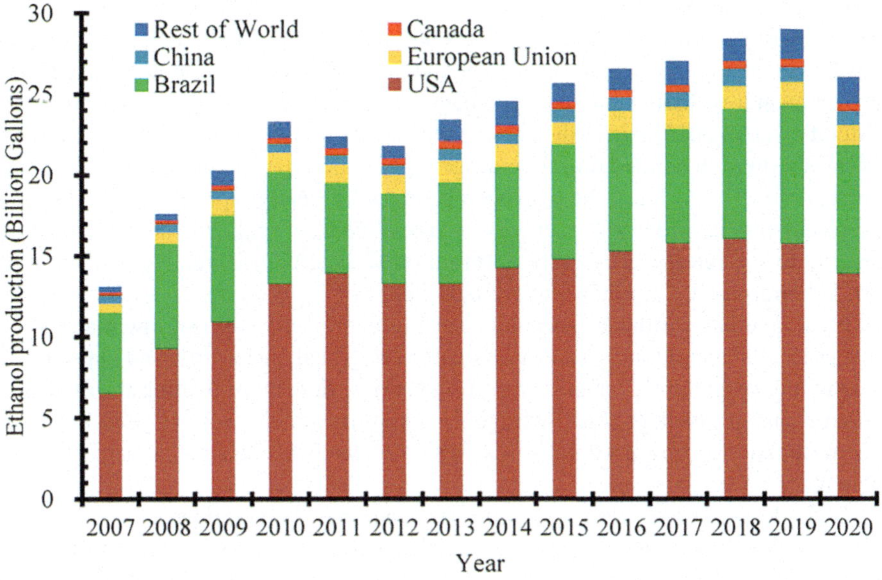

Fig. 2.6 Global ethanol production (billion gallons) from 2007 to 2020 (Awogbemi et al. 2022). (Derived from open-access article distributed under the terms and conditions of the Creative Commons Attribution (CC BY) license (https://creativecommons.org/licenses/by/4.0/))

45 °C, crushed using a ball mill, and sieved to a particle size ranging between 0.5 and 1.0 mm. In addition to these physical pretreatment measures, various chemical, biological, and ultrasonic techniques are used. These techniques involve using acids, alkalis, wet oxidation, ozonolysis, and hydrogen peroxide with metal salts (Meenakshi and Kumaresan 2014; Hijosa-Valsero et al. 2018; Soltaninejad et al. 2022). These pretreatment methods effectively improve product yield by increasing the surface area of the reactants (Awogbemi et al. 2021). Different techniques such as hydrolysis, saccharification, and fermentation have been used to convert PPW into ethanol (Chintagunta et al. 2016), bioethanol (Taher et al. 2017; Chohan et al. 2020) or biobutanol (Hijosa-Valsero et al. 2018) with production rates ranging from 7.6 to 40 g/L. The results of these studies indicate that PPW is a feasible and readily available feedstock that can be used for bioethanol production at a low cost.

The α-amylase, cellulase, Termamylamyloglucosidase, and *Clostridium acetobutylicum* are all used to produce PPW-based bioethanol (Khawla et al. 2014; Sujeeta et al. 2018; Chavez et al. 2020). To enhance the conversion of PPW to acetone, butanol, and ethanol via fermentation using *C. acetobutylicum*, uses the powdered orange peel (*Citrus sinensis*) extract has been used (Kamboj and Dhanya 2021). According to a report, adding orange peel extract boosted butanol production. *C. acetobutylicum* is being employed for the fermentation-based conversion of PPW to biobutanol. Hence, PPW is successfully used as a cheap and environmentally benign feedstock in a low-input manufacturing process (Abedini et al. 2020).

Biohydrogen Production

The growth of hydrogen production is being propelled by various factors, such as advancements in hydrogen separation methods, a rise in hydrogen usage, and government regulations to reduce greenhouse gas emissions and desulfurization (Moodley et al. 2023). Nevertheless, the high storage cost and the significant amount of carbon dioxide emitted during hydrogen production have hindered its widespread application. Almost all of the hydrogen generated globally (approximately 95%), is derived from non-renewable energy sources such as coal, oil, and natural gas, leading to the release of roughly 830 million tons of CO_2 annually (Atilhan et al. 2021; Ferraren-De Cagalitan and Abundo 2021).

Hydrogen has various applications, such as acting as a fuel and raw material for industries. Its utilization as a fuel is crucial in the global energy mix and plays a vital role in reducing carbon emissions from industries, thereby mitigating the effects of climate change. Besides this, hydrogen's versatility enables it to be used as a fuel, chemical, and feedstock for industrial processes, leading to a constant rise in global demand.

Biohydrogen production is a more environmentally sustainable, secure, cost-effective, and less energy-demanding alternative to hydrogen production from fossil fuel sources (Ferraren-De Cagalitan and Abundo 2021; Islam et al. 2021). Various waste vegetables and biomasses are utilized to effectively generate biohydrogen, capitalizing on the high content of carbohydrates, total lipids, glucose, sucrose, and starch (Sinha and Pandey 2014; Marone et al. 2014). Biohydrogen is also produced from untreated and hydrolyzed potato peels (Mars et al. 2010). Several methods employed for its production include biological, thermochemical, and electrochemical methods. A flow chart showing the production process of biohydrogen is shown in Fig. 2.7.

Biohydrogen from PPW is also used to complement Portuguese transportation fuel (Ferreira et al. 2011). In addition to feedstock composition, availability, cost-effectiveness, substrate purity, ease of hydrolysis and biodegradation are critical considerations in biohydrogen production. Thus, novel investigations are necessary to leverage the benefits of feedstock composition, availability, and purity to meet the growing demand for biohydrogen.

2.3.3.5 Animal Feed

Recently, multiple studies have indicated that the waste produced from processing fruits and vegetables can be utilized efficiently in making high-quality animal feed (Ray 2022). One such waste product is PP, which is rich in polyphenols and can be repurposed for animal feed production (Kasapidou et al. 2015). Traditionally, farmers feed PP to cows and pigs due to its high starch content. The inclusion of PP in beef cattle diets has been shown to enhance milk production and improve the yield and quality of meat to a significant degree. However, some purveyors, meat vendors, and chefs in Japan and North America have expressed concerns that PP-fed beef is less desirable than corn-fed beef, as it is believed to be softer, more rigid, more watery, inferior in color and less flavorful (Hossain et al. 2014).

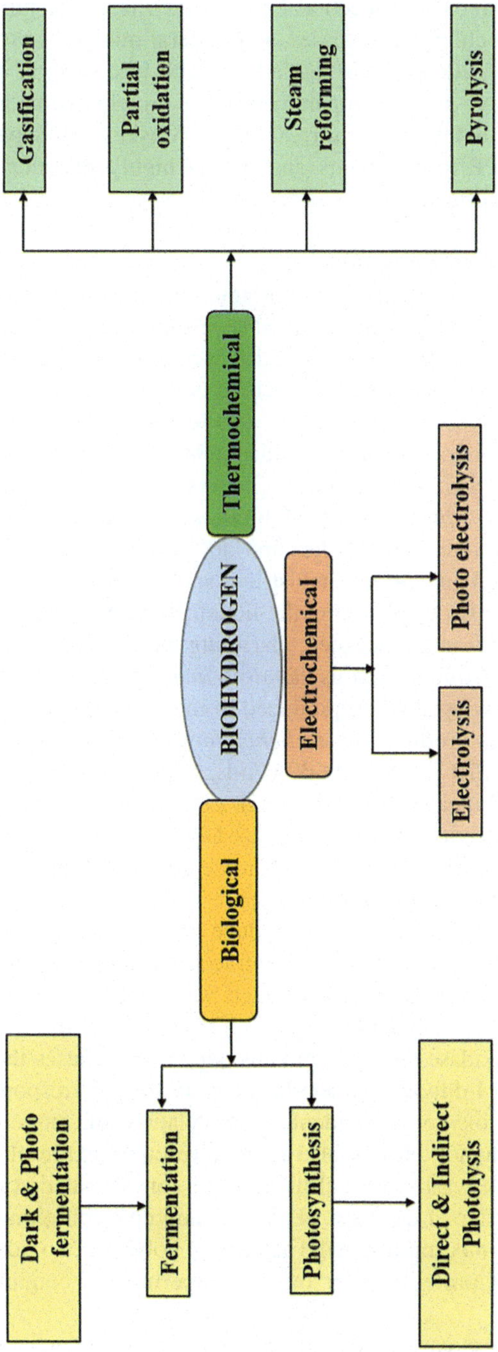

Fig. 2.7 The production process of biohydrogen

Nevertheless, PP can replace up to 50% of the concentrate feed mixture (which includes ingredients like cottonseed meal, rice bran, wheat bran, limestone, vitamins, yellow corn, sodium chloride, molasses, and mineral mixture) in sheep rations with success (Tawila et al. 2008). PP is also a beneficial feed material for fish, as it has been shown to improve nutritional parameters like protein and fat in liver tissue and fish muscles when fed to them for 30 days (Maske and Satyanarayan 2012). Therefore, feeding PP to ruminants and fish is highly beneficial to their health without any harmful effects.

2.3.3.6 Production of Enzymes

Potato peels can serve as a medium for microorganisms to grow in solid-state fermentation (SSF), making them a valuable resource for producing enzymes and other beneficial products (Mahmood et al. 1998; Shukla and Kar 2006). Mannanases enzyme can reduce the viscosity of extracts in coffee production, extract vegetable oil from leguminous seeds, and aid in textile and paper production. *Bacillus amyloliquefaciens* has been seen to produce using PP as a carbon source, with an optimal dose of 14 g/L. The resulting process yielded a maximum activity of mannanase with 105.7 U/mg protein (Mabrouk and El Ahwany 2008).

Microorganisms were among the first to produce amylases commercially. Amylases are utilized in various industries, such as textiles, food, baking, and detergents. They are also employed in the liquefaction or saccharification of starch, improving digestibility in animal feed, clarifying haze in fruit juices and beer, and sizing textile fibers. Strains such as *Bacillus licheniformis*, *Bacillus subtilis*, and *Aspergillus niger* grew on amylase produced from PP (Shukla and Kar 2006; Jadhav et al. 2013). Enzymes produced by *B. licheniformis* and *B. subtilis* strains showed enzyme activities of 270 U/mL and 600 U/mL, respectively (Shukla and Kar 2006; Behera et al. 2019). Protease obtained from *B. subtilis* DM-04 using PP as a substrate displayed enzyme activity of 400.0 ± 13.0 U/gds and is used as an additive in detergent industries. SSF production of protease does not require additional supplementation for co-carbon and co-nitrogen.

A. niger and *B. subtilis* strains can produce cellulolytic enzymes like xylanase, carboxymethyl cellulose, cellulase, and extracellular enzymes such as α-amylase, protease, and polygalacturonase lyase through SSF of PP. These enzymes have been shown to possess high enzymatic activities (Mahmood et al. 1998; Santos et al. 2012). Polyphenol oxidase enzyme contains copper that causes the oxidation of *o*-diphenols (including l-dihydroxyphenylalanine) to their corresponding *o*-quinones and the molecular oxygen-dependent hydroxylation of monophenols to their corresponding *o*-diphenols. It can also remove reactive textile phenolic pollutants and textile dyes from wastewater, effluents, and contaminated soil. PP is blended with a phosphate buffer of different pH values and concentrations to extract polyphenol oxidase. The maximum specific activity of polyphenol oxidase was achieved as 3572.74 U/mg of protein, under optimal experimental conditions (Mukherjee et al. 2012).

Polyphenol oxidase can be extracted from PP using an aqueous two-phase extraction method. It is an attractive alternative to chemical extraction since it is a one-step process and easy to scale up, allowing for the clarification, concentration, and partial purification of polyphenol oxidase. The researchers have achieved a maximum yield of 77.8% for polyphenol oxidase using this method (Niphadkar et al. 2015). Conventional three-phase partitioning and ultrasound-assisted three-phase partitioning methods, to obtain maximum extraction and purification of polyphenol oxidase from PP have been proposed (Niphadkar and Rathod 2015).

2.3.3.7 Production of Biofertilizer

Organic fertilizers play a critical role in maintaining sustainable agriculture. PP has proven to be a successful source for producing biofertilizers. Due to its high protein and starch content, PP can be degraded by soil microorganisms to produce fertilizers with high nitrogen content. Vermicompost is derived by earthworms (*Pheretima elongata*) by feeding on PP and exhibited a higher bacterial count than the surrounding soil, indicating its effectiveness as a biofertilizer (Pandit et al. 2012). The slurry obtained from PP biogas plants (anaerobic digesters) is also a valuable biofertilizer for land application that replenishes nutrients in the soil (Muhondwa et al. 2015).

Furthermore, a liquid biofertilizer can be produced by fermenting a mixture of cow dung, PP, legume peels, carrot peels, neem leaves, and tulsi leaves in water for 45 days, resulting in increased vegetative growth, yield, and physicochemical characteristics of strawberry fruits (Tiwari et al. 2016). In addition, the residue left after ethanol production via SSF was inoculated with different microbes such as *Nostoc muscorum*, *Fischerella muscicola*, *Aulosira fertilissima*, *Anabaena variabilis*, *Cylindrospermum muscicola*, *Azotobacter chroococcum* and *Azospirillum lipoferum* in equal ratios. This process improved the nitrogen, phosphorus, and potassium ratio (Chintagunta et al. 2016). Hence, PP represents an efficient resource for the production of biofertilizers.

2.3.3.8 Applications in Biofilms/Biocomposites

Due to some difficulties in adequately disposing of synthetic and petroleum-based plastics, there has been substantial interest in bioplastics made from renewable resources (Ray and Swain 2011). Researchers used cellulose nanocrystals obtained from PP through alkali treatment and acid hydrolysis and added them to polymer films made from polyvinyl alcohol and starch at concentrations of 0–2% (w/w). This addition of cellulose nanocrystals resulted in a polymer film with improved resistance to water permeability for the polyvinyl alcohol composite (Chen et al. 2012).

Making biopolymer films involves combining PP and glycerol using high-pressure homogenization and heat treatment. This process creates smooth films. Even without starch, PP can still form films, which may be due to soluble pectin. However, films reinforced with cellulose have more potent mechanical properties and can completely block grease (Rommi et al. 2016). The adhesive and binder obtained from PP are utilized in packaging materials. To make biodegradable bioplastics with elastic properties, PP fermentation residue fibers are blended with polyhydroxybutyrate. Although the resulting biocomposites had poor mechanical properties, their biodegradation rate is substantially faster than pure polyhydroxybutyrate, mainly when the fiber content is over 15%. The degradation rate of the PP-polyhydroxybutyrate biocomposite increased with increasing fiber content and biocomposites with a 50% fiber content required 8 months to degrade fully (Wei et al. 2015). These biocomposites have potential applications in horticulture and agriculture as rapidly degrading biomaterials. PP-based biopolymer films are also effective against *Listeria monocytogenes* on cold-smoked salmon, with the edible film exhibiting antimicrobial activity by providing an additional barrier to the growth of *L. monocytogenes* on the surface of the salmon (Tammineni et al. 2013).

2.3.3.9 Antioxidants from PP

Loss of food quantity in the food industry mainly occurs due to a chain reaction called lipid oxidation which involves free radicals. One effective way to extend the shelf life of processed foods and prevent the production of harmful oxidation molecules is by adding synthetic antioxidants. However, synthetic antioxidants, such as butylated hydroxytoluene, butylatedhydroxyanisole, tertubutylhydroquinone, and propyl gallate have potential health risks. As a result, consumers are increasingly interested in using natural antioxidants that offer more health benefits with fewer side effects. Fruit and vegetable waste is a good source of natural antioxidants inhibiting oxidation and maintaining product quality. Potatoes are particularly rich in antioxidants such as phenolic compounds. The recovery of these valuable compounds from potato waste could be an economically viable alternative to synthetic antioxidants for food-processing businesses (Onyeneho and Hettiarachchy 1993; Singh et al. 2011; Amado et al. 2014).

The section of a potato that consists of 10 mm of skin and cortex, known as the PP section, contains more phenolic compounds than other parts of the potato, such as the vascular ring and medullary tissue (Al-Saikhan et al. 1995). These phenolic compounds can exist in either bound or free form, including ferulic acid, *p*-coumaric acid, chlorogenic acid, and caffeic acid. Ferulic acid plays a role in enhancing the thermal stability of food plants and improving their texture and cell adhesion. The PP section contains 1.26 and 3.66 mg of ferulic acid equivalent per gram of dry matter inbound and accessible form, respectively (Nara et al. 2006). HPLC analysis has identified several primary phenolic acids in the PP section, including gallic acid, protocatechulcgentlsic acid, *p*-hydroxybenzoic acid, chlorogenic acid, ferulic acid, vanillic acid, syringic acid, caffeic acid, salicylic acid, and *p*-coumaric acid (Onyeneho and Hettiarachchy 1993; Habeebullah et al. 2010).

The traditional method of extracting antioxidants from PP involves using organic solvents such as methanol, ethanol, ethyl acetate, and acetone (Sooch et al. 2022). Although these solvents are highly effective at extracting antioxidants, they have some drawbacks. For instance, they require longer extraction times, consume large amounts of solvent, pose a high toxicity risk, and can degrade heat-sensitive components. However, ethanol is preferred over other solvents due to its higher efficiency, lower toxicity, and reduced cost. In addition to organic solvents, different extraction methods such as pressurized liquid extraction, microwave-assisted extraction and subcritical water extraction (also known as pressurized polarity water extraction, superheated water extraction, high-temperature water extraction, or hot liquid water extraction) can also be used to extract polyphenols from PP (Singh et al. 2011; Amado et al. 2014).

Phenolic acids such as chlorogenic acid, caffeic acid, gallic acid, protocatechuic acid, *p*-hydroxybenzoic acid, and syringic acid are commonly extracted using subcritical water extraction. At a temperature of 180 °C and pressure of 6 MPa, it takes approximately 60 min to recover these acids and the amounts recovered are presented as weight percentages. The quantities of these acids recovered are chlorogenic acid (14.59 ± 0.98 mg/100 g), caffeic acid (9.23 ± 0.73 mg/100 g), gallic acid (29.56 ± 0.79 mg/100 g), protocatechuic acid (13.58 ± 0.13 mg/100 g), *p*-hydroxybenzoic acid (11.18 ± 0.16 mg/100 g), and syringic acid (3.69 ± 0.06 mg/100 g) (Singh and Saldaña 2011).

As per the solid-liquid extraction methods, the antioxidant activity of PP is measured against 2,2-diphenyl-1-picrylhydrazyl (DPPH) using 75% ethanol at a temperature of 80 °C for 22 min indicating an antioxidant activity of 352 mg Trolox equivalents/100 g dried PP. On the other hand, pressurized liquid extraction using 70% ethanol at 125 °C indicated an antioxidant activity of 339 mg Trolox equivalents/100 g dried PP. Although pressurized liquid extraction did not improve antioxidant recovery compared to solid-liquid extracts, however, using aqueous ethanol as a solvent for extraction resulted in a higher recovery of polyphenols instead of using 100% methanol (Wijngaard et al. 2012). Different extracts of PP were found to have a total polyphenolic content ranging from 2.9 to 4.2 mg/g PP powder. These extracts have the potential to act as potent antioxidants and protect erythrocytes from oxidative damage. Experiments using rat RBCs (red blood cells) confirmed that PP extracts effectively inhibited H_2O_2-induced morphological changes. Additionally, PP extracts substantially protected human erythrocyte membrane proteins against oxidative damage caused by ferrous ascorbate (Singh and Rajini 2008). PP extracts can be safely used as natural antioxidants in oils, fats, and other food products to prevent lipid oxidation (Habib and Shah 2004).

PP contains polysaccharides, which can serve as a natural antioxidant and can be added to pharmaceutical (specifically antitumor), food, and cosmetic products. This polysaccharide displays scavenging activity against DPPH and other radicals. Additionally, it exhibits strong ABTS (2,2′-azino-bis(3-ethylbenzothiazoline-6-sulfonic acid)) radical-scavenging and β-carotene-bleaching inhibition activities. Based on these observations, it can be inferred that PP extracts possess efficient antioxidant properties in vitro, with the effect being dependent on the dosage (Jeddou et al. 2016). Thus, PP can be a valuable source of natural antioxidants for inhibiting lipid oxidation.

2.4 Conclusion and Future Perspective

Potato peel, a significant byproduct of potato-processing factories can create valuable compounds through various techniques. These compounds could be used to create useful products. The biorefinery approach can be employed to maximize the use of chemicals, utilities, and timing, resulting in more significant economic benefits for the potato peel-producing industry. Due to its high phenolic content, potato peel is considered a good source of antioxidants, making it essential for dietary fiber and nutrition. However, potato peel contains various valuable bioactive compounds, including polyphenols (phenolic acids and flavonoids), glycoalkaloids (α-chaconineand α-solanine), and polysaccharides (starch and pectin). Recovering these bioactive compounds can mitigate environmental problems and increase the profitability of the food industry. Innovative technologies are necessary to achieve this objective and to obtain higher extraction yields while preserving the properties of the extracts. Industrial enzymes and biofuels can be produced through fermentation, providing promising energy sources. Implementing a biorefinery that uses potato skins as the primary source can lead to future industrial applications and maximize the use of this byproduct. Managing potato peel waste is a significant concern for the potato industry, but utilizing the peel can lead to various nutritionally beneficial applications, such as the production of bio-hydrogen, biogas, and bio-fertilizer. These activities can help developing countries boost their economy, reduce poverty, and establish industry linkages. Sharing new ideas, technologies and future trials for sustainable economic growth can be based on these activities. However, there is still a need to develop the engineering design of potato peel biorefineries, which should be a focus of future research.

Acknowledgments The financial support from the Department of Biotechnology, Ministry of Science and Technology, Govt. of India, to Department of Biotechnology, Himachal Pradesh University, Shimla, India, is thankfully acknowledged.

Consent of Publication The authors give their consent for the publication of this chapter.

Conflict of Interest The authors declare that they have no conflict of interest.

References

Abd-El-Magied MM (1991) Effect of dietary fiber of potato peel on the rheological and organoleptic characteristics of biscuits. Egypt J Food Sci 19:293–300

Abedini A, Amiri H, Karimi K (2020) Efficient biobutanol production from potato peel wastes by separate and simultaneous inhibitors removal and pretreatment. Renew Energy 160:269–277

Adeleye T, Yeo H, Seth R, Hafez H, Biswas N (2022) Influence of mix ratio of potato peel and pig manure on reaction kinetics and methane recovery from anaerobic co-digestion. Can J Civ Eng 49(5):675–682

Al-Saikhan MS, Howard LR, Miller JC Jr (1995) Antioxidant activity and total phenolics in different genotypes of potato (Solanum tuberosum, L.). J Food Sci 60(2):341–343

Al-Weshahy A, Rao VA (2012) Potato peel as a source of important phytochemical antioxidant nutraceuticals and their role in human health—a review. IntechOpen, pp 207–224

Amado IR, Franco D, Sánchez M, Zapata C, Vázquez JA (2014) Optimisation of antioxidant extraction from Solanum tuberosum potato peel waste by surface response methodology. Food Chem 165:290–299

Arapoglou D, Varzakas T, Vlyssides A, Israilides CJWM (2010) Ethanol production from potato peel waste (PPW). Waste Manag 30(10):1898–1902

Arora A, Camire ME (1994) Performance of potato peels in muffins and cookies. Food Res Int 27(1):15–22

Atilhan S, Park S, El-Halwagi MM, Atilhan M, Moore M, Nielsen RB (2021) Green hydrogen as an alternative fuel for the shipping industry. Curr Opin Chem Eng 31:100668. https://doi.org/10.1016/j.coche.2020.100668

Atitallah IB, Antonopoulou G, Ntaikou I, Alexandropoulou M, Nasri M, Mechichi T, Lyberatos G (2019) On the evaluation of different saccharification schemes for enhanced bioethanol production from potato peels waste *via* a newly isolated yeast strain of *Wickerhamomyces anomalus*. Bioresour Technol 289:121614. https://doi.org/10.1016/j.biortech.2019.121614

Awogbemi O, Von Kallon DV, Aigbodion VS (2021) Trends in the development and utilization of agricultural wastes as heterogeneous catalyst for biodiesel production. J Energy Inst 98:244–258

Awogbemi O, Kallon DVV, Owoputi AO (2022) Biofuel generation from potato peel waste: current state and prospects. Recycling 7(2):23. https://doi.org/10.3390/recycling7020023

Bancal V, Ray RC (2022) Overview of food loss and wastes in fruits and vegetables: from issues to resources. In: Ray RC (ed) Fruits and vegetable wastes: valorization to bioproducts and platform chemicals. Springer, Singapore, pp 3–32

Barampouti EM, Christofi A, Malamis D, Mai S (2023) A sustainable approach to valorize potato peel waste towards biofuel production. Biomass Convers Biorefinery 13:8197–8208

Behera SS, Ray RC, Das U, Panda SK, Saranraj P (2019) Microorganisms in fermentation. In: Berenjian A (ed) Essentials in fermentation technology. Springer, Switzerland, pp 1–39

Camire ME, Camire AL (1994) Enzymatic starch hydrolysis of extruded potato peels. Starch-Stärke 46(8):308–311

Camire ME, Violette D, Dougherty MP, McLaughlin MA (1997) Potato peel dietary fiber composition: effects of peeling and extrusion cooking processes. J Agric Food Chem 45(4):1404–1408

Chavez AY, Morales R, Gonzalez C, Moya FV (2020) Production of ethanol from two varieties of potato peel waste through cellulolytic and amylolytic enzymes. Int J Energy Clean Environ 21(1):41–58

Chen D, Lawton D, Thompson MR, Liu Q (2012) Biocomposites reinforced with cellulose nanocrystals derived from potato peel waste. Carbohydr Polym 90(1):709–716

Chen X, Ma X, Peng X, Lin Y, Yao Z (2018) Conversion of sweet potato waste to solid fuel via hydrothermal carbonization. Bioresour Technol 249:900–907

Chen X, Peng X, Ma X (2020) Investigation of Mannich reaction during co-liquefaction of microalgae and sweet potato waste: combustion performance of bio-oil and bio-char. Bioresour Technol 317:123993. https://doi.org/10.1016/j.biortech.2020.123993

Chintagunta AD, Jacob S, Banerjee R (2016) Integrated bioethanol and biomanure production from potato waste. Waste Manag 49:320–325

Chohan NA, Aruwajoye GS, Sewsynker-Sukai Y, Kana EG (2020) Valorisation of potato peel wastes for bioethanol production using simultaneous saccharification and fermentation: process optimization and kinetic assessment. Renew Energy 146:1031–1040

Cristóbal J, Caldeira C, Corrado S, Sala S (2018) Techno-economic and profitability analysis of food waste biorefineries at European level. Bioresour Technol 259:244–252

Daimary N, Eldiehy KS, Boruah P, Deka D, Bora U, Kakati BK (2022) Potato peels as a sustainable source for biochar, bio-oil and a green heterogeneous catalyst for biodiesel production. J Environ Chem Eng 10(1):107108. https://doi.org/10.1016/j.jece.2021.107108

Dhingra D, Michael M, Rajput H (2012) Physico-chemical characteristics of dietary fibre from potato peel and its effect on organoleptic characteristics of biscuits. J Agric Eng 49(4):25–32

Ebrahimian F, Karimi K (2020) Efficient biohydrogen and advanced biofuel coproduction from municipal solid waste through a clean process. Bioresour Technol 300:122656. https://doi.org/10.1016/j.biortech.2019.122656

FAO et al. (2020) The state of food security and nutrition in the world 2020. In: Transforming food systems for affordable healthy diets, https://doi.org/FAO, Rome

Faruk MO, Mumtaz T, Rashid HO, Zhang Y, Liu S, Wu H (2014) Supplementary effect of potato peel hydrolysate on the citric acid production by *Aspergillus niger* CA16. Biotechnology 9(8): 308–310

Ferraren-De Cagalitan DDT, Abundo MLS (2021) A review of biohydrogen production technology for application towards hydrogen fuel cells. Renew Sust Energ Rev 151:111413. https://doi.org/10.1016/j.rser.2021.111413

Ferreira AF, Ribau JP, Silva CM (2011) Energy consumption and CO_2 emissions of potato peel and sugarcane biohydrogen production pathways, applied to Portuguese road transportation. Int J Hydrog Energy 36(21):13547–13558

Fu BR, Sun Y, Wang YX (2013) Construction of biogas fermentation system based on feedstock of potato residue. J Northeast Agric Univ 44:102–106

Gil A, Siles JÁ, Serrano A, Martín MÁ (2015) Mixture optimization of anaerobic co-digestion of tomato and cucumber waste. Environ Technol 36(20):2628–2636

Habeebullah K, Nielsen NS, Jacobsen C (2010) Antioxidant activity of potato peel extracts in a fish-rapeseed oil mixture and in oil-in-water emulsions. J Am Oil Chem Soc 87(11):1319–1332

Habib F, Shah WH (2004) Utilization of potato peels extract as a natural antioxidant in soy bean oil. Food Chem 85(2):215–220

Hijosa-Valsero M, Paniagua-García AI, Díez-Antolínez R (2018) Industrial potato peel as a feedstock for biobutanol production. New Biotechnol 46:54–60

Hossain MB, Tiwari BK, Gangopadhyay N, O'Donnell CP, Brunton NP, Rai DK (2014) Ultrasonic extraction of steroidal alkaloids from potato peel waste. Ultrasonics Sonochem 21(4): 1470–1476

Islam AK, Dunlop PS, Hewitt NJ, Lenihan R, Brandoni C (2021) Bio-hydrogen production from wastewater: a comparative study of low energy intensive production processes. Clean Technol 3(1):156–182

Jadhav SA, Kataria PK, Bhise KK, Chougule SA (2013) Amylase production from potato and banana peel waste. Int J Curr Microbiol App Sci 2(11):410–414

Jeddou KB, Chaari F, Maktouf S, Nouri-Ellouz O, Helbert CB, Ghorbel RE (2016) Structural, functional, and antioxidant properties of water-soluble polysaccharides from potato peels. Food Chem 205:97–105

Kamboj R, Dhanya MS (2021) Orange peel extract enhanced sugar recovery and butanol production from potato peel by *Clostridium acetobutylicum*. Int J Green Energy 18(10):987–997

Kasapidou E, Sossidou E, Mitlianga P (2015) Fruit and vegetable co-products as functional feed ingredients in farm animal nutrition for improved product quality. Agriculture 5(4):1020–1034

Keswani MH, Patil AR (1985) The boiled potato peel as a burn wound dressing: a preliminary report. Burns 11(3):220–224

Khawla BJ, Sameh M, Imen G, Donyes F, Dhouha G, Raoudha EG, Oumèma NE (2014) Potato peel as feedstock for bioethanol production: a comparison of acidic and enzymatic hydrolysis. Ind Crop Prod 52:144–149

Kumari P, Mehta A, Sehgal R, Ray RC, Gupta R (2022) Microbial enzymes and organic acids production from vegetables and fruit wastes and their applications. In: Ray RC (ed) Fruits and vegetable wastes: valorization to bioproducts and platform chemicals. Springer, Singapore, pp 237–257

Liang S, McDonald AG (2014) Chemical and thermal characterization of potato peel waste and its fermentation residue as potential resources for biofuel and bioproducts production. J Agric Food Chem 62(33):8421–8429

Liang S, McDonald AG, Coats ER (2014) Lactic acid production with undefined mixed culture fermentation of potato peel waste. Waste Manag 34(11):2022–2027

Liang S, Han Y, Wei L, McDonald AG (2015a) Production and characterization of bio-oil and bio-char from pyrolysis of potato peel wastes. Biomass Convers Biorefinery 5:237–246

Liang S, McDonald AG, Coats ER (2015b) Lactic acid production from potato peel waste by anaerobic sequencing batch fermentation using undefined mixed culture. Waste Manag 45:51–56

Mabrouk ME, El Ahwany AM (2008) Production of 946-mannanase by *Bacillus amylolequifaciens* 10A1 cultured on potato peels. Afr J Biotechnol 7(8)

Macias-Corral M, Samani Z, Hanson A, Smith G, Funk P, Yu H, Longworth J (2008) Anaerobic digestion of municipal solid waste and agricultural waste and the effect of co-digestion with dairy cow manure. Bioresour Technol 99(17):8288–8293

Mahmood AU, Greenman J, Scragg AH (1998) Orange and potato peel extracts: analysis and use as Bacillus substrates for the production of extracellular enzymes in continuous culture. Enzym Microb Technol 22(2):130–137

Maldonado AFS, Mudge E, Gänzle MG, Schieber A (2014) Extraction and fractionation of phenolic acids and glycoalkaloids from potato peels using acidified water/ethanol-based solvents. Food Res Int 65:27–34

Marone A, Izzo G, Mentuccia L, Massini G, Paganin P, Rosa S, Signorini A (2014) Vegetable waste as substrate and source of suitable microflora for bio-hydrogen production. Renew Energy 68:6–13

Mars AE, Veuskens T, Budde MA, Van Doeveren PF, Lips SJ, Bakker RR, Claassen PA (2010) Biohydrogen production from untreated and hydrolyzed potato steam peels by the extreme thermophiles *Caldicellulosiruptor saccharolyticus* and *Thermotoga neapolitana*. Int J Hydrog Energy 35(15):7730–7737

Maske NS, Satyanarayan S (2012) Effect of special fish feed prepared using potato peels on fresh water fish *Labeo rohita*. J Ind Pollut Contr 29(1)

Meenakshi A, Kumaresan R (2014) Ethanol production from corn, potato peel waste and its process development. Int J ChemTech Res 6(5):2843–2853

Mohapatra S, Ramachandran S, Ray RC (2018) Bioethanol from biorenewablefeedstocks: technology, economics and challenges. In: Ray RC, Ramachandran S (eds) Bioethanol production from food crops. Academic, London, pp 3–27

Moodley P, Ray RC, Kana EBG (2023) Advances in lignocellulosic biofuel production systems. Elsevier. 316p

Muhondwa JP, Martienssen M, Burkhardt M (2015) Feasibility of anaerobic digestion of potato peels for biogas as mitigation of greenhouse gases emission potential. Int J Environ Res 9(2): 481–488

Mukherjee S, Bandyopadhayay B, Basak B, Mandal N, Apurba DEY, Mondal B (2012) An improved method of optimizing the extraction of polyphenol oxidase from potato (*Solanum tuberosum* L.) peel. Notulae Scientia Biologicae 4(1):98–107

Nara K, Miyoshi T, Honma T, Koga H (2006) Antioxidative activity of bound-form phenolics in potato peel. Biosci Biotechnol Biochem 70(6):1489–1491

Niphadkar SS, Rathod VK (2015) Ultrasound-assisted three-phase partitioning of polyphenol oxidase from potato peel (Solanum tuberosum). Biotechnol Prog 31(5):1340–1347

Niphadkar SS, Vetal MD, Rathod VK (2015) Purification and characterization of polyphenol oxidase from waste potato peel by aqueous two-phase extraction. Prep Biochem Biotechnol 45(7):632–649

Onyeneho SN, Hettiarachchy NS (1993) Antioxidant activity, fatty acids and phenolic acids compositions of potato peels. J Sci Food Agric 62(4):345–350

Panda SK, Sahu L, Behera SK, Ray RC (2020) Research and production of organic acids and their potential for industry. In: Molina G, Gupta VK, Singh BN (eds) Bioprocessing for food ingredients production. Wiley-Blackwell, New Jersey. https://doi.org/10.1002/9781119434436.ch9

Pandey AK, Mishra BK, Arora A, Singh S, Lata, Ray RC (2011) Bioaugmentation and biovalorization of agro-food and beverage industry effluents. In: Singh A, Parmar N, Kuhad RC (eds) Bioaugmentation, biostimulation and biocontrol, Part 1, Varma AK (series eds.) Soil biology series. Springer, Heidelberg, pp 85–126

Pandit NP, Ahmad N, Maheshwari SK (2012) Vermicomposting biotechnology an eco-loving approach for recycling of solid organic wastes into valuable biofertilizers. J Biofertil Biopestic 2:113

Patel SK, Gupta RK, Kalia VC, Lee JK (2021) Integrating anaerobic digestion of potato peels to methanol production by methanotrophs immobilized on banana leaves. Bioresour Technol 323: 124550

Pathak PD, Mandavgane SA (2015) Preparation and characterization of raw and carbon from banana peel by microwave activation: application in citric acid adsorption. J Environ Chem Eng 3(4):2435–2447

Pathak PD, Mandavgane SA, Puranik NM, Jambhulkar SJ, Kulkarni BD (2018) Valorization of potato peel: a biorefinery approach. Crit Rev Biotechnol 38(2):218–230

Rasheed H, Ahmad D, Bao J (2022) Genetic diversity and health properties of polyphenols in potato. Antioxidants 11:603

Ray RC (2020) Sustainable biofuels: challenges and opportunities. Academic. 502p

Ray RC (2022) Fruits and vegetable wastes: valorization into bioproducts and platform chemicals. Springer. 448p

Ray RC, Ramachandran S (2018) Bioethanol production from food crops: sustainable sources, interventions, and challenges. Academic (Elsevier)

Ray RC, Swain MR (2011) Bio-ethanol, bioplastics and other fermented industrial products from cassava starch and flour. In: Pace CM (ed) Cassava: farming, uses and economic impacts. Nova Science Publishers Inc., Hauppauge, NY, pp 1–32

Rommi K, Rahikainen J, Vartiainen J, Holopainen U, Lahtinen P, Honkapää K, Lantto R (2016) Potato peeling streams as raw materials for biopolymer film preparation. J Appl Polym Sci 133(5):42862

Sanaei-Moghadam A, Abbaspour-Fard MH, Aghel H, Aghkhani MH, Abedini-Torghabeh J (2014) Enhancement of biogas production by co-digestion of potato pulp with cow manure in a CSTR system. Appl Biochem Biotechnol 173:1858–1869

Santos TC, Gomes DPP, Bonomo RCF, Franco M (2012) Optimisation of solid state fermentation of potato peel for the production of cellulolytic enzymes. Food Chem 133(4):1299–1304

Shukla J, Kar R (2006) Potato peel as a solid state substrate for thermostable α-amylase production by thermophilic Bacillus isolates. World J Microbiol Biotechnol 22:417–422

Singh N, Rajini PS (2008) Antioxidant-mediated protective effect of potato peel extract in erythrocytes against oxidative damage. Chem Biol Interact 173(2):97–104

Singh PP, Saldaña MD (2011) Subcritical water extraction of phenolic compounds from potato peel. Food Res Int 44(8):2452–2458

Singh N, Kamath V, Rajini PS (2005a) Attenuation of hyperglycemia and associated biochemical parameters in STZ-induced diabetic rats by dietary supplementation of potato peel powder. Clin Chim Acta 353(1–2):165–175

Singh N, Kamath V, Rajini PS (2005b) Protective effect of potato peel powder in ameliorating oxidative stress in streptozotocin-diabetic rats. Plant Foods Hum Nutr 60(2):49–54

Singh N, Kamath V, Narasimhamurthy K, Rajini PS (2008) Protective effect of potato peel extract against carbon tetrachloride-induced liver injury in rats. Environ Toxicol Pharmacol 26(2): 241–246

Singh A, Sabally K, Kubow S, Donnelly DJ, Gariepy Y, Orsat V, Raghavan GSV (2011) Microwave-assisted extraction of phenolic antioxidants from potato peels. Molecules 16(3): 2218–2232

Sinha P, Pandey A (2014) Biohydrogen production from various feedstocks by *Bacillus firmus* NMBL-03. Int J Hydrog Energy 39(14):7518–7525

Soltaninejad A, Jazini M, Karimi K (2022) Sustainable bioconversion of potato peel wastes into ethanol and biogas using organosolv pretreatment. Chemosphere 291:133003

Sooch BS, Mann MK, Sharma P, Ray RC (2022) Phenolics and other bioactive compounds from vegetables and fruit wastes: extraction methods and their possible utilizations. In: Ray RC (ed) Fruits and vegetable wastes: valorization to bioproducts and platform chemicals. Springer, Singapore, pp 81–114

Sujeeta KM, Mehta S, Sihag K (2018) Optimization of conditions for bioethanol production from potato peel waste. Int J Chem Stud 6(1):2021–2024

Swain MR, Ray RC, Patra JK (2012) Citric acid: microbial production and applications in food and pharmaceutical applications. In: Vargas DA, Medina JV (eds) Citric acid: synthesis, properties and applications. Nova Science Publishers Inc., Hauppauge, NY, pp 97–118

Taher BI, Fickers P, Chniti S, Hassouna M (2017) Optimization of enzymatic hydrolysis and fermentation conditions for improved bioethanol production from potato peel residues. Biotechnol Prog 33(2):397–406

Tammineni N, Ünlü G, Min SC (2013) Development of antimicrobial potato peel waste-based edible films with oregano essential oil to inhibit *Listeria monocytogenes* on cold-smoked salmon. Int J Food Sci Technol 48(1):211–214

Tawila MA, Omer HA, Gad SM (2008) Partial replacing of concentrate feed mixture by potato processing waste in sheep rations. Cellulose 20(2.50):26–73

Tiwari V, Maji S, Kumar S, Prajapati G, Yadav R (2016) Use of kitchen waste based bio-organics for strawberry (*Fragaria* × *ananassa* Duch) production. Afr J Agric Res 11(4):259–265

Ude MU, Oluka I, Eze PC (2020) Optimization and kinetics of glucose production via enzymatic hydrolysis of mixed peels. J Bioresour Bioprod 5(4):283–290

Uslu H, İnci I, Bayazit SS (2010) Adsorption equilibrium data for acetic acid and glycolic acid onto Amberlite IRA-67. J Chem Eng Data 55(3):1295–1299

Wei L, Liang S, McDonald AG (2015) Thermophysical properties and biodegradation behavior of green composites made from polyhydroxybutyrate and potato peel waste fermentation residue. Ind Crop Prod 69:91–103

Wijngaard HH, Ballay M, Brunton N (2012) The optimisation of extraction of antioxidants from potato peel by pressurised liquids. Food Chem 133(4):1123–1130

Potato Peel Enrichment in Functional Food and Feed

3

Harsh Kumar, Shivani Guleria, Rajni Dhalaria, Richard Cimler, Vandana Guleria, Ruchi Sharma, Rajni Choudhary, Sivakumar Manickam, and Kamil Kuča

Abstract

The potato is the fourth most common crop cultivated globally every year. Potato peel is the major waste released by potato-processing industries and poses a global environmental threat. Potato peel contains beneficial nutritional rich

H. Kumar · R. Cimler
Centre of Advanced Technologies, Faculty of Science, University of Hradec Kralove, Hradec Kralove, Czech Republic
e-mail: richard.cimler@uhk.cz

S. Guleria
Department of Biotechnology, TIFAC-Centre of Relevance and Excellence in Agro and Industrial Biotechnology (CORE), Thapar Institute of Engineering and Technology, Patiala, India

R. Dhalaria
School of Biological and Environmental Sciences, Shoolini University of Biotechnology and Management Sciences, Solan, India

V. Guleria · R. Sharma
School of Bioengineering and Food Technology, Shoolini University of Biotechnology and Management Sciences, Solan, India

R. Choudhary
School of Health Sciences & Technology, University of Petroleum and Energy Studies, Dehradun, Uttarakhand, India

S. Manickam
Petroleum and Chemical Engineering, Faculty of Engineering, Universiti Teknologi Brunei, Bandar Seri Begawan, Brunei
e-mail: manickam.sivakumar@utb.edu.bn

K. Kuča (✉)
Department of Chemistry, Faculty of Science, University of Hradec Kralove, Hradec Kralove, Czech Republic
e-mail: kamil.kuca@uhk.cz

R. C. Ray (ed.), *Roots, Tubers, and Bulb Crop Wastes: Management by Biorefinery Approaches*, https://doi.org/10.1007/978-981-99-8266-0_3

organic and bioactive compounds that can be used in a variety of ways. Aside from serving as a natural antioxidant in food, phenolic compounds found in potato peel can be specifically beneficial in chronic disease treatment. Increased demand for food products based on animals has had a direct impact on feed supply and price. This has sparked the interest of researchers in the potential of low-cost, less competitive, and locally available material as feed ingredients. Therefore, the current chapter discusses the health-promoting and nutritional value of potato peel. Furthermore, a thorough discussion of the food and feed industries' current knowledge of this bio-residue.

Keywords

Potato peel · Enrichment · Nutritional profile · Functional food · Feed

Abbreviations

ABTS	2,2′-Azino-bis(3-ethylbenzothiazoline-6-sulfonic acid)
BHA	Butylated hydroxyanisole
BHT	Butylated hydroxytoluene
BWG	Body weight gain
DPPH	2,2-Diphenyl-1-picrylhydrazyl
DW	Dry weight
FAO	Food and Agriculture Organization
FCR	Feed conversion ratio
FI	Feed intake
FRAP	Ferric reducing antioxidant power
GHGs	Greenhouse gases
IPPM	Irish potato peel meal
MDA	Malondialdehyde
PhE	Phenolic extract
PP	Potato peels
PPE	Potato peel extract
PPF	Potato peel fibers
PPWs	Potato peel wastes
SPPM	Sun-dried Irish potato peel
TBA	Thiobarbituric acid
TBARS	2-Thiobarbituric acid reactive substances
TBHQ	*tert*-Butyl hydroquinone
TEAC	Trolox equivalent antioxidant capacity
TPC	Total phenolic contents

3.1 Introduction

Potato (*Solanum tuberosum* L.), belongs to the Solanaceae family and genus *Solanum*. It is one of the oldest crops harvested for food purposes. After wheat, rice, and maize, potato is the fourth most important food crop (Verma et al. 2021). The history of potatoes began in the Andes Mountains of South America (between 8000 and 5000 BC) near Lake Titicaca which is located in the border provinces of Bolivia and Peru at 12,500 ft above sea level (Verma et al. 2021). The potato and its wild relatives can be traced back to Western South America where they have been consumed for at least 13,000 years. Potato gained importance in the 1530s when Spanish conquistadors came in search of gold in Peru. The Spanish carried potatoes to Europe in the years 1570 and 1593 and from there potatoes reached other continents (Verma et al. 2021). In India, the potato was first introduced by a Portuguese sailor in the early seventeenth century. In the twentieth century, potatoes became one of the highest and most extensively cultivated food crops on the planet. Worldwide, around 100 countries where potato is now grown widely including Europe, North America, former Soviet Union countries, Asia, Africa, and Latin America (Verma et al. 2021).

Many food industries use potatoes for starch, chips, French fries, flour, and canning. The processing of potatoes such as cutting, trimming, and peeling released a huge amount of potato peel wastes (PPWs) (Ebrahimian et al. 2022). Depending on the peeling method used (abrasive, steam, or lye), the residue stream accounts for 15–40% of the primary potato weight (Barampouti et al. 2021). To manage the PPWs, the landfilling system is used, but it has a huge impact on the environment due to the release of GHGs (Green House Gases) and poisonous leachates (Chohan et al. 2020; Ebrahimian and Karimi 2020) (Fig. 3.1).

However, these cost-free wastes are used as raw material for the production of bio-based products because it contains large amounts of starch and lignocelluloses (Taher et al. 2017). The PPWs can be reduced by applying techniques like biorefinery for the production of a wide range of bioproducts as well as the reduction of end waste residue, thereby supporting the long-term advancement of a circular bio-economy. Biological and thermochemical technologies can be used to process PPWs. The techniques like fermentation or anaerobic digestion referred to as biological processing involve a group of microorganisms to convert PPWs into valuable products (Abedini et al. 2020; Chohan et al. 2020; Patel et al. 2021; Adeleye et al. 2022). Thermochemical processing, instead, refers to the thermal breakdown of PPWs into valuable bioproducts via, hydrothermal liquefaction, carbonization, and pyrolysis (Chen et al. 2018, 2020; Daimary et al. 2022). PPWs have been extensively studied in the manufacturing of a broad range of valuable products, such as enzymes, biopolymers, biochemicals, biosorbents, biochar, biofuels, and organic acids. Therefore, the current chapter summarises the application of potato peel enrichment in the manufacturing of animal feeds, and functional foods.

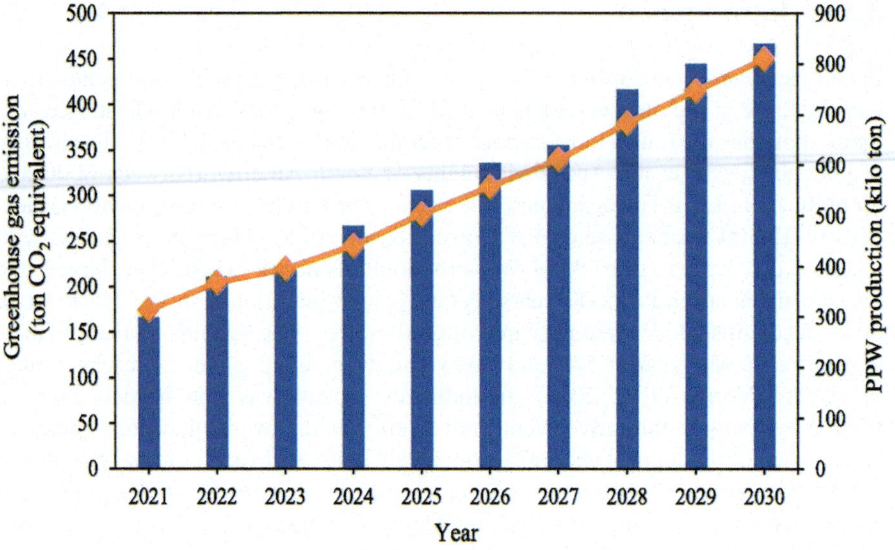

Fig. 3.1 GHG emissions and worldwide production of potato peel wastes (PPWs). (Source: Ebrahimian et al. (2022), (Licence Number: 5546500480123))

3.2 Global Production of Potato

In the year 2021, a total of 376 million tonnes of potatoes were produced globally according to a recent report by the Food and Agriculture Organisation (FAO). China and India are the major potato-producing countries with an overall production of 94 million tonnes and 54 million tonnes, respectively followed by Ukraine (21 million tonnes), and Russia (18.2 million tonnes) (PND 2023). In 2021, the total harvested area was estimated at around 18,132,694 ha, producing a global average of about 21 tonnes per ha according to FAO data (PND 2023). Germany was the largest producer in the European Union, making up 11,312,100 metric tonnes. In the North American region, the United States accounts for 18,582,370 metric tonnes. In the African region, Egypt produced 6,902,816 metric tonnes of potatoes in 2021. Peru was South America's leading producer, with a total production of 56,661,443 metric tonnes (PND 2023).

3.3 Agronomic Features of Potato Peel: Taxonomy, Morphology, and Biodiversity

The tuber periderm comprises peel or skin as an outer or protecting layer against water loss and pest and pathogen attacks. However, it also helps to repair the damaged epidermis and thus acts as a wound-healing tissue (Barel and Ginzberg 2008). Under conditions such as tuber development, maturation, and postharvest handling epidermis becomes damaged, and the periderm along with phellogen, phelloderm cells, and phellem (or skin), work together to replace the epidermis (Barel and Ginzberg 2008). In the central section of the periderm, a meristematic layer phellogen is found that provides phellem cells through outward and phelloderm through inward cell divisions across tuber development (Fogelman et al. 2015). Many cell layers containing phellem have suberized cell walls due to the accumulation of a complex macromolecule known as suberin which is in charge of the protective properties and structural strength of skin (Fogelman et al. 2015; Tanios et al. 2020). Suberin is made up of two main components, one is aromatic and the other is aliphatic. These two domains help in distinguishing the suberin polymers of potatoes from other species. It is responsible for potato peels (PP's) impermeability to soluble compounds, water, and gases (Sampaio et al. 2020).

The formation of suberin takes place under normal environmental conditions, but wounding (biotic factors or mechanical injury) and environmental stress factors (e.g., aridity, salinity, etc.) stimulate its biosynthesis through the upregulation of suberin-related genes differentially expressed in the skin (*StKCS20*-like, *StCYP86A22*, *StFAR*3, and *StPOD72*-like genes) (Schreiber et al. 2005; Yang and Bernards 2006; Vulavala et al. 2017). Periderm has been further subdivided into two categories. ie. native and wound periderm. The native one is formed naturally during tuber development and wound periderm is formed when tubers are injured and skin abrasion is noticed (Sabba and Lulai 2002). Both periderm types are hydrophobic due to their high lipid content; however, their water permeability differs. The permeability of wound periderm is greater than that of native periderm (Schreiber et al. 2005; Yang and Bernards 2006). The chemical constituents of periderm, thickness, and strength of cell walls of phellogen and biosynthesized of the suberin in phellem cells and deposition of waxes occurred during maturation of tuber. As a result, the tissues of the periderm give essential protection to tubers against external factors (Lulai and Freeman 2001).

The accumulation of pigments in the periderm and peripheral tissues of tuber cortices is attributed to the variability in tuber skin color (red, purple, and white color) in various genotypes of potatoes (Zhao et al. 2015). The most familiar pigments such as pelargonidin and peonidin (anthocyanins) located in the vacuoles are extracted from colored skin tubers (Fogelman et al. 2015). The color stability and saturation of pigment are influenced by a variety of factors which include the presence of vacuolar peroxidases, co-pigmentation, interactions with metallic ions, and soil pH (Sampaio et al. 2020). Agronomic factors may also have an impact on the tuber skin quality and development. According to Vulavala et al. (2016), Si-fertilization leads to enhanced skin cell area, upregulation of suberin biosynthetic genes, and delay skin maturation as a result, improved lignification, and suberization also enhances resistance of the skin to injury and tuber storage ability. The process of skin-set is accelerated in potato crops by mechanical, physical, or chemical vine killing, permitting the early harvest without any increased mechanical wounding risk associated with unripe skin (Sampaio et al. 2020). The farming system also influences skin color, for example, the skin color of the Arinda cultivar was improved by organic farming, as reported by Lombardo et al. (2017). Furthermore, the after-harvest synergistic action of light and wounding on the deposition of steroidal glycoalkaloids in tuber skins reported by Nie et al. (2019), while in storage sprouting and α-solanine content, inhibited by the use of nano-silica which has been associated to adverse effects on living beings (Zhang et al. 2018). Although the amount of glycoalkaloids may be influenced not just after harvesting but also by environmental conditions and genetic factors, growth and farming practices (Sampaio et al. 2020).

3.4 Nutritional Composition of Potato Peel

Potato peel (PP) has a high nutritional value, as it is a rich source of protein, starch, and dietary fiber (Jeddou et al. 2017). Based on genotype there is great variability in the metabolite profile of PP (Inostroza-Blancheteau et al. 2018). Carbohydrates are the most plentiful macronutrient in PP, accounting for 69–88 g/100 g dw (dry weight) (Table 3.1). Starch accounts for 30–52% dw percent of whole carbohydrate content (Sampaio et al. 2020). PP contains a good amount of soluble and insoluble dietary fibers (Jeddou et al. 2016), which can provide numerous health benefits such as lowering blood cholesterol, controlling diabetes, and regulating intestinal health (Jeddou et al. 2017). Elkahoui et al. (2018) investigated the total dietary fiber content of the Russet variety of both organic and non-organic PP and found similar amounts of dietary fiber in both farming types (21.4% and 22.39% dw respectively). On the other hand, a gravimetric enzymatic method was used by Jeddou et al. (2017) and described a higher dietary fiber in the variety Spunta (commercially unknown). Another study compared the dietary fiber content (insoluble and soluble) of commercial and Spunta varieties in PP, yielding 19.23% and 19.59% insoluble dietary fiber, respectively, and 9.21% and 10.14% soluble dietary fiber, respectively. Kumari et al. (2017) described a 51% higher total dietary fiber content in the PP of the Lady Claire variety. The total content of soluble sugars in PP is low, ranging between 0.9% and 1.0% dw (Amado et al. 2014; Arapoglou et al. 2010), whereas 0.6% dw fermentable reducing sugars are present (Arapoglou et al. 2010). PP has been reported to contain glucose, fructose, sucrose galactose, rhamnose, and arabinose, with glucose being the most plentiful soluble sugar (Choi et al. 2016; Jeddou et al. 2016). Choi et al. (2016) examined the PP of genotypes Atlantic, Goun, Superior, K1, K20, and K30 (Korean potato genotypes) for soluble sugar content, the results showed the presence of sucrose (290–427 mg/100 g dw), glucose (566–723 mg/100 g dw), and fructose (433–683 mg/100 g dw). In another study, the PP of the Spunta variety is reported to contain 76.25% glucose, 3.84% galactose, 0.51% rhamnose, and 0.19% arabinose (Jeddou et al. 2016) analyzed. The physicochemical properties of water-soluble polysaccharides derived from PP waste were analyzed by Jeddou et al. (2016). The polysaccharide-rich extracts showed high water-holding (4.097 ± 0.537 (g H_2O/g)) and fat-binding (4.398 ± 0.04 (g oil/g)) capacities. Additionally, the low molecular weight oligosaccharides are produced by the hydrolysis of water-soluble polysaccharides by the enzyme (Jeddou et al.

Table 3.1 Nutritional value (mean values ± standard deviation) of *Solanum tuberosum* peels, as reported in the literature over the last 10 years. Values are expressed in g/100 g dw (Sampaio et al. 2020, License Number: 5546590714352)

	Potato peel variety and origin								
Proximate	Organic Russet—United States	Non-organic Russet—United States	Red potato—United States	Gold potato—United States	Lady Rosetta—Ireland	Lady Claire—Ireland	Spunta—Tunisia	Agria—Spain	Unknown variety—Greece
Moisture	3.67	3.78	4.46	5.66	6.98 ± 0.05	4.08 ± 0.04	7.3 ± 0.3	7.30 ± 0.23	NR
Carbohydrates	76	71	72	70	72.53 ± 0.08	77.38 ± 0.65	88.0 ± 4.4	86.97 ± 0.43	68.7
Protein	11.98	17.19	15.99	14.17	11.17 ± 0.03	12.44 ± 0.09	2.099 ± 0.105	6.47 ± 0.23	8
Fat	1.12	1.1	0.81	1.17	2.09 ± 0.01	1.27 ± 0.38	0.733 ± 0.037	0	2.6
Ash	7.32	7.34	6.69	9.12	7.24 ± 0.02	4.83 ± 0.13	0.906 ± 0.006	5.46 ± 0.17	6.34

NR not reported

2018). In this study, the oligosaccharide-rich extracts showed great foaming and emulsion properties.

The other macronutrient is protein, the second most common present in PP, despite the amount varies between studies, in the range of 2–17 g/100 dw (Table 3.1). A total crude protein, essential amino acid, free amino acid, and an asparagine content in the range of 9.52–10.58 g/100 g dw,429–666 mg/100 g dw, 1383–2077 mg/100 g dw, 90.4–115.8 mg/g dw respectively present in PP reported by Choi et al. (2016). The fat content of PP varies greatly between varieties, with some authors reporting no fat and others reporting values up to 2.6 g/100 g dw (Table 3.1). The presence of two unsaturated fatty acids omega-6 (9,10,11-trihydroxy-12(Z)-octadecenoic acid) and an omega-3 (9,10,11-trihydroxy-12(Z),15(Z)-octadecadienoic acid) found in PP of the Chinese variety reported by Wu et al. (2012). Moreover, the existence of quinic acid (0.63–0.71 mg/g dw) is reported by the same authors (Wu et al. 2012); which aids in diabetes treatment, by increasing pancreatic beta cell insulin secretion (Heikkilä et al. 2019). Singh et al. (2011) reported the presence of ascorbic acid (1.44 ± 0.5 mg/g dw) in the PP of Russet Burbank (Canadian cultivar). As per the reports available, PP appears to be a good source of nutritious substances, especially dietary fiber and can be included in the human diet for increasing the nutritional value of foods. Although, there is a difference in the chemical and nutritional composition of PP that can be associated with a variety of factors such as varietal differences in addition to agronomic and environmental aspects (Kumari et al. 2017).

3.5 Health-Promoting Activities of Potato Peel

Because of its health-promoting properties, potato peel has sparked a lot of interest. The presence of phenolic compounds in PP contributes to its antioxidant properties, and including it in a human diet has health benefits (Nara et al. 2006). Phenolics found in PP extract have been studied for their medicinal properties such as antibacterial, anti-diabetic, hepatoprotective, and antioxidant activities (Singh et al. 2020). The study carried out by de Sotillo et al. (1998) stated that PP extract is safe and can be utilized as a natural antioxidant in food preservation as it is found to be non-mutagenic in *Escherichia coli* and *Salmonella typhimurium* microsome assay. Furthermore, potato peel extract (PPE) can be used in food as a dietary antioxidant component to suppress oxidative stress and slow the progression of degenerative and chronic diseases (Singh and Rajini 2004, 2008; Singh et al. 2005). Studies have shown that PP can help in improving nutritional value and in retarding the oxidation process when added to food preparations (Al-Weshahy and Rao 2009; Pathak et al. 2017).

A great free radical-scavenging activity of the PPE was carried out by Singh and Rajini (2004) under in vitro conditions. There was a strong correlation among both antioxidant activity and total phenolic content (TPC) of PP as reported by Singh et al. (2008). Phenolic compounds in PP play an important role as lipid peroxidation inhibitors and scavengers. They significantly prevent the morphological changes in rat red blood cells induced by exogenous oxidants. Al-Weshahy and Rao (2009) reported the Trolox equivalent antioxidant capacity (TEAC) of six potato variety's methanolic extracts of PP (grown in Ontario) in a range of 0.28–0.55 mM. The lowest was recorded in the potato variety Yukon Gold (light skin) and the highest in Purple Majesty and Siecle (red skin). The highest content of free phenolics and the highest antioxidant capacity in the purple peel extract (835.59 mol Trolox equivalents/g), when compared to the Russet, Yellow, and Innovator varieties. Albishi et al. (2013) reported the highest free phenolic content and antioxidant capacity (835.59 μmol Trolox equivalents/g) of the purple variety's peel extract when compared to Innovator, Yellow, and Russet varieties. The peel's DPPH (2,2-Diphenyl-1-picrylhydrazyl) radical scavenging activity is high, indicating that PP has the potential to be a valuable source of antioxidative compounds. The presence of caffeic and chlorogenic acid in free phenolics of PP was found to be significantly related to DPPH radical-scavenging activity (Nara et al. 2006). These acids (caffeic and chlorogenic acid) are reported to have 3.95 and 7.87 μmol of Trolox equivalents/g of dry matter DPPH radical-scavenging activity, respectively. These two acids (free phenolic acid) in PP were found to have a total scavenging activity of 57% (Nara et al. 2006). Another study found that using pressurized liquid and solid-liquid extraction processes, the value of 42.52 and 35.32 mg $FeSO_4$/g FRAP (Ferric reducing antioxidant power) respectively was observed in PP extract (Alvarez et al. 2014). Because of the high content of phenolics and flavonoids, the methanolic extract of PP showed high antioxidant capacity in all three assays, DPPH radical-scavenging activity, β-Carotene/linoleic acid bleaching, 2,2′-azino-bis (3-ethylbenzothiazoline-6-sulfonic acid) (ABTS) radical-scavenging activity (Mohdaly et al. 2010a, b). There was a reduction in oxidative stress, attenuated high blood sugar, and lipid-lowering effect as well as protective action against diabetic eye lens damage in streptozotocin-induced diabetic rats, dietary supplemented with PP powder (10%) reported by Singh et al. (2005). PP feeding remarkably reduced blood sugar levels and kidney and liver hypertrophy (diabetes-induced) in diabetic rats. Arun et al. (2015) described the highest phenolic content in PP extract and also valuable anti-diabetic and antioxidant properties. Yang et al. (2015) described the inhibitory effect of PP's ethanol extract of Jayoung cultivar (colored potato) in a mouse model against atopic dermatitis. It can be used to treat atopic dermatitis as an effective and safe therapeutic tool. It has been reported that PP extract protects erythrocyte membrane proteins from oxidative damage by decreasing the degree of lipid peroxidation (Singh and Rajini 2008). The high concentration of PP extract (10^5 μg/mL) revealed antibacterial activity against Gram-positive (*Bacillus cereus*) and Gram-negative (*S. typhimurium*, *E. coli*) bacteria (Sotillo et al. 1998). Singh et al. (2008) studied the hepatoprotective effect of PP extract in rats against CCl_4-induced liver damage and oxidative stress. It was observed that PP extract protected the liver by reducing the formation of hepatic malondialdehyde, lowering serum alanine transaminase and aspartate transaminase levels, and by the reduction in glutathione depletion. Moreover, PP extracts also normalized the altered activities induced by CCl_4 of antioxidant enzymes superoxide dismutase, catalase, glutathione peroxidase, and glutathione-*S*-transferase in rats to control levels.

3.6 Concept of Food Enrichment

Nutrients are progressively being added to improve the quality of food because many nutrients, especially micronutrients, are lost during the production process. White bread, for example, is the most commonly eaten fortified food, various vitamins are added to it after the process of bleaching, reducing its levels significantly (Kumar et al. 2022). The addition of nutrients to food is termed fortification and enrichment. A practice of increasing the content of one or more micronutrients such as vitamins and minerals intentionally in food items to enhance the nutritional value and provide public health benefits with minimal health risks is termed fortification (WHO, FAO). On the other hand, enrichment refers to the addition of micronutrients that are lost during processing conditions (Kumar et al. 2022).

3.6.1 Enrichment Applications in Cereal-Based Foods

Dietary fibers from various sources have been combined with wheat flour, and dietary fiber extracted from potato peel can be a valuable source as it contains 50% of dietary fiber (w/w) (Javed et al. 2019). Potato peel fiber (PPF) can help an economy boost by serving as a low-cost material with better chemical and physical properties compared to wheat bran in the baking industry (Javed et al. 2019). As per the reports, PPF has a lower starch content with higher water-holding capacity. Additionally, a few quality defects such as a musty smell due to the use of PPF can be eliminated by using extruded PP (Alonso et al. 2006). Furthermore, Abd-El-Magied (1991) reported the effect of PP in biscuit processing by replacing wheat flour with 5% and 10% (w/w) PP, which resulted in a smaller stack weight with an acceptable sensory score. Curti et al. (2016) studied the ability of the PP fiber in reducing the stealing of bread (0.4 g fiber/100 g flour). The authors reported the increase in frozen water content by adding potato fiber to bread and which resulted in soft bread crumbs after a storage of 7 days. At the end of the storage day, the crumb hardness values were lower for the formulations supplemented with PP (3.7 ± 0.6 N) than that of control formulations (4.5 ± 0.7 N).

Jeddou et al. (2017) reported the use of PP powder in cakes besides bread. In this study, the 2%, 5%, and 10% commercial unknown and Spunta variety's PP powder was supplemented in wheat flour. The PP powders contained a large level of protein and dietary fiber. The resulting cakes had a better texture overall, due to PP powder's fat absorption and water binding abilities. The 5% PP formulation of the Spunta variety showed 30.24% reduced cake hardness when compared with the control cake. Furthermore, substituting PP powder for wheat flour increased the elasticity-to-extensibility ratio and strength of the dough. A sensory evaluation based on consumer approval found that there is no significant difference in the formulations of control and fortified cake. However, increased insoluble fiber content (21.4%) was recorded in PP fortified cake to that of control (15.9%), and the same was reported for soluble dietary fiber (4.8% in fortified and 3.3%, in control cake). Finally, when compared to the control, the PP-fortified cakes had a darker color with lower L* values (lightness) and higher A* values (red color). These color changes could be associated with the color properties of PP powders, but they could also be due to the non-enzymatic browning effect of sugar present in PP. Other applications for PP powder include its use in biscuits as a wheat flour substitute, contributing to increased water absorption, stability of dough, and development time of dough, while also increasing the fiber content and antioxidant activities at the same time (Hallabo et al. 2018).

3.6.2 Meat-Based Foods

The PP extracts are reported to control meat lipid oxidation caused by radiation (Kanatt et al. 2005). In this study, the ethanolic extract of PP from Kufri Chandra Mukhi (Indian Potato variety) was added to lamb meat before irradiation (4 mL of PP extract (1%) in 100 g of meat, 5 kGy). Lipid peroxidation was evaluated in terms of carbonyl value and thiobarbituric acid (TBA) number after 7 days of cold storage (0–3 °C). A 54% lower TBA number and 30% lower carbonyl content was found in meat containing PP extract compared to meat containing no PP, highlighting the ability of PP extract in reducing meat product's lipid oxidation. Mansour and Khalil (2000) investigated the ability of PP extract in reducing lipid oxidation in ground beef patties (for 12 days, 5 °C). The antioxidant activity of the ethanolic extract of PP was evaluated using TBARS (2-thiobarbituric acid reactive substances), which revealed a 59.5% inhibition compared to the control (maximum at pH range 5.0–6). Furthermore, when compared to control samples, the extracts were able to control color change and lipid oxidation during cold storage in beef patties. In another study, Albishi et al. (2013) added four different PP methanolic-acetonic extracts derived from Innovator, Yellow, Purple, and Russet potato varieties in a fish meat model system to study the PP potential of inhibition of oxidation. Over a 7 days storage period, the extracts were used to delay the emergence of oxidative rancidity in cooked comminuted salmon. The results showed that the PP extracts of Innovator, Yellow, Purple, and Russet inhibited the TBARS formation by 31.4%, 9.48%, 39.7%, and 83.4%, respectively. Farvin et al. (2012) investigated the role of PP as a natural antioxidant source for retarding protein and lipid oxidation in another fish model (minced horse mackerel). The authors used both ethanol and water as solvents and compared two concentrations of the Sava variety's PP extracts (4.8 or 2.4 g/kg). When compared to control and water extracts, the ethanolic extracts of PP were better in inhibiting protein and lipid oxidation, resulting in lower levels of volatiles, carbonyl compounds, and peroxide values. For example, the addition of ethanolic extract of 4.8 or 2.4 g/kg resulted in 0.41 and 0.58 mEq/kg peroxide values, whereas the control and water extract samples resulted in 3.45 and 5.78 mEq/kg peroxide values respectively. The PP ethanolic extracts also protected fish meat from the loss of, tryptophan, tocopherol, and tyrosine residues. It was observed that ethanolic extracts constituted more phenolic compounds than water extracts.

3.6.3 Enrichment in Edible Oils

Lipid oxidation is a free radical chain reaction that reduces the shelf-life of several processed foods. The incorporation of synthetic antioxidants is the most effective strategy employed by the food industry to reduce the generation of toxic oxidation molecules and rancidity in food products. However, consumer concerns about the safety of synthetic food additives are increasing, as well as their growing preference for natural alternatives (Carocho and Ferreira 2013). Following this trend, some authors have used PP as an antioxidant to prevent the lipid oxidation of soybean oil. Some measurable factors can indicate the oxidation incidence, such as conjugated dienes, trienes, the production of peroxides (the primary products of lipid peroxidation), and *p*-anisidine values. According to Franco et al. (2016), incorporation of ethanolic extract of PP reduces the lipid oxidation indices in soybean oil such as conjugated dienes, peroxide, and *p*-anisidine values. The stability of soybean oil was examined over a 15 days storage period under accelerated oxidation conditions (60 ° C). Three concentrations of PP ethanolic extracts of the Agria variety were examined based on chlorogenic acid levels such as 31.94, 20.37, and 14.01 ppm. At the end of the storage period, the control oil had a 180.78 mEq/kg higher peroxide values than the PP enriched oils (below 162 mEq/kg oil).In terms of conjugated dienes, 52% of higher inhibition was recorded in oil treated with the PP's highest concentration (31.94 ppm chlorogenic acid equivalent), and 41% was recorded in a sample treated with BHT (butylated hydroxytoluene). This indicates that extracts of PP were more effective than BHT at reducing polyunsaturated fatty acid loss. Finally, in control samples, the levels of *p*-anisidine, associated with carbonyls (secondary oxidation products), were higher than in the other treatments. Mohdaly et al. (2010a, b) did an antioxidant comparison study of methanolic extracts of sugar beet and PP in soybean and sunflower oil at 70 °C, 72 h (accelerated oxidation conditions). The peroxide values of sunflower oil with sugar beet pulp, PP (200 ppm), BHA (butylated hydroxyanisole), BHT (butylated hydroxytoluene), and *tert*-butyl hydroquinone (TBHQ) were 13.8 ± 0.26, 11.9 ± 0.19, 13.9 ± 0.24, 13.4 ± 0.26 and 11.0 ± 0.26 mEq/kg, respectively whereas higher peroxide value (25.8 ± 0.24 mEq/kg) was observed in control oil at the end of storage time. The *p*-anisidine values for sugar beet pulp, PP, BHA, BHT, and TBHQ, containing samples were recorded as 9.40 ± 0.31, 9.10 ± 0.24, 8.84 ± 0.14, 9.36 ± 0.14, and 9.42 ± 0.14, respectively, whereas a value of 14.2 ± 0.14 was recorded in the untreated control sample. These findings indicate that PP extracts have lower antioxidant activity than TBHQ but higher than BHT and BHA. Similar findings were recorded for soyabean oil samples, with PP extracts at 100 and 200 ppm having a higher antioxidant activity than BHT and BHA at 200 ppm. Furthermore, vegetable oil with PP extracts appears to be a safe natural antioxidant substitute to inhibit lipid oxidation.

Another study examined the PP extract's antioxidant potential on the storage stability of a 5% oil-in-water emulsion, and fish-rapeseed oil mixture (Koduvayur Habeebullah et al. 2010). The antioxidant effect of Sava and Bintje's aqueous and ethanolic PP extract (Danish potato varieties) was evaluated using peroxide and anisidine and peroxide values, as well as sensory evaluation. It was observed that the Sava variety's PP-ethanolic extract (2400 mg/kg concentration) was 9 times more efficient in reducing anisidine value, 5.7 times more efficient in reducing peroxide value than BHT, while aqueous extract showed no antioxidant effect. According to the sensory analysis, rancidity intensities were higher in oil-water extracts than in oil-PP ethanolic extracts.

3.6.4 Films and Coatings

The development of biodegradable films and coatings derived from natural bio-based polymers such as proteins, polysaccharides, and lipids has sparked a new interest. These materials are less costly, recyclable, and most importantly biodegradable than conventional petroleum-based polymers (Zhang et al. 2021). Some packaging films and coatings that are biobased may have properties different from chemically synthesized plastic-based packaging films and coatings. Moreover, recent progress in the research and development sector to develop natural bio-based polymers has resulted in the generation of new materials with improved properties with the ability to extend food shelf life. Composite films and coatings with antibacterial and antioxidant properties are increasingly being used in the food packaging industry. Bioactive packaging materials may strengthen quality and food safety by delaying the process of oxidation and inhibiting the growth of harmful microbes responsible for spoilage (Khan et al. 2012). Biobased packaging films and coatings that have bioactive agents such as plant extracts, antimicrobial peptides, and oil are commonly referred to as active packaging materials. Based on the presence of bioactive agents, these are also categorized as antioxidant, antibacterial packaging films/coatings, or other active packaging materials.

Merino et al. (2021) reported that the addition of potato peel with polyglycerol-3 plasticizer resulted in decreased water vapor permeability, increased thermal resistance, and improved mechanical properties of the film. The films were prepared with polyglycerol-3 plasticizer (20 wt%) and their suitability as packaging for dry foods was estimated by measuring its oxygen permeability and the total migration of its components in the Tenax food simulant. The migration results (8.6 mg/dm^2) are as per the current food contact material regulation. Kang et al. (Kang and Min 2010; Kang et al. 2015) reported that by controlling the glycerol concentration and homogenization parameters, the moisture sensitivity of the film (potato peel-based) can be altered. Borah et al. (2017) reported that ultrasound-treated PP (60 min) required minimal PP for film development and had better characteristics. Water solubility, water vapor permeability, moisture absorption, elongation capacity, and breakage strength of the developed film were 38.92 ± 0.702%, 7.25 × 10^{-9} g/Pa h m, 12.88 ± 0.348%, 7.61 ± 0.824 mm, 242.01 ± 3.074 g respectively. Moreover adding 1.5% clove oil (antimicrobial agent) with film forming solution of selected film and wrapping it on the bread resulted in reducing the hardness, lowering the weight loss, and inhibiting microbial load on the surface of the bread after 5 days of storage. According to Xie et al. (2020), PP films containing curcumin (1–5%) in fresh pork led to a reduction in lipid oxidation with lower MDA (malondialdehyde) content during storage. The impact of Phenolic extract (PhE) of potato peel on films (starch-based) for mechanical and physiochemical properties was investigated by Lopes et al. (2021). PhE gave the neat films a yellowish color while maintaining their transparency. Furthermore, starch/PhE-based films revealed antioxidant activity (56–85% $ABTS^{•+}$ inhibition after 7 days), water tolerance, elasticity, and stretchability in contrast to the pristine materials. A sensory panel praised the smoked seabream fillet packaged containing films for its pleasant aroma and taste. The judges praised the golden color increase of the fillets, as well as their lower succulence and higher stiffness in comparison to the fillets packed with commercial plastic. The golden color increase of fillets was highly valued by the judges, as well as their higher stiffness and lower succulence compared to the fillets packed with commercial plastic.

3.7 Enrichment of Potato Peel in Animal Feeds

The expense of livestock feed is rising due to the increase in prices of fertilizer and extreme weather. Food waste can be used as a substitute for feed ingredients. This can reduce the cost of feed, disposal, and environmental pollution (Gebrechristos and Chen 2018). Food waste has a great nutritive value for animal feed (Myer and Johnson 2010). Moreover, the demand for farm animals is increasing in developing countries. Because feed scarcity is the primary issue, unconventional feed resources can play a vital role. The vegetable and fruit processing industry produces a large amount of waste. Such unusual resources are a great source of nutrients for livestock (Gebrechristos and Chen 2018). Being a natural source of fiber with a low level of protein and energy, PP is the most valuable food waste among others and could be used as an animal feed substitute (Chimonyo 2017). Moreover, food waste has traditionally been fed to swine without processing. As a result, the use of food waste as residual waste as animal feed ingredients and additives is essential.

3.7.1 Pig Feed

PP is produced in large quantities and has a higher moisture content, due to which putrefaction occurs in a very short period. This will facilitate their use as a source of dietary fiber and energy for pigs (Chimonyo 2017). Before giving food waste to pigs, it should be heat-treated to avoid the risk of animal diseases and harmful pathogens, thus protecting the pig meat consumers (Gebrechristos and Chen 2018). Potato waste feeding holds an antimicrobial activity by lowering coliform bacteria and improving weanling pig performance (Jin et al. 2008). Differences in the processing management, steaming temperature and time, starch proportion, and potato variety, therefore, influence the incorporation of correct levels of steamed PP. In finishing pigs, Edwards et al. (1986) and Nicholson et al. (1988) suggested the addition of 250 g/kg of steamed PP in the diet while Lunen et al. (1989) recommended up to 300 g/kg.

According to Edwards et al. (1986), a decrease in body weight gain (BWG) and dressing percentage in pigs fed with steamed PP could be due to high crude fiber (60.1 g/kg dry matter) and low dietary energy digestibility.

3.7.2 Chicken Feed

Abdel-Hafeez et al. (2018) found that combining 15% potato peel with enzymes in broiler feed reduces total cholesterol and lipoprotein cholesterol (low-density) serum while also improving body weight. In another study, it was observed that the addition of 5% PP with a mixture of cellulase, xylanase, protease, and mannanase (exogenous enzymes) had similar body weight gain, feed intake (FI), and nutrient digestibility to the control (Rahman et al. (2021). Moreover, peel at 10%, 15%, and 20% concentration along with exogenous enzymes when added to the broiler diet, had a negative effect on feed conversion ratio, body weight gain, and nutrient digestibility ($p < 0.05$). However, Dietary treatments did not affect carcass parameters ($p > 0.05$). Kpanja et al. (2020) investigated the blood biochemical and hematological profiles of broiler chickens, fed diets containing graded levels of IPPM (Irish potato peel meal) for maize. Based on 20% and 23% crude protein in finisher and starter respectively, four experimental diets were developed and formulated to have 0%, 5%, 10%, and 15% IPPM for maize grain replacement. In a completely randomized design, each treatment was replicated three times. Data on blood biochemical, hematological profiles, and growth indices were collected. During 7–28 days (starter period), feeding the test ingredient had no negative effects on feed cost/kg gain or daily feed intake, but feed conversion ratio (FCR) and weight gain were reduced above 5% maize replacement with the test material. During 28–63 days (finisher phase), feed intake increased significantly on the 15% replacement diet to that of control however in treatment-based diets it was not statistically different. The maximum daily gain was observed on a 5% replacement diet, but no significant difference was observed to that of 10%, 15%, and control diets. With IPPM, the feed conversion ratio increased significantly above 10% maize replacement. The test material had no negative effects on blood chemical or hematological parameters, as their values were all within normal limits. According to Wafar et al. (2017), SPPM (sun-dried Irish potato peel) can replace up to 75% maize in diets of broiler chicken without affecting, carcass characteristics, growth serum, or serum biochemistry.

3.7.3 Fish Feed

Ansari and Singh (2022) studied the effect of 5%, 10%, and 20% potato peel addition in the diet of *Catla catla* fish. A long-term dietary study found that a 20% potato peel diet resulted in a high percentage of weight gain, followed by 10%, 5% potato peel, and control. A similar trend was also observed in the gain of fish length with the highest percentage in 20%,10%, 5% potato peel, and control. The fish remained healthy throughout the study, and the survival rate was 100%. Sreeja et al. (2015) investigated the efficacy of different potato peel powder (feed additive) with *Bacillus* (probionts) on *Etroplus suratensis*, formulating two different pelleted diets with 40% protein and supplementing Diet A with 2% potato peel powder + 1% *Bacillus*, while diet B was supplemented with 2% CMC (carboxymethyl cellulose) with no probionts (control). During the 91-day experiment period, the growth rate was higher in *E. suratensis* fed with Diet A than in the control. El-Nadi et al. (2017) reported that the common carp (*Cyprinus carpio*) could endure up to 60% potato peal addition in their diet as a substitute for corn. According to Verma and Satyanarayan (2016), *Labeo rohita* feed with potato peel resulted in bright body scales and showed fingerling's growth response higher as compared to commercial fish feed.

3.8 Conclusion

During the processing of potatoes for human consumption, huge amounts of potato peel are generated as agro-waste. There is an urgent need of reinstating by-products and waste in the food supply chain and to associated industries from the perspective of the circular economy and the growing environmental impact of agro-industry. Potato peel is a potential candidate to be used as a nutraceutical constituent in food, which will increase the value of the potato industry. Several studies have shown that retrieving antioxidants and bioactive compounds from potato peel has a high potential to be used as antimicrobial or active ingredients in food products and supplements. The potency of potato peel meals has sparked increased interest as feed ingredients in animal diets. Although potato peel has been incorporated into animal feed, reviews have been mixed. The abundance of commercial feed additives (enzyme mixture) at reasonable prices and expertise in processing technologies and crop breeding to decrease antinutritional compounds will improve the animal utilization of these by-products. The anticipated increase in peel availability necessitates additional research into feed technologies that would help enhance their utilization by animals to minimize feed costs and potential environmental risks. The envision of an increase in the availability of peel requires thorough research into feed technologies that would aid in improving the utilization of feed by animals to reduce the cost of feed and related environmental issues.

References

Abdel-Hafeez HM, Saleh ES, Tawfeek SS, Youssef IM, Abdel-Daim AS (2018) Utilization of potato peels and sugar beet pulp with and without enzyme supplementation in broiler chicken diets: effects on performance, serum biochemical indices and carcass traits. J Anim Physiol Anim Nutr 102(1):56–66

Abd-El-Magied MM (1991) Effect of dietary fibre of potato peel on the rheological and organoleptic characteristics of biscuits. Egypt J Food Sci 19:293–300

Abedini A, Amiri H, Karimi K (2020) Efficient biobutanol production from potato peel wastes by separate and simultaneous inhibitors removal and pretreatment. Renew Energy 160:269–277

Adeleye T, Yeo H, Seth R, Hafez H, Biswas N (2022) Influence of mix ratio of potato peel and pig manure on reaction kinetics and methane recovery from anaerobic co-digestion. Can J Civ Eng 99(999):1–8

Albishi T, John JA, Al-Khalifa AS, Shahidi F (2013) Phenolic content and antioxidant activities of selected potato varieties and their processing by-products. J Funct Foods 5:590–600

Alonso A, Beunza JJ, Bes-Rastrollo M, Pajares RM, Martínez-González MA (2006) Vegetable protein and fiber from cereal are inversely associated with the risk of hypertension in a Spanish cohort. Arch Med Res 37:778–786

Alvarez VH, Cahyadi J, Xu D, Saldaña MD (2014) Optimization of phytochemicals production from potato peel using subcritical water: experimental and dynamic modeling. J Supercrit Fluids 90:8–17

Al-Weshahy A, Rao AV (2009) Isolation and characterization of functional components from peel samples of six potato varieties growing in Ontario. Food Res Int 42:1062–1066

Amado IR, Franco D, Sánchez M, Zapata C, Váazquez JA (2014) Optimisation of antioxidant extraction from *Solanum tuberosum* potato peel waste by surface response methodology. Food Chem 165:290–299

Ansari M, Singh N (2022) Growth surveillance of *Catla catla* fingerlings fed on potato peel diet. Ecol Environ Conserv 2022:S268–S272

Arapoglou D, Varzakas T, Vlyssides A, Israilides C (2010) Ethanol production from potato peel waste (PPW). Waste Manag 30(10):1898–1902

Arun KB, Chandran J, Dhanya R, Krishna P, Jayamurthy P, Nisha P (2015) A comparative evaluation of antioxidant and antidiabetic potential of peel from young and matured potato. Food Biosci 9:36–46

Barampouti EM, Christofi A, Malamis D, Mai S (2021) A sustainable approach to valorize potato peel waste towards biofuel production. Biomass Convers Bioref 13:8197–8208

Barel G, Ginzberg I (2008) Potato skin proteome is enriched with plant defense components. J Exp Bot 59(12):3347–3357

Borah PP, Das P, Badwaik LS (2017) Ultrasound treated potato peel and sweet lime pomace based biopolymer film development. Ultrason Sonochem 6:11–19

Carocho M, Ferreira ICFR (2013) A review on antioxidants, prooxidants, and related controversy: natural and synthetic compounds, screening and analysis methodologies and future perspectives. Food Chem Toxicol 51(1):15–25

Chen X, Ma X, Peng X, Lin Y, Yao Z (2018) Conversion of sweet potato waste to solid fuel via hydrothermal carbonization. Bioresour Technol 249:900–907

Chen X, Peng X, Ma X (2020) Investigation of mannich reaction during co-liquefaction of microalgae and sweet potato waste: combustion performance of bio-oil and bio-char. Bioresour Technol 317:123993

Chimonyo M (2017) A review of the utility of potato by-products as a feed resource for smallholder pig production. Anim Feed Sci Technol 227:107–117

Chohan NA, Aruwajoye G, Sewsynker-Sukai Y, Kana EG (2020) Valorisation of potato peel wastes for bioethanol production using simultaneous saccharification and fermentation: process optimization and kinetic assessment. Renew Energy 146:1031–1040

Choi SH, Kozukue N, Kim HJ, Friedman M (2016) Analysis of protein amino acids, non-protein amino acids and metabolites, dietary protein, glucose, fructose, sucrose, phenolic, and flavonoid content and antioxidative properties of potato tubers, peels, and cortexes (pulps). J Food Compos Anal 50:77–87

Curti E, Carini E, Diantom A, Vittadini E (2016) The use of potato fiber to improve bread physico-chemical properties during storage. Food Chem 195:64–70

Daimary N, Eldiehy KS, Boruah P, Deka D, Bora U, Kakati BK (2022) Potato peels as a sustainable source for biochar, bio-oil and a green heterogeneous catalyst for biodiesel production. J Environ Chem Eng 10(1):107108

Ebrahimian F, Karimi K (2020) Efficient biohydrogen and advanced biofuel coproduction from municipal solid waste through a clean process. Bioresour Technol 300:122656

Ebrahimian F, Denayer JF, Karimi K (2022) Potato peel waste biorefinery for the sustainable production of biofuels, bioplastics, and biosorbents. Bioresour Technol 360:127609

Edwards SA, Fairbairn CB, Capper AL (1986) Liquid potato feed for finishing pigs: feeding value, inclusion rate and storage properties. Anim Feed Sci Technol 15:129–139

Elkahoui S, Bartley GE, Yokoyama WH, Friedman M (2018) Dietary supplementation of potato peel powders prepared from conventional and organic Russet and non-organic gold and red potatoes reduces weight gain in mice on a high-fat diet. J Agric Food Chem 66(24):6064–6072

El-Nadi ASM, Abozaid H, Zeinhom MM (2017) Graded levels of potato peels effects as partial replacement for corn in diet of common carp (*Cyprinus carpio*). Int J ChemTech Res 10(2): 566–572

Farvin KHS, Grejsen HD, Jacobsen C (2012) Potato peel extract as a natural antioxidant in chilled storage of minced horse mackerel (*Trachurus trachurus*): effect on lipid and protein oxidation. Food Chem 131(3):843–851

Fogelman E, Tanami S, Ginzberg I (2015) Anthocyanin synthesis in native and wound periderms of potato. Physiol Plant 153(4):616–626

Franco D, Pateiro M, Rodr I, Zapata C (2016) Antioxidant ability of potato (*Solanum tuberosum*) peel extracts to inhibit soybean oil oxidation. Eur J Lipid Sci Technol 118(12):1891–1902

Gebrechristos HY, Chen W (2018) Utilization of potato peel as eco-friendly products: a review. Food Sci Nutr 6(6):1352–1356

Hallabo SAS, Helmy SA, Elhassaneen Y, Shaaban M (2018) Utilization of mango, onion and potato peels as sources of bioactive compounds in biscuits processing. Biosci Res 15(4): 3647–3657

Heikkilä E, Hermant A, Thevenet J, Bermont F, Kulkarni SS, Ratajczak J, Santo-Domingo J, Dioum EH, Canto C, Barron D, Wiederkher A, Marchi UD (2019) The plant product quinic acid activates Ca^{2+}-dependent mitochondrial function and promotes insulin secretion from pancreatic beta cells. Br J Pharmacol 176:3250–3323

Inostroza-Blancheteau C, de Oliveira Silva FM, Durán F, Solano J, Obata T, Machado M, Ferine AR, Reyes-Díaz M, Nunes-Nesi A (2018) Metabolic diversity in tuber tissues of native Chiloé potatoes and commercial cultivars of *Solanum tuberosum* ssp. *tuberosum* L. Metabolomics 14(10):1–15

Javed A, Ahmad A, Tahir A, Shabbir U, Nouman M, Hameed A (2019) Potato peel waste—its nutraceutical, industrial and biotechnological applications. AIMS Agric Food 4(3):807–823

Jeddou KB, Chaari F, Maktouf S, Nouri-ellouz O, Helbert CB, Ghorbel RE (2016) Structural, functional, and antioxidant properties of water-soluble polysaccharides from potatoes peels. Food Chem 205:97–105

Jeddou KB, Bouaziz F, Zouari-Ellouzi S, Chaari F, Ellouz-Chaabouni S, Ellouz- Ghorbel R, Nouri-Ellouz O (2017) Improvement of texture and sensory properties of cakes by addition of potato peel powder with high level of dietary fiber and protein. Food Chem 217:668–677

Jeddou KB, Bouaziz F, Helbert CB, Nouri-Ellouz O, Maktouf S, Ellouz-Chaabouni S, Ellouz-Ghorbel R (2018) Structural, functional, and biological properties of potato peel oligosaccharides. Int J Biol Macromol 112:114–1155

Jin Z, Yang YX, Choi JY, Shinde PL, Yoon SY, Hahn TW, Lim HT, Park Y, Hahm KS, Joo JW, Chae BJ (2008) Potato (*Solanum tuberosum* L. cv. *Gogu Valley*) protein as a novel antimicrobial agent in weanling pigs. J Anim Sci 86(7):1562–1572

Kanatt SR, Chander R, Radhakrishna P, Sharma A (2005) Potato peel extracts a natural antioxidant for retarding lipid peroxidation in radiation processed lamb meat. J Agric Food Chem 53:1499–1504

Kang HJ, Min SC (2010) Potato peel-based biopolymer film development using high-pressure homogenization, irradiation, and ultrasound. LWT Food Sci Technol 43(6):903–909

Kang HJ, Won MY, Lee SJ, Min SC (2015) Plasticization and moisture sensitivity of potato peel-based biopolymer films. Food Sci Biotechnol 24:1703–1710

Khan A, Khan RA, Salmieri S, Tien CL, Riedl B, Bouchard J, Chauve G, Tan V, Kamal MR, Lacroix M (2012) Mechanical and barrier properties of nanocrystalline cellulose reinforced chitosan based nanocomposite films. Carbohydr Polym 90:1601–1608

Koduvayur Habeebullah SF, Nielsen NS, Jacobsen C (2010) Antioxidant activity of potato peel extracts in a fish-rapeseed oil mixture and in oil-in-water emulsions. J Am Oil Chem Soc 87(11): 1319–1332

Kpanja EJ, Duru S, Kotso JA, Gonjoh PT (2020) Effect of diets containing graded levels of Irish potato peel meal (IPPM) on carcass quality, haematological and blood biochemical profile of broiler chickens. Nig J Anim Sci 2(3):242–249

Kumar H, Bhardwaj K, Kuča K, Sharifi-Rad J, Verma R, Machado M, Kumar D, Cruz-Martins N (2022) Edible mushrooms' enrichment in food and feed: a mini-review. Int J Food Sci Technol 57(3):1386–1398

Kumari B, Tiwari BK, Hossain MB, Rai DK, Brunton NP (2017) Original article ultrasound-assisted extraction of polyphenols from potato peels: profiling and kinetic modelling. Int J Food Sci Technol 52(6):1432–1439

Lombardo S, Pandino G, Mauromicale G (2017) The effect on tuber quality of an organic versus a conventional cultivation system in the early crop potato. J Food Compos Anal 62:189–196

Lopes J, Goncalves I, Nunes C, Teixeira B, Mendes R, Ferreira P, Coimbra MA (2021) Potato peel phenolics as additives for developing active starch-based films with potential to pack smoked fish fillets. Food Packag Shelf Life 28:100644

Lulai EC, Freeman TP (2001) The importance of phellogen cells and their structural characteristics in susceptibility and resistance to excoriation in immature and mature potato tuber (*Solanum tuberosum* L.) periderm. Ann Bot 88(4):555–561

Lunen TV, Anderson DM, Laurent AMST, Nicholson JWG, Dean PR (1989) The feeding value of potato steam peel for growing-finishing pigs. Can J Anim Sci 69:225–234

Mansour EH, Khalil AH (2000) Evaluation of antioxidant activity of some plant extracts and their application to ground beef patties. Food Chem 69(2):135–141

Merino D, Paul UC, Athanassiou A (2021) Bio-based plastic films prepared from potato peels using mild acid hydrolysis followed by plasticization with a polyglycerol. Food Packag Shelf Life 29: 100707

Mohdaly AAA, Sarhan MA, Mahmoud A, Ramadan MF, Smetanska I (2010a) Antioxidant efficacy of potato peels and sugar beet pulp extracts in vegetable oils protection. Food Chem 123(4): 1019–1026

Mohdaly AA, Sarhan MA, Smetanska I, Mahmoud A (2010b) Antioxidant properties of various solvent extracts of potato peel, sugar beet pulp and sesame cake. J Sci Food Agric 90:218–226

Myer R, Johnson H (2010) Feeding food wastes to livestock. J Anim Sci 1(11):1–2

Nara K, Miyoshi T, Honma T, Koga H (2006) Antioxidative activity of bound-form phenolics in potato peel. Biosci Biotechnol Biochem 70:1489–1491

Nicholson JWG, Snoddon PM, Dean PR (1988) Digestibility and acceptability of potato steam peel by pigs. Can J Anim Sci 68:233–239

Nie X, Li C, Zhang G, Shao Z, Wang X, Shi H, Guo H (2019) Light exposure and wounding: synergistic effects on steroidal glycoalkaloid accumulation in potato tubers during storage. Int J Food Sci Technol 54(10):2939–2948

Patel SK, Gupta RK, Kalia VC, Lee JK (2021) Integrating anaerobic digestion of potato peels to methanol production by methanotrophs immobilized on banana leaves. Bioresour Technol 323: 124550

Pathak PD, Mandavgane SA, Puranik NM, Jambhulkar SJ, Kulkarni BD (2017) Valorization of potato peel: a biorefinery approach. Crit Rev Biotechnol 38:218–230

Potato News Today (2023) Global potato statistics: latest FAO data published. https://www.potatonewstoday.com/2023/01/21/global-potato-statistics-latest-fao-data-published/. Accessed 15 Mar 2023

Rahman MAU, Jamal U, Anwar U, Bilal MQ, Riaz M, Hussain M, Ahmad S (2021) Effects of potato peels inclusion with exogenous enzymes in broiler diet on growth performance, nutrients digestibility, and carcass characteristics. Sci Prog 104(4):00368504211061972

Sabba RP, Lulai EC (2002) Histological analysis of the maturation of native and wound periderm in potato (*Solanum tuberosum* L.) tuber. Ann Bot 90(1):1–10

Sampaio SL, Petropoulos SA, Alexopoulos A, Heleno SA, Santos-Buelga C, Barros L, Ferreira ICFR (2020) Potato peels as sources of functional compounds for the food industry: a review. Trends Food Sci Technol 103:118–129

Schreiber L, Franke R, Hartmann K (2005) Wax and suberin development of native and wound periderm of potato (*Solanum tuberosum* L.) and its relation to peridermal transpiration. Planta 220(4):520–530

Singh N, Rajini PS (2004) Free radical scavenging activity of an aqueous extract of potato peel. Food Chem 85:611–616

Singh N, Rajini PS (2008) Antioxidant-mediated protective effect of potato peel extract in erythrocytes against oxidative damage. Chem Biol Interact 173:97–104

Singh N, Kamath V, Rajini PS (2005) Attenuation of hyperglycemia and associated biochemical parameters in STZ-induced diabetic rats by dietary supplementation of potato peel powder. Clin Chim Acta 353:165–175

Singh N, Kamath V, Narasimhamurthy K, Rajini PS (2008) Protective effect of potato peel extract against carbon tetrachloride-induced liver injury in rats. Environ Toxicol Pharmacol 26:241–246

Singh A, Sabally K, Kubow S, Donnelly DJ, Gariepy Y, Orsat V (2011) Microwave-assisted extraction of phenolic antioxidants from potato peels. Molecules 16(3):2218–2232

Singh B, Singh J, Singh JP, Kaur A, Singh N (2020) Phenolic compounds in potato (*Solanum tuberosum* L.) peel and their health-promoting activities. Int J Food Sci Technol 55(6): 2273–2281

Sotillo DRD, Hadley M, Wolf-Hall C (1998) Potato peel extract a nonmutagenic antioxidant with potential antimicrobial activity. J Food Sci 63(5):907–910

Sreeja SA, Selvi AT, Kumari PN, Palavesam A (2015) Effect of potato peel powder with *Bacillus* on growth and biochemical changes of fish *Etroplus suratensis*. Int J Innov Sci Eng Technol 3(9):169–173

Taher IB, Fickers P, Chniti S, Hassouna M (2017) Optimization of enzymatic hydrolysis and fermentation conditions for improved bioethanol production from potato peel residues. Biotechnol Prog 33(2):397–406

Tanios S, Thangavel T, Eyles A, Tegg RS, Nichols DS, Corkrey R, Wilson CR (2020) Suberin deposition in potato periderm: a novel resistance mechanism against tuber greening. New Phytol 225(3):1273–1284

Verma SK, Deka B, Bordoloi R, Sharma N, Sahariah BJ (2021) Prospects of medicinal and commercial utilization of potato peel waste. Int J Pharm Res Health Sci 9(5):3336–3341

Verma SR, Satyanarayan S (2016) Effect of special fish feed prepared using food industrial waste on *Labeo rohita*. In: Fisheries and aquaculture in the modern world. IntechOpen, pp 1–9. https://doi.org/10.5772/62736

Vulavala VKR, Elbaum R, Yermiyahu U, Fogelman E, Kumar A, Ginzberg I (2016) Silicon fertilization of potato: expression of putative transporters and tuber skin quality. Planta 243(1):217–229

Vulavala VKR, Fogelman E, Rozental L, Faigenboim A, Tanami Z, Shoseyov O, Ginzberg I (2017) Identification of genes related to skin development in potato. Plant Mol Biol 94(4–5):481–494

Wafar RJ, Ojinnaka PE, Tarimbuka LI, Iliya DS, Shehu II (2017) Growth performance, carcass characteristics and serum biochemistry of broiler chicken fed graded levels of sun-dried Irish potato peel meal. Turkish J Agric Food Sci Technol 5(5):525–529

Wu Z, Xu H, Ma Q, Cao Y, Ma J, Ma C (2012) Isolation, identification and quantification of unsaturated fatty acids, amides, phenolic compounds and glycoalkaloids from potato peel. Food Chem 135(4):2425–2429

Xie Y, Niu X, Yang J, Fan R, Shi J, Ullah N, Feng X, Chen L (2020) Active biodegradable films based on the whole potato peel incorporated with bacterial cellulose and curcumin. Int J Biol Macromol 150:480–491

Yang WL, Bernards MA (2006) Wound-induced metabolism in potato (*Solanum tuberosum*) tubers: biosynthesis of aliphatic domain monomers. Plant Signal Behav 1(2):59–66

Yang G, Cheon SY, Chung KS, Lee SJ, Hong CH, Lee KT, Jang DS, Jeong JC, Kwon OK, Nam JH, An YJ (2015) *Solanum tuberosum* L. cv Jayoung epidermis extract inhibits mite antigen-induced atopic dermatitis in NC/Nga mice by regulating the Th1/Th2 balance and expression of filaggrin. J Med Food 18:1013–1021

Zhang L, Li M, Zhang G, Wu L, Cai D, Wu Z (2018) Inhibiting sprouting and decreasing α-solanine amount of stored potatoes using hydrophobic nano-silica. ACS Sustain Chem Eng 6(8): 10517–10525

Zhang L, Yu D, Regenstein JM, Xia W, Dong J (2021) A comprehensive review on natural bioactive films with controlled release characteristics and their applications in foods and pharmaceuticals. Trends Food Sci Technol 112:690–707

Zhao CL, Wen GS, Mao ZC, Xu SZ, Liu ZJ, Zhao MF, Lin C (2015) Molecular structures of the stem tuber anthocyanins of colored potatoes and their coloring effects on the tubers. Nat Prod Commun 10(3):461–466

Management of Potato Peel Waste Through Biorefinery Approaches

4

Omojola Awogbemi and Daramy VandiVon Kallon

Abstract

Management of the huge waste generated is among the challenges confronting humanity in recent decades. Waste generated from food residues has continued to increase owing to population explosion and increased commercial and industrial activities. Potato tuber is one of the most consumed food, globally, due to its composition and health benefits. Management of the huge waste generated from the peeling of potatoes has elicited genuine concerns among stakeholders. Inappropriate disposal and management of potato peel wastes (PPWs) constitutes an eyesore, attracts flies and other pathogens, exacerbates water pollution and contamination, and impedes efforts toward environmental sustainability. The biorefinery approach is an effective waste management strategy for the conversion and utilization of PPWs. The chapter focuses on the biorefinery approaches for the management of PPWs for environmental and economic sustainability. The techniques for the production of biofuels, bioadsorbents, biopolymers, energy storage devices, and other value-added products from PPWs are discussed. The adoption of biorefinery techniques for the management of PPWs contributes to energy security, renewable energy penetration, waste conversion, and recovery of raw materials for industries. More multidisciplinary studies are needed to discover and escalate more cost-effective and sustainable techniques for the conversion and utilization of PPWs.

Keywords

Potato peel waste · Biofuel · Waste management · Waste conversion · Biorefinery · Organic acids · Environmental sustainability

O. Awogbemi (✉) · D. V. Kallon
Department of Mechanical and Industrial Engineering Technology, University of Johannesburg, Johannesburg, South Africa

R. C. Ray (ed.), *Roots, Tubers, and Bulb Crop Wastes: Management by Biorefinery Approaches*, https://doi.org/10.1007/978-981-99-8266-0_4

Abbreviations

ABE	Acetone butanol, ethanol
AD	Anaerobic digestion
ICE	Internal combustion engine
PPWs	Potato peel wastes
SGGs	Sustainable development goals
TS	Total solids
VS	Volatile solids

4.1 Introduction

Management of waste is one of the major challenges confronting humanity in this generation. The problem of waste generation, disposal, and management is further exacerbated by consistent population explosion, changes in lifestyle, urbanization, and increased socioeconomic activities. For example, The United Nations (UN) has projected that the global population which was 7.51 billion in 2016 will become 8.55 billion in 2030 and further increase to 9.71 billion in 2050 (Roser and Rodés-Guirao 2019). In the same vein, global waste generation has been estimated to rise to 3.40 billion tons by 2050. East Asia and the Pacific are projected to continue to dominate global waste generation (Fig. 4.1) (WorldBank 2018). The market value for global waste management which was about USD 1.61 trillion in 2020 has been projected to

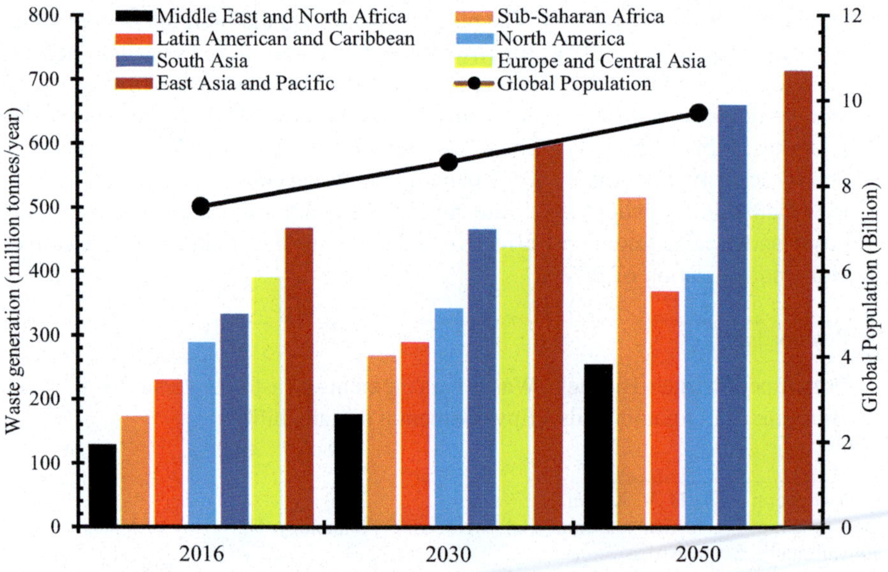

Fig. 4.1 Global waste generation million tons/year and global population (Billion)

rise to USD 2.48 trillion in 2030 (Tiseo 2022). This shows the huge financial resources that will continue to be expended on waste management in the immediate future. Effective management of global waste will contribute to ensuring a clean and healthy environment, improved air quality, maintenance of good health, and environmental sustainability. Inappropriate waste disposal coupled with ineffective waste management strategies adversely exacerbates environmental degradation and climate change, hampers efforts at producing clean water, pollutes terrestrial and aquatic habitats, and provides habitations for fertile breeding of mosquitoes, pests, rodents, and other disease-causing pathogens (Abubakar et al. 2022; Awogbemi and Kallon 2022). This, perhaps, explains why effective and sustainable waste management is one of the priorities in meeting some of the UN's Sustainable Development Goals (SDGs), including ensuring clean water and hygiene, building sustainable cities and inclusive communities, and mitigating climate change, which are SDG 6, 11, and 13, respectively (UN 2015).

Generally, waste is defined as any material that has become useless or unable to meet its expected production purpose. Since they are no longer useful, they are discarded, dumped, and thrown away. Waste can be generated in the form of raw materials, semi-finished, finished products, or a product of human activities (Awogbemi and Kallon 2023; Awogbemi et al. 2022a, b). Waste can occur in solid, semi-solid, liquid, and gaseous states. Solid wastes include glass, plastic, bones, etc. while semisolid wastes are animal waste, animal fats, slurry, etc. Common examples of liquid waste are wastewater, blood, chemicals, etc. Waste can be categorized according to the source of generation as industrial wastes (industrial waste water, plastic, glass, etc.), domestic wastes (food leftovers, vegetables, peels, eggshells, etc.), agricultural wastes (weed, husks, wood, crop residues, etc.), commercial wastes (papers, plastic, etc.), and electronic wastes (TV casing, used bulbs, wires, and cables, etc.). In terms of biodegradability, waste can also be categorized as biodegradable (food leftovers, leaves, peels, etc.) and non-biodegradable (plastic, bones, glass, chemicals, etc.) (Awogbemi et al. 2022a, b).

Waste management is a strategic process involving waste collection, transportation, and treatment before disposal of the remaining harmless residues into the appropriate location. Among the various waste management approaches in operation, the concept of reduce, reuse, and recycle (3R) is arguably the most cost-effective and environmentally sustainable (Awogbemi et al. 2022a, b). Waste reduction entails minimizing the quantity of waste produced and discarded to dumpsites by ensuring adequate utilization of materials before disposal. Waste reuse is extending the useful life of the product in another form or for another purpose without necessarily reprocessing or modification of the material. The modification, conversion, and reprocessing of waste into a new product or another useful form is the hallmark of waste recycling (Saad et al. 2022). The 3R waste management concept aims to maximize material utilization, improve sanitation, create a safe and healthier environment, reduce ecological pollution, eradicate contamination of terrestrial and aquatic habitats, and ensure effective resource utilization towards sustainable development.

Just like other categories of waste, the generation of household wastes and agricultural residues like potato peel wastes (PPWs) has increased in recent decades owing to increased food production to feed the ever-increasing population and provide raw materials for industries. Sadly, most of these wastes are neither properly disposed and nor recycled. Available data shows that only about 13.5% of global waste generation is recycled, annually. To correct this scenario, governments from various jurisdictions have continued to commit huge financial resources to waste recycling and sustainable waste management strategies. Also, various innovative technologies have been adopted and collaborative research commissioned to convert waste materials to useful forms and safeguard the environment from the impact of inappropriate waste management (Ray 2022). For example, in recent research, Zorpas (2020), Khan et al. (2022), and Esparza et al. (2020) studied the process, framework, challenges, and future perspectives for effective waste management to ameliorate the impact of waste on humanity and environment. They recommended the strengthening of existing waste management systems and the introduction and adoption of innovative strategies and scientific approaches for effective waste management toward environmental sustainability.

More specifically, Wu (2016), Galhano dos Santos et al. (2016), and Javed et al. (2019) investigated and cataloged various techniques and technologies for the conversion and recycling of PPW into useful forms for various applications. Similarly, in a series of recent studies, Abdelraof et al. (2019), Lu et al. (2020), and Soltaninejad et al. (2022) reported the practicality of converting PPWs into cellulose, biogas, and ethanol, respectively, using low-cost technologies. The techniques and technologies for producing biofuels, bioplastics, biosorbents, cellulose nanocrystals, metallic nanoparticles, and other products from PPWs were investigated. The outcome of the research revealed PPWs as a readily available, low-cost, and environmentally friendly feedstock for diverse applications (Ebrahimian et al. 2022; Olad et al. 2020; Yang et al. 2022). Bearing in mind the effectiveness of the techniques for recycling and conversion of PPWs for various applications as a feasible waste management approach, Kurniawan et al. (2023), Sanjaya and Abbas (2023), and Caferra et al. (2023) suggested effective adoption and utilization of the huge PPWs generated annually into the circular economy platform to escalate attainment of resource recovery, environmental sanitation, and sustainable ecosystem.

Despite the trajectory of cutting-edge research in the waste management domain, the relevant question to ask, which forms the motivation for this study, is whether the subject of waste management through biorefinery approaches has been exhausted. In the current intervention, the various biorefinery approaches to potato peel waste management will be investigated to come up with strategies for converting the waste to various useful products. The layout of this chapter includes the generation, composition, and pretreatment of potato peel waste, avenues for the utilization of PPWs for biofuel, bioadsorbent, biopolymer, energy storage devices production, and other uses. The challenges facing the conversion of PPWs to other useful forms through biorefinery approaches are exposed while future research areas and other recommendations are suggested. The study is limited to a desktop review of recent advances in the management of PPW through biorefinery approaches.

4.2 Generation, Composition, and Pretreatment of Potato Peel Wastes

Potato is a starchy food and a root tuber harvested from the swollen end of an underground *Solanum tuberosum* plant stem. The origin of potato can be traced to about 2500 BC in present southern Peru and northwestern Bolivia. There are about 4000 varieties of potatoes cultivated and consumed globally with russet potato, white potato, red potato, yellow potato, purple potato, and sweet potato as the most popular. Currently, potato is grown in virtually all types of soils and consumed in more than 100 countries (Awogbemi et al. 2022a, b). The global production of potatoes which was about 328 million tons (Mt) in 2010 increased to 368.8 Mt, 371 Mt, and 376 Mt in 2019, 2020, and 2021, respectively. The Asia continent dominated production in 2021 contributing about 51% (189 Mt) of the total global production followed by Europe, Americas, Africa, and Oceania with 107 Mt, 45 Mt, 20 Mt, and 1.7 Mt, in that order (FAOSTAT 2022). At the country level, China, India, and Ukraine led the production chart in 2019, 2020, and 2021, as shown in Fig. 4.2. The continuous potato production is driven by the increased popularity of potato as low-cost nutritional food, health benefits, and as raw materials for diverse industries.

Uncooked potato is made up of about 79% water, 17% carbohydrates, 2% protein, and 2% fats. Potato is also rich in amino acids, potassium, phosphorous, magnesium, sodium, and other essential minerals (Table 4.1). When consumed either in mashed, fried, baked, or boiled forms or as soup, potato offers numerous nutritional and health benefits. For example, Awogbemi et al. (2022a, b) report that the consumption of potatoes boosts immunity, lowers blood sugar, and strengthens bones and teeth. Apart from their nutritional and health benefits, potato is also used as raw materials for various industrial processes.

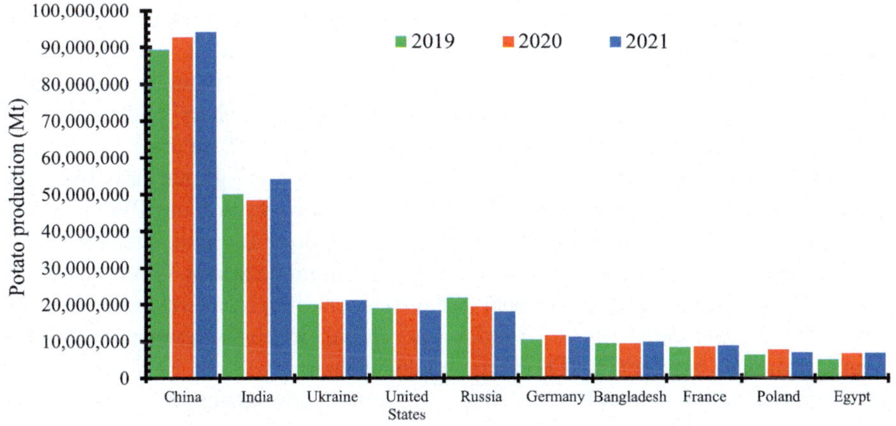

Fig. 4.2 Ten biggest producers of potatoes in 2019, 2020, and 2021

Table 4.1 Composition of potato (FAOSTAT 2022)

Element	Composition (per 100 g)	Element	Composition (per 100 g)
Water content	78.6	Magnesium	23 mg
Energy	79 kcal	Phosphorus	55 mg
Energy	332 kJ	Potassium	417 mg
Protein	2.14 g	Sodium	5 mg
Total lipid (fat)	0.08 g	Zinc	0.29 mg
Ash	1.13 g	Copper	0.103 mg
Carbohydrate	18.1 g	Manganese	0.157 mg
Total dietary fiber	1.3 g	Selenium	0.4 µg
Total sugars	0.62 g	Vitamin C	5.7 mg
Sucrose	0.13 g	Thiamine	0.082 mg
Glucose	0.25 g	Riboflavin	0.033 mg
Fructose	0.23 g	Niacin	1.04 mg
Starch	15.9 g	Pantothenic acid	0.301 mg
Calcium	13 mg	Folate	14 µg
Iron	0.86 mg	Saturated fatty acids	0.14 g
Beta-carotene	5 µg	Polyunsaturated fatty acids	0.19 g

Generally, PPWs are produced when the exterior layer of the potato tuber is removed. This usually takes place in households, restaurants, and other potato processing outlets to gain access to the edible part of the potato either for boiling, frying, mashing, or soup preparation. PPWs constitute about 15–40% by mass of an average potato tuber and are usually thrown away as waste after peeling. This shows that a huge quantity of PPWs is generated globally going by the millions of tons of potato tubers harvested and consumed annually. Indiscriminate disposal of PPWs makes the environment unclean, produces an unpleasant odor, and attracts flies, cockroaches, rats, and other pathogens. Inappropriate disposal of PPWs pollutes the atmosphere, contaminates underground and surface water bodies, and exacerbates environmental degradation. Due to the presence of glycoalkaloid compounds, improper management of PPWs aggravates ecological toxicity and constitutes grave health hazards to humans and animals (Joshi et al. 2020). For every 100 g of raw potato tuber, the peel is made up of about 83 g water, 8.7 g total carbohydrates, 7.8 g starch, and other components (Table 4.2). The composition of PPWs is one of the major factors in determining the appropriate technique for its conversion.

Certain operations are usually carried out on the lignocellulosic biomass, including PPWs, to alter their compositional and physicochemical makeup before conversion. These operations are to subdue the initial intrinsic resistance to their degradation and conversion. The pretreatment techniques are to increase the ease of degradation and thereby improve the conversion efficiency. During the pretreatment process, the complex molecules in the PPWs are decomposed and fragmented into cellulose, hemicellulose, and lignin, preparatory to their final

Table 4.2 Chemical composition of raw potato peel waste

Parameter	Dry weight composition (%)				
Water	83.3–85.1	12.4 ± 2.4	NS	NS	5.26
TS (%)	NS	96.4 ± 0.7	84.40 ± 0.10	NS	NS
VS (g/100 g TS)	NS	NS	90.40 ± 0.20	NS	NS
Protein (N_{tot} 6.25)	1.2–2.3	NS	NS	17.1 ± 0.3	10.73
Total lipids	0.1–0.4	NS	NS	1.2 ± 0.0	NS
Total carbohydrate	8.7–12.4	NS	NS	63.2 ± 4.2	43.2
Starch (g/100 g TS)	7.8	24.4 ± 3.2	44.80 ± 0.70	34.3.2 ± 2.7	23.01
Ash (g/100 g TS)	0.9–1.6	NS	11.00 ± 0.10	9.6 ± 0.1	7.45
Total phenolic content	1.02–2.92	NS	NS	NS	2.5 mg/g
Total flavonoids	0.51–0.96	NS	NS	NS	NS
Lignin (g/100 g TS)	NS	NS	4.30 ± 0.13	NS	32.88
Cellulose (g/100 g TS)	NS	28.5 ± 1.2	34.30 ± 0.60	NS	NS
Hemicellulose (g/100 g TS)	NS	18.1 ± 1.6	5.90 ± 1.90	NS	NS
Acid soluble	NS	1.1 ± 0.9	NS	6.2 ± 0.2	NS
Acid insoluble	NS	29.2 ± 1.5	NS	4.1 ± 0.0	NS
Reference	Javed et al. (2019)	Barampouti et al. (2021)	Ben Atitallah et al. (2019)	Liang et al. (2014)	Hijosa-Valsero et al. (2018)

TS total solid, *VS* volatile solids, *NS* not stated

conversion to other forms. The pretreatment process helps in reducing energy, time, and cost of decomposition ensures the formation and preservation of simple sugar and other intermediate products, and minimizes the construction of inhibitory products during the actual conversion process (Areepak et al. 2022; Awogbemi and Kallon 2022). Various pretreatment techniques including physical, chemical, biological, physicochemical, thermal, and green solvent-based have been used in the recent past, with having their benefits and drawbacks (Achinas et al. 2019; Ebrahimian et al. 2022; Procentese et al. 2018; Soltaninejad et al. 2022). Table 4.3 compiles some of the most widely used pretreatment techniques for PPWs.

Table 4.3 Major pretreatment techniques for PPWs (Achinas et al. 2019; Awogbemi and Kallon 2022; Soltaninejad et al. 2022)

Technique	Description	Examples	Benefits	Drawbacks
Physical	The reduction in particle size and increment in surface area of PPW	• Mechanical • Ultrasonic • Thermal	• Safe, non-toxic, and easily achievable • Saves time, energy, and cost • Significant size reduction • Improved surface area	• Cannot degrade lignin • Must be combined with other processes
Biological	Application of aerobic bacteria, enzymes, or fungi for biomass decomposition	• Fungal • Bacterial • Termite • Microbial consortium • Enzymes	• Nonhazardous • Energy efficient • Low cost • Environmentally friendly • Mild reaction process	• Note efficient • Slow process • More research needed
Chemical	The use of special chemicals to break the crystal structure	• Acid • Alkali • Oxidation • Organic solvent	• Fast and inexpensive • Environmentally friendly • Easy lignin removal • Reliable	• Toxic and corrosive • High energy consumption • Time-consuming • Expensive • Contaminate the environment
Physicochemical	Combination of physical and chemical pretreatments techniques	• Steam explosion • Alkali-heat • Ammonia fiber explosion • Extrusion	• Cost-effective • No pollution • Low capital cost • Simple, safe, and achievable • Generates no inhibition	• High energy consumption • Requires high temperature and pressure • Costly infrastructure
Green solvent-based	Use of sustainable methods	• Biochemical • Ionic liquids • Supercritical fluids • Deep	• Low cost and safe • Good chemical and thermal stability • High efficiency • Nonpolluting	• Special equipment needed • Needs more investigation • Low yield

(continued)

Table 4.3 (continued)

Technique	Description	Examples	Benefits	Drawbacks
		eutectic solvents	• Eco-friendly and biodegradable	

PPW potato peel wastes

4.3 Biorefinery Approaches for PPWs Utilization

The biorefinery approach is one of the sustainable conversion and utilization platforms for PPWs. Through this approach, PPWs are used for the production of biofuels, bioadsorbents, biopolymers, energy storage devices, and other useful products. This section is dedicated to providing updated information on the topic.

4.3.1 Biofuel Production

The investigation and production intensification of biofuel as an alternative and low-cost fuel for internal combustion (ICE) has gained traction in recent decades. Human and financial resources have been committed to sustained efforts aimed at developing low-cost feedstocks, catalysts, reactors, and other intensification processes for biofuel production to meet up with the increased demand. Increased demand for biofuels has triggered a significant rise in global biofuel production in recent decades. The global biofuel production which was about 59 million Tons of Oil Equivalent (Mtoe) in 2010 became 100 Mtoe in 2020 and has been projected to rise to 116 Mtoe in 2025 (IEA 2022). The global biodiesel market value which was about USD 120 billion in 2010 has been estimated to become USD 141 billion in 2025 and USD 201 billion by 2030 (Sönnichsen 2022). The predicted growth is due to increased investment in biofuel production as a result of policies in favor of increased biofuel utilization for ICE applications, energy security, and reduction in greenhouse gas emissions. PPW has been used in various forms to improve biofuel production for diverse applications.

There is verifiable evidence that PPWs have been dried, pulverized, and deployed to enhance biogas yield in an anaerobic digestion (AD) process. Suntikunaporn et al. (2014), Sheikh and Sandeep (2019), and Liang and McDonald (2015) deployed PPWs as feedstock for biogas production through AD. The high carbohydrate content of PPWs acts as an impetus to simulate AD and subsequent biogas yield. In a similar study, Lu et al. (2020), Lahbab et al. (2021), and Awosusi et al. (2021) co-digested PPWs and animal wastes for improved biomethane production. The authors were unanimous on the application of PPWs as a cheap, readily available, and easily digestible feedstock for biogas production.

Recent investigations also showed the use of PPWs as feedstock for the production of ethanol, bioethanol, or biobutanol by the process of hydrolysis, saccharification, and fermentation. Hijosa-Valsero et al. (2018), Meenakshi and Kumaresan (2014), and Soltaninejad et al. (2022) deployed the use of various enzymes including Alpha-amylase, and cellulose from *Saccharomyces cerevisiae*, *Clostridium*, etc. for accelerated bioethanol production. In similar studies, PPW was pretreated and converted to Acetone Butanol, Ethanol (ABE), and Biobutanol via fermentation using *Clostridium acetobutylicum* MTCC 11274 for various applications. The authors demonstrated PPW as a cheap, easily fermented, and high-yield feedstock to synthesize ABE and biobutanol (Abedini et al. 2020). PPW has also been processed and converted to bio-oil, biohydrogen, biochar, and syngas to meet up with the increased demand and utilization potentials of these biofuels for various applications (Awogbemi et al. 2022a, b). Recent studies indicated the possibility of using PPW as a feedstock for pyrolysis to generate solid, liquid, and gaseous fuels. In different studies, Daimary et al. (2022) and Frantzi and Zabaniotou (2021) pyrolyzed PPW under various process parameters to yield biochar, bio-oil, and syngas in varying proportions. The conversion yield was comparable to other high-yielding food agricultural wastes.

Apart from the use of PPWs as feedstock, PPWs have been deployed as catalysts to enhance biodiesel production. This is achieved when PPWs are converted to powder and used as a heterogeneous catalyst for the transesterification process. This helps to lower production expenditure and ensure the conversion of waste to useful catalysts (Awogbemi et al. 2021; Daimary et al. 2022). Catalyst derived from PPWs and other agro-wastes are more efficient, biodegradable, contributes to waste to energy, and are an effective feedstock for the circular economy.

4.3.2 Production of Bioadsorbents

Pollution and contamination of surface and underground water have greatly impacted the availability and accessibility of clean and portable water for domestic and industrial uses. Also, the huge volume of wastewater generated in households, agricultural, and industrial sectors are released into the environment without adequate treatment and remediation measures. The use of commercial and chemical-based adsorbents is not only expensive, and toxic to the human body but also environmentally unsustainable (Muhammad et al. 2023). These have shifted attention to the use of adsorbents fabricated from wastes and other renewable sources. The utilization of PPWs and other wastes offers cost-effective, readily available, easily accessible, biodegradable, and environmentally benign materials as bioadsorbents for the removal of contaminants. They are also effective in the removal of heavy metals, dyes, organic compounds, and other contaminants from wastewater. Specifically, activated carbon constructed from pyrolyzed PPWs has been used for the adsorption of heavy metals such as cobalt, copper, nickel, cadmium, iron, lead, arsenic, chromium, etc. dyes, toxic chemicals, and contaminants from wastewater and other aqueous solutions (El-Azazy et al. 2019).

The application of PPWs as an effective and low-cost adsorbent for the removal of chromium (Mohammed and Salim 2017), malachite green (Guechi and Hamdaoui

2016), anionic dye (Bouhadjra et al. 2021), arsenic and fluoride (Bibi et al. 2017), toxic metalloids (Anastopoulos et al. 2019) and other pollutants from wastewater and aqueous solutions have been demonstrated. Similar studies have confirmed the use of biochar or activated carbon derived from PPW for the adsorption of arsenic, cadmium, mercury, lead, chromium, manganese, dyes, dorzolamide, pramipexole, and other contaminants from wastewater from agricultural, pharmaceutical, and industrial sectors (Figueira et al. 2022; Gogoi et al. 2023; Salih et al. 2021). The outcomes of these studies show that the process of using PPWs to take away dyes, heavy metals, and other contaminants from wastewater and other aqueous solutions is straightforward, effective, economically feasible, and ecologically safe. They unanimously concluded that PPWs can be converted into bioadsorbents for cheap, eco-friendly, and a substitute for the costly commercial adsorbents commonly used to purify wastewater and other aqueous solutions.

4.3.3 Production of Biopolymer

Plastics, composites, and polymers have become indispensable in our daily lives including households, commercial, agricultural, automobiles, construction, and other sectors. Synthetic and petroleum-based plastics, composites, and polymers are for the production of many household utensils like packaging materials, shopping bags, plastic bottles, automobile parts, electrical insulators, electronic casings, pipes, and other containers. However, their disposal and end-of-life usage has continued to create a great challenge to environmentalists and other stakeholders due to their non-biodegradability. These challenges have resulted in the fabrication of plastic and other polymers from waste materials and other renewable biomass. Plastic, cellulose, and other polymers produced from PPWs and other renewable sources are cheaper to produce, exhibit better mechanical properties, are biodegradable, and are eco-friendly (Asyraf et al. 2022; Ferrari et al. 2022).

Recent research by Sugumaran et al. (2015, 2018) have demonstrated the fabrication of polypropylene and linear low-density polyethylene from PPWs. The authors reported that the biocomposites derived from PPWs are biodegradable, eco-friendly, and exhibit better physicochemical, mechanical, and thermal properties than petroleum-based composites. The deployment of ultrasonic-treated PPWs for the production and development of biopolymer films resulted in products with enhanced water vapor permeability, breakage strength, elongation capacity, and better mechanical properties (Borah et al. 2017). In recent years, starch has been extracted from PPWs for the production of bioplastics to replace petroleum-based plastic. The bioplastics are found to be biodegradable, harmless to the environment, and demonstrate sufficient physical, mechanical, and chemical properties (Chaffa et al. 2022; Taha et al. 2022). Generally, PPWs are used as reinforcement and additive for the production of packaging materials, bioactive shopping bags, bioplastic, pot handles, pipes, electronic casing, and other composite-based materials for commercial, food packaging, medical, industrial, horticultural, and agricultural applications.

4.3.4 Energy Storage Applications

Inadequate energy for the use of the ever-growing population and to drive social, economic, and industrial growth is one of the banes of the modern age. The most commonly used available energy storage device is the battery system which is expensive, non-renewable, not environmentally friendly, and suffers from disposal problems. When batteries become old, they corrode, their chemicals soak into the soil, contaminate underground and surface water, and impact the ecosystem. Leaching, disintegration, and degradation of the batteries cause photochemical smog pollution and exacerbate climate change and global warming. The production of batteries also requires a large amount of lithium, nickel, cobalt, lead, and other raw materials that are poisonous, and flammable, and trigger environmental pollution and degradation processes. This has shifted the attention of researchers to the use of renewable biomass such as agricultural residues and food wastes for the production of energy storage materials.

In recent research, Wesley et al. (2021) constructed a copper(II)phthalocyanine compound composite from biochar derived from PPWs and activated with a KOH solution. The copper(II)phthalocyanine compound composite was used to produce a supercapacitor for energy storage applications. Earlier, Carbon–Sulfur composite has been produced from biochar derived from the carbonization of PPWs and used as cathode materials for Lithium-Sulfur batteries (Andreas Arie et al. 2021) while carbon electrodes were fabricated from PPWs biochar generated by carbonization for supercapacitor by Pertiwi et al. (2022). A similar study by Khalafallah et al. (2021) demonstrated the construction of supercapacitors from heteroatoms doped with porous carbon obtained from PPWs for supercapacitors. The supercapacitors derived from PPW were found to possess large surface area which substantially promotes the electrochemical and capacitive performance, improves its porous features, and enhances charge storage capabilities. When compared with supercapacitors constructed from nonrenewable materials, PPW-derived supercapacitors demonstrated better energy density, power density, and longer cycle life. Supercapacitors manufactured from PPWs also exhibit simplicity and economic, sanitary, and environmental benefits.

4.3.5 Other Biorefinery Approaches to PPWs Utilization

Researchers have continued to exploit the unique properties and composition of PPWs for their conversion into value-added products and other diverse applications. Notable pathway for PPWs application includes animal feed production, biofertilizer production, organic acids production, and medical and pharmaceutical applications.

4.3.5.1 Animal Feed

In recent years, researchers in animal nutrition have exploited the polyphenol content of PPWs to boost the quality of cow, beef, fish, etc. feeds. The addition of PPWs to beef feed boosts the starch and energy contents of their meal thereby improving the meat quality and daily milk production. In a research, Kpanja et al. (2020) reported that the addition of PPWs to the meal of broiler chicken improves weight gain, hematological, and blood biochemical parameters. The addition of raw PPW powder as an additive to chicken feed elongates the shelf life of the chicken feed by about 4 months (Iegorov et al. 2021). Feeding fishes with shredded PPWs for 1 month improves the protein and fat levels, strengthen the muscles and liver tissue, and ensure healthier growth without any side effect (Maske and Satyanarayan 2012). PPWs have been able to replace more than half of the concentrate feed mixture in sheep ration resulting in lower feed costs, more rapid growth, and better meat production (Pathak et al. 2018). Consequently, farmers and animal nutritionists have exploited the use of PPWs as an additive to animal feed for cost reduction, more nutritious animal feed, and increased profit for small-holder farmers.

4.3.5.2 Biofertilizer Production

Since PPWs contain large percentages of starch and proteins which can be decomposed by microorganisms to produce nitrogen. The nitrogen generated nourishes the soils and serves as a major ingredient of biofertilizers and a sustainable replacement for chemical fertilizers. The slurry obtained from the AD of PPWs replenishes the lost nutrients in the soil and serves as a liquid biofertilizer for farmers. Tiwari et al. (2016) reported that the slurry obtained from the AD of PPWs for biogas production was used as a liquid biofertilizer for the cultivation of strawberry fruit. After the application of the biofertilizer, the was a significant improvement in vegetative growth, yield, fruit quality, and better physical and chemical characteristics of harvested strawberries. In another instance, the residue obtained from the fermentation of PPWs for bioethanol production was used as liquid manure to increase the content of nitrogen, phosphorous, and potassium in the soil and ensure better crop yield. Thus confirming the application of PPWs slurry as a low-cost and effective biofertilizer (Chintagunta et al. 2016).

4.3.5.3 Organic Acids Production

Organic acids are one of the most important and widely used materials for various industrial processes and production. However, the high cost of commercial organic acids has necessitated their generation from waste and other biomass sources. For example, acetic and lactic acids were produced from the enzymatic fermentation of PPWs. The reaction yielded 0.22 g/g lactic acid and 0.06 g/g acetic acid when the PPWs substrate was mixed with hydrolytic enzymes, α-amylase, glucoamylase, and cellulose (Liang et al. 2014). Lactic and acetic acids are widely used in the pharmaceutical, food, and cosmetic industries. Similarly, Faruk et al. (2014) and Miller et al. (2022) synthesized citric acid from PPW hydrolysate. Citric acid is an important raw material in the food and beverage, pharmaceutical, cosmetic, metallurgy, and metal processing industries. Calcio Gaudino et al. (2020) and Singh et al. (2020) also reported the synthesis of phenolic acids from PPWs. The authors exploited the high concentration of phenolic compounds in PPWs to extract phenolic acid during the fermentation of PPWs to achieve a high yield of phenolic acid for use in food and allied industries. The process replaced the detrimental, costly, and unsustainable process of using methanol, acetone, and other organic solvents to extract phenolic acid from plants. The production of organic acids from PPW is cost-effective and environmentally friendly.

4.3.5.4 Medical and Pharmaceutical Applications

The presence of glycoalkaloid PPWs has been exploited and escalated in the medical and pharmaceutical industries. Secondary metabolites such as steroidal alkaloids (a-solanine, a-chaconine, solanidine, and demissidine) extracted from PPWs demonstrate antifungal, anti-inflammatory, and antibacterial properties (Wahyudi et al. 2020). These unique properties of PPWs make them useful to fight against insect attacks and healing burns. Recent research shows that bandages made from PPWs are easy to apply and remove, more comfortable, and ensures rapid epithelial regeneration. Pathak et al. (2018) reported that bandages made from PPWs exhibit better survival of superficial skin cells, guarantee fast epithelial renewal, and drastically decrease bacterial infection during the healing process when compared with gauze and other commercial bandages. Advancement in medical research has revealed the diverse health-benefiting effects of PPWs not only in nutritional, biotechnological, and pharmaceutical sectors but also in curative or therapeutic care. Kowalczewski et al. (2022) reported the presence of some bioactive compounds in PPWs makes them a potential candidate for the management and treatment of cancer, diabetes, obesity, hypertension, and hyperlipidemia.

4.4 Challenges, Prospects, and Future Research

The concept of biorefinery as discussed in this chapter is the conversion of PPWs into biofuels, bioadsorbents, biopolymers, energy storage devices, and an array of other products. The adoption of biorefinery approaches in the management of PPWs allows the generation of diverse valuable products through various techniques. This will ensure optimal utilization of materials, enhance waste conversion technologies, reduce the cost of waste management, and contributes towards achieving waste minimization and zero waste. PPWs management through biorefinery approaches also contributes towards material recovery from waste, reduced cost of production, socioeconomic growth, cleaner environment, and bioeconomy. Though additional costs including capital cost, overhead cost, operational cost, etc. are incurred in PPW conversion through biorefinery approaches, the overall benefits make the option a worthwhile adventure.

Not withstanding the high potential of PPWs to be converted to biofuels, biocomposites, biopolymers, and several other value-added products through a biorefinery approach, there are many challenges confronting the process. As shown in Table 4.1, the moisture content of PPWs is high. This is a major setback in the biorefinery approach to its management. For example, the high moisture content makes it attractive to flies, rodents, cockroaches, etc., generates an offensive odor, and contributes to its rapid decay. Also, the collection, handling, separation, transportation, and storage of PPWs are impacted by the high water content. In most cases, huge costs, energy, and infrastructure are required to dry PPWs before storage and further processing. To overcome this drawback, low-cost, easy-to-operate, and low-energy consumption drying equipment and technologies to dehydrate PPWs and other wet wastes are needed.

Though the global production of potato peel has been rising in recent decades, the product is not yet available all year round. Also, the yield per hectare of potation is not encouraging and has been discouraging farmers to cultivate the crop. These together with the high cost of storage of potato has led to an inadequate supply of PPWs regularly. There is a need to encourage farmers to cultivate potatoes in large quantities. Potato growers can be organized into groups and granted soft loans and other incentives to encourage them. High-yielding potato species need to be developed to increase yield per hectare. Most of the technologies and techniques for PPWs conversion and utilization are still at the infant or laboratory stages. There is an urgent need to scale up to an industrial level to fully tap the benefit is PPWs conversion.

Going forward, there is a need to institute and fund collaborative research into more low-cost avenues for PPWs conversion and utilization. To improve potato yield, more studies should be conducted on the soil type, pests, species, and the influence of other environmental conditions. The application of PPWs for energy storage needs to be further investigated and upgraded to an industrial scale. Further investigations are needed to optimize capital investment, energy consumption, the nature of the solvent, yield, and integration into current processing lines. The use of innovative technologies for the storage, conversion, and utilization of PPWs needs further investigation.

4.5 Conclusion

Potato is one of the most consumed vegetables, globally. The unique composition and constituents of potatoes influence their applications in diverse sectors. The management of the huge PPWs generated annually is a source of concern for environmentalists, governments, and other stakeholders. Inappropriate disposal of PPWs constitutes an eyesore, attracts flies, rodents, and other disease-causing pathogens, oozes offensive odor, and contributes to environmental pollution and contamination. Biorefinery approaches to PPWs management are a feasible pathway to ensuring effective and sustainable conversion and utilization of PPWs by the application of appropriate conversion technologies.

Through biorefinery approaches, PPWs are converted to biofuels, bioadsorbents, biopolymers, energy storage devices, and a wide range of other value-added products such as biofertilizers, animal feed, organic acids, etc. These ensure resource recovery from PPWs, contributes to sanitation, provide the much-needed raw materials for the industries, and ensure a sustainable environment. Though biorefinery approaches to PPWs management involve additional cost and infrastructure for waste collection, sorting, transportation, storage, and processing, the overall benefits of this approach far outweigh the drawbacks.

Going forward, more financial, logistics, infrastructural, and policy investments are needed to further escalate the multidisciplinary investigations into biorefinery conversion of PPWs and other agricultural residues. More sensitization campaigns are needed to suppress some social, religious, and cultural biases against waste handling, conversion, and utilization among the population. Governments at all levels and across various jurisdictions should implement existing policies and enact new laws to support waste conversion owing to its multifaceted benefits for humanity. Farmers and other small-scale entrepreneurs involved in waste conversion should be encouraged and incentivized. Youths and women groups should be encouraged to develop a sustained interest in potato cultivation and PPW conversion.

References

Abdelraof M, Hasanin MS, El-Saied H (2019) Ecofriendly green conversion of potato peel wastes to high productivity bacterial cellulose. Carbohydr Polym 211:75–83. https://doi.org/10.1016/j.carbpol.2019.01.095

Abedini A, Amiri H, Karimi K (2020) Efficient biobutanol production from potato peel wastes by separate and simultaneous inhibitors removal and pretreatment. Renew Energy 160:269–277. https://doi.org/10.1016/j.renene.2020.06.112

Abubakar IR, Maniruzzaman KM, Dano UL, AlShihri FS, AlShammari MS et al (2022) Environmental sustainability impacts of solid waste management practices in the global South. Int J Environ Res Public Health 19(19):12717. https://doi.org/10.3390/ijerph191912717

Achinas S, Li Y, Achinas V, Euverink GJW (2019) Biogas potential from the anaerobic digestion of potato peels: process performance and kinetics evaluation. Energies 12(12):2311. https://doi.org/10.3390/en12122311

Anastopoulos I, Pashalidis I, Hosseini-Bandegharaei A, Giannakoudakis DA, Robalds A, Usman M et al (2019) Agricultural biomass/waste as adsorbents for toxic metal decontamination of aqueous solutions. J Mol Liq 295:111684. https://doi.org/10.1016/j.molliq.2019.111684

Andreas Arie A, Lenora S, Kristianto H, Frida Susanti R, Kee Lee J (2021) Potato peel-based carbon–sulfur composite as cathode materials for lithium-sulfur battery. J Nanosci Nanotechnol 21(12):6243–6247. https://doi.org/10.1166/jnn.2021.19288

Areepak C, Jiradechakorn T, Chuetor S, Phalakornkule C, Sriariyanun M, Raita M et al (2022) Improvement of lignocellulosic pretreatment efficiency by combined chemo-mechanical pretreatment for energy consumption reduction and biofuel production. Renew Energy 182:1094–1102. https://doi.org/10.1016/j.renene.2021.11.002

Asyraf MRM, Ishak MR, Syamsir A, Nurazzi NM, Sabaruddin FA, Shazleen SS et al (2022) Mechanical properties of oil palm fibre-reinforced polymer composites: a review. J Mater Res Technol 17:33–65. https://doi.org/10.1016/j.jmrt.2021.12.122

Awogbemi O, Kallon DVV (2022) Pretreatment techniques for agricultural waste. Case Stud Chem Environ Eng 6:100229. https://doi.org/10.1016/j.cscee.2022.100229

Awogbemi O, Kallon DVV (2023) Achieving affordable and clean energy through the conversion of waste plastic to liquid fuel. J Energy Inst 106:101154. https://doi.org/10.1016/j.joei.2022.101154

Awogbemi O, Kallon DVV, Aigbodion VS (2021) Trends in the development and utilization of agricultural wastes as heterogeneous catalyst for biodiesel production. J Energy Inst 98:244–258. https://doi.org/10.1016/j.joei.2021.06.017

Awogbemi O, Kallon DVV, Bello KA (2022a) Resource recycling with the aim of achieving zero-waste manufacturing. Sustainability 14(8):4503. https://doi.org/10.3390/su14084503

Awogbemi O, Kallon DVV, Owoputi AO (2022b) Biofuel generation from potato peel waste: current state and prospects. Recycling 7(2):23. https://doi.org/10.3390/recycling7020023

Awosusi A, Sethunya V, Matambo T (2021) Synergistic effect of anaerobic co-digestion of South African food waste with cow manure: role of low density-polyethylene in process modulation. Mater Today Proc 38:793–803. https://doi.org/10.1016/j.matpr.2020.04.584

Barampouti EM, Christofi A, Malamis D, Mai S (2021) A sustainable approach to valorize potato peel waste towards biofuel production. Biomass Convers Biorefin 13:8197. https://doi.org/10.1007/s13399-021-01811-4

Ben Atitallah I, Antonopoulou G, Ntaikou I, Alexandropoulou M, Nasri M et al (2019) On the evaluation of different saccharification schemes for enhanced bioethanol production from potato peels waste via a newly isolated yeast strain of *Wickerhamomyces anomalus*. Bioresour Technol 289:121614. https://doi.org/10.1016/j.biortech.2019.121614

Bibi S, Farooqi A, Yasmin A, Kamran MA, Niazi NK (2017) Arsenic and fluoride removal by potato peel and rice husk (PPRH) ash in aqueous environments. Int J Phytoremediation 19(11): 1029–1036. https://doi.org/10.1080/15226514.2017.1319329

Borah PP, Das P, Badwaik LS (2017) Ultrasound treated potato peel and sweet lime pomace based biopolymer film development. Ultrason Sonochem 36:11–19. https://doi.org/10.1016/j.ultsonch.2016.11.010

Bouhadjra K, Lemlikchi W, Ferhati A, Mignard S (2021) Enhancing removal efficiency of anionic dye (Cibacron blue) using waste potato peel powder. Sci Rep 11(1):2090. https://doi.org/10.1038/s41598-020-79069-5

Caferra R, D'Adamo I, Morone P (2023) Wasting energy or energizing waste? The public acceptance of waste-to-energy technology. Energy 263:126123. https://doi.org/10.1016/j.energy.2022.126123

Calcio Gaudino E, Colletti A, Grillo G, Tabasso S, Cravotto G (2020) Emerging processing technologies for the recovery of valuable bioactive compounds from potato peels. Foods 9(11):1598. https://doi.org/10.3390/foods9111598

Chaffa TY, Meshesha BT, Mohammed SA, Jabasingh SA (2022) Production, characterization, and optimization of starch-based biodegradable bioplastic from waste potato (*Solanum tuberosum*) peel with the reinforcement of false banana (*Ensete ventricosum*) fiber. Biomass Convers Biorefin. https://doi.org/10.1007/s13399-022-03426-9

Chintagunta AD, Jacob S, Banerjee R (2016) Integrated bioethanol and biomanure production from potato waste. Waste Manage 49:320–325. https://doi.org/10.1016/j.wasman.2015.08.010

Daimary N, Eldiehy KSH, Boruah P, Deka D, Bora U, Kakati BK (2022) Potato peels as a sustainable source for biochar, bio-oil and a green heterogeneous catalyst for biodiesel production. J Environ Chem Eng 10(1):107108. https://doi.org/10.1016/j.jece.2021.107108

Ebrahimian F, Denayer JFM, Karimi K (2022) Potato peel waste biorefinery for the sustainable production of biofuels, bioplastics, and biosorbents. Bioresour Technol 360:127609. https://doi.org/10.1016/j.biortech.2022.127609

El-Azazy M, El-Shafie AS, Issa AA, Al-Sulaiti M, Al-Yafie J, Shomar B, Al-Saad K (2019) Potato peels as an adsorbent for heavy metals from aqueous solutions: eco-structuring of a green adsorbent operating Plackett–Burman design. J Chem 2019:4926240. https://doi.org/10.1155/2019/4926240

Esparza I, Jiménez-Moreno N, Bimbela F, Ancín-Azpilicueta C, Gandía LM (2020) Fruit and vegetable waste management: conventional and emerging approaches. J Environ Manag 265: 110510. https://doi.org/10.1016/j.jenvman.2020.110510

FAOSTAT (2022) World's production and harvested areas of potatoes 1970–2021. https://www.fao.org/faostat/en/. Accessed 20 Jan 2023

Faruk MO, Mumtaz T, Rashid HO, Zhan Y, Liu S, Wu H (2014) Supplementary effect of potato peel hydrolysate on the citric acid production by Aspergillus niger CA16. J Biotechnol 9(8): 308–310

Ferrari F, Striani R, Fico D, Alam MM, Greco A, Esposito Corcione C (2022) An overview on wood waste valorization as biopolymers and biocomposites: definition, classification,

production, properties and applications. Polymers 14(24):5519. https://doi.org/10.3390/polym14245519
Figueira P, Henriques B, Teixeira F, Afonso N, Pinto J, Tavares D et al (2022) Potentialities of agro-based wastes to remove Cd, Hg, Pb, and As from contaminated waters. Water Air Soil Pollut 233(3):78. https://doi.org/10.1007/s11270-022-05543-7
Frantzi D, Zabaniotou A (2021) Waste-based intermediate bioenergy carriers: syngas production via coupling slow pyrolysis with gasification under a circular economy model. Energies 14(21): 7366. https://doi.org/10.3390/en14217366
Galhano dos Santos R, Ventura P, Bordado JC, Mateus MM (2016) Valorizing potato peel waste: an overview of the latest publications. Rev Environ Sci Biotechnol 15(4):585–592. https://doi.org/10.1007/s11157-016-9409-7
Gogoi N, Samanta P, Dahutia P (2023) Agro-wastes as low-cost biosorbent for dyes removal from wastewater. In: Agriculture waste management and bioresource: the circular economy perspective, pp 149–172. https://doi.org/10.1002/9781119808428.ch7
Guechi EK, Hamdaoui O (2016) Sorption of malachite green from aqueous solution by potato peel: kinetics and equilibrium modeling using non-linear analysis method. Arab J Chem 9:S416–S424. https://doi.org/10.1016/j.arabjc.2011.05.011
Hijosa-Valsero M, Paniagua-García AI, Díez-Antolínez R (2018) Industrial potato peel as a feedstock for biobutanol production. New Biotechnol 46:54–60. https://doi.org/10.1016/j.nbt.2018.07.002
IEA (2022) Global biofuel production 2010–2025 compared to consumption in the Sustainable Development Scenario. https://www.iea.org/data-and-statistics/charts/global-biofuel-production-2010-2025-compared-to-consumption-in-the-sustainable-development-scenario. Accessed 20 Jan 2023
Iegorov B, Tsiundyk O, Lapinska A, Fihurska L (2021) Using raw potato peel in the production of extruded feed additive. Food Sci Technol 15(2):85–92
Javed A, Ahmad A, Tahir A, Shabbir U, Nouman M, Hameed A (2019) Potato peel waste—its nutraceutical, industrial and biotechnological applications. AIMS Agric Food 4(3):807–823. https://doi.org/10.3934/agrfood.2019.3.807
Joshi A, Sethi S, Arora B, Azizi AF, Thippeswamy B (2020) Potato peel composition and utilization. In: Raigond P, Dutt S, Chakrabarti SK (eds) Potato. Springer, Singapore, pp 229–245
Khalafallah D, Quan X, Ouyang C, Zhi M, Hong Z (2021) Heteroatoms doped porous carbon derived from waste potato peel for supercapacitors. Renew Energy 170:60–71. https://doi.org/10.1016/j.renene.2021.01.077
Khan S, Anjum R, Raza ST, Ahmed Bazai N, Ihtisham M (2022) Technologies for municipal solid waste management: current status, challenges, and future perspectives. Chemosphere 288: 132403. https://doi.org/10.1016/j.chemosphere.2021.132403
Kowalczewski PŁ, Olejnik A, Świtek S, Bzducha-Wróbel A, Kubiak P, Kujawska M, Lewandowicz G (2022) Bioactive compounds of potato (*Solanum tuberosum* L.) juice: from industry waste to food and medical applications. Crit Rev Plant Sci 41(1):52–89. https://doi.org/10.1080/07352689.2022.2057749
Kpanja E, Duru S, Kotso J, Gonjoh P (2020) Effect of diets containing graded levels of Irish potato peel meal (IPPM) on carcass quality, haematological and blood biochemical profile of broiler chickens. Niger J Anim Sci 22(3):242–249
Kurniawan TA, Meidiana C, Dzarfan Othman MH, Goh HH, Chew KW (2023) Strengthening waste recycling industry in Malang (Indonesia): lessons from waste management in the era of Industry 4.0. J Clean Prod 382:135296. https://doi.org/10.1016/j.jclepro.2022.135296
Lahbab A, Djaafri M, Kalloum S, Benatiallah A, Atelge MR, Atabani AE (2021) Co-digestion of vegetable peel with cow dung without external inoculum for biogas production: experimental

and a new modelling test in a batch mode. Fuel 306:121627. https://doi.org/10.1016/j.fuel.2021.121627

Liang S, McDonald AG (2015) Anaerobic digestion of pre-fermented potato peel wastes for methane production. Waste Manag 46:197–200. https://doi.org/10.1016/j.wasman.2015.09.029

Liang S, McDonald AG, Coats ER (2014) Lactic acid production with undefined mixed culture fermentation of potato peel waste. Waste Manag 34(11):2022–2027. https://doi.org/10.1016/j.wasman.2014.07.009

Lu Y, Zhang Q, Wang X, Zhou X, Zhu J (2020) Effect of pH on volatile fatty acid production from anaerobic digestion of potato peel waste. Bioresour Technol 316:123851. https://doi.org/10.1016/j.biortech.2020.123851

Maske NS, Satyanarayan S (2012) Effect of special fish feed prepared using potato peels on fresh water fish *Labeo rohita*. J Ind Pollut Control 29:33–38

Meenakshi A, Kumaresan R (2014) Ethanol production from corn, potato peel waste and its process development. Int J ChemTech Res 6(5):2843–2853

Miller K, Reichert CL, Schmid M, Loeffler M (2022) Physical, chemical and biochemical modification approaches of potato (peel) constituents for bio-based food packaging concepts: a review. Foods 11(18):2927. https://doi.org/10.3390/foods11182927

Mohammed NMS, Salim HAM (2017) Adsorption of Cr(VI) ion from aqueous solutions by solid waste of potato peels. Sci J Univ Zakho 5(3):254–258. https://doi.org/10.25271/2017.5.3.392

Muhammad S, Yahya EB, Abdul Khalil H, Marwa M, Albadn YM (2023) Recent advances in carbon and activated carbon nanostructured aerogels prepared from agricultural wastes for wastewater treatment applications. Agriculture 13(1):208. https://doi.org/10.3390/agriculture13010208

Olad A, Doustdar F, Gharekhani H (2020) Fabrication and characterization of a starch-based superabsorbent hydrogel composite reinforced with cellulose nanocrystals from potato peel waste. Colloids Surf A Physicochem Eng Asp 601:124962. https://doi.org/10.1016/j.colsurfa.2020.124962

Pathak PD, Mandavgane SA, Puranik NM, Jambhulkar SJ, Kulkarni BD (2018) Valorization of potato peel: a biorefinery approach. Crit Rev Biotechnol 38(2):218–230. https://doi.org/10.1080/07388551.2017.1331337

Pertiwi D, Yanti N, Taslim R (2022) High potential of yellow potato (*Solanum tuberosum* L.) peel waste as porous carbon source for supercapacitor electrodes. J Phys Conf Ser 2193:012019. https://doi.org/10.1088/1742-6596/2193/1/012019

Procentese A, Raganati F, Olivieri G, Russo ME, Rehmann L, Marzocchella A (2018) Deep eutectic solvent pretreatment of agro-industrial food waste. Biotechnol Biofuels 11(1):1–12. https://doi.org/10.1186/s13068-018-1034-y

Ray RC (2022) Fruits and vegetable wastes: valorization into bioproducts and platform chemicals. Springer, Singapore. 448p

Roser M, Rodés-Guirao L (2019) Future population growth. https://ourworldindata.org/future-population-growth. Accessed 25 Jan 2023

Saad A, Bal M, Khatib J (2022) The need for a proper waste management plan for the construction industry: a case study in Lebanon. Sustainability 14(19):12783. https://doi.org/10.3390/su141912783

Salih SM, Hussein SA, Ajeel AA, Al Maliki AA, Owen G (2021) Removal of Cr and Mn from wastewater using food wastes as low-cost biosorbents. Afr J Environ Econ Manage 9(3):1–8

Sanjaya E, Abbas A (2023) Plasma gasification as an alternative energy-from-waste (EFW) technology for the circular economy: an environmental review. Resour Conserv Recycl 189:106730. https://doi.org/10.1016/j.resconrec.2022.106730

Sheikh AY, Sandeep NM (2019) Sustainable low-thermal pretreatment enhances substrate solubilization and biogas production: a case study with potato peel waste. Int J Innov Sci Eng Technol 12(2):31–38

Singh B, Singh J, Singh JP, Kaur A, Singh N (2020) Phenolic compounds in potato (*Solanum tuberosum* L.) peel and their health-promoting activities. Int J Food Sci Technol 55(6): 2273–2281. https://doi.org/10.1111/ijfs.14361

Soltaninejad A, Jazini M, Karimi K (2022) Sustainable bioconversion of potato peel wastes into ethanol and biogas using organosolv pretreatment. Chemosphere 291:133003. https://doi.org/10.1016/j.chemosphere.2021.133003

Sönnichsen N (2022) Global biofuels market size 2020–2030. https://www.statista.com/statistics/217179/global-biofuels-market-size/. Accessed 20 Jan 2023

Sugumaran V, Vimal KK, Kapur GS, Narula AK (2015) Preparation and morphological, thermal, and physicomechanical properties of polypropylene-potato peel biocomposites. J Appl Polym Sci 132(34). https://doi.org/10.1002/app.42445

Sugumaran V, Kapur GS, Narula AK (2018) Sustainable potato peel powder–LLDPE biocomposite preparation and effect of maleic anhydride-grafted polyolefins on their properties. Polym Bull 75:5513–5533. https://doi.org/10.1007/s00289-018-2340-z

Suntikunaporn M, Echaroj S, Rimpikul W, Kaewprapha P, Puttarak N, Geethalakshmi B, Yiarayong P (2014) Evaluation of agricultural wastes for biogas production. Thammasat Int J Sci Technol 19(1):1–8

Taha TH, Abu-Saied M, Elnouby M, Hashem M, Alamri S, Desouky EE, Morsy K (2022) Profitable exploitation of biodegradable polymer including chitosan blended potato peels' starch waste as an alternative source of petroleum plastics. Biomass Convers Biorefin. https://doi.org/10.1007/s13399-021-02244-9

Tiseo I (2022) Global waste management market value 2020–2030. https://www.statista.com/statistics/246178/projected-global-waste-management-market-size/. Accessed 20 Jan 2023

Tiwari V, Maji S, Kumar S, Prajapati G, Yadav R (2016) Use of kitchen waste based bio-organics for strawberry (*Fragaria* × *ananassa* Duch) production. Afr J Agric Res 11(4):259–265. https://doi.org/10.5897/AJAR2015.10349

United Nations (2015) Millennium development goals. https://sdgs.un.org/goals. Accessed 20 Jan 2023

Wahyudi IA, Ramadhan FR, Wijaya RIK, Ardhani R, Utami TW (2020) Analgesic, anti-inflammatory and anti-biofilm-forming activity of potato (*Solanum tuberosum* L.) peel extract. Indones J Cancer Chemoprev 11(1):30–35. https://doi.org/10.14499/indonesianjcanchemoprev11iss1pp30-35

Wesley RJ, Durairaj A, Ramanathan S, Obadiah A, Justinabraham R, Lv X, Vasanthkumar S (2021) Potato peels biochar composite with copper phthalocyanine for energy storage application. Diam Relat Mater 115:108360. https://doi.org/10.1016/j.diamond.2021.108360

WorldBank (2018) What a Waste 2.0. https://datatopics.worldbank.org/what-a-waste/trends_in_solid_waste_management.html. Accessed 22 Jan 2023

Wu D (2016) Recycle technology for potato peel waste processing: a review. Procedia Environ Sci 31:103–107. https://doi.org/10.1016/j.proenv.2016.02.014

Yang G, Xu C, Varjani S, Zhou Y, Wc Wong J, Duan G (2022) Metagenomic insights into improving mechanisms of Fe^0 nanoparticles on volatile fatty acids production from potato peel waste anaerobic fermentation. Bioresour Technol 361:127703. https://doi.org/10.1016/j.biortech.2022.127703

Zorpas AA (2020) Strategy development in the framework of waste management. Sci Total Environ 716:137088. https://doi.org/10.1016/j.scitotenv.2020.137088

Bioprocessing Cassava Bagasse: Part I—Bioproducts and Biochemicals

5

Nizzy Albert Mariathankam and Kannan Suruli

Abstract

Cassava is an essential food source to satisfy the demands of a growing population in tropical and subtropical countries. However, the cassava field and cassava-based industrial activities generate a variety of wastes during harvest and processing. The improper disposal and burning of these wastes result in various detrimental effects on environmental quality and human health. Dumping and burning of these wastes contribute to leachate formation and greenhouse gas generation. However, the composition of cassava waste could be a significant source for the production of products of value addition. Moreover, the conversion of cassava wastes/residues into value-added products is the major route to sustainable development. This chapter highlights the status of cassava production and waste generation and discusses the valorization of cassava wastes for the generation of value-added bioproducts and biochemicals.

Keywords

Valorization · Bioadsorbents · Bioplastics · Nanomaterials · Biofilms · Biochar · Activated carbon

Abbreviations

AC	Activated carbon
AD	Anaerobic digestion
CBN	Cassava bagasse nanofibers

N. Albert Mariathankam (✉) · K. Suruli
Department of Environmental Studies, School of Energy Sciences, Madurai Kamaraj University, Madurai, Tamil Nadu, India

R. C. Ray (ed.), *Roots, Tubers, and Bulb Crop Wastes: Management by Biorefinery Approaches*, https://doi.org/10.1007/978-981-99-8266-0_5

CD	Carbon dots
CMC	Carboxymethyl cellulose
CNWS	Cellulose nanowhiskers
CPDs	Carbonized polymer dots
CQDs	Carbon quantum dots
GQDS	Graphene quantum dots
PHAs	Polyhydroxyalkanoates
PHBs	Polyhydroxybutyrate
PLA	Polylactic acid
PUFs	Polyurethane foams
SC	Super capacitors

5.1 Introduction

Cassava (*Manihot esculenta* Crantz) is the major staple food crop in tropical and sub-tropical countries. It is the third most significant food carbohydrate source after rice and maize in the tropics and the fifth most common starchy crop produced worldwide (Kiangkitiwan and Srikulkit 2013). Cassava is a dicotyledonous plant, that belongs to the Euphorbiaceae family, which is also referred to by various regional names, including Yuca, Manioc, Tapioca, Aypu, Camioc, and Boniato. Commonly, either the cooked or baked tuber or food preparations made from cassava are consumed as dietary. Cooked cassava root has 191 calories/100 g; the majority of them (84%) come from carbohydrates (40 g), with the remainder coming from protein (1.5 g), fat (3 g), fiber (2 g), and other trace elements (USDA 2020). Due to its wide range of uses in the agricultural economy, cassava is regarded as a significant commercial crop.

Apart from direct food, the agro-industries based on cassava help to produce value-added goods such as sago, starch, chips, flour, and sago wafers (Nizzy et al. 2022). The most lucrative product made from cassava tubers is starch and flour. Starch extracted from cassava tubers is used as an alternative for wheat and corn starch products due to its ease of availability, low cost, and high yield. There were approximately 8 million tonnes of cassava starch produced globally in 2018, with a compound annual growth rate of 4.73% between 2011 and 2018. So, it is anticipated to approach 10 million tonnes in 2024, with an average yearly increase of 3.76% from 2019 to 2024 (FAO 2019). Cassava starch is typically produced by a series of mechanical processing procedures, including peeling and rasping of cassava root, separation of fiber, and dewatering and drying of starch (Padi and Chimphango 2020). Cassava starch can be used in a variety of industries, including food, textile, pharmaceutical, and detergent industries, and can be used as the adhesive in the paper, plywood, and cement industries (Nizzy and Kannan 2022; Serpa-Fajardo et al. 2022). In addition, various fermented foods and beverages have been prepared from cassava tubers (Ray and Sivakumar 2009).

However, the cassava-based industry produces large amounts of cassava bagasse (cassava pulp) every day, which ferments uncontrollably and emits foul odors. About 0.38 tonnes of cassava peel and 1.4 tonnes of bagasse are produced for every tonne of cassava starch produced (Edama et al. 2014). Bagasse has a very high level of perishability due to its high moisture content, which is over 85% when it leaves the processing. Starch and fibrous compounds are present in nearly equal amounts in their dry matter, along with trace amounts of proteins, lipids, and minerals, which together constitute less than 5% of its dry matter (Teixeira et al. 2009). The primary difficulty in managing cassava bagasse with high moisture content is the lack of usage by other industrial sectors as a result of processing, storage, transport, and conservation issues. They need to dry out to accomplish this (Polachini et al. 2016). Hence, the agro-industrial sector has made extensive use of various drying alternatives, including solar drying, hot air drying, spray drying, or osmosis dehydration, in addition to more recently developed options like high hydrostatic pressure, ultrasound, electric pulses, vacuum, centrifugal force, etc., to increase the effectiveness of osmotic dehydration and drying process to shorten drying times while using less energy (Serpa-Fajardo et al. 2022).

Many innovative techniques are currently being used to recycle agricultural-based industrial waste into advantageous goods for use in various commercial and industrial operations. Even though cassava bagasse has several negative effects, its starch content has emerged as a key ingredient in the production of high-value products that can be made inexpensively, such as biocomposites, bioadsorbents, bioplastics, nanomaterials, carbon dots, antimicrobials, compost, and biofertilizers. The innovations in practices and goods created for the environmental value addition of cassava bagasse are highlighted in this book chapter.

5.2 World Production of Cassava

Cassava is the indigenous root food crop of South America and in the eighteenth to nineteenth century, it was introduced to Asia, mainly in India, Java, and the Philippines (Onwueme 2002). Cassava is the most widely grown crop in many African countries due to its advantageous agronomic qualities, such as its tolerance to low soil fertility, resistance to drought and diseases, high adaptability to any climatic conditions, ease of cultivation, extreme starch-yielding capacity, and trouble-free placement on local markets (Amelework et al. 2021; Tabaglio et al. 2023). Africa accounted for over 61% of the world's total 278 million metric tonnes of cassava production in 2018 (FAOSTAT 2020). Additionally, cassava has significantly aided Africa's development and offers small-scale farmers a means of subsistence. FAOSTAT (2023) reported that Nigeria leads the globe in both the production and consumption of cassava. Nigeria produces the most cassava in Africa (61.03 million tonnes from an area of 9.08 lakh ha of land) while Thailand and Brazil produce the most cassava in Asia and the Americas, respectively, with 30.1 million tonnes and 18.09 million tonnes, from an area of 1.46 lakh ha and 1.2 lakh ha. Figure 5.1 provides an overview of the top-producing nations for cassava tubers in Africa, Asia, and America.

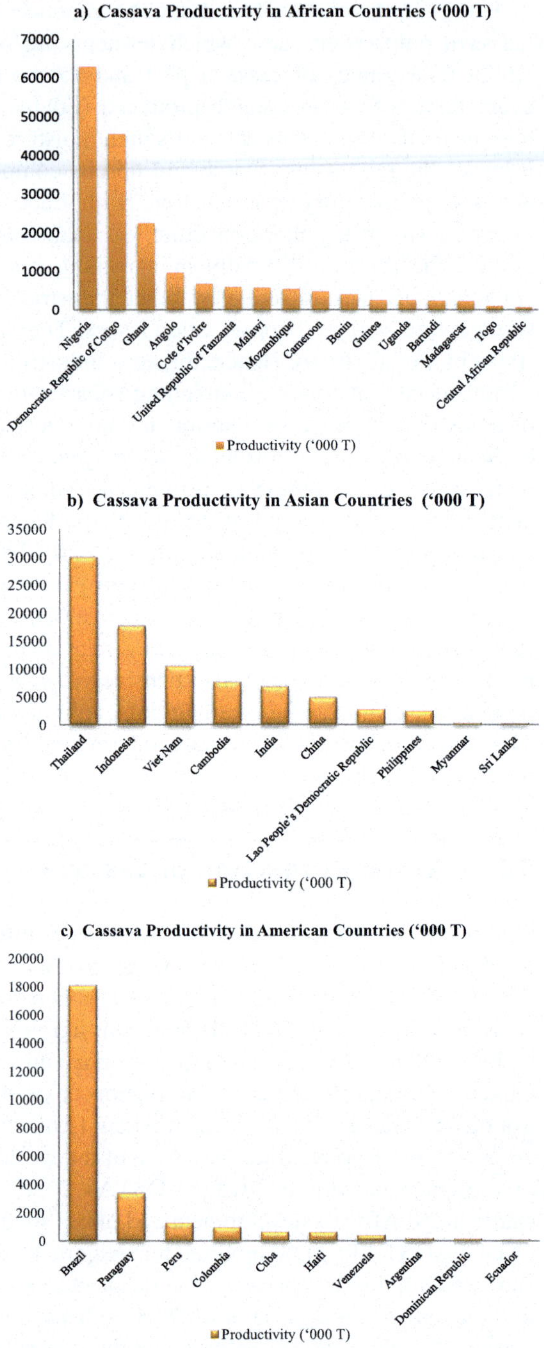

Fig. 5.1 Cassava production in the major countries of (**a**) Africa, (**b**) Asia, and (**c**) Latin America

5.3 Types of Waste Generated

Two types of waste are produced by cassava processing industries: Solid waste and liquid waste.

5.3.1 Solid Waste

Cassava stalks, leaves, and straw make up the majority of the solid wastes left in the cassava cultivating field, and the by-products of cassava processing industries include bagasse, bark, bran, peel, and fiber, as well as a significant amount of spent water. Due to the widespread cultivation of cassava around the globe, suitable use for the by-products must be identified to reduce any negative effects on the environment. The nature and the market demand of the final product affected the chemical composition and nature of the solid trash (peels and bagasse) produced during cassava processing. In comparison to toasted cassava granules, cassava paste after fermentation produces more liquid waste and baggase (Achi et al. 2015). The cassava peels are part of the waste, which are plentiful and cause the most environmental concerns. It might range from 20% to 35% of the total weight of processed cassava tubers (Ukwuru and Egbonu 2013). After the manufacturing of starch and sago, an enormous quantity of the fibrous by-product of the cassava-processing industry known as cassava bagasse or thippi is created, which has a detrimental effect on the land, water, and air. Cassava bagasse is frequently dumped into the environment without being treated, which is a severe concern for locations with starch factories. Other than a limited amount of use as feed or combustible, it finds no beneficial use (Jyothi et al. 2005). Cassava-based industries produce bagasse/thippi in quantities of 40–60 tonnes annually, endangering both the environment and the health of those living nearby (ICAR 2021).

Cassava bagasse needs to be treated before it can be returned to the environment. During the processing of 250–300 tonnes of cassava tubers, roughly 1.6 tonnes of solid peel and 280 tonnes of cassava bagasse with high moisture content are produced (Pandey et al. 2000). It can be utilized as animal feed despite having low protein content because of the high proportion of carbohydrates. On a dry weight basis, it has between 30% and 50% starch. It can serve as an excellent substrate for microbial processes for the manufacture of value-added goods because of its rich organic composition and low ash content. Leftover starches in cassava peel and bagasse shared structural similarities with commercial cassava starch, although peel and bagasse, were significantly abundant in potassium (8570 $\pm$ 56 mg/kg) and calcium (5300 $\pm$ 147 mg/kg), respectively (Thuppahige et al. 2023).

However, cassava bagasse is composed of fiber, residual starch with hemicelluloses, cellulose, and lignin. The composition of cassava bagasse is furnished in Table 5.1. Hence, creating a simple and eco-friendly process for turning cassava residue into high-value products is essential from an economic point of view.

Table 5.1 Composition of cassava bagasse

Cellulose (%)	Hemicellulose (%)	Lignin (%)	Starch (%)	Reference
28.5	31.6	22.2	–	Zanatta et al. (2016)
38	37	–	–	Edhirej et al. (2017)
26.85	8.33	4.5	37.29	Yun et al. (2019)
1.45	1.19	1.03	73.79	Prasertsung et al. (2019)
27	30	2.8	–	Travalini et al. (2019)
35.57	5.36	4.53	–	Ortega et al. (2018)
14.75	3.31	0.59	51.01	Bunterngsook et al. (2017)
10.04	29.26	–	–	Edhirej et al. (2017)

5.3.2 Liquid Waste

In addition, for every tonne of cassava tubers used in the production of sago and starch, 20,000–30,000 L of water is required. The process ends with nearly equal proportions of the effluent, which is a highly organic and acidic substance (Nizzy et al. 2019). When this effluent is released into the environment without proper treatment, it pollutes the soil, stream, and groundwater. Additionally, cassava wastewater left in open or stagnant channels serves as a breeding habitat for various pathogenic microbes, mosquitoes, and other parasitic insects and emits an unpleasant odor (Oghenejoboh 2015).

5.4 Bioproducts and Biochemicals

5.4.1 Bioproducts

Thippi and peel from the cassava-based industry have been directly utilized as the primary components of cattle feed preparation on the animal feed market. Along with cow feed preparations, poultry, fish, and swine feed preparations also employ cassava thippi flour. In addition, cassava leaves that have been partially dried can be fed to animals without causing harm (Nizzy and Kannan 2022). At present, the solid and liquid wastes generated from the cassava-based industry have been converted to various value-added products in all need-based sectors such as water treatment, pharmaceuticals, chemical industry, Agriculture, and biorefinery (Fig. 5.2).

5.4.1.1 Biocomposites

Composites are made up of a wide range of organic and/or inorganic elements, including metals, nanocarbons, proteins, natural and manufactured polymers, and polysaccharides. Biocompatible and environmentally friendly composites prepared from biological sources are referred to as biocomposites. Natural cellulosic fiber-based composites have several benefits over traditional materials, including significant toughness, flexibility, ease of production, recyclability, and environmental friendliness. There are many different types of biocomposites, including films,

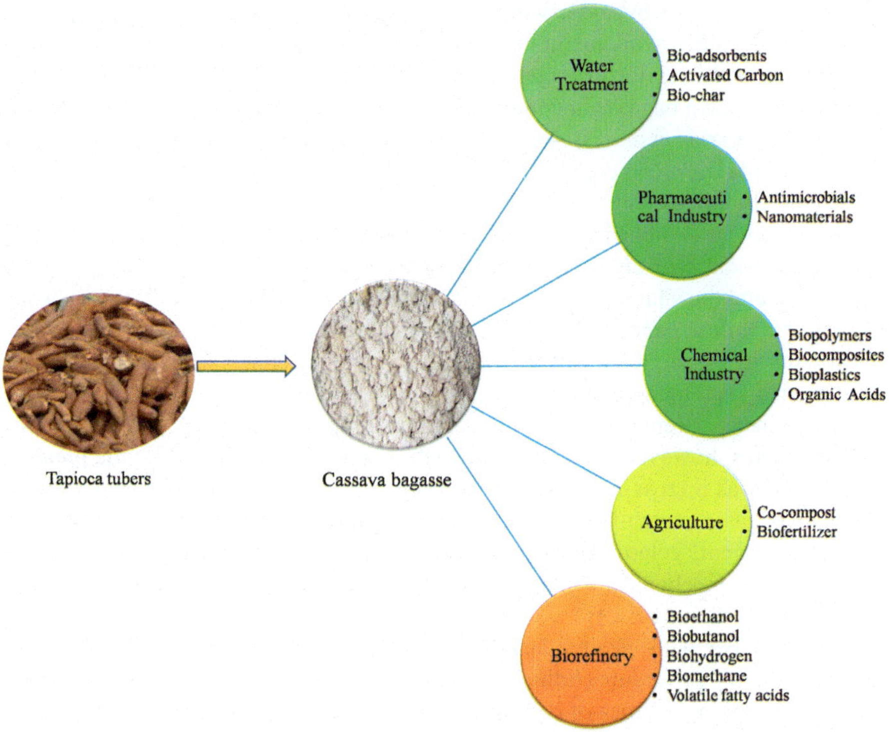

Fig. 5.2 Value-added products recovered from cassava bagasse

membranes, moldings, coatings, fibers, and foams (Haraguchi 2014). Biocomposites from natural fibers are highly valued in a variety of industrial sectors because they are biodegradable, renewable, and produced sustainably. Improved impact resistance and thermal re-formability are two benefits of the bio-composites, which are also less expensive than comparable products manufactured solely from plastic (Corradini et al. 2008). Cassava bagasse was used to create a composite that resembles cardboard. These composites were created using a blend of 90% cassava bagasse and 10% Kraft paper. To enhance the material's mechanical qualities, kraft paper was added as a source of long fibers. The created material resembles the molded fiber packaging used in egg boxes, which is made using recycled paper. However, because it comes from dependable and renewable sources, cassava bagasse has advantages over recycled paper (Matsui et al. 2003).

Moreover, in composites and polymers, natural fibers are being used as filling components more frequently. Cellulose and hemicelluloses present in cassava bagasse are considered potential fillers for polymers, it also contain a significant amount of starch, a natural polymer with a high polarity that interacts with lignocellulosic fibers to improve mechanical properties. Florencia et al. (2019) investigated that cassava peels and bagasse are suitable fillers for thermoplastic corn starch film preparation. Cassava bagasse, which has a high starch content and little protein, hence, it is a readily available large source of biocomposite material. Cassava starch is also used as the reinforcing agent in nature-based biocomposites. The viability of employing cassava bagasse in composites to reinforce low-density polyethylene was evaluated by Farias et al. (2014). They revealed that cassava bagasse could be used as a structural reinforcement and also have benefits in the environment when used in packaging, building, and automotive parts. The modifications in starch films and their mechanical and barrier properties can be explained by the chemical composition and particle size distribution of cassava bagasse (Versino and Garcia 2018). Malic acid and the silane coupling agent KH-550 were used to modify the hydroxyl groups of cellulose nanofibrils (CNFs), which were then added to cassava starch as a reinforcing agent to create a film with outstanding mechanical, hydrophobic, and barrier qualities (Huang et al. 2020).

Low tensile strength and water resistance are drawbacks of starch-based biocomposites, however, the composite matrix can be strengthened by the presence of citric acid as a cross-linker along with the addition of plasticizers and gelatine. Cassava bagasse starch-based biocomposites prepared using bamboo fiber as a reinforcing agent, gelatine from chicken feet, and lime juice as a cross-linker to increase their mechanical strength and water stability (Silviana and Dzulkarom 2019). In addition, a hybrid composite was prepared from cassava bagasse and sugar palm fiber with cassava starch as the matrix and fructose as the plasticizer using the casting procedure (Edhirej et al. 2017). Moreover, jute or kapok fibers were used as two different types of cellulosic fibers to enhance the thermoplastic cassava starch matrix (Prachayawarakorn et al. 2013).

5.4.1.2 Biopolymers

A biopolymer is a type of polymer that is created by or derived from a living entity, such as a plant or a microbe, as opposed to petroleum, which is the usual source of polymers. Biopolymers mostly come from renewable sources (Christian 2016). Polymeric materials are serving as vital for industry and society because of their adaptability, toughness, and resistance to deterioration. However, there are numerous environmental issues associated with the use of polymers made from petrochemicals. These materials are difficult to decompose; their continued use has become a significant contributor to environmental pollution. Due to the poisonous gases produced, the process of degrading and burning became constrained. Therefore, there have been concerns and inquiries in the scientific community on the creation of biodegradable plastics as alternatives to synthetic polymers that are not biodegradable. Contrary to synthetic plastics derived from petroleum, the synthesis of biodegradable biopolymers from agro-based industrial waste is an important tactic for lowering production and waste management costs. In the packaging industry, they are utilized as coatings to lessen major environmental concerns by reducing the need for plastic packaging. Starch-based biopolymers can be generated without the aid of advanced methods, such as fermentation or microbiological processes. Polyhydroxyalkanoates (PHAs) are one such biodegradable form of polymer, which can be synthesized from starchy agriculture residues by different bacteria.

PHAs are polyesters made from R-hydroxy alkanoic acid monomers that are thermoplastic or elastomeric. These polymers are categorized structurally according to the number of carbon atoms in their monomeric units, which can be between 3 and 14 (Vega-Castro et al. 2021). As a carbon source, hydrolyzed cassava peel from agricultural industry waste has been used to make PHAs.

The most crucial factor concerning the restrictions of cassava starch by-products as a substrate for PHA-producing bacteria is water solubility. As a result, the amylase hydrolytic activity of microorganisms may be able to inhibit PHA production. But, alkaline or acidic chemical hydrolysis, high pressure, or high temperature could promote the release of carbohydrates from cellulosic material and promote the synthesis of PHA (Krueger et al. 2012). Currently, it is recognized and understood that certain gram-negative and gram-positive bacteria, photosynthetic cyanobacteria, yeasts, transgenic plants, and Archaea create and store various forms of Polyhydroxy butyrates (PHBs) under altered and stressful situations (Mourao et al. 2021). *Bacillus megaterium* is an amylolytic strain of bacteria that can biotransform the cassava starch by-product into a biodegradable PHAs biopolymer (Krueger et al. 2012). After that many researchers have worked on the synthesis of PHAs from cassava wastes using various bacteria such as *Bacillus tequilensis* (Chaleomrum et al. 2014), *Bacillus safensis* (Babaniyi et al. 2020), *Cupriavidus necator* (Vega-Castro et al. 2021). Mourao et al. (2021) explored the potential for using cassava peel hydrolysate, as a carbon source for the production and characterization of PHBs by the Amazonian microalga *Stigeoclonium* sp. B23.

Cellulose is a polymer present in the cell wall of all plants, which has been extracted and applied in various sectors. But, due to the insoluble nature of cellulose in water, many applications are restricted. The carboxymethylation of cellulose produces water-soluble cellulose derivatives known as carboxymethylcellulose (CMC), which can be used to boost the utility of cellulose (Rachtanapun et al. 2012). CMC is a biodegradable, nontoxic, inexpensive, and renewable polymer, which has been prepared from cellulose. Commonly, CMC has been widely used in the food, medicine, petroleum, paper, and textile industries. Cellulose in the cassava residues was used as the attractive raw material for the synthesis of Carboxymethyl cellulose and carboxymethyl starch to maximize the value and minimize the waste (Yun et al. 2019). Yuvano et al. used centrifugation, ultrasonication, freeze drying, and acid hydrolysis to separate the nitrocellulose from the cassava waste. After that, CMC was created from cassava nitrocellulose utilizing the etherification and alkalization processes (Yuwono et al. 2020). Similarly, cassava stem was used to extract the cellulose, which was subsequently transformed into sodium-CMC; here the CMC was produced from the extracted cellulose from the cassava stem using isopropyl alcohol as a solvent and the addition of monochloride acetic acid and 20% NaOH (Obele et al. 2017).

5.4.1.3 Bioplastics

Conventional plastic waste is currently one of the most prevalent environmental pollutants, due to its inability to degrade, it can easily collect in the environment and release harmful substances like microplastics. It is also a matter of concern because of its effects on the environment, including flooding, build-up of trash in landfills, and even air pollution if burned. These substances have the potential to get into the food chain and impact on human health. Due to their detrimental effects on the environment, numerous researches are being conducted to develop biodegradable plastics (bioplastics) that can replace these conventional plastics. Bioplastics are biodegradable and environmentally benign, prepared from biomass as a substitute for conventional biopolymers out of concern for the environment and public safety (Maulida et al. 2016). Bioplastics is one of the types of biopolymer, which are mostly created using substances derived from plants such as starch and cellulose. Since it is generally possible to generate clear, non-toxic, odorless, good gas barrier, and biodegradable, the creation of starch-based films has received substantial study. However, issues with global hunger are genuine, and the packaging sector should refrain from using starches, a well-known source of carbohydrates. Therefore, the utilization of various residual starches can reduce production costs, encourage the development of novel biomaterials, and eliminate rivalry with the food chain (Luchese et al. 2021).

Starch present in agro-based wastes is a naturally occurring polymer that can be used to create bioplastics since it is readily available, inexpensive, and biodegradable. Starch-rich cassava bagasse has been utilized as biodegradable plastic, because of its widespread availability and low cost (Maulida et al. 2016; Adhani et al. 2019; Abel et al. 2021). Since starch-based bioplastics still have some drawbacks such as weak mechanical characteristics and significant moisture absorption, which make them more brittle than synthetic thermoplastics, hence a substitute is utilized to enhance its properties (Cao et al. 2008). The most promising substance that has proven to be filled with reinforcement is cellulose (Maulida et al. 2016). And also, by adding plasticizers such as sorbitol, glycerol, and xylitol, this brittleness has been reduced, resulting in flexible starch-based bioplastics (Maulida et al. 2016; Singh et al. 2018; Abel et al. 2021; Edhirej et al. 2016; Abidin et al. 2020).

In addition, the fiber in the cassava bagasse served as reinforcement in the thermoplastic matrix and enhanced the maximum tensile strength. To create low-cost composite materials with good performance, blends of polylactic acid (PLA) and thermoplastic starch were also created. Although PLA is a biodegradable polymer, its uses are constrained by its high price and generally subpar qualities when compared to polymers derived from petroleum. One of the most effective methods is adding starch to PLA because it is a plentiful and affordable biopolymer. Cassava starch powder surface modified with soybean oil maleate was added to PLA as filler to enhance the mechanical strength (Kiangkitiwan and Srikulkit 2013). Asrofi et al. (2021) revealed that the addition of sugarcane bagasse fiber into the cassava substrate strengthened the structural integrity of composite bioplastics. In addition, cellulose nanowhiskers (CNWs) can improve the mechanical and water barrier qualities of starch polymers. CNWs are promising fillers for starch matrices due to their comparable chemical structures, which tend to improve the filler interfacial compatibility through hydrogen bonding. de Campos et al. (2017) developed cassava starch films reinforced with CNWs made from oil palm mesocarp fibers by casting, which can improve the mechanical and water barrier characteristics of starch-based plastics.

5.4.1.4 Carbon Dots

Carbon dots (CDs), a group of zero-dimensional luminescence carbon-based nanomaterials with a diameter <10 nm, have come to light as an outstanding contender for a variety of uses, including bioimaging, catalysis, sensing, and drug administration. They are highly biocompatible, have excellent optical characteristics, need straightforward low-cost synthesis techniques, and have little toxicity (Kurian and Paul 2021; Pete et al. 2023). According to their various formation mechanisms, micro-/nanostructures, and features, CDs are primarily divided into graphene quantum dots (GQDs), carbon quantum dots (CQDs), and carbonized polymer dots (CPDs) (Liu et al. 2020). There are two common ways to create CDs made from biomass. The top-down route is the first one. These techniques use ultrasonic synthesis, laser ablation, chemical oxidation, and other processes to break down bulk carbon compounds into carbon nanoparticles. The second method is a bottom-up approach in which micromolecules are heated to high temperatures and then pyrolyzed and carbonized to create CDs, which go through processes like thermal decomposition, hydrothermal treatment, microwave radiation, electrochemistry, etc. (Wang et al. 2020). Generally, the carbon dots are produced by cassava waste by carbonization followed by dissolving in water and the clear solution can be obtained by centrifugation or filtration process. The illumination of CDs can be visualized by UV 405 nm laser (Putro et al. 2019). The illusion of carbon dot production based on Putro et al. (2019) and Prasannan and Imae (2013) is presented in Fig. 5.3.

CDs were produced from various biomass residues such as strawberry (Huang et al. 2013), sugarcane molasses (Huang et al. 2017), grape pulp (Varisco et al. 2017), orange peel (Fatimah et al. 2018), mango kernels (Saini et al. 2022) etc. Putro et al. (2019) demonstrated that CDs were produced utilizing the hydrothermal technique and Poly Ethylene Glycol passivation from cassava peel, which may increase fluorescence when used, and revealed that the prepared CDs can be applied in the metal ion and salt biosensing as well as bioimaging. Using sugarcane bagasse pulp as the precursor, hydrothermal synthesis of highly fluorescent carbon dots with an efficient yield of 17.98% was carried out. Moreover, in antibacterial testing, CDs may serve as a significant replacement for traditional antibiotic medications (Pandiyan et al. 2020).

5.4.1.5 Bioadsorbents

Adsorption is the process by which molecules of a substance are drawn to the surface of a solid or liquid and adhered there. Adsorption is a widely utilized alternative method for eliminating organic contaminants from aqueous solutions (Schwantes et al. 2016). The goal of bioadsorption is to remove or recover organic and inorganic chemicals from aqueous solutions by adsorbing them onto a biological medium. Bioadsorbents are inexpensive and are mostly made from bio-based materials, such as organic wastes, plant residues, and agricultural biomass, and can be utilized to

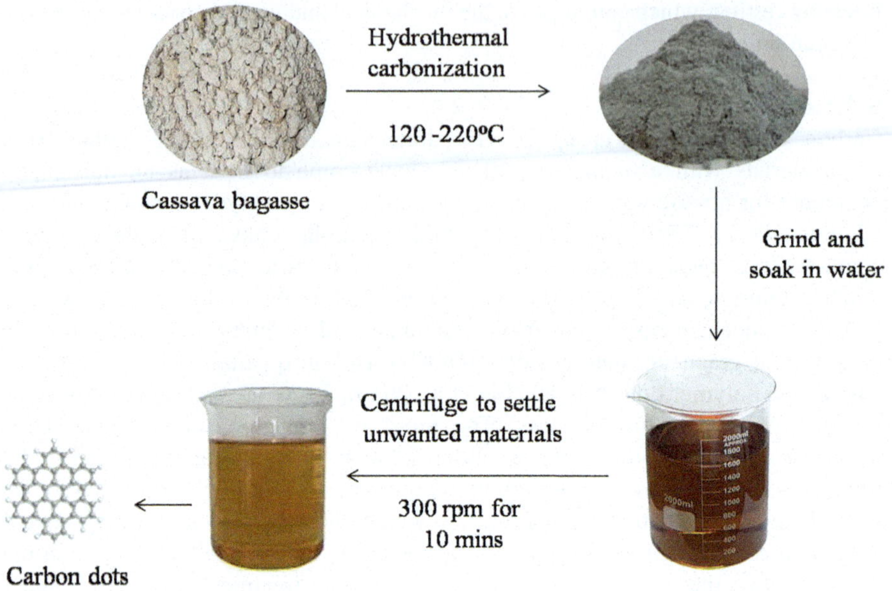

Fig. 5.3 Carbon dots prepared from cassava bagasse

treat wastewater pollutants like heavy metals, pesticides, dyes, etc. in a sustainable manner. Activated carbon, biochar, and polyurethane foam are the bioadsorbents that can be derived from biomass. Biosorbents derived from the cassava peel were used to remove the most widely used pesticide, chlorpyrifos in the water system (Schwantes et al. 2022a). Schwantes et al. (2022b) examined the potential use of solid wastes from the cassava-based industry in the production of adsorbents and their application in the removal of the heavy metals Cd^{2+}, Pb^{2+}, and Cr^{total} from water.

Activated carbon (AC) is still used primarily for adsorption in industrial and domestic processes. AC made from biomass has recently gained popularity as a powerful adsorbent because of its wide availability and high functional group content. In physical activation, activated carbon is often heated to a high temperature and activated with steam or carbon dioxide, but in chemical activation methods, activated carbon is impregnated with chemical reagents such as NaOH, KOH, H_3PO_4, and $ZnCl_2$. The activated carbon was then thermally degraded by carbonization at about 800 °C (Alongamo et al. 2021; Chaisit et al. 2020). Cassava tuber peelings were chemically activated to produce ACs for the batch adsorption of nickel (II) ions from an aqueous solution using potassium hydroxide and phosphoric acid at impregnation ratios of 2:1 and 1:1, respectively (Alongamo et al. 2021). The AC developed from cassava wastes impregnated with H_3PO_4 is activated to remove dyes and metal ions (Rajeshwarisivaraj et al. 2001). To remove naphthol blue-black, Astuti et al. (2020) produced a microwave-induced H_3PO_4 AC from cassava peel. Moreover, AC from cassava peels was investigated for use in supercapacitors (SC),

Table 5.2 Adsorption of pollutants using activated carbon prepared from cassava wastes

Cassava wastes	Activation	Toxic chemicals	Efficiency	Reference
Cassava leaves	Activated at 600 °C for 60 min	Rhodamine-B	77% in 45 min with 25 g/L as the dose	Louis and Thomas (2013)
Cassava residue	Heating at 700 °C for 90 min and carbonization with KOH	Formaldehyde and iodine	Adsorption of 295 mg/g formaldehyde and 1387 mg/g iodine	Wang et al. (2014)
Cassava peel	Carbonized in a muffle furnace for 120 min at 380 °C activated with 150 g 1.0 M zinc chloride	Lead present in the oil refinery wastewater	100% reduction	Oghenejoboh et al. (2016)
Cassava peel	Chemical activation using H_3PO_4	Naphthol blue-black	99.7% of naphthol blue-black was adsorbed	Astuti et al. (2020)
Cassava root	Activation at 700 °C with the chemical activation of phosphoric acid	Cadmium from aqueous solution	100% at 400 °C	Poolprasert and Chorchong (2020)
Cassava peel	Chemical activation using H_3PO_4 and KOH	Nickel(II) ions from aqueous solution	47.39 mg/g for phosphoric acid and 41.15 mg/g for KOH	Alongamo et al. (2021)
Cassava sludge	Chemical activation with $ZnCl_2$	Hexavalent chromium from potable water	95.44% with the dosage of 2.5 g/L	Li et al. (2021)
Cassava husk	$ZnCl_2$ and cassava husk were combined and pyrolyzed for 80 min at 650 °C	Diuron (herbicide) from water	68%	Georgin et al. (2022)
Cassava stem	Chemical activation with NaOH and ZnCl	Ofloxacin from simulated pharmaceutical wastewater	58.82 mg/g	Sulaiman et al. (2022)
Cassava bagasse		Methylene blue from aqueous medium	170.13 mg/g	Diehl et al. (2023)

this showed a cheap and environmentally acceptable source of carbon materials for prospective utilization as electrode materials for SC (Amakoromo et al. 2021). Moreover, a novel composite that effectively removed colors from an aqueous solution by activated carbon graphene oxide with extracts from cassava flour solution was investigated (Ndagijimana et al. 2021) (Table 5.2).

Polyurethane foams (PUF) have been used as one of the sorbents for the decontamination of water polluted with oil and biodiesel spilling (Ambrosio et al. 2017). Commonly, the implementation of this material on a practical scale is constrained because polyurethane synthesis frequently develops into synthetic reactants. Recently, research has focus on the synthesis of polyurethane using starch, a cost-effective and sustainable production process (Zia et al. 2015). In this regard, Keller et al. (2020) synthesized PUF from starch-rich cassava wastes, which demonstrated high biodiesel removal, and 96.8% of biodiesel sorption in 5 min. Cassava waste was used to synthesize polyurethane, a renewable raw material that lowers the expense and negative environmental effects of foam manufacture. In addition, PUF impregnated with cassava bagasse starch hydrolysate as the main carbon source was employed to produce L-lactic acid using *Lactocaseibacillus casei* in a solid-state environment (John et al. 2007).

5.4.1.6 Biochar

Biochar is a porous carbonaceous substance and is a very effective adsorbent, made from the hydrothermal carbonization and pyrolysis of biomass, such as plants, food wastes, and animal wastes. It has fixed carbon content and high surface area (Fu et al. 2019; Nizzy and Kannan 2022). The three main methods for producing biochar from biomass are thermochemical, biochemical, and mechanical extraction. Moreover, biochar can be used as a soil amendment, which results in less compacted soil, greater soil moisture and aeration, and less loss of plant nutrients from the rhizosphere. Wijitkosum and Sriburi (2021) applied biochar made from the pyrolysis of cassava stems to enhance the cultivation of maize in low-fertility soil. When cassava-based biochar was applied to the soil, corn production and yield were higher than when fertilizer was used. Additionally, biochar can be used to restore heavy metal-contaminated soil, minimize environmental toxin pollution, increase soil carbon storage, and increase carbon sequestration in plant biomass to combat long-term climate change (Guo et al. 2020; Li et al. 2017; Spokas et al. 2012; Sohi et al. 2009).

The largest concentration of nutrients, particularly C, Ca, Mg, organic matter, ash, and pH, are found in biochar made from cassava pulp, whereas K, Na, and Al concentrations are the lowest (Zemriyetti et al. 2022). Some specially designed reactors are also used in the efficient conversion of cassava waste into biochar. The conversion of cassava peels into biochar was accomplished by using a top-lit updraft reactor with retort heating. The reactor is fueled by biomass and is reasonably affordable, easy to operate, ecologically benign, and modified for the production of biochar, which is an alternative to an electrical reactor. The 160-min carbonization procedure resulted in a high reactor temperature of 338 °C and an output of 55.13% biochar (Odeyemi et al. 2023). Different methods of biochar preparation and their application is furnished in Table 5.3.

According to Liu et al. (2018), hydrothermal carbonization is a revolutionary method that uses a variety of chemical reactions to transform biomass into carbon-rich material, accomplished at temperatures between 150 and 350 °C in an enclosed system. These chemical reactions include hydrolysis, dehydration, decarboxylation, demethanation (removal of methane from biochar), and aromatization. For

Table 5.3 Methods and application of biochar prepared from biochar

Cassava wastes	Method	Application	Reference
Cassava stem	Pyrolysis at 500 °C	Malachite green adsorption	Zaid et al. (2017)
Cassava residues	Carbonization at 600 °C	Organic resources	Fru et al. (2019)
Cassava stump	Pyrolysis at 500 °C	Methylene Blue dye adsorption	Meili et al. (2019)
Cassava stem	Pyrolysis slowly at 200–400 °C under argon atmosphere	Adsorption of Drimarene Turquoise and Reactive red	Navya et al. (2020)
Cassava slag	Hydrothermal carbonization	Adsorption of Rhodamine B	Wu et al. (2020)
Cassava ethanol sludge	$NaHCO_3$ activation	Adsorption of tetracycline	Zhang et al. (2021)
Cassava waste	Ammonium modified	Mono and co-biosorption of Sulfamerazine, oxytetracycline, and Norfloxacin	Luo et al. (2021)
Cassava straw	Carbonization at 400 °C	Malachite blue dye adsorption	Deng et al. (2021)
Cassava pulp	Pyrolysis at 300–350 °C	Soil amendment	Zemriyetti et al. (2022)
Cassava tubers	Pyrolysis at 450–500 °C	Soil amendment in the agriculture field	Wijitkosum (2022)
Cassava peel	Carbonization for 160 min at the peak temperature of 338 °C	Adsorbent	Odeyemi et al. (2023)

rhodamine B adsorption from the aqueous media, Wu et al. (2020) developed a new cassava slag biochar using the hydrothermal technique.

5.4.1.7 Compost

Cassava thippi has major and micronutrient content, excellent water-holding capacity, good porosity, low bulk density, high starch, fiber, low protein, and cyanide; hence, it can be composted to nutrient-rich organic manure. Cassava thippi composted with various raw materials, microbial cultures, and earthworms revealed that it had the highest levels of nutrients and the narrowest carbon-to-nitrogen (C:N) ratio (Syamala et al. 2017). Ogbo (2010) revealed that two phosphate solubilizing fungi, *Aspergillus fumigatus*, and *A. niger* were isolated from decomposing cassava peels were employed to convert cassava wastes into phosphate biofertilizer using the semi-solid fermentation process.

Generally, the water pollution brought on by fertilizer leaching, particularly as a result of fertilizer overuse, is one of the key issues associated with intensive agriculture. Biocomposites made from cassava bagasse were utilized as urea fertilizer controlled-released solutions for enhancing plant growth (Versino et al. 2020).

Moreover, the majority of the urea is wasted in the environment due to the over usage of fertilizer in high quantities. Hence, urea-controlled release hydrogel is designed. Distantina et al. (2021) developed cassava bagasse-based bead hydrogels and used them to regulate the rate of urea release.

5.4.1.8 Mushrooms

Currently, agriculture residues rich in organic compounds including coconut residue, rice, and wheat straw, coffee waste, cassava peel, and sawdust have been utilized as the substrate for edible mushroom cultivation. Cassava bagasse was also used as the growth substrate for the mushrooms, *Lentinus squarrosulus*, and *Pleurotus ostreatus* (Kupradi et al. 2017). But the poor nitrogen content of cassava waste is the only bottleneck in mushroom cultivation when compared with other agricultural wastes, but it can be overcome by the addition of some nitrogen-based agriculture waste like rice bran, wheat bran, coconut waste, etc. Elsisura and Figueroa (2022) revealed that the number of fruiting bodies and the fresh weight of oyster mushrooms was improved by cassava peel waste (80%) combined with coconut waste (10%) and coffee waste (10%).

5.4.2 Platform Chemicals

5.4.2.1 Organic Acids

Lactic acid is an essential organic acid, widely used in industries based on chemical, food, and pharmaceuticals. Recently, there has been a lot of interest in the creation of biodegradable plastic using lactic acid. It can be produced from starch by lactic acid bacteria and fungi. Previously, L-lactic acid was produced by *La. casei* in a solid-state environment employing polyurethane foam impregnated with cassava bagasse starch hydrolysate as the principal carbon source (John et al. 2007). Ray et al. produced lactic acid using *Lactoplantibacillus plantarum* from cassava bagasse as the substrate (Ray et al. 2008). The fungus *Rhizopus oryzae* utilized the partially hydrolyzed cassava starch to make lactic acid, with a maximum yield of 68.32 g/L (Naranong and Poocharoen 2001). By using an innovative simultaneous saccharification and co-fermentation process with *Bacillus coagulans* and *Lactocaseibacillus rhamnosus*, lactic acid was produced from cassava bagasse with a high yield and productivity. The synergistic effects of cellulase and glucoamylase content of cassava bagasse on sugar yield in this fermentation process greatly enhanced lactic acid production (Chen et al. 2020). Similarly, the fungi, *Rhizopus microsporus* has been used for the production of lactic acid from cassava starch (Yuwa-amornpitak and Chookietwattana 2014).

Brevibacterium divaricatum is used to maximize the production of glutamic acid from leftovers at a cassava starch mill using submerged fermentation, and the highest glutamate yield was obtained at the pH of 7.0 at 30 °C (Jyothi et al. 2005).

5.4.2.2 Nanomaterials

The high fiber content of the cassava bagasse has been used to produce nanofibers, which are utilized as the reinforcing agent in cassava starch film preparation (Travalini et al. 2019). Due to their great rigidity and strength, both the production of nanoscale cellulose fibers and their use in composite materials have drawn growing attention. Cassava bagasse nanofibers (CBN) can be prepared by mechanical, enzymatic, and chemical treatment. CBN was generated mechanically to enhance the thermal, mechanical, and barrier properties of biopolymer films, and can range in size from 1 to 100 nm in one dimension (Pham and Nguyen 2014). Previously, Wicaksono et al. (2013) revealed that CBN can be prepared from cassava bagasse using a combination of chemical (alkali treatment, bleaching, and acid hydrolysis) and mechanical (high-velocity mixer) processes with estimated lengths of several micrometers and widths between 5 and 8 nm. The addition of cellulose nanofibers will increase the tensile strength and lower the elongation of films at the break. CBN has various advantages, including widely available, non-toxic, biodegradable, and resistant.

Nanowhiskers (crystalline cellulose) can be extracted from cassava bagasse using acid hydrolysis. Cellulose whiskers were directly isolated from cassava bagasse and used to produce nanocomposite films with a natural rubber matrix. After the addition of cellulose whiskers, the rubbery modulus increased (Pasquini et al. 2010). Cellulose nanocrystals recovered from cassava stem cellulose via acid hydrolysis were used to investigate the removal of cationic methylene blue dye from an aqueous solution (Obele et al. 2021).

5.4.2.3 Sorbitol and Polyols

Starch modification is essential to enhance the role of functional groups in diverse applications. Crosslink reaction, etherification, esterification, radical graft copolymerization, and oxidation reaction are the main chemical modifications. Cassava starch was modified with glutaraldehyde to be employed as a binder in the preparation of formaldehyde-free particleboards from wood chip wastes (Akinyemi et al. 2019). Moreover, cassava bagasse has a lot of promise for being used directly in the creation of biodegradable food packaging. In association with food packaging, powdered cassava bagasse, water, and glycerol were combined to create film-forming combinations (Thuppahige et al. 2023).

Sorbitol (D-glucitol) is a sugar polyol, which has been used as an additive in the food industry. Most commonly, glucose is hydrogenated to produce sorbitol. Li et al. (2023) revealed that cassava dreg and cassava starch were converted into sorbitol using a biomass-derived bifunctional catalyst. The maximum sorbitol yield of 11.43% was obtained with cassava dreg as the reactant, whereas, using cassava starch as the reactant, the maximum sorbitol yield was 82.51%. The combination of cassava pulp hydrolysate and sugarcane bagasse hydrolysate was used as the feedstock for sorbitol production with the aid of *Zymomonas mobilis*, a thermotolerant strain (Chamnipa et al. 2022).

Polyether polyol, a liquefaction by-product of cassava residue using ethylene glycol as a liquefaction reagent and sulphuric acid as a catalyst, could be utilized to make polyurethane material or alkyd resins (Kang et al. 2019).

5.4.2.4 Xanthan Gum and Other Microbial Polysaccharides

Xanthan gum is a microbial polysaccharide that has been used as a thickening and stabilizer in food additives. It is a water-soluble polysaccharide produced by *Xanthomonas* sp. Sujithra et al. investigated the microbial fermentation of enzyme-hydrolyzed cassava bagasse and was employed to produce xanthan gum using *X. campestris* (Sujithra et al. 2019).

Pullulan is another microbial polysaccharide, produced by the yeast-like fungus, *Aureobasidium pullulans* using cassava bagasse. It is suitable for various commercial applications, especially in food and biomedical fields. An amount of 2.75 g pullulan/100 g cassava bagasse hydrolysate supplemented with a basal medium was produced using *A. pullulans* (Ray and Moorthy 2007). Sugumaran et al. (2014) found that a high yield of cassava bagasse-based pullulan was obtained at suitable conditions at 5.5 initial pH, 1:2 moisture ratio, sodium nitrite as nitrogen source, mannose as supplemented carbons source, and 4 days fermentation time. Moreover, using cassava bagasse and Asian palm kernel as solid substrates, *A. pullulans* produced pullulan in a lab-scale solid-state fermentor, and the pullulan was extracted from the substrate using a solid-liquid method (Sugumaran and Ponnusami 2017).

5.4.2.5 Bio-Oil

Pyrolysis of cassava residues is used to produce bio-oil. In general, the aromatics content is a crucial criterion for assessing the quality of bio-oils, and those with high aromatics content have the potential to be added to gasoline (Zhang et al. 2022). Aromatics like benzene, toluene, and xylenes are significant platform chemicals that can be used widely as organic solvents or building blocks for the synthesis of a variety of functional polymer materials. Yuan et al. (2023) investigated that the pyrolysis of cassava wastes can be converted into aromatics using zeolite-based catalysts. Cassava bagasse is also used to produce neutral lipids and fatty acids in batch and fed-batch fermentations with high yields using heterotrophic *Chlorella protothecoides*, which have been used to make biodiesel (Chen et al. 2015).

5.4.2.6 Microbial Enzymes

The thermostable enzymes are having significant importance in various industries for the processing of starch, paper, leather, food, pharmaceuticals, detergents, textile, and the removal of toxic waste. Generally, these enzymes exist in thermophilic microorganisms. Thermophilic *Bacillus* spp. are the major group of industrially used microbes for enzyme production from carbon sources. Cassava bagasse was used as the substrate in an investigation into the production of amylase by *Bacillus brevis* MTCC 7521 in solid-state fermentation (Ray and Kar 2009). Sugumaran et al. (2013), studied the production of thermotolerant xylanase enzyme from cassava bagasse, wheat flour, and wheat bran using *B. subtilis* in submerged fermentation, and demonstrated that the use of cassava bagasse as a carbon source was found to maximize enzyme synthesis and activity. Gois et al. (2020) revealed that the maximum amylase yield (437.76 U/g) was obtained from cassava bagasse at a pH of 6, temperature 37 °C in 72 h of solid-state fermentation using *Bacillus* spp.

5.4.2.7 Biocolor

Biocolours are the colors/pigments obtained from natural sources, which are added to coloring drugs, food, cosmetics, etc., to replace synthetic colors. In general, biocolors are produced by microbes by fermentation using various natural substrates. To synthesize inexpensive biocolors, agricultural residues can be utilized as the substrates like wheat bran, rice bran, cassava and sugarcane bagasse, oil cake, rice flour, etc. Manimala and Murugesan (2017) identified the pigment carotenoid-producing yeast *Rhodotorula musilaginosa* by utilizing cassava bagasse as the natural cost-effective substrate, and the results revealed that the maximum biocolor production was obtained with cassava bagasse and more stability at pH 7. In addition, cassava wastewater is also used as the alternative substrate and sole source of nutrients for the growth of fungi, *R. glutinis* to produce carotenoid pigment (Ribeiro et al. 2019).

5.5 Conclusion

The main product produced in the cassava-based industry is sago and starch. However, the most significant challenges are the disagreeable amount of cassava residues and wastewater. These cassava residues need to be properly treated and disposed of because they could cause a negative impact on the environment and human health, by polluting the air, oil, and water. Agricultural waste can be transformed into high-value goods like biofuels, biochemicals, and biomaterials, which can reduce the need for non-renewable resources and address environmental issues with agricultural waste disposal. The waste from cassava contains essential elements like starch, cellulose, hemicelluloses, lignin, etc. that can be used to make goods with additional value, particularly in the fields of biorefineries, pharmaceuticals, agriculture, and chemical industries.

References

Abel OM, Chinelo AS, Chidioka NR (2021) Enhancing cassava peels starch as feedstock for biodegradable plastic. J Mater Environ Sci 12(2):169–182

Abidin NDZ, Azhar NS, Sarip MN, Hamid HA, Nasir NAHA (2020) Production of bioplastic from cassava peel with different concentrations of glycerol and $CaCO_3$ as filler. 4th international sciences, technology and engineering conference (ISTEC). AIP Conf Proc 2332:020004-1–020004-8. https://doi.org/10.1063/5.0043482

Achi CG, Coker AO, Sridhar MKC (2015) Utilization of cassava processing waste in Ibadan as a viable and sustainable option to meet cassava processing energy needs—a case study from Ibadan City, Nigeria. In: The thirtieth international conference on solid waste technology and management

Adhani L, Sofian M, Kustiyah E (2019) Synthetic bioplastics from cassava skin using the nata method and the addition of plasticizers. AIP Conf Proc 2085:020076. https://doi.org/10.1063/1.5095054

Akinyemi BA, Olamide O, Oluwasogo D (2019) Formaldehyde free particle boards from wood chip wastes using glutaraldehyde modified cassava starch as binder. Case Stud Construct Mater 11:e00236. https://doi.org/10.1016/j.cscm.2019.e00236

Alongamo BAA, Ajifack LD, Ghogomu JN, Nsami NJ, Ketcha JM (2021) Activated carbon from the peelings of cassava tubers (*Manihot esculenta*) for the removal of Nickel(II) ions from aqueous solution. J Chem 5545110:14. https://doi.org/10.1155/2021/5545110

Amakoromo TE, Abumere O, Amusan JA, Anye V, Bello A (2021) Porous carbon from *Manihot esculent* (cassava) peels waste for charge storage applications. Curr Opin Green Sustain Chem 4: 100098. https://doi.org/10.1016/j.crgsc.2021.100098

Ambrosio E, Lucca DL, Garcia MHB, de Souza MTF, de S Freitas TKF, de Souza RP, Visentainer JV, Garcia JC (2017) Optimization of photocatalytic degradation of biodiesel using TiO_2/H_2O_2 by experimental design. Sci Total Environ 581–582:1–9

Amelework AB, Bairu MW, Maema O, Venter SL, Laing M (2021) Adoption and promotion of resilient crops for climate risk mitigation and import substitution: a case analysis of cassava for South African agriculture. Front Sustain Food Syst 5:617783

Asrofi M, Sapuan SM, Ilyas RA, Ramesh M (2021) Characteristic of composite bioplastics from tapioca starch and sugarcane bagasse fiber: effect of time duration of ultrasonication (Bath-Type). Mater Today Proc 46(4):1626–1630. https://doi.org/10.1016/j.matpr.2020.07.254

Astuti W, Hidayah M, Fitriana L, Mahardhika MA, Irchamsyah EF (2020) Preparation of activated carbon from cassava peel by microwave-induced H_3PO_4 activation for naphthol blue-black removal. AIP Conf Proc 2243:020003. https://doi.org/10.1063/5.0001464

Babaniyi BR, Chidozie OV, Ademola B (2020) Characterization and blending of polyhydroxyalkanoate (PHA) produced by *Bacillus safensis* FO-366(T) on cassava peels. GSC Adv Res Rev 05(02):151–158. https://doi.org/10.30574/gscarr.2020.5.2.0084

Bunterngsook B, Laothanachareon T, Natrchalayuth S, Lertphanich S, Fujii T, Inoue H, Youngthong C, Chantasingh D, Eurwilaichitr L, Champreda V (2017) Optimization of a minimal synergistic enzyme system for hydrolysis of raw cassava pulp. RSC Adv 7:48444–48453. https://doi.org/10.1039/C7RA08472B

Cao X, Chen Y, Chang PR, Muir AD, Falk G (2008) Starch-based nanocomposites reinforced with flax cellulose nanocrystals. Express Polym Lett 2(7):502–510. https://doi.org/10.3144/expresspolymlett.2008.60

Chaisit S, Chanlek N, Khajonrit J, Sichumsaeng T, Maensiri S (2020) Preparation, characterization, and electrochemical properties of KOH-activated carbon from cassava root. Mater Res Express 7:105605

Chaleomrum N, Chookietwattana K, Dararat S (2014) Production of PHA from cassava starch wastewater in sequencing batch reactor treatment system. APCBEE Proc 8:167–172. https://doi.org/10.1016/j.apcbee.2014.03.021

Chamnipa N, Klanrit P, Thanonkeo S, Thanonkeo P (2022) Sorbitol production from a mixture of sugarcane bagasse and cassava pulp hydrolysates using thermotolerant *Zymomonas mobilis* TISTR548. Ind Crop Prod 88:115741. https://doi.org/10.1016/j.indcrop.2022.115741

Chen J, Liu X, Wei D, Chen G (2015) High yields of fatty acid and neutral lipid production from cassava bagasse hydrolysate (CBH) by heterotrophic *Chlorella protothecoides*. Bioresour Technol 191:281–290. https://doi.org/10.1016/j.biortech.2015.04.116

Chen H, Chen B, Su Z, Wang K, Wang B, Wang Y, Si Z, Wu Y, Cai D, Qin P (2020) Efficient lactic acid production from cassava bagasse by mixed culture of Bacillus coagulans and Lactobacillus rhamnosus using stepwise pH controlled simultaneous saccharification and co-fermentation. Ind Crop Prod 146:112175. https://doi.org/10.1016/j.indcrop.2020.112175

Christian SJ (2016) Natural fibre-reinforced noncementitious composites (biocomposites). In: Harries KA, Sharma B (eds) Nonconventional and vernacular construction materials. Woodhead Publishing, pp 111–126. https://doi.org/10.1016/B978-0-08-100038-0.00005-6

Corradini E, Agnelli JAM, Morais LC, Mattoso LHC (2008) Estudo das propriedades de compósitos biodegradáveis de amido/glúten de milho/glicerol reforçados com fibras de sisal. Polímeros 18(4):353–358

de Campos A, de Sena Netoa AR, Rodrigues VB, Luchesi BR, Moreiraa FKV, Correaa AC, Mattosoa LHC, Marconcini JM (2017) Bionanocomposites produced from cassava starch and oil palm mesocarp cellulose nanowhiskers. Carbohydr Polym 175:330–336. https://doi.org/10.1016/j.carbpol.2017.07.080

Deng H, Li YF, Tao SQ, Li AN, Li QY, Hu LN (2021) Efficient adsorption capability of banana and cassava biochar for malachite green: removal process and mechanism exploration. Environ Eng Res 27(3):200575. https://doi.org/10.4491/eer.2020.575

Diehl M, Silva LFO, Schnorr CM, Netto S, Bruckmann FS, Dotto GL (2023) Cassava bagasse as an alternative biosorbent to uptake methylene blue environmental pollutant from water. Environ Sci Pollut Res 30:51920–51931. https://doi.org/10.1007/s11356-023-26006-4

Distantina S, Gunawan L, Damai RA, Kaavessina M (2021) Bead gels based on carrageenan-cassava bagasse as a urea controlled release fertilizer. AIP Conf Proc 2382(1):030008. https://doi.org/10.1063/5.0060999

Edama NA, Sulaiman A, Rahim SNA (2014) Enzymatic saccharification of tapioca processing wastes into biosugars through immobilization technology (Mini Review). Biofuel Res J 1(1): 2–6. https://doi.org/10.18331/BRJ2015.1.1.3

Edhirej A, Sapuan SM, Jawaid M, Zahari NI (2016) Effect of various plasticizers and concentration on the physical, thermal, mechanical, and structural properties of cassava starch based films. Starch 69(2):1500366. https://doi.org/10.1002/star.201500366

Edhirej A, Sapuan SM, Jawaid M, Zahari NI (2017) Cassava/sugar palm fiber reinforced cassava starch hybrid composites: physical, thermal and structural properties. Int J Biol Macromol 101: 75–83. https://doi.org/10.1016/j.ijbiomac.2017.03.045

Elsisura IB, Figueroa MAG (2022) Growth and yield performance of oyster mushroom cultivated in combined cassava peels, coconut residue and coffee waste substrates. Am J Environ Clim 1(1): 1–11. https://doi.org/10.54536/ajec.v1i1.206

Farias FO, Jasko AC, Colman TAD, Pinheiro LA, Schnitzler E, Barana AC, Demiate IM (2014) Characterisation of cassava bagasse and composites prepared by blending with low-density polyethylene. Braz Arch Biol Technol 57(6):821–830. https://doi.org/10.1590/S1516-8913201402506

Fatimah S, Isnaeni, Tahir D (2018) Assisted surface-state recombination of orange-peel carbon nanodots in various matrices. Makara J Sci 22(1):29–34. https://doi.org/10.7454/mss.v22.i1.8301

Florencia V, Lopez O, Garcia M (2019) Exploitation of by-products from cassava and ahipa starch extraction as filler of thermoplastic corn starch. Composites B 182:107653. https://doi.org/10.1016/j.compositesb.2019.107653

Food and Agriculture Organization (FAO) (2019) Food outlook—biannual report on global food markets

FAOSTAT (2020) Food and agriculture organization. Food and agriculture data

FAOSTAT (2023) Food and agriculture organization. Food and agriculture data

Fru BS, Angwafo TE, Ngome AF (2019) Agro-environmental characterization of biochar issued from crop wastes in the humid forest zone of Cameroon. Int J Recycl Org Waste Agric 8:1–13. https://doi.org/10.1007/s40093-018-0223-9

Fu MM, Mo CH, Li H, Zhang YN, Huang WX, Wong MH (2019) Comparison of physicochemical properties of biochars and hydrochars produced from food wastes. J Clean Prod 236:117637. https://doi.org/10.1016/j.jclepro.2019.117637

Georgin J, Pinto D, Franco DSP, Schadeck Netto M, Lazarotto JS, Allasia DG, Tassi R, Silva LFO, Dotto GL (2022) Improved adsorption of the toxic herbicide diuron using activated carbon obtained from residual cassava biomass (*Manihot esculenta*). Molecules 27(21):7574. https://doi.org/10.3390/molecules27217574

Gois IM, Santos AM, Silva CF (2020) Amylase from *Bacillus* sp. produced by solid state fermentation using cassava bagasse as starch source. Braz Arch Biol Technol 63:e20170521. https://doi.org/10.1590/1678-4324-2020170521
Guo M, Weiping S, Jing T (2020) Biochar-facilitated soil remediation: mechanisms and efficacy variations. Front Environ Sci 8:183
Haraguchi K (2014) Biocomposites. In: Kobayashi S, Müllen K (eds) Encyclopedia of polymeric nanomaterials. Springer, Berlin. https://doi.org/10.1007/978-3-642-36199-9_316-1
Huang H, Lv JJ, Zhou DL, Bao N, Xu Y, Wang AJ, Feng JJ (2013) One-pot green synthesis of nitrogen-doped carbon nanoparticles as fluorescent probes for mercury ions. RSC Adv 3(44): 21691–21696. https://doi.org/10.1039/c3ra43452d
Huang G, Chen X, Wang C, Zheng H, Huang Z, Chen D, Xie H (2017) Photoluminescent carbon dots derived from sugarcane molasses: synthesis, properties, and applications. RSC Adv 7: 47840–47847. https://doi.org/10.1039/C7RA09002A
Huang L, Zhao H, Yi T, Qi M, Xu H, Mo Q, Huang C, Wang S, Liu Y (2020) Preparation and properties of cassava residue cellulose nanofibril/cassava starch composite films. Nanomaterials 10(4):755. https://doi.org/10.3390/nano10040755
ICAR—Central Tuber Crops Research Institute (2020–2021) SCSP Programme
John RP, Nampoothiri KM, Pandey A (2007) Polyurethane foam as an inert carrier for the production of L (+)-lactic acid by *Lactobacillus casei* under solid-state fermentation. Lett Appl Microbiol 44:582–587. https://doi.org/10.1111/j.1472-765X.2007.02137.x
Jyothi AN, Sasikiran K, Nambisan B, Balagopalan C (2005) Optimisation of glutamic acid production from cassava starch factory residues using *Brevibacterium divaricatum*. Process Biochem 40(11):3576–3579. https://doi.org/10.1016/j.procbio.2005.03.046
Kang J, Chen W, Yao Y, Jin Y, Cheng X, Lu Q (2019) Optimisation of bio-polyol production from cassava residue using ethylene glycol as the liquefaction reagent. Mater Sci Ed 34(4):945–949. https://doi.org/10.1007/s11595-019-2142-7
Keller M, Ambrosio E, Oliveira VM, Goes MM, Carvalho GM, Batistela VR, Garcia JC (2020) Polyurethane foams synthesis with cassava waste for biodiesel removal from water bodies. Bioresour Technol Rep 10:100396. https://doi.org/10.1016/j.biteb.2020.100396
Kiangkitiwan N, Srikulkit K (2013) Poly(lactic acid) filled with cassava starch-g-soybean oil maleate. Sci World J 2013:7. https://doi.org/10.1155/2013/860487
Krueger CL, Radetski CM, Bendia AG, Oliveira IM, Castro-Silva MA, Rambo CR, Antonio RV, Lima AOS (2012) Bioconversion of cassava starch by-product into *Bacillus* and related bacteria polyhydroxyalkanoates. Electron J Biotechnol 15(3):8–18. https://doi.org/10.2225/vol15-issue3-fulltext-6
Kupradi C, Khongla C, Musika S, Ranok A, Tamaruay K, Woraratphoka J, Mangkalanan S (2017) Cultivation of *Lentinus squarrosulus* and *Pleurotus ostreatus* on cassava bagasse based substrates. Int J Agric Technol 13(6):883–892
Kurian M, Paul A (2021) Recent trends in the use of green sources for carbon dot synthesis—a short review. Carbon Trends 3:100032. https://doi.org/10.1016/j.cartre.2021.100032
Li H, Dong X, da Silva EB, de Oliveira LM, Chen Y, Ma LQ (2017) Mechanisms of metal sorption by biochars: biochar characteristics and modifications. Chemosphere 178:466–478
Li L, Li Y, Liu Y, Ding L, Jin X, Lian H, Zheng J (2021) Preparation of a novel activated carbon from cassava sludge for the high-efficiency adsorption of hexavalent chromium in potable water: adsorption performance and mechanism insight. Water 13:3602. https://doi.org/10.3390/13243602
Li Z, Lan W, Liu C (2023) Biomass derived bifunctional catalyst for the conversion of cassava dreg into sorbitol. Ind Crop Prod 197:116493. https://doi.org/10.1016/j.indcrop.2023.116493
Liu F, Yu R, Ji X, Guo M (2018) Hydrothermal carbonization of holocellulose into hydrochar: structural, chemical characteristics, and combustion behaviour. Bioresour Technol 263:508–516. https://doi.org/10.1016/j.biortech.2018.05.019
Liu J, Li R, Yang B (2020) Carbon dots: a new type of carbon-based nanomaterial with wide applications. ACS Cent Sci 6(12):2179–2195. https://doi.org/10.1021/acscentsci.0c01306

Louis NM, Thomas S (2013) Activated carbon from tapioca leaves for treating dye waste water. Toxicol Environ Heal Sci 5(3):131–137. https://doi.org/10.1007/s13530-013-0166-5

Luchese CL, Rodrigues RB, Tessaro IC (2021) Cassava starch-processing residue utilization for packaging development. Int J Biol Macromol 183:2238–2247

Luo J, Li X, Ge C, Muller K, Yu H, Deng H, Shaheen SM, Tsang DCW, Bolan NS, Rinklebe J, SikOk Y, Gao B, Wang H (2021) Preparation of ammonium-modified cassava waste-derived biochar and its evaluation for synergistic adsorption of ternary antibiotics from aqueous solution. J Environ Manag 298:113530. https://doi.org/10.1016/j.jenvman.2021.113530

Manimala MRA, Murugesan R (2017) Studies on physiochemical analysis and applicability of carotenoid pigment production from *Rhodotorula mucilaginosa* as food colourant. Int J Agric Stat Sci 13(2):507–511

Matsui KN, Larotonda FDS, Paes SS, Luiz DB, Pires ATN, Laurindo JB (2003) Cassava bagasse-Kraft paper composites: analysis of influence of impregnation with starch acetate on tensile strength and water absorption properties. Carbohydr Polym 55(3):237–243. https://doi.org/10.1016/j.carbpol.2003.07.007

Maulida, Siagian M, Tarigan P (2016) Production of starch based bioplastic from cassava peel reinforced with microcrystalline cellulose Avicel PH101 using sorbitol as plasticizer. J Phys Conf Ser 710:012012. https://doi.org/10.1088/1742-6596/710/1/012012

Meili L, Godoy RPS, Soletti JI, Carvalho HV, Ribeiro LMO, Silva MGC, Vieira MGA, Gimenes ML (2019) Cassava (*Manihot esculenta* Crantz) stump biochar: physical/chemical characteristics and dye affinity. Chem Eng Commun 206(7):829–841. https://doi.org/10.1080/00986445.2018.1530991

Mourao MM, Xavier LP, Urbatzka R, Figueiroa LB, Costa CEFD, Dias CGBT, Schneider MPC, Vasconcelos V, Santos AV (2021) Characterization and biotechnological potential of intracellular polyhydroxybutyrate by *Stigeoclonium* sp. B23 using cassava peel as carbon source. Polymers 13(5):687. https://doi.org/10.3390/polym13050687

Naranong N, Poocharoen D (2001) Production of L-lactic acid from raw cassava starch by *Rhizopus oryzae* NRRL 395. Nat Sci 35:164–170

Navya A, Nandhini S, Sivamani S, Vasu G, Sivarajasekar N, Hosseini-Bandegharaei A (2020) Preparation and characterization of cassava stem biochar for mixed reactive dyes removal from simulated effluent. Desalin Water Res 189:440–451. https://doi.org/10.5004/dwt.2020.25635

Ndagijimana P, Liu X, Xu Q, Lai D, Wang G, Pan B, Wang Y (2021) Cassava flour extracts solution to induce gelatin cross-linked activated carbon-graphene oxide composites: the adsorption performance of dyes from aqueous media. Environ Adv 5:100079. https://doi.org/10.1016/j.envadv.2021.100079

Nizzy AM, Kannan S (2022) A review on the conversion of cassava wastes into value-added products towards a sustainable environment. Environ Sci Pollut Res 29:69223–69240. https://doi.org/10.1007/s11356-022-22500-3

Nizzy AM, Kannan S, Anand SB (2019) Identification and characterization of bacteria isolated from sago industrial effluent. Indian J Exp Biol 57:427–434

Nizzy AM, Kannan S, Anand SB (2022) Feasibility study of biohydrogen production using *Clostridium sartagoforme* NASGE 01 from cassava industry effluent. Int J Environ Sci Technol 19:7217–7228. https://doi.org/10.1007/s13762-022-03995-2

Obele CM, Ewulonu CM, Ogbuagu JO (2017) Synthesis and characterization of carboxymethyl cellulose from cassava stem (*Manihot esculenta*). Lignocellulose 6(2):109–118

Obele CM, Ejimofor MI, Atuanya CU, Ibenta ME (2021) Cassava stem cellulose (CSC) nanocrystal for optimal methylene blue biosorption with response surface design. Curr Res Green Sustain Chem 4:100067. https://doi.org/10.1016/j.crgsc.2021.100067

Odeyemi SO, Iwuozor KO, Emenike EC, Odeyemi OT, Adeniyi AG (2023) Valorization of waste cassava peel into biochar: an alternative to electrically-powered process. Total Environ Res Themes 6:100029. https://doi.org/10.1016/j.totert.2023.100029

Ogbo FC (2010) Conversion of cassava wastes for biofertilizer production using phosphate solubilizing fungi. Bioresour Technol 101(11):4120–4124. https://doi.org/10.1016/j.biortech.2009.12.057

Oghenejoboh KM (2015) Effects of cassava wastewater on the quality of receiving water body intended for fish farming. Curr J Appl Sci Technol 6(2):164–171. https://doi.org/10.9734/BJAST/2015/14356

Oghenejoboh KM, Otuagoma M, Ohimor EO (2016) Application of cassava peels activated carbon in the treatment of oil refinery wastewater—a comparative analysis. J Ecol Eng 17(2):52–58. https://doi.org/10.12911/22998993/62290

Onwueme IC (2002) Cassava in Asia and Pacific. In: Cassava: biology, production and utilization. CAB International, p 55

Ortega MV, Rhenals LR, Salcedo JG (2018) Effects of enzymatic hydrolysis on the physicochemical and structural properties of cassava bagasse (*Manihot esculenta* Cranz). Adv J Food Sci Technol 15:158–165. https://doi.org/10.19026/ajfst.15.5889

Padi RK, Chimphango A (2020) Commercial viability of integrated waste treatment in cassava starch industries for targeted resource recoveries. J Clean Prod 265:121619. https://doi.org/10.1016/j.jclepro.2020.121619

Pandey A, Soccol CR, Nigam PS, Soccol VT, Vandenberghe LP, Mohan R (2000) Biotechnological potential of agro-industrial residues. II: Cassava bagasse. Bioresour Technol 74:81–87. https://doi.org/10.1016/S0960-8524(99)00143-1

Pandiyan S, Arumugam L, Srirengan SP, Pitchan R, Sevugan P, Kannan K, Pitchan G, Hegde TA, Gandhirajan V (2020) Biocompatible carbon quantum dots derived from sugarcane industrial wastes for effective nonlinear optical behavior and antimicrobial activity applications. ACS Omega 5(47):30363–30372. https://doi.org/10.1021/acsomega.0c03290

Pasquini D, Teixeira EM, Curvelo AAS, Belgacem MN, Dufresne A (2010) Extraction of cellulose whiskers from cassava bagasse and their applications as reinforcing agent in natural rubber. Ind Crop Prod 32(3):486–490. https://doi.org/10.1016/j.indcrop.2010.06.022

Pete AM, Ingle PU, Raut RW, Shende SS, Rai M, Minkina TM, Rajput VD, Kalinitchenko VP, Gade AK (2023) Biogenic synthesis of fluorescent carbon dots (cds) and their application in bioimaging of agricultural crops. Nanomaterials 13:209. https://doi.org/10.3390/nano13010209

Pham H, Nguyen QP (2014) Effect of silica nanoparticles on clay swelling and aqueous stability of nanoparticle dispersions. J Nanopart Res 16:2137

Polachini T, Leonardo L, Lopes-Filho J, Telis-Romero J (2016) Water adsorption isotherms and thermodynamic properties of cassava bagasse. Thermochim Acta 632:79–85. https://doi.org/10.1016/j.tca.2016.03.032

Poolprasert P, Chorchong T (2020) Preparation of activated carbon from cassava root for cadmium removal in aqueous solution sample. J Renew Energy Smart Grid Technol 15(02):26–38. https://ph01.tci-thaijo.org/index.php/RAST/article/view/243208

Prachayawarakorn J, Chaiwatyothin S, Mueangta S, Hanchana A (2013) Effect of jute and kapok fibers on properties of thermoplastic cassava starch composites. Mater Des 47:309–315. https://doi.org/10.1016/j.matdes.2012.12.012

Prasannan A, Imae T (2013) One-Pot synthesis of fluorescent carbon dots from orange waste peels. Ind Eng Chem Res 52(44):15673–15678. https://doi.org/10.1021/ie402421s

Prasertsung I, Aroonraj K, Kamwilaisak K, Saito N, Damrongsakkul S (2019) Production of reducing sugar from cassava starch waste (CSW) using solution plasma process (SPP). Carbohydr Polym 205:472–479. https://doi.org/10.1016/j.carbpol.2018.10.090

Putro PA, Roza L, Isnaeni I (2019) The effect of poly (ethylene glycol) on the photoluminescence properties of carbon dots from cassava peels synthesized by hydrothermal methods. SPEKTRA Jurnal Fisika Dan Aplikasinya 4(1):11–20. https://doi.org/10.21009/SPEKTRA.041.02

Rachtanapun P, Luangkamin S, Tanprasert K, Suriyatem R (2012) Carboxymethyl cellulose film from durian rind. LWT Food Sci Technol 48:52–58. https://doi.org/10.1016/j.lwt.2012.02.029

Rajeshwarisivaraj, Sivakumar S, Senthilkumar P, Subburam V (2001) Carbon from cassava peel, an agricultural waste, as an adsorbent in the removal of dyes and metal ions from aqueous solution. Bioresour Technol 80(3):233–235

Ray R, Kar S (2009) Statistical optimization of α-amylase production by *Bacillus brevis* MTCC 7521 in solid-state fermentation using cassava bagasse. Biologia 64(5):864–870. https://doi.org/10.2478/s11756-009-0160-1

Ray RC, Moorthy SN (2007) Exopolysaccharide (pullulan) production from cassava starch residue by *Aureobasidium pullulans* strain MTTC 1991. J Sci Ind Res 66:252–255

Ray RC, Sivakumar PS (2009) Traditional and novel fermented foods and beverages from tropical root and tuber crops: review. Int J Food Sci Technol 44(6):1073–1087. https://doi.org/10.1111/j.1365-2621.2009.01933.x

Ray RC, Mohapatra S, Panda S, Kar S (2008) Solid substrate fermentation of cassava fibrous residue for production of α-amylase, lactic acid and ethanol. J Environ Biol 29(1):111–115

Ribeiro JES, Sant'Ana AMS, Martini M, Sorce C, Andreucci A, de Melo dJN, da Silva FLH (2019) *Rhodotorula glutinis* cultivation on cassava wastewater for carotenoids and fatty acids generation. Biocatal Agric Biotechnol 22:101419. https://doi.org/10.1016/j.bcab.2019.101419

Saini S, Kumar K, Saini P, Mahawar DK, Rathore KS, Kumar S, Dandiaa A, Parewa V (2022) Sustainable synthesis of biomass-derived carbon quantum dots and their catalytic application for the assessment of α,β-unsaturated compounds. RSC Adv 12:32619–32629. https://doi.org/10.1039/D2RA05201F

Schwantes D, Gonçalves AC, Coelho GF, Campagnolo MA, Dragunski DC, Tarley CRT, Miola AJ, Leismann EAV (2016) Chemical modifications of cassava peel as adsorbent material for metals ions from wastewater. J Chem 15:3694174. https://doi.org/10.1155/2016/3694174

Schwantes D, Gonçalves AC Jr, Fuentealba D, Carneiro MFH, Tarley CRT, Prete MC (2022a) Removal of chlorpyrifos from water using biosorbents derived from cassava peel, crambe meal, and pinus bark. Chem Eng Res Des 188:142–165. https://doi.org/10.1016/j.cherd.2022.09.033

Schwantes D, Gonçalves AC Jr, Perina HA, Tarley CRT, Dragunski DC, Conradi E Jr, Zimmermann J (2022b) Ecofriendly biosorbents produced from cassava solid wastes: sustainable technology for the removal of Cd^{2+}, Pb^{+}, and Cr^{total}. Adsorpt Sci Technol 5935712:18. https://doi.org/10.1155/2022/5935712

Serpa-Fajardo JG, Hernandez-Ramos EJ, Fernández-Lambert G, Sandoval-Herazo LC, Andrade-Pizarro RD (2022) Post-industrial context of cassava bagasse and trend of studies towards a sustainable industry: a scoping review—Part I. F1000Research 11:562. https://doi.org/10.12688/f1000research

Silviana S, Dzulkarom MC (2019) Synthesis of cassava bagasse starch-based biocomposite reinforced woven bamboo fibre with lime juice as crosslinker and epoxidized waste cooking oil (EWCO) as bioplasticizer. J Phys Conf Ser 1295:012076. https://doi.org/10.1088/1742-6596/1295/1/012076

Singh K, Madhav S, Appro N (2018) Smart bio-polymer from tapioca sago possessing novel in-built film ability by synergistic approach using film initiator. Drug Des Dev 4(2):NAPDD.MS.ID.555635

Sohi SP, Lopez-Capel E, Krull E, Bol R (2009) Biochar, climate change and soil: a review to guide future research. CSIRO Land Water Sci Rep 5:17–31

Spokas KA, Cantrell KB, Novak JM, Archer DW, Ippolito JA, Collins HP, Boateng AA, Lima IM, Lamb MC, McAloon AJ et al (2012) Biochar: a synthesis of its agronomic impact beyond carbon sequestration. J Environ Qual 41:973–989

Sugumaran KR, Ponnusami V (2017) Conventional optimization of aqueous extraction of pullulan in solid-state fermentation of cassava bagasse and Asian palm kernel. Biocatal Agric Biotechnol 10:204–208. https://doi.org/10.1016/j.bcab.2017.03.010

Sugumaran KR, Kumar BK, Mahalakshmi M, Ponnusami V (2013) Cassava bagasse—low cost substrate for thermo-tolerant xylanase production using *Bacillus subtilis*. Int J Chem Tech Res 5(1):394–400

Sugumaran K, Jothi PA, Ponnusami V (2014) Bioconversion of industrial solid waste—cassava bagasse for pullulan production in solid state fermentation. Carbohydr Polym 99:22–30. https://doi.org/10.1016/j.carbpol.2013.08.039

Sulaiman NS, Mohamad Amini MH, Danish M, Sulaiman O, Hashim R, Demirel S, Demirel GK (2022) Characterization and ofloxacin adsorption studies of chemically modified activated carbon from cassava stem. Materials 15(15):5117. https://doi.org/10.3390/ma15155117

Sujithra B, Deepika S, Akshaya K, Ponnusami V (2019) Production and optimization of xanthan gum from three-step sequential enzyme treated cassava bagasse hydrolysate. Biocatal Agric Biotechnol 21:101294. https://doi.org/10.1016/j.bcab.2019.101294

Syamala C, Kuzhivilayil SJ, Manikantan Nair M, Sreekumar J (2017) Management of cassava starch factory solid waste (Thippi) through composting to a nutrient-rich organic manure. Commun Soil Sci Plant Anal 48(6):595–607. https://doi.org/10.1080/00103624.2016.1243700

Tabaglio V, Fiorini A, Ndayisenga V, Ndereyimana A, Minuti A, Nyembo Nyembo R, Nyembo Ngoy D, Bertoni G (2023) Sustainable intensification of cassava production towards food security in the Lomami Province (DR Congo): role of planting method and landrace. Agronomy 13(1):228. https://doi.org/10.3390/agronomy13010228

Teixeira EM, Pasquini D, Curvelo AAS, Corradini E, Belgacem MN, Dufresne A (2009) Cassava bagasse cellulose nanofibrils reinforced thermoplastic cassava starch. Carbohydr Polym 78(3): 422–431

Thuppahige VTW, Moghaddam L, Welsh ZG, Wang T, Karim A (2023) Investigation of critical properties of Cassava (*Manihot esculenta*) peel and bagasse as starch-rich fibrous agro-industrial wastes for biodegradable food packaging. Food Chem 422:136200. https://doi.org/10.1016/j.foodchem.2023.136200

Travalini AP, Lamsal B, Magalhães WLE, Demiate IM (2019) Cassava starch films reinforced with lignocellulose nanofibers from cassava bagasse. Int J Biol Macromol 139:1151–1161. https://doi.org/10.1016/j.ijbiomac.2019.08.115

Ukwuru MU, Egbonu SE (2013) Recent development in cassava-based products research. Acad J Food Res 1(1):001–013. https://doi.org/10.15413/ajfr.2012.0105

United States Department of Agriculture (USDA) (2020) Agricultural Research Service. FoodData Central Search Results. https://www.ars.usda.gov/northeast-area/beltsville-md-bhnrc/beltsville-human-nutrition-researchcenter/food-surveys-research-group/docs/fndds-download-databases/

Varisco M, Zufferey D, Ruggi A, Zhang Y, Erni R, Mamula O (2017) Synthesis of hydrophilic and hydrophobic carbon quantum dots from waste of wine fermentation. R Soc Open Sci 4(12): 170900. https://doi.org/10.1098/rsos.170900

Vega-Castro O, Leon E, Arias M, Teresa Cesario M, Ferreira F, da Fonseca MMR, Segura A, Valencia P, Simpson R, Nunez H, Contreras-Calderon J (2021) Characterization and production of a polyhydroxyalkanoate from cassava peel waste: manufacture of biopolymer microfibers by electrospinning. J Polym Environ 29:187–200. https://doi.org/10.1007/s10924-020-01861-1

Versino F, Garcia MA (2018) Particle size distribution effect on cassava starch and cassava bagasse biocomposites. ACS Sustain Chem Eng 7(1):1052–1060. https://doi.org/10.1021/acssuschemeng.8b04700

Versino F, Urriza M, Garcia MA (2020) Cassava-based biocomposites as fertilizer controlled-release systems for plant growth improvement. Ind Crop Prod 44:112062. https://doi.org/10.1016/j.indcrop.2019.112062

Wang X, Wang S, Yin X, Chen J, Zhu L (2014) Activated carbon preparation from cassava residue using a two-step KOH activation: preparation, micropore structure and adsorption capacity. J Biobaased Mater Bioenergy 8(1):35–42. https://doi.org/10.1166/jbmb.2014.1408

Wang Y, Sun J, He B, Feng M (2020) Synthesis and modification of biomass derived carbon dots in ionic liquids and their application: a mini review. Green Chem Eng 1(2):94–108. https://doi.org/10.1016/j.gce.2020.09.010

Wicaksono R, Syamsu K, Yuliasih I, Nasir M (2013) Cellulose nanofibers from cassava bagasse: characterization and application on tapioca-film. Chem Mater Res 3(13):79–88

Wijitkosum S, Sriburi T (2021) Applying cassava stems biochar produced from agronomical waste to enhance the yield and productivity of maize in unfertile soil. Fermentation 7:277. https://doi.org/10.3390/fermentation7040277

Wijitkosum S (2022) Biochar derived from agricultural wastes and wood residues for sustainable agricultural and environmental applications. Int Soil Water Conserv Res 10(2):335–341. https://doi.org/10.1016/j.iswcr.2021.09.006

Wu J, Yang J, Huang G, Xu C, Lin B (2020) Hydrothermal carbonization synthesis of cassava slag biochar with excellent adsorption performance for Rhodamine B. J Clean Prod 251:119717. https://doi.org/10.1016/j.jclepro.2019.119717

Yuan H, Li C, Shan R, Zhang J, Chen Y (2023) Aromatics production from catalytic pyrolysis of waste cassava residue using La and P modified ZSM-5: experimental and kinetic study. Ind Crop Prod 198:116753. https://doi.org/10.1016/j.indcrop.2023.116753

Yun T, Pang B, Lu J, Lv Y, Cheng Y, Wang H (2019) Study on the derivation of cassava residue and its application in surface sizing. Int J Biol Macromol 128:80–84. https://doi.org/10.1016/j.ijbiomac.2019.01.115

Yuwa-amornpitak T, Chookietwattana K (2014) L-Lactic acid production from cassava starch by thermotolerant *Rhizopus microspores* LTH23. J Biol Sci 14(4):284–291. https://doi.org/10.3923/jbs.2014.284.291

Yuwono SD, Wahyuningsih E, Noviany, Kiswandono AA, Simanjuntak W, Hadi S (2020) Characterization of carboxymethyl cellulose (CMC) synthesized from microcellulose of cassava peel. Mater Plast 57(4):225–235. https://doi.org/10.37358/MP.20.4.5422

Zaid MKZM, Jamion NA, Omar Q, Yong K (2017) Sorption of malachite green (MG) by cassava stem biochar (CSB) kinetic and isotherm studies. J Fundam Appl Sci 9:273–287. https://doi.org/10.4314/jfas.v9i6s.22

Zanatta ER, Reinehr TO, Awadallak JA et al (2016) Kinetic studies of thermal decomposition of sugarcane bagasse and cassava bagasse. J Therm Anal Calorim 125:437–445. https://doi.org/10.1007/s10973-016-5378-x

Zemriyetti, Mayly S, Refnizuida, Girsang R, Hidayat B (2022) Characterization biochar properties from waste of cassava, sugarcane, and coconut. J Environ Agric Stud:37–41. https://doi.org/10.32996/jeas.2022.3.1.4

Zhang Z, Li Y, Ding L, Yu J, Zhou Q, Kong Y, Ma J (2021) Novel sodium bicarbonate activation of cassava ethanol sludge derived biochar for removing tetracycline from aqueous solution: performance assessment and mechanism insight. Bioresour Technol 330:124949. https://doi.org/10.1016/j.biortech.2021.124949

Zhang J, Li C, Yuan H, Chen Y (2022) Enhancement of aromatics production via cellulose fast pyrolysis over Ru modified hierarchical zeolites. Renew Energy 84:280–290. https://doi.org/10.1016/j.renene.2021.11.103

Zia F, Zia KM, Zuber M, Kamal S, Aslam N (2015) Starch based polyurethanes: a critical review updating recent literature. Carbohydr Polym 134:784–798

Bioprocessing of Cassava Bagasse: Part II—Potential for Renewable Biofuels

6

Yeshona Sewsynker-Sukai and Edson L. Meyer

Abstract

Dwindling fossil fuel reserves combined with the exponentially growing population has raised concerns over energy sustainability. Biofuel production from microbial bioprocesses has appeared as the potential supplemental energy source to current non-renewable ones. The former is renewable, sustainable, and environmentally friendly. Notwithstanding, biofuel processes are currently not feasible, attributable to the limited supply and high costs associated with food-based crops as feedstocks. Interestingly, the agricultural industry generates large quantities of biomass waste residues in the form of lignocellulosic biomass (LCB) and lignocellulose-starch biomass (LCSB) that consist of leaves, peels, and bagasse, amongst various others, that are usually landfilled or disposed of, resulting in detrimental effects on the environment. Both LCB and LCSB consist of cellulose, hemicellulose, and lignin in their structural composition, however, the latter contains starch, which is considered to be a major advantage over the former as a feedstock for microbial bioprocesses. Particularly, cassava bagasse (CB), a waste generated by the cassava starch industry, has demonstrated significant potential as a feedstock for biofuel production. The present chapter outlined the potential of CB as a feedstock for biofuel production. The status of global cassava production, waste generation, and disposal of CB is presented. Then, the general composition of CB and the significance of its structural composition in industrial biotechnology is highlighted. Subsequently, the different stages during LCSB bioprocesses are briefly described followed by the application of CB for the production of bioethanol, biohydrogen, biogas, and biobutanol. To conclude,

Y. Sewsynker-Sukai (✉) · E. L. Meyer
Fort Hare Institute of Technology, University of Fort Hare, Alice, South Africa
e-mail: yeshonasukai@gmail.com

R. C. Ray (ed.), *Roots, Tubers, and Bulb Crop Wastes: Management by Biorefinery Approaches*, https://doi.org/10.1007/978-981-99-8266-0_6

the challenges and prospects of using LCSB such as CB in bioprocessing were noted.

Keywords

Lignocellulosic biomass · Lignocellulose-starch biomass · Cassava bagasse · Pretreatment · Biofuels · Bioprocessing

Abbreviations

ABE	Acetone-butanol-ethanol
AD	Anaerobic digestion
CB	Cassava bagasse
CSTR	Continuous stirred tank reactor
HRT	Hydraulic retention time
LCB	Lignocellulosic biomass
LCSB	Lignocellulosic starch biomass
PSSF	Prehydrolysis and simultaneous saccharification and fermentation
SBR	Sequencing batch reactor
SHF	Separate hydrolysis and fermentation
SSF	Simultaneous saccharification and fermentation
UASB	Up-flow anaerobic sludge blanket

6.1 Introduction

The depletion of fossil fuel reservoirs coupled with the consequent increase in fuel prices have skewed research towards the development of renewable energy using bioprocesses (Laltha et al. 2022; Behera et al. 2022). Of particular interest are the high quantities of agricultural biomass waste residues that are generated annually. It has been estimated that approximately 50% of the global biomass is agricultural and this equates to a projected 181.5 billion tons of lignocellulosic biomass, of which, only 8.2 billion tons are utilized for different applications (Singh et al. 2022), with remaining waste released into the environment. Channeling these waste-based substrates for microbial biofuel bioprocesses has gained significant attention. Lignocellulosic-based bioprocessing for biofuel production currently falls in line with global sustainable development goals (Sewsynker-Sukai et al. 2020; David et al. 2022; Laltha et al. 2022) that aim to alleviate wastes, which generally pose hazardous to the environment since it is typically landfilled or burnt, releasing harmful gases. With this in mind, the race for sustainable energy from renewable sources is presently underway (Sewsynker-Sukai et al. 2023a, b). Microbial fermentative biofuel production can be achieved by the utilization of simple sugars recovered from the disintegration of polymeric carbohydrates embedded in the biomass (Laltha et al. 2022). Two types of lignocellulosic-based biomass exist and include lignocellulosic biomass (LCB) and lignocellulose-starch biomass (LCSB). Several wastes have been assessed as feedstocks for biofuel production and include sugarcane leaves (Moodley and Gueguim Kana 2017), corn cobs (Sewsynker-Sukai and Gueguim Kana 2018), and sorghum leaves (Rorke et al. 2017), as well as LCSB, substrates such as cassava bagasse (CB) (Ray et al. 2008), cassava peels (Aruwajoye et al. 2020) and potato peels (Sanusi et al. 2022), to name a few.

In the present time, LCBs have been the preferred agricultural waste substrate for biofuel production, even though these systems remain in their infancy. LCB has been characterized as by-products generated from agricultural activities such as food preparation, farming activities, and the remains of crops post-harvesting, and mainly consists of cellulose, hemicellulose, and lignin (Sewsynker-Sukai et al. 2023a). Inclination towards utilizing LCB as compared to sugar and starch crops stems from its characterization as waste, thereby avoiding the food versus fuel debacle that has caused significant tension globally (Ray and Ramachandran 2018; Sewsynker-Sukai et al. 2023b). Despite the benefit of being a waste, LCB majorly comprises recalcitrant lignocellulosic structures that present several challenges for microbial fermentation (David et al. 2022). Conversely, LCSB contains starch as an additional benefit and major polysaccharide component as part of their structural composition. LCSB consists of 2–70% starch, 4–50% cellulose, 7–30% hemicellulose, and 6–42% lignin (Aruwajoye et al. 2020). The main structural component of LCSB, starch, is a biopolymer of monomeric sugars, usually in the form of amylose and amylopectin chains. The α-1–4 glycosidic bonds link the amylose polymer while the amylopectin contains α-1–6 bonds as well as the presence of α-1–4. Recently, LCSB has garnered interest as substrates for the production of supplemental energy sources such as bioethanol, biohydrogen, biogas, and biobutanol. In particular, CB has gained prominence as a suitable LCSB feedstock for biofuel production (Escaramboni et al. 2018; Balbin et al. 2022; Ruttithiwapanich et al. 2022; Johnravindar et al. 2019).

However, the major structural component in LCSB substrates is starch which is semi-crystalline, presenting some challenges for utilization within industrial bioprocesses. Starch crystallinity is conferred by the amylopectin-to-amylose ratio within the biomass. Generally, a larger amylopectin content indicates higher crystallinity and vice-versa (Alcázar-Alay and Meireles 2015). In the case of cassava starch, the amylose proportion has shown to be relatively low, ranging from 8% to 28% (Wheatley et al. 2003), thereby resulting in a high amylopectin content. Therefore, to process starch-based substrates with high crystallinity such as cassava bagasse for use in biofuel processing, they must first be subjected to pretreatment, liquefaction, and enzymatic conditions that modify their structure and alter their physicochemical properties (Liu et al. 2017).

LCSB-based bioprocesses typically comprise four major processes that include: pretreatment, starch liquefaction, enzymatic saccharification, and fermentation process (Aruwajoye et al. 2020). Pretreatment is carried out to break down the resistant lignocellulosic structures and typically includes chemicals such as acids or alkali. For the liquefaction process, α-amylase enzymes are added to the biomass at high temperatures (≤90 °C) to gelatinize the starchy portion of the chemical-pretreated biomass. The gelatinization produces shorter oligosaccharide chains such as dextrins, maltose, and maltotriose (Aruwajoye et al. 2020). After the liquefaction process, amyloglucosidase and cellulase, also referred to as the saccharifying enzymes are added to the vessel and result in the conversion of dextrins to glucose at temperatures ranging from 50 to 60 °C. Once the liquefaction and saccharification steps are completed, the inoculum is introduced into the fermentation vessel and is incubated at optimal temperatures that facilitate the metabolic conversion of glucose to the product of interest (Aruwajoye et al. 2020).

This chapter focuses on CB and its potential use as a feedstock for biofuel production. First, the status of global cassava production, waste generation, and disposal of CB is highlighted. Subsequently, the general composition of CB and the significance of its structural composition in industrial biotechnology is presented. Afterward, the different stages during LCSB bioprocesses are outlined followed by the application of CB for renewable biofuels such as bioethanol, biohydrogen, biogas, and biobutanol. Finally, the challenges and prospects of using LCSB such as CB in bioprocessing are presented.

6.2 Status of Global Production and Waste Disposal of Cassava Waste

Cassava (*Manihot esculenta* Crantz) a tropical root crop is tagged as the third most vital food source in the tropics, after rice and corn (John 2009). This starchy, tuberous root crop is a staple food for an estimated 800 million people worldwide with Africa being the leading producer due to some 500 million people who depend on the root as their main staple (Statista 2023). This crop yields the highest quantity of starch per hectare of any crop, ranking it as the fourth crop in worldwide production after other major crops like rice, wheat, and maize. In 2021, the total worldwide cassava production reached just under 233 million metric tons with the top three producers being Nigeria (27.11%), the Democratic Republic of Congo (19.64%), and Thailand (12.95%) as depicted in Fig. 6.1 (Statista 2023). The global production of cassava is expected to reach approximately 335 million metric tons by 2026 (CMO 2022–2026).

Cassava processing is conducted industrially to isolate the flour and starch components, thereby generating liquid and solid residues. Flour processing leads to the generation of solid residues (bagasse) (Fig. 6.2), whereas starch processing results in the production of liquid residues. To put this in context, the processing of approximately 250–300 t of cassava tubers generates an estimated 1.6 t of solid peels and 280 t of bagasse with an 85% moisture content (Escaramboni et al. 2018; Pandey et al. 2000). The bagasse is typically made up of fibrous root material and contains starch that could not be processed during extraction. Additionally, when poor processing conditions are present, higher concentrations of starch content are left-over in the resultant CB.

CB, also commonly referred to as cassava pulp, has very rarely been used as animal feed due to its low nitrogen content (2.3%) (John 2009). Apart from its carbohydrate content, CB is considered nutrient deficient with calcium, potassium, lipid, and phosphorous concentrations of less than 1% (John 2009; Pandey et al.

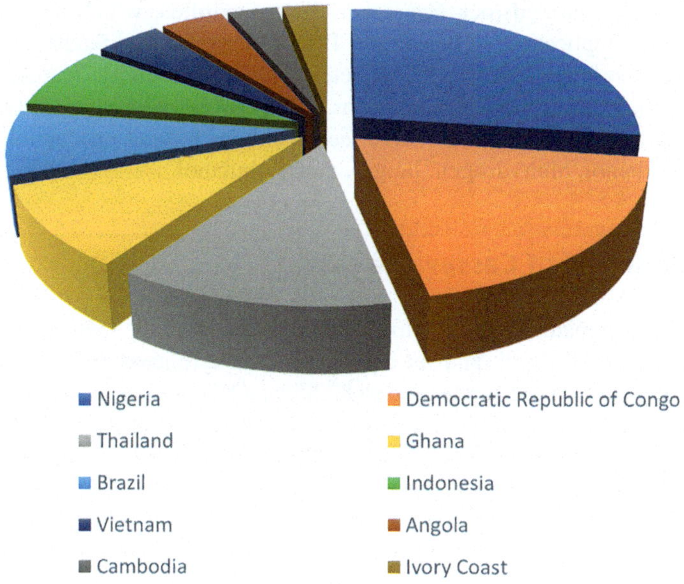

Fig. 6.1 Global top ten cassava producing countries in 2021. (Modified from Statista 2023)

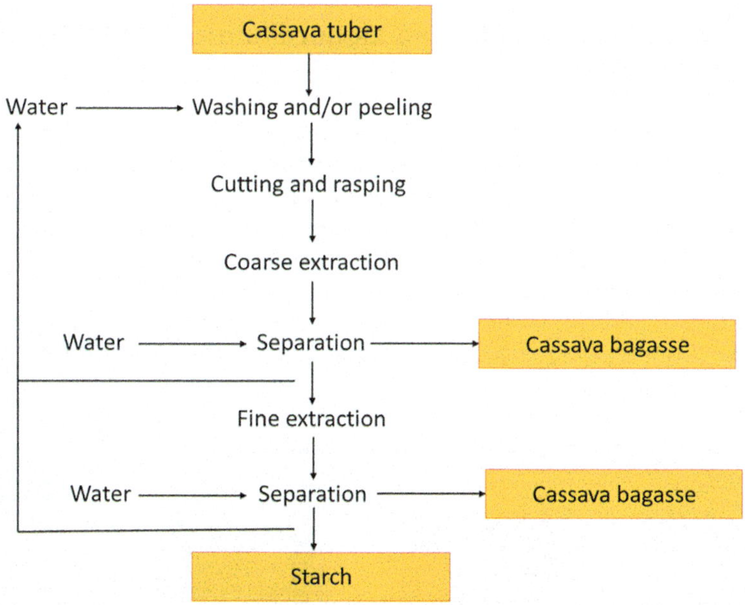

Fig. 6.2 Cassava starch extraction towards the generation of cassava bagasse

2000). To date, this feedstock has largely been landfilled, which results in severe environmental implications. Therefore, CB represents an ideal feedstock for microbial bioprocesses that does not compete with food security and also negates landfilling, thereby resolving the problem of environmental pollution. LCSBs such as CB constitute promising alternative feedstocks for biofuel production processes such as bioethanol, biohydrogen, biogas, and biobutanol.

6.3 Potential of Cassava Bagasse

CB is a fibrous material with a high moisture content of 85% and consists of relatively low protein, ash, and fat content, and an estimated 16% dry matter that majorly comprises carbohydrates (up to 82.85%). Of this 82.85%, greater than 50% is present in the form of starch as the main carbohydrate component (Polachini et al. 2016; De Souza et al. 2018; Travalini et al. 2019), and fiber in the range of 15–50% (Chen et al. 2015). Even with these beneficial components, bagasse is not being utilized by the cassava starch production industry presently, thus generating high volumes of waste that have posed negative environmental effects due to the resulting pollution (Ray and Swain 2011; Grasso 2020). LCSB feedstocks such as CB have gained popularity for use in bioprocessing in line with the "waste to wealth" concept (Sewsynker-Sukai et al. 2023a).

6.4 Different Stages During Lignocellulose-Starch Biomass Bioprocesses

As highlighted in Sect. 6.1, LCSB-based bioprocesses consist of four key processes including pretreatment, liquefaction, enzymatic saccharification, and the microbial fermentation process (Fig. 6.3), and these are briefly explained in this section. The pretreatment of starch-based substrates has been well-documented (Aruwajoye et al. 2017; Chohan et al. 2020; Phowan and Danvirutai 2014; Huang et al. 2019) and include physical methods such as heat, pressure, and moisture (Jyothi et al. 2010; Martínez et al. 2015; Zhu 2015), chemical methods (acid and alkali) (Falade and Ayetigbo 2015; Ptaszek et al. 2013; Wang et al. 2020; Zhu 2015), enzymatic

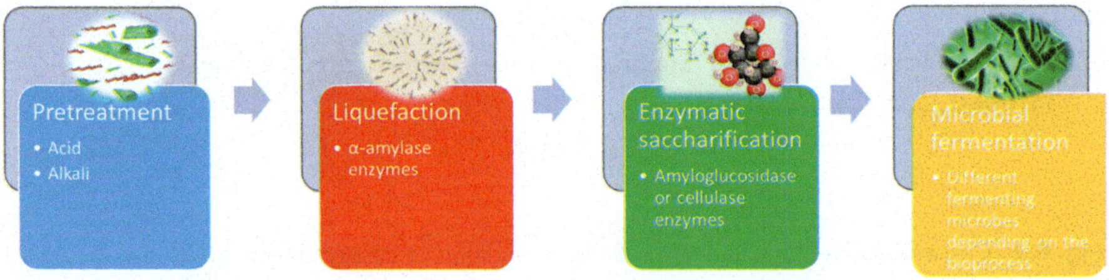

Fig. 6.3 Schematic showing different stages during lignocellulose-starch biomass bioprocesses

methods (Azmi et al. 2017; Zhang et al. 2021; Zhu 2015), or combined methods (Gatt et al. 2018; Zhu 2015).

The bulk of starch pretreatment studies available in the scientific domain has utilized acid hydrolysis (such as HCl and H_2SO_4). Although effective for starch hydrolysis, acids are corrosive and therefore require specialized pretreatment reactors, which have shown to be costly. Additionally, acid usage negatively impacts the environment due to the end products formed. Enzymatic hydrolysis has emerged as a more suitable pretreatment alternative to acids since it is environmentally benign as a result of the milder reaction conditions that generate biodegradable end products (Jujjavarapu and Dhagat 2019). In recent years, starch hydrolysis has been conducted using a combination of α-amylase and amyloglucosidase enzymes (Azmi et al. 2017; Zhang et al. 2021). Owing to the semicrystalline nature of starch granules, they are hydrolyzed very slowly by α-amylase and amyloglucosidase (Martínez et al. 2015; Uthumporn et al. 2012), rendering them ineffective for industrial biofuel processing (Zhang et al. 2021). Thus, the granular structure of starch must first be disrupted before the enzymatic hydrolysis stage almost always subjecting the LCSB feedstock to either hydrothermal or chemical methods such as acid (Martínez et al. 2015).

Starch enzymatic hydrolysis is typically performed in three steps including:

1. Gelatinization, a process carried out by heating in excess water, that unwinds the starch structure and improves the accessibility of enzymes to starch polysaccharides (Gatt et al. 2018; Jujjavarapu and Dhagat 2019; Ruiz et al. 2011).
2. Liquefaction of the resultant gelatinized starch and dextrinization by α-amylase enzymes at 80–90 °C for about 30–90 min, decreasing the viscosity of the solution and circumventing starch retrogradation (Aruwajoye et al. 2017, 2020; Gatt et al. 2018; Jujjavarapu and Dhagat 2019; Ruiz et al. 2011).
3. Saccharification is conducted using amyloglucosidases at 50–60 °C for about 12–96 h, resulting in the breakdown of the sugars in the liquefied starch to maltose or glucose (Aruwajoye et al. 2020; Gatt et al. 2018; Jujjavarapu and Dhagat 2019; Ruiz et al. 2011).

Once the substrate is broken down into a simpler, utilizable form, the microbial fermentative process occurs and is performed using microorganisms that possess metabolic capabilities based on the required target product. Nevertheless, it must be noted that dark fermentation and anaerobic digestion processes for biohydrogen and biogas production respectively, sometimes follow a slightly different biofuel production trajectory since the microbes used during these processes possess metabolic pathways that produce the required enzymes for starch degradation, thus avoiding separate processes.

Microbial biofuel production by fermentative processes is carried out in different bioprocess types that include:

- Separate hydrolysis and fermentation (SHF).
- Simultaneous saccharification and fermentation (SSF).

- Prehydrolysis and simultaneous saccharification and fermentation (PSSF) (Sewsynker-Sukai et al. 2023a, b).

SHF entails separate hydrolysis and fermentation steps under different optimal conditions (temperature) for the enzyme and microbe. Conversely, the SSF bioprocess comprises a single, combined hydrolysis and fermentation stage with conditions that use conditions that can accommodate but are not necessarily optimal for the enzyme and microbe. PSSF processes include a short hydrolysis step, termed prehydrolysis (usually 6–24 h) under optimal enzymatic conditions, followed by the introduction of the inoculum and fermentation nutrients and thereafter switching to SSF mode by adjusting the temperature to suit both the enzyme and microbe. The different bioprocess types have mainly been documented for bioethanol production (David et al. 2020; Sewsynker-Sukai and Gueguim Kana 2018; Jugwanth et al. 2020; Aruwajoye et al. 2020). Section 6.5 below summarizes the different studies that have been performed on different biofuels (bioethanol, biohydrogen, biogas, and biobutanol) using CB. Additionally, the effect of different pretreatment conditions and input factors relating to the fermentation process using this feedstock are also highlighted.

6.5 Application of CB for Renewable Biofuels

Various studies on biofuels including bioethanol, biohydrogen, biogas, and biobutanol produced from CB are briefly discussed below and summarized in Table 6.1.

6.5.1 Bioethanol

Bioethanol can be used as a fuel on its own or in various ratios with other fossil fuels (Ray and Swain 2011). Compared to traditional fossil fuels, bioethanol is renewable, sustainable, can easily be stored, and possesses a higher oxygen content and higher octane number, among others (Putra et al. 2015). CB has shown to be promising as an ideal substrate for producing bioethanol. For instance, the largest cassava-producing country, Nigeria, has the potential to produce 7.56 million cubic meters of ethanol production on an annual basis, which far exceeds the local governmental mandate that stipulated that 10% renewable fuel must be created in addition to processing facilities that are dedicated to the utilization of a single feedstock (Iye and Bilsborrown 2013). In this regard, various efforts have been directed at generating high ethanol yield from cassava industrial wastes by hydrolysis and fermentation of sugars. There has been an abundance of studies focusing on the use of CB for bioethanol production as shown in Table 6.1. In a recent study, Escaramboni et al. (2018) investigated ethanol biosynthesis by fast hydrolysis of CB using crude fungal amylases produced under optimized conditions and observed an 89.2% ethanol yield, which translated to 247.32 L/t of ethanol using a fed-batch

Table 6.1 Various biofuels produced from cassava bagasse

Inoculum	Biofuel	Pretreatment and/or enzymatic conditions	Fermentation process conditions	Biofuel production	Reference
Rhizolus oligosporus (CCT 3762)	Bioethanol	Gelatinization (80 °C per 30 min); amylolytic crude extract (1.5 U/mL starch solution)	9 g of yeasts, 10.8% (w/v) of reducing sugars, 32 °C, 80 rpm	247.32 L/t ethanol yield	Escaramboni et al. (2018)
Saccharomyces cerevisae	Bioethanol	Dextrinization (Liquozymeat 90 °C for 60 min), amyloglucosidases 60 °C for 60 min	10% inoculum, filtrate after pretreatment, 30 °C, 100 rpm	0.21 g/g ethanol yield	Martinez et al. (2018)
Candida tropicalis BCC7755	Bioethanol	Multi-activity enzyme (cellulase = 10 FPU/g, pectinase = 14 IU/g and hemicellulose = 13 IU/g based on xylanase), 40 °C for 48 h with rotary shaking at 200 rpm	50% inoculum, 4% SL, 40 °C, 200 rpm	14.3 g/L ethanol concentration	Rattanachomsri et al. (2009)
S. cerevisiae	Bioethanol	15% H_2SO_4 at 100 °C with the ratio of 1:1 (cassava peel: CB) for 120 min	10% inoculum, filtrate after pretreatment, 30 °C	6.2% bioethanol	Heriyanti et al. (2020)
S. cerevisiae SHY08-3	Bioethanol	SSF process	6% inoculum, pH 5, 12% SL (initial), 4% feeding (fed-batch), 30 °C, 100 rpm	43.25 g/L Bioethanol concentration	Zhu et al. (2012)
S. cerevisiae TISTR 5596	Bioethanol	2% H_2SO_4 for 30 min, cellulase (40 °C, 9 h)	10% inoculum, pH 4.5, hydrolysates from 30 g cassava pulp and 60 g H_2SO_4, 30 °C, 150 rpm	11.9 g/L Bioethanol concentration	Akaracharanya et al. (2011)
Pichia kudriavzevii	Bioethanol	20% SL, 120 °C for 10 min	Cells inoculated to optical density of 0.6, Pretreatment hydrolysate, pH 6, 42 °C, 120 rpm	42 g/L Bioethanol concentration	Ndubuisi et al. (2018)

(continued)

Table 6.1 (continued)

Inoculum	Biofuel	Pretreatment and/or enzymatic conditions	Fermentation process conditions	Biofuel production	Reference
Pretreated cow dung	Biohydrogen	1 kg substrate with 1 L diluted 1% HCl, 100 °C for 90 min	0.25 kg heatshocked inoculum, 30–35 °C, pH 6.09, HRT: 83.24 h	12.39 L H_2	Balbin et al. (2022)
			10% SL, pH 7, 60 °C		
Soil and sediments (*Clostridium thermopalmarium, Clostridium. isatidis, Thermo anaerobacterium*)	Biohydrogen	–		760 mL of H_2/L	Pason et al. (2020)
UASB granules from brewery wastewater	Biohydrogen	0.5% H_2SO_4 at 121 °C with the ratio of 1:15 (dry wt.: volume) for 30 min	3.0 g/L biomass, 25 g COD/L, pH 5.5, 35 °C	342 mL H_2 g/$COD_{reduced}$	Phowan and Danvirutai (2014)
Clostridium butyricum TISTR 1032 and *Enterobacter. aerogenes* TISTR 1468	Biohydrogen	0.5% H_2SO_4 at 121 °C with the ratio of 1:15 (dry wt.: volume) for 30 min	4 mL *C. butyricum*, 2 mL *E. aerogenes*, 25 g COD/L, pH 5.5, 36 °C, 150 rpm	345.8 mL H_2 g/$COD_{reduced}$	Phowan et al. (2010)
Fresh pig manure	Biogas	–	250 g VS inoculum concentration, biomass-water ratio 1:1.22 (kg/L) where biomass was 1 kg, Ambient temperature, manual agitation daily	7.43 L biogas volume per kg of cassava bagasse	Jaro et al. (2021)
Municipal wastewater sludge	Biogas	–	27.4 g CB and 58.3 g total solids of inocula, 55 °C, 120 rpm	0.395 m^3/kg VS_{added} (batch) and 0.36 m^3/kg VS_{added} (semi-continous)	Zhang et al. (2016)
Sludge	Biogas	–	8 kg-COD/m^3-d organic loading rate, 100 rpm	1.87 m^3-CH_4/ton-$pulp_{fresh}$-d (SBR) and	Ruttithiwapanich et al. (2022)

				0.22 m^3-CH_4/ton-$pulp_{fresh}$-d (CSTR)	
Clostridium sp. strain AS3	Acetone, Butanol and Ethanol	25 g CB, 2 N HCl	60 g/L CB, 5% inoculum, pH 6, 33 °C, 85 rpm	10.5 g/L acetone, 42 g/L butanol and 115.5 g/L bioethanol	Johnravindar et al. (2019)
Clostridium. tyrobutyricum CtΔack-adhE2	Butanol	10% SL, 121 °C for 30 min, pH 5.0–5.5, glucoamylase and cellulase, 50 °C, 150 rpm for 72 h	Immobilized cells, pH 6, 37 °C	15.0 g/L biobutanol	Huang et al. (2019)
Clostridium. acetobutylicum	Acetone, Butanol and Ethanol	10% SL, 121 °C for 30 min, glucoamylase at 65 °C, pH 4.2 and 200 rpm for 24 h, cellulase at 50 °C, pH 5.0, and 200 rpm for 24 h	Immobilized cells, pH 5, 37 °C	27.0 g/L acetone, 76.4 g/L butanol, 5.1 g/L ethanol	Lu et al. (2012)

SL solid loading

process. In the same year, Martinez et al. (2018) assessed ethanol production from CB using alpha-amylase and amyloglucosidases to evaluate its effect on the Brazilian energy matrix and achieved an ethanol yield of 0.21 g/g. In that same year, Valeriano et al. (2018) investigated optimal operational conditions of the starch hydrolysis process from CB followed by ethanol production using *Saccharomyces cerevisiae* Hansen BY4741. The optimized starch conditions (0.517 mL amyloglucosidase/g starch and 0.270 mL Termamyl/g starch) gave a maximum bioethanol concentration of 68 g/L corresponding to a yield of 0.48 g/g (Valeriano et al. 2018).

In an earlier study, Rattanachomsri et al. (2009) studied a multi-activity enzyme (cellulase, pectinase, and hemicellulase) for sugar release from CB also referred to as pulp and produced 14.3 g/L ethanol (Rattanachomsri et al. 2009). Later, Heriyanti et al. (2020) assessed a combined 50:50 of cassava peel and bagasse in addition to the effect of acid type (HCl or H_2SO_4) and acid concentration (7% and 15%) on bioethanol production using *S. cerevisiae*. The maximum ethanol production of 6.2. % was observed using 15% H_2SO_4 (Heriyanti et al. 2020). With the biofuel process cost rocketing, studies have been focused on dual-product systems. For example, Li and Zhu (2011) explored a novel biphasic, sequential-culture system for ethanol production from cassava bagasse by eliminating pretreatment and enzyme processes coupled with biohydrogen as a profitable by-product. The findings from that study demonstrated superior glucose (1.75- and 1.17-fold) with the co-cultures (*Clostridium thermocellum* and *Thermoanaerobacterium aotearoense*) compared to their individual *C. thermocellum* or *T. aotearoense* monoculture, respectively. The maximum bioethanol production achieved was 8.83 g/L using the co-culture. The co-culture produced a hydrogen yield of 4.06 mmol, translating to 1.54- and 2.09-fold higher compared to the mono-cultures *C. thermocellum* and *T. aotearoense*, respectively (Li and Zhu 2011). Similarly, Thongchul et al. (2010) evaluated a dual-product bioprocess by generating lactic acid and bioethanol from CB using *Rhizopus oryzae* NRRL395 and reported maximum concentrations of 21.82 and 12.09 g/L respectively. Several other bioethanol production studies on CB using *S. cerevisiae* (Zhu et al. 2012; Akaracharanya et al. 2011) and *Pichia kudriavzevii* (not detailed here) have also been reported (Ndubuisi et al. 2018) as shown in Table 6.1.

6.5.2 Biohydrogen

Biohydrogen production has emerged as a potential supplemental energy source since it is renewable, environmentally friendly (Giang et al. 2019), possesses a high calorific value of (141 kJ/g) (Sarangi and Nanda 2020); allowing it to produce a high amount of energy per unit mass (Hay et al. 2013). This fuel has the potential to be transformed into hydrocarbon fuels or electricity by incorporating fuel cell technologies (Sarangi and Nanda 2020). Biohydrogen production may be generated by several biological processes (Giang et al. 2019) including (1) dark fermentation, (2) photo-fermentation, (3) direct/indirect biophotolysis, and (4) microbial electrolysis cell (MEC) (Hay et al. 2013). Compared to the other processes, the dark fermentation process offers the advantage of using ambient temperatures and pressures, reducing the energy input needed (Hay et al. 2013), demonstrating its feasibility in the near future, and is, therefore, the only method focused on in this chapter. Biohydrogen production from LCSB feedstocks such as CB entails a multi-step process that may include but is not limited to pretreatment, enzymatic hydrolysis, and fermentation (Patil et al. 2021). Bottlenecks of the dark fermentation process are attributed to the fermentation costs. For instance, a suitable, cost-effective feedstock that can provide sufficient carbon and energy sources for fermenting microorganisms is imperative. During dark fermentative hydrogen production, hydrolytic microorganisms break down complex organic compounds into simpler ones under anaerobic conditions in a phase called hydrolysis, resulting in the production of organic acids (such as propionic acid, butyric acid, and acetic acid). Following hydrolysis, acetogenesis results in the production of hydrogen and carbon dioxide. Since carbohydrates are a frequent component of plant tissues and are the primary substrate for biohydrogen generation during fermentation, plant biomass with a high carbohydrate content may hold promise for this purpose. There have been numerous studies on starch waste residues for biohydrogen production (Vendruscolo 2014). However, the use of CB has been scantily reported as depicted in Table 6.1. Recently, Balbin et al. (2022) investigated key variables (pH and hydraulic retention time: HRT), substrate and inoculum pretreatment (heat and acid), and gas dissolution for biohydrogen production using the utilization of CB. Findings from the aforementioned study showed that at a pH of 6.09 and HRT of 83.24 h, 1 kg of CB can produce 12.39 L H_2 (Balbin et al. 2022). Earlier, Pason et al. (2020) examined novel bacterial consortium addition to the effects of pH (5–7) and solid loading (1–15%) for one-step hydrogen production from cassava bagasse without any pretreatment and enzyme saccharification. The microbial consortium designated as BK17 achieved a maximum yield of 760 mL of H_2/L under pH and solid loading of 7% and 10% respectively, with dominant microbes identified as starch-degrading *Clostridium thermopalmarium* (61.2%) and *Clostridium isatidis* (25.3%) that co-existed with acetogenic bacteria, *Thermoanaerobacterium* (6.7%) and *Fonticellatunisiensis* (5.06%) (Pason et al. 2020). Interestingly, Pason et al. (2020) detected polysaccharide-degrading enzymes such as amylase, cellulase, and xylanase which are responsible for breaking down the various starch and lignocellulosic structures to release fermentable sugars for microbial fermentation.

Likewise, Phowan and Danvirutai (2014) assessed the effects of initial pH (5–8), substrate (5–40 g COD/L), and biomass concentrations (0.5–5 g/L) from CB using three different mixed seed cultures obtained from different industrial wastewater plants in Thailand (anaerobic sludge from anaerobic wastewater treatment plant, Up-flowAnaerobic Sludge Blanket (UASB) granules from the UASB wastewater treatment plant of a brewery company, and the UASB granules obtained from the UASB wastewater treatment plant of a starch industry). Findings from that study showed that the optimal substrate concentration of 25 g COD/L with a biomass concentration of 3 g/L and an initial medium pH of 5.5 gave a peak hydrogen yield of 342 mL H_2 g/$COD_{reduced}$ using the UASB granules from the brewery wastewater as the inoculum source. Over a decade ago, Phowan et al. (2010) studied the presence and absence of a reducing agent (*Enterobacter aerogenes* TISTR 1468) for hydrogen production from CB using *Clostridium butyricum* TISTR 1032. Additionally, the impact of initial pH (5–8) and glucose concentration (5–40 g COD/L) on hydrogen production was determined (Phowan et al. 2010).

Major results from that study indicated that the addition of the reducing agent (*E. aerogenes*) together with a glucose concentration of 25 g COD/L and initial pH of 5.5 led to a maximum hydrogen yield of 345.8 mL H_2 g/COD reduced. Furthermore, the presence of the reducing agent decreased the lag phase and generated a 14.5% higher cumulative hydrogen production (Phowan et al. 2010).

6.5.3 Biogas

Biogas is a renewable and sustainable biofuel that can be generated globally for supplemental energy, particularly for domestic uses. It is generated from organic wastes (such as LCSB or LCB) or industrial wastewater, mainly using a process called anaerobic digestion (AD) (Moghaddam et al. 2019; Li et al. 2017). Biogas is mainly composed of a mixture of methane (CH_4) (55–75%), carbon dioxide (CO_2) (25–45%), and hydrogen (H_2) (5–10%) (de Mes et al. 2003; Kamali et al. 2016). This fuel can be used in a domestic setting for heating, or it can be upgraded to a higher quality (such as natural gas) or utilized in the co-generation of electricity and heat (de Mes et al. 2003). AD may be subdivided into four key phases that encompass hydrolysis, acidogenesis, acetogenesis, and methanogenesis (Angelidaki et al. 2011; Manyi-Loh et al. 2013). With the potential of direct conversion of LSCB such as cassava bagasse to biogas, this fuel shows promising results as the search for alternative energy continues. Table 6.1 outlines a few studies that have examined biogas production from cassava bagasse and different microbial sources. Recently, Jaro et al. (2021) investigated the effects of inoculum concentration (fresh pig manure) and biomass-water ratio on the biogas production from cassava bagasse. Maximum biogas production of 7.43 L/kg of CB was achieved under the optimum conditions (250 g VS inoculum concentration and biomass-water ratio 1:1.22 (kg/L), where biomass was 1 kg) (Jaro et al. 2021). In a separate study, Zhang et al. (2016) studied the effectiveness of municipal wastewater sludge as an inoculum for different fermentation modes (batch and semi-continuous) and recorded maximum specific methane yields of 0.395 m^3/kg VS_{added} (batch) and 0.36 m^3/kg VS_{added} (semi-continuous). Lately, Ruttithiwapanich et al. (2022) assessed biogas production from cassava bagasse in a 50-m^3 pilot-scale sequencing batch reactor (SBR) and continuous stirred tank reactor (CSTR) using sludge from a biogas plant treating starch-wastewater. The SBR generated a higher methane production of 1.87 m^3-CH_4/ton-$pulp_{fresh}$-d compared to 0.22 m^3-CH_4/ton-$pulp_{fresh}$-d by the CSTR (Ruttithiwapanich et al. 2022). Other studies documented using cassava waste consisting of bagasse but did not specify, produced biogas with maximum yields of 0.837 m^3 kg/TVS (Boontian et al. 2023) and 63.1% CH_4 (Kohmuean et al. 2020).

6.5.4 Biobutanol

Industrial solvent biobutanol is regarded as a cutting-edge transportation fuel due to its similarities to regular petrol in terms of vaporization heat, energy density, and octane number. Moreover, the mechanical design of current automobiles and the transportation pipeline are compatible with butanol is compatible with, thus making this fuel a suitable alternative to conventional fuels like gasoline. Butanol production is produced from the popular acetone–butanol–ethanol (ABE) fermentation, considered to be one of the oldest fermentation processes known to man dating back to as early as 1861 (Jones and Woods 1986). Furthermore, butanol outperforms the well-known ethanol since it possesses a higher energy content, lower vapor pressure, and a similar air-to-fuel ratio to gasoline (Lee et al. 2008). Solventogenic Clostridia are typically used in ABE fermentation and metabolize an array of sugars such as glucose, xylose, arabinose, cellobiose, and mannose (Ezeji and Blaschek 2008). Nevertheless, major limitations of biobutanol production stem from the use of food-based substrates including corn starch and molasses, which have been shown to have a limited supply and are uneconomical since they contribute to more than 50% of the product cost (Ezeji and Blaschek 2008). Therefore, alternative feedstocks such as CB, that are abundant, inexpensive, and not food-based must be sought. Some studies on biobutanol production using CB are summarized in Table 6.1.

Biobutanol was produced utilizing CB and an engineered *Clostridium tyrobutyricum* CtΔack-adhE2 strain and observed a maximum production of 15 g/L (Huang et al. 2019). In the same year, Johnravindar et al. (2019) explored biobutanol production CB using a newly isolated solvent-tolerant *Clostridium* species strain AS3 through the ABE fermentation pathway and obtained 42 g/L butanol. The feasibility of producing butanol from concentrated CB hydrolysate as the substrate in ABEfermentation with a high butanol tolerant strain of *Clostridium acetobutylicum* in a fed-batch fermentation process was examined by Lu et al. (2012). Maximum butanol production of 76.4 g/L was achieved with acetone (27.0 g/L) and ethanol (5.1 g/L) also formed as products during ABE fermentation (Lu et al. 2012).

6.6 Conclusion, Challenges, and Future Prospects

Unlike LCB which consists of resistant structures, LCSB substrates are composed of starch as an additional and major polysaccharide component in their structure. It is this starch portion that renders them valuable for bioprocesses such as biofuel production. Particularly, CB, a waste generated from the starch industry has gained traction over the years as a suitable feedstock for use in several different biofuel processes. Biofuel production from microbial fermentation processes is exhaustively being investigated in search of an economically viable process for industrial-scale processes as the race to reduce the carbon footprint continues. Despite the benefits of biofuels, current production methods are costly, generate low product yields, and require high energy input, thus making these unfeasible on an industrial scale. This chapter highlighted previous reports that used CB as a substrate for the production of various biofuels including bioethanol, biohydrogen, biogas, and biobutanol. Even though quite a few studies exist on the use of CB for the production of different biofuels, there has been a scarcity of reports on kinetic modeling using CB, and therefore must be explored. Additionally, pretreatments that employ industrial wastes should be evaluated in order to curb costs. Another aspect to be considered includes the genetic engineering of microorganisms to achieve superior microbial co-cultures that have high substrate conversion efficiencies, can produce enzymes for starch conversion to simple sugars, and have a high biofuel production capability. Taking the cost and environmental factors into consideration, research is presently moving towards integrated biorefineries that generate biofuels in addition to other bioproducts from cassava bagasse to enhance the process economics, all while recycling the waste effluent produced from each bioprocess, thereby achieving a circular bioeconomy that is in line with global sustainable development goals.

References

Akaracharanya A, Kesornsit J, Leepipatpiboon N, Srinorakutara T, Kitpreechavanich V, Tolieng V (2011) Evaluation of the waste from cassava starch production as a substrate for ethanol fermentation by *Saccharomyces cerevisiae*. Ann Microbiol 61:431–436

Alcázar-Alay SC, Meireles MAA (2015) Physicochemical properties, modifications, and applications of starches from different botanical sources. Food Sci Technol 35(2):215–236

Angelidaki I, Karakashev D, Batstone DJ, Plugge CM, Stams AJM (2011) Chapter 16—Biomethanation and its potential. Methods Enzymol 494:327–351

Aruwajoye GS, Faloye FD, Gueguim Kana EB (2017) Soaking assisted thermal pretreatment of cassava peels wastes for fermentable sugar production: process modeling and optimization. Energy Convers Manag 150:558–566

Aruwajoye GS, Sewsynker-Sukai Y, Gueguim Kana EB (2020) Valorisation of cassava peels through simultaneous saccharification and ethanol production: effect of prehydrolysis time, kinetic assessment and preliminary scale up. Fuel 278:118351

Azmi AS, Malek MIA, Puad NIM (2017) A review on acid and enzymatic hydrolyses of sago starch. Int Food Res J 24(12):265–273

Balbin AJFT, Gonzales HA, Bornillo LAM, Tagupa JHA, Jaro RH, Ido AL, Arazo RA (2022) High biohydrogen yield from fresh cassava waste pulps via gas dissolution in a dark fermentative environment. Cleaner Waste Syst 3:100039

Behera SS, Saranraj P, Ray RC (2022) Microbial bioethanol fermentation technologies—recent trends and future prospects. In: Gómez Castro FI, Guti-Errez-Antonio C (eds) Biofuels and biorefining. Elsevier, pp 75–107

Boontian N, Yingchon U, Padri M (2023) pH and heat pretreatments with zero-valent iron addition to enhance biogas production from cassava pulp wastewater: optimization and comparison of mathematical models. Fermentation 9:622

Chen J, Liu X, Wei D, Chen G (2015) High yields of fatty acid and neutral lipid production from cassava bagasse hydrolysate (CBH) by heterotrophic Chlorella protothecoides. Bioresour Technol 191:281–290

Chohan NA, Aruwajoye G, Sewsynker-Sukai Y, Gueguim Kana EB (2020) Valorisation of potato peel wastes for bioethanol production using simultaneous saccharification and fermentation: process optimization and kinetic assessment. Renew Energy 146:1031–1040

CMO (Cassava Market Outlook) (2022–2026) Daily updated industry statistics and top emerging trends about the cassava industry. https://www.reportlinker.com/clp/global/3978. Accessed 13 Aug 2023

David AN, Sewsynker-Sukai Y, Sithole B, Gueguim Kana EB (2020) Development of a green liquor dregs pretreatment for enhanced glucose recovery from corn cobs and kinetic assessment on various bioethanol fermentation types. Fuel 274:117797

David AN, Sewsynker-Sukai Y, Gueguim Kana EB (2022) Co-valorization of corn cobs and dairy wastewater for simultaneous saccharification and lactic acid production: process optimization and kinetic assessment. Bioresour Technol 348:126815

de Mes TZD, Stams AJM, Reith JH, Zeeman G (2003) Methane production by anaerobic digestion of wastewater and solid wastes. In: Reith JH, Wijffels RH, Barten H (eds) Biomethane and biohydrogen: status and perspectives of biological methane and hydrogen production. Dutch Biological Hydrogen Foundation, Petten, Netherlands, pp 58–102

De Souza CB, Jonathan M, Isay SM, Schols HA, Venema K (2018) Characterization and in vitro digestibility of by-products from Brazilian food industry: cassava bagasse, orange bagasse and passion fruit peel. Bioact Carbohydr Diet Fibre 16:90–99

Escaramboni B, Núñez EGF, Carvalhoa AFA, Netoa PDO (2018) Ethanol biosynthesis by fast hydrolysis of cassava bagasse using fungal amylases produced in optimized conditions. Ind Crop Prod 112:368–377

Ezeji TC, Blaschek HP (2008) Fermentation of dried distillers' grains and solubles (DDGS) hydrolysates to solvents and value-added products by solventogenic clostridia. Bioresour Technol 99:5232–5242

Falade KO, Ayetigbo OE (2015) Effects of annealing, acid hydrolysis and citric acid modifications on physical and functional properties of starches from four yam (*Dioscorea* spp.) cultivars. Food Hydrocoll 43:529–539

Gatt E, Rigal L, Vandenbossche V (2018) Biomass pretreatment with reactive extrusionusing enzymes: a review. Ind Crop Prod 122:329–339

Giang TT, Lunprom S, Liao Q, Reungsang A, Salakkam A (2019) Enhancing hydrogen production from *Chlorella* sp. biomass by prehydrolysis with simultaneous saccharification and fermentation (PSSF). Energy 12:908

Grasso S (2020) Extruded snacks from industrial by-products: a review. Trends Food Sci Technol 99:284–294

Hay JXW, Wu TY, Juan JC, Jahim MJ (2013) Biohydrogen production through photo fermentation or dark fermentation using waste as a substrate: overview, economics, and future prospects of hydrogen usage. Bioprod Bioref 7:334–352

Heriyanti AP, Fibriana F, Tirtasari NL (2020) Bioethanol production from cassava-based industrial wastes using acid hydrolysis and simple fermentation. J Phys Conf Ser 1567:022024

Huang J, Du Y, Bao T, Lin M, Wang J, Yang S (2019) Production of n-butanol from cassava bagasse hydrolysate by engineered *Clostridium tyrobutyricum* overexpressing *adhE2*: kinetics and cost analysis. Bioresour Technol 292:121969

Iye E, Bilsborrown P (2013) Cellulosic ethanol production from agricultural residues in Nigeria. Energy Policy 63:207–214

Jaro RH, Icalina MAA, Talemporos RT, Napinas PM, Potato DNC, Manuel LG, Cubio GM, Ido AL, Arazo RO (2021) Biogas production from waste pulps of cassava (*Manihot esculenta* Crantz) via anaerobic digestion. Energy Ecol Environ 6:204–212

John RP (2009) Biotechnological potentials of cassava bagasse. In: Singh nee' Nigam P, Pandey A (eds) Biotechnology for agro-industrial residues utilisation. Springer, Dordrecht. https://doi.org/10.1007/978-1-4020-9942-7_11

Johnravindar D, Elangovan N, Gopal NO, Muthaiyan A, Fei Q (2019) Biobutanol production from cassava waste residue using *Clostridium* sp. AS3 in batch culture fermentation. Biofuels. https://doi.org/10.1080/17597269.2019.1608671

Jones DT, Woods DR (1986) Acetone–butanol fermentation revisited. Microbiol Rev 50:484–524

Jugwanth Y, Sewsynker-Sukai Y, Gueguim Kana EB (2020) Valorization of sugarcane bagasse for bioethanol production through simultaneous saccharification and fermentation: optimization and kinetic studies. Fuel 262:116552

Jujjavarapu SE, Dhagat S (2019) Evolutionary trends in industrial production of α-amylase. Recent Pat Biotechnol 13(1):4–18

Jyothi AN, Sajeev MS, Sreekumar JN (2010) Hydrothermal modifications of tropical tuber starches. 1. Effect of heat-moisture treatment on the physicochemical, rheological and gelatinization characteristics. Starch-Stärke 62(1):28–40

Kamali M, Gameiro T, Costa MEV, Capela I (2016) Anaerobic digestion of pulp and paper mill wastes—an overview of the developments and improvement opportunities. Chem Eng J 298: 162–182

Kohmuean P, Boonrod N, Wongkoblap A (2020) Biogas production from cassava waste: effect of concentration. IOP Conf Ser Mater Sci Eng 778:012115

Laltha M, Sewsynker-Sukai Y, Gueguim Kana EB (2022) Simultaneous saccharification and citric acid production from paper wastewater pretreated banana pseudostem: optimization of fermentation medium formulation and kinetic assessment. Bioresour Technol 361:127700

Lee SY, Park JH, Jang SH, Nielsen LK, Kim J, Jung KS (2008) Fermentative butanol production by Clostridia. Biotechnol Bioeng 101:209–228

Li P, Zhu M (2011) A consolidated bio-processing of ethanol from cassava pulp accompanied by hydrogen production. Bioresour Technol 102:10471–10479

Li H, Mehmood D, Thorin E, Yu Z (2017) Biomethane production via anaerobic digestion and biomass gasification. Energy Procedia 105:1172–1177

Liu Y, Chen J, Luo S, Li C, Ye J, Liu C, Gilbert RG (2017) Physicochemical and structural properties of pregelatinized starch prepared by improved extrusion cooking technology. Carbohydr Polym 175:265–272

Lu C, Zhao J, Yang ST, Wei D (2012) Fed-batch fermentation for n-butanol production from cassava bagasse hydrolysate in a fibrous bed bioreactor with continuous gas stripping. Bioresour Technol 104:380–387

Manyi-Loh CE, Mamphweli SN, Meyer EL, Okoh AI, Makaka G, Simon M (2013) Microbial anaerobic digestion (bio-digesters) as an approach to the decontamination of animal wastes in pollution control and the generation of renewable energy. Int J Environ Res Public Health 10: 4390–4417

Martínez MM, Pico J, Gómez M (2015) Physicochemical modification of native and extruded wheat flours by enzymatic amylolysis. Food Chem 167:447–453

Martinez DG, Feiden A, Bariccatti R, de Freitas Zara KR (2018) Ethanol production from waste of cassava processing. Appl Sci 8:2158

Moghaddam EA, Ericsson N, Hansson PA, Nordberg Å (2019) Exploring the potential for biomethane production by willow pyrolysis using life cycle assessment methodology. Energy Sustain Soc 9:1–18

Moodley P, Gueguim Kana EB (2017) Microwave-assisted inorganic salt pretreatment of sugarcane leaf waste: effect on physiochemical structure and enzymatic saccharification. Bioresour Technol 235:35–42

Ndubuisi IA, Nweze JE, Onoyima NJ, Yoshinori M, Ogbonna JC (2018) Ethanol production from cassava pulp by a newly isolated thermotolerant *Pichia kudriavzevii* LC375240. Energy Power Eng 10:457–474

Pandey A, Soccol CR, Nigam P, Soccol VT, Vandenberghe LPS, Mohan R (2000) Biotechnological potential of agro-industrial residues. II: Cassava bagasse. Bioresour Technol 74(1):81–87

Pason P, Tachaapaikoon C, Panichnumsin P, Ketbot P, Waeonukul R, Kosugi A, Ratanakhanokchai K (2020) One-step biohydrogen production from cassava pulp using novel enrichment of anaerobic thermophilic bacteria community. Biocatal Agric Biotechnol 27: 101658

Patil PD, Tiwari MS, Bhange VP, Marghade DT, Kumaran S (2021) Bioconversion of lignocellulosic residues into hydrogen. In: Inamuddin, Khan A (eds) Sustainable bioconversion of waste to value added products. Springer International Publishing, Cham, Switzerland, pp 59–80

Phowan P, Danvirutai P (2014) Hydrogen production from cassava pulp hydrolysate by mixed seed cultures: effects of initial pH, substrate and biomass concentrations. Biomass Bioenergy 64:1–10

Phowan N, Reungsang A, Danvirutai P (2010) Bio-hydrogen production from cassava pulp hydrolysate using co-culture of *Clostridium butyricum* and *Enterobacter aerogenes*. Biotechnolnology 9(3):348–354

Polachini T, Leonardo L, Lopes-Filho J et al (2016) Water adsorption isotherms and thermodynamic properties of cassava bagasse. Thermochim Acta 632:79–85

Ptaszek A, Lukasiewicz M, Bednarz S (2013) Environmental friendly polysaccharide modification—rheological properties of oxidized starches water systems. Starch-Stärke 65(1–2):134–145

Putra MD, Abasaeed AE, Atiyeh HK, Al-Zahrani SM, Gaily MH, Sulieman AK, Zeinelabdeen MA (2015) Kinetic modeling and enhanced production of fructose and ethanol from date fruit extract. Chem Eng Commun 202:1618–1627

Rattanachomsri U, Tanapongpipat S, Eurwilaichitr L, Champreda V (2009) Simultaneous non-thermal saccharification of cassava pulp by multi-enzyme activity and ethanol fermentation by Candida tropicalis. J Biosci Bioeng 107(5):488–493

Ray RC, Ramachandran S (2018) Bioethanol production from food crops: sustainable sources, interventions, and challenges. Academic. 460p

Ray RC, Swain MR (2011) Bio-ethanol, bioplastics and other fermented industrial products from cassava starch and flour. In: Pace CM (ed) Cassava: farming, uses and economic impacts. Nova Science Publishers Inc., Hauppauge, NY, pp 1–32

Ray RC, Mohapatra S, Panda S, Kar S (2008) Solid substrate fermentation of cassava fibrous residue for production of α-amylase, lactic acid, and ethanol. J Environ Biol 29(1/2):111–115

Rorke DCS, Suinyuy TN, Gueguim Kana EB (2017) Microwave-assisted chemical pre-treatment of waste sorghum leaves: process optimization and development of an intelligent model for determination of volatile compound fractions. Bioresour Technol 224:590–600

Ruiz MI, Sanchez CI, Torrres RG, Molina DR (2011) Enzymatic hydrolysis ofcassava starch for production of bioethanol with a Colombian wild yeast strain. J Braz Chem Soc 22(12): 2337–2343

Ruttithiwapanich T, Waewsak C, Laopitinan O, Saengchan K, Lueangwattanapong L, Songkasiri W (2022) Comparative study of biogas production from cassava pulp in a sequencing batch reactor and a continuous stirred tank reactor. IOP Conf Ser Earth Environ Sci 1050:012013

Sanusi IA, Aruwajoye A, Revaprasadu N, Sewsynker-Sukai Y, Meyer E, Gueguim Kana EB (2022) A novel autoclave-assisted nanoparticle pre-treatment for improved sugar recovery from potato peel waste: process optimisation, nanoparticle recyclability and bioethanol production. Biomass Convers Biorefin. https://doi.org/10.1007/s13399-022-03574-y

Sarangi PK, Nanda S (2020) Biohydrogen production through dark fermentation. Chem Eng Technol 43:601–612

Sewsynker-Sukai Y, Gueguim Kana EB (2018) Simultaneous saccharification and bioethanol production from corn cobs: process optimization and kinetic studies. Bioresour Technol 262: 32–41

Sewsynker-Sukai Y, David AN, Gueguim Kana EB (2020) Recent developments in the application of Kraft pulping alkaline chemicals for lignocellulosic pretreatment: potential beneficiation of green liquor dregs waste. Bioresour Technol 306:123225

Sewsynker-Sukai Y, David AN, Rorke DCS, Sanusi IA, Aruwajoye G, Meyer EL, Gueguim Kana EB (2023a) Integrated biorefineries: the path forward (Chapter 17). In: Advances in lignocellulosic biofuel production systems. Elsevier. (ISBN: 978-0-323-91192-4)

Sewsynker-Sukai Y, David AN, Laltha M, Meyer EL, Gueguim Kana EB (2023b) Lignocellulosic biofuel production: insight into microbial factories (Chapter 15). In: Advances in lignocellulosic biofuel production systems. Elsevier. (ISBN: 978-0-323-91192-4)

Singh N, Singhania RR, Nigam PS, Dong C, Patel AK, Puri M (2022) Global status of lignocellulosic biorefinery: challenges and perspective. Production of n-butanol from cassava bagasse hydrolysate by engineered *Clostridium tyrobutyricum* overexpressing adhE2: kinetics and cost analysis. Bioresour Technol 344:126415

Statista (2023). https://www.statista.com/statistics/1391572/global-leading-cassava-producing-countries/. Accessed 13 Aug 2023

Thongchul N, Navankasattusas S, Yang S (2010) Production of lactic acid and ethanol by Rhizopus oryzae integrated with cassava pulp hydrolysis. Bioprocess Biosyst Eng 33:407–416

Travalini A, Lamsal B, Esteves W et al (2019) Cassava starch films reinforced with lignocellulose nanofibers from cassava bagasse. Int J Biol Macromol 139:1151–1161

Uthumporn U, Shariffa YN, Karim AA (2012) Hydrolysis of native and heat-treated starches at sub-gelatinization temperature using granular starch hydrolyzing enzyme. Appl Biochem Biotechnol 166(5):1167–1182

Valeriano IH, Marques GLL, Freitas SP, Couri S, Penha ED, Gonçalves MMM (2018) Cassava pulp enzymatic hydrolysate as a promising feedstock for ethanol production. Braz Arch Biol Technol 61:18161214. ISSN 1678-4324 Online Edition

Vendruscolo F (2014) Biohydrogen production from starch residues. WASET 8(12). https://doi.org/10.1002/er.3224

Wang X, Huang L, Zhang C, Deng Y, Xie P, Liu L, Cheng J (2020) Research advances in chemical modifications of starch for hydrophobicity and its applications: a review. Carbohydr Polym 240: 116292

Wheatley CC, Chuzel G, Zakhia N (2003) CASSAVA. The nature of the tuber. In: Caballero (ed) Encyclopedia of food sciences and nutrition. Academic, pp 964–969. ISBN 9780122270550

Zhang J, Xu JY, Wang DQ, Ren NQ (2016) Anaerobic digestion of cassava pulp with sewage sludge inocula. Bioresources 11(1):451–465

Zhang Y, Li W, Huang M, Xu X, Jiang M, Hu H, Huang Z, Liang J, Qin Y (2021) Non-digesting strategy for efficient bioconversion of cassava to bioethanol via mechanical activation and metal salts pretreatment. Renew Energ 169:95–103

Zhu F (2015) Composition, structure, physicochemical properties, and modifications of cassava starch. Carbohydr Polym 122:456–480

Zhu M, Li P, Gong X, Wang J (2012) A comparison of the production of ethanol between simultaneous saccharification and fermentation and separate hydrolysis and fermentation using unpretreated cassava pulp and enzyme cocktail. Biosci Biotechnol Biochem 76(4): 671–678

Bio-Valorization of Sweet Potato Bagasse into Food Additives, Feeds, and Fuels

7

Ivi Chakraborty, Arup Chattopadhyay, Pinky Maity, Aditi Pahari, Subhradeep Pramanik, Hirak Banerjee, Rajib Nath, and Ramesh C. Ray

Abstract

Sweet potato is superior to other carbohydrate sources and rich in vitamins, minerals, dietary fiber, and protein content. The crop generates much bagasse that currently has little utilization and goes waste into landfills and water bodies. Leaves, vines, and tuber peels contain polyphenols and carotenoids that could be used as antioxidants and food colorants. Several useful bioactive compounds, including polyphenols, vitamins, fibers, enzymes, oils, and others, can be recovered from the sweet potato wastes produced during the starch manufacturing process. The current chapter discusses various waste streams produced during the stages of harvesting, post-harvest handling, starch extraction, and alcoholic fermentation, the presence of phenolic and bioactive compounds in these waste streams, various extraction techniques, and the potential for converting these valuable materials into various value-added products with higher consumer acceptance.

I. Chakraborty · P. Maity
Department of Post Harvest Management, Faculty of Horticulture, Bidhan Chandra Krishi Viswavidyalaya, Nadia, West Bengal, India

A. Chattopadhyay · S. Pramanik
Department of Vegetable Science, Faculty of Horticulture, Bidhan Chandra Krishi Viswavidyalaya, Nadia, West Bengal, India

A. Pahari · H. Banerjee · R. Nath
Department of Agronomy, Faculty of Agriculture, Bidhan Chandra Krishi Viswavidyalaya, Nadia, West Bengal, India

R. C. Ray (✉)
ICAR-Central Tuber Crops Research Institute (Regional Centre), Bhubaneswar, Odisha, India

R. C. Ray (ed.), *Roots, Tubers, and Bulb Crop Wastes: Management by Biorefinery Approaches*, https://doi.org/10.1007/978-981-99-8266-0_7

Keywords

Sweet potato · Bagasse · Waste generation · Value-added products · Polyphenols · Sochu

Abbreviations

DM	Dry matter
DW	Dry weight
SPPS	Sweet potato peels
SPW	Sweet potato wastes

7.1 Introduction

Sweet potatoes are the seventh-most significant food crop in the world (FAO 2009; Ray and Tomlins 2010). The crop outperforms all other main food crops in terms of edible energy production (194 MJ/ha/day) (Woolfe 1992), is more productive over a shorter time period on marginal lands (Ray and Ravi 2005), and contributes tremendously to food and nutritional security (Nedunchezhiyan and Ray 2010; Mukhopadhyay et al. 2011).

A significant portion of the plants are wasted in landfills and aquatic bodies as a result of the processing of this root tuber into various value-added products (Akoetey et al. 2017). According to Trevisan et al. (2020), 10% of greenhouse gas emissions are attributable to food waste. Nearly 45% of the food produced by tubers is wasted globally each year. About 7% of the crop was wasted on sweet potatoes in the world in 2011 (FAO 2016). Very little can be done with vines and leaves besides feeding farm animals. Starting at the farm level, waste is produced from removing vines and leaves as well as from tubers that don't match size grading standards or are harmed as a result of harvesting methods. The by-products of processing are a valuable source of carbohydrates, protein, anthocyanins, and vitamins that can be used as raw materials to create products with additional value (Islam et al. 2002). Sweet potato waste (SPW) is a byproduct of starch manufacturing that is typically regarded as unusable and polluting. According to Yang and Ling (1989), SPW can be classified as a material that is rich in carbon because it primarily contains crude protein, crude fiber, and carbs. Therefore, it's important to consider how to utilize the SPW to create value-added products that are more widely accepted.

Table 7.1 Composition of 100 g raw sweet potato tuber

Energy	194 MJ/ha/day
Carbohydrates	27.0 g
Sugars	4.2 g
Dietary fiber	3.0 g
Fat	0.2 g
Protein	1.5 g
Vitamin A equivalents	709 μg
β-Carotene	8509 μg
Thiamine (Vitamin B_1)	0.1 mg
Riboflavin (Vitamin B_2)	0.1 mg
Niacin (Vitamin B_3)	0.61 mg
Pantothenic acid (Vitamin B_5)	0.8 mg
Vitamin B_6	0.2 mg
Folate (Vitamin B_9)	11 μg
Vitamin C	24 mg
Calcium	46.0 mg
Iron	0.8 mg
Magnesium	25.0 mg
Phosphorus	49.0 mg
Potassium	337 mg
Sodium	55 mg
Zinc	0.3 mg

USDA National Nutrient Database for Standard Reference, Release # 22 (2009), NBD No. 11507

7.2 Proximate Composition of Sweet Potato

The crop is a nutritional powerhouse, scoring higher than most underground crops in terms of vitamins, minerals, dietary fiber, and protein content (Table 7.1) (Chattopadhyay et al. 2002a, 2002b, 2006a; Lebot 2009; Ray and Tomlins 2010). However, the proximate composition of the tuber varies with cultivar, maturity, and storage time (Woolfe 1992; Nedunchezhiyan and Ray 2010). Yellow and light-fleshed varieties are both strong sources of vitamin C, while orange- and yellow-fleshed varieties are rich sources of carotene and lutein, respectively (Mitra et al. 2010; Bechoff et al. 2011; Burri 2011). Moreover, anthocyanin-rich purple-fleshed types have become more popular because of their unique color and rich antioxidant properties (Lila 2004; Choi et al. 2010; Panda et al. 2012). Compared to dark, leafy greens, orange-fleshed varieties currently on the market have higher concentrations of β-carotene (3000–16,000 μg/100 g fresh weight (FW)) (Bovell-Benjamin 2007; Kidmose et al. 2007; Ofori et al. 2009) and retinol equivalents (Chattopadhyay et al. 2006a), which appear to be more bio-available. In many food-insecure nations and regions, including the tribal-dominated regions of India, such varieties could combat vitamin A deficiency (Nath et al. 2005; Burri 2011).

7.3 Waste Generation During Harvesting and Postharvest Handling

After being cooked, the polyphenol and carotenoid-rich leaves and vines may be utilized as food for people. After harvest, the quality of roots rapidly deteriorates due to weevil infestation, pathological decay, sprouting, weight loss, physical damage, and weight loss (Chattopadhyay et al. 2006b; Thankappan 1989; Ray and Ravi 2005; Ray et al. 2010; Abano et al. 2011).

Sweet potatoes are industrially processed into a staggering number of byproducts that are now used for relatively little. Trims and peels from the stage of raw material preparation that comes before further processing make up these by-products. Although these byproducts are frequently dumped on the ground or dumped into bodies of water, they can also be fed to animals.

According to a significant body of recent research, functional proteins, phenolic compounds, soluble fiber, anthocyanins, and carotenoids may be recovered from sweet potato by-products (Ray and Tomlins 2010). These wastes could be used as feedstock for the anaerobic digestion of biogas or the ethanol and lactic acid manufacturing process. It is possible to turn leaf and vine, peel, discarded root, and leftovers from different sweet potato processing processes into food, feed, and market goods. Various methods of using the solid and liquid wastes produced in a sweet potato processing factory were suggested by Akoetey et al. (2017). The valorization options for sweet potato bagasse and wastes are given in Fig. 7.1.

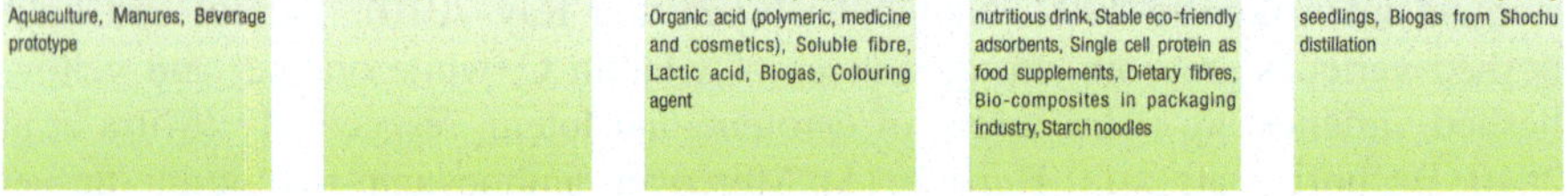

Fig. 7.1 Valorization options for sweet potato wastes and bagasse

7.3.1 Utilization of Leaves, Vines, and Discarded Tubers

Although sweet potato leaves and vines are incredibly nutritious and can be cooked (Sun et al. 2014), they are typically thrown away following the harvest of the tubers, especially in North America. As important feed for ruminants and other livestock species, sweet potato leaves, and vines are also employed (Backer et al. 1980; Gonzalez et al. 2003; Giang et al. 2004). The leaves are a great source of fiber, vitamins A and C, iron, protein, and carbs (Smart and Simmonds 1995). According to An et al. (2003), sweet potato vines have a high crude protein content that ranges from 16% to 29% of the dry matter (DM). In addition to a base diet of other fodder, feeding vines to cows enhanced milk production (Etela et al. 2008). According to Anastacio et al. (2016), sweet potato leaf and peel extract can be used to create a beverage prototype that is comparable to popular vegetable and fruit drinks. Silage that is nutrient-rich can also be made from sweet potato tops (leaves and vines) (Fasuyi 2006). Li et al. (2017) investigated the fermentation-based enhancement of the nutritional content of rice straw silage with sweet potato vines. Additionally, researchers have looked into the possibility of using root residues which are typically discarded instead of traditional animal diets for pigs (Dom et al. 2017), rabbits (Chakrabarti et al. 2017), and poultry (Khan 2017). Mussoline and Wilkie (2015) suggested that around 45% of secondary outlets from the total yield of sweet potatoes might be used for ethanol production and the remaining 55% of the culls for biogas recovery; this strategy could provide the facility with the energy inputs needed. Genetic engineering can be used to improve the starch content of sweet potato biomass for fuel production, hence reducing the costs associated with the enzyme process (El Sheikha and Ray 2017).

According to Huang et al. (2010), a water extract from sweet potato leaves reduced the generation of nitric oxide in macrophages and shielded mouse liposomes from oxidative damage. *Escherichia coli* and *Bacillus* bacteria's growth was inhibited by lyophilized powder made from sweet potato leaves.

The Suioh cultivar of sweet potatoes contained more lutein (31.5–42.6 mg/100 g FW), a non-provitamin A yellow to orange color carotenoid, than did 120 other fruits and vegetables (Isghiguro and Yoshimoto 2006). In order to prevent eye problems, sweet potato leaves should be taken into account (Isghiguro and Yoshimoto 2006). Additionally, tartrazine (FD & C Yellow 5) may be replaced by the lutein present in sweet potato leaves. Using typical agricultural waste, sweet potato leaves, Hue et al. (2014) were able to extract yellow color.

When chickens were fed up to 160 g/kg of sweet potato leaves against Lucerne meal in a trial evaluating the utilization efficiency of the two feed sources, there was no change in growth rate, food intake, or conversion (Farrell et al. 2000). Sweet potato vines supported greater weight growth in goats when combined with Batiki grass (*Ischaemum aristatum* var. *indicum*), a low-nutrient fodder crop (Aregheore 2004). Sweet potato vines supported the production of goat meat when they were supplemented with an energy-protein concentrate, and they were said to be especially helpful when feed was limited (Katongole et al. 2009). As a protein supplement for pigs in the tropics, sweet potato leaves are still potentially extremely beneficial (Régnier et al. 2012, 2013). According to Nguyen et al. (2012), ensiling the leaves increased the nutrients' digestion compared to drying the leaves.

Even though the phytic acid content (3.29%) is high for fish growth, sweet potato leaves can also be employed to sustain small-scale aquaculture (Dongmeza et al. 2009). Animals fed with sweet potato leaves produced more manure than the control animals in a rice/sweet potato cropping combination. According to Oke and Workneh (2013), leaf manure can be utilized as fertilizer for growing rice.

Yang et al. (2019) devised a straightforward process to simultaneously extract peroxidases, phenolics, and flavonoids using the Column chromatographic extraction method with gradient elution from old sweet potato stems. With water and 50% ethanol, peroxidase and phenolics in dried or fresh sweet potato stems were successfully removed, with extraction rates of 91% and 98.7%, respectively. The activity of the purified peroxidases was strong in producing theaflavins from catechins. Following extraction, the leftover material can be utilized to create organic fertilizer, fish and animal feed, dietary fiber, and other products. The bulk sweet potato trash can be used to make useful functional items, which has the potential to boost the crop's economic profit, make better use of its resources, and reduce environmental waste.

Low-topping sweet potato types that are intended for both food and fodder allow for the distribution of fodder availability throughout the year without considerably impacting tuberous root production (Leon-Verlarde 2000). To produce the most herbage for fodder without sacrificing the yield of tuberous roots, Ahmed et al. (2012) found that planting sweet potatoes on ridges and harvesting the vines 105 days after planting (when roughly 60% of the plant's growth phase was complete) were key strategies. The year-round need for feed poses a productivity barrier in mixed crop-livestock production systems (Leon-Verlarde 2000). Pastoralists from Ethiopia have started growing sweet potatoes using irrigation to help with the severe livestock feed shortage and food insecurity issues (Solomon 1999).

Agriculture waste is used as a catalyst in the manufacturing of biodiesel from inexpensive and readily accessible feedstocks. In a recent study, a green heterogeneous catalyst was produced from waste sweet potato leaves and used to produce biodiesel from waste soybean cooking oil and *Scenedesmus obliquus* oil. Due to the predominance of potassium in carbonates and oxides form, the green heterogeneous catalyst demonstrated strong catalytic activity for converting the waste soybean cooking oil and *S. obliquus* oil to biodiesel. The maximum *S. obliquus* oil and waste soybean cooking oil conversion of 99.50% and 98.0% were obtained using 7 and 5 wt% catalyst loading and 15:1, 9:1 methanol to oil molar ratio for 90 and 120 min at 60 °C, respectively (Eldiehy et al. 2022).

7.3.2 Utilization of Peels

Waste produced by industrial procedures using sweet potato peels is massive. According to Schaub and Leonard (1996), 30% of the raw material used in canning operations for sweet potatoes is abandoned and not used for human consumption. The peels are normally utilized as either fertilizer or animal food if they are used at all. Sweet potato peels (SPPs) have several essential nutrient compositions and they are cheap and readily available. The use of SPPs to produce biogas for domestic and industrial use provides a larger volume of methane, controls environmental pollution, and prevents exposure to health (Adeyosoye et al. 2010).

Maloney Katherine et al. (2012) looked at the protein composition of sweet potato puree SPPs produced at two different stages of processing. The principal peels of the sweet potato came from the initial peeling. Sweet potatoes were blanched, developed secondary peels, and then peeled a second time. The additional peels were said to make protein extraction easier. Through an improved procedure that made use of saline and calcium chloride, about 32% of the protein in the peels was recovered.

SPPs are utilized as fish feed in aquaculture systems for subsistence. According to Omoregie et al. (2009), fish fed with sweet potato peels showed reduced development when compared to fish given with a control diet (no sweet potato peels). However, there was no growth depression when the amount of peels in the diet was kept to 5–20%.

The phenolic chemicals found in sweet potato peels are in large amounts and have up to three times the antioxidant activity of other plant tissues. According to studies on the extraction of phenolics from sweet potato peels, the amount of phenolics produced was influenced by the solvent: solid ratio and the depth of peel-cut (Anastacio and Carvalho 2013). Total phenolic compounds ranged from 1.02 to 6.21 gallic acid equivalents/g of dry weight (DW), and Trolox equivalent antioxidant activity ranged from 0.71 to 8.11 TE (trolox equivalent)/g of dry weight (DW).

For usage as dietary supplements, Oluyori et al. (2016) proposed a workable isolation approach for several anti-cancer components from sweet potato peels. Sweet potato peels have also been utilized to separate organic acids (such as limonenic acid), which have been used in polymeric, pharmaceutical, and cosmetic applications (Omojasola and Adeniran 2014). Additionally, SPPs have about three times as much total phenolics as the flesh and whole root, according to Steed and Truong's (2008) research. Extraction of phenolic compounds from peels, which are produced as by-products during processing, would be a smart method to reduce waste while creating a product with additional value and a positive economic impact. Sweet potato peels can be used to create soluble fiber, an essential dietary supplement. A fresh source of affordable dietary fiber may be created by turning sweet potatoes into low-cost soluble fiber.

Using *Saccharomyces cerevisiae*, sweet potato peels which are typically regarded as non-valuable waste could be effectively turned to ethanol. As a result, SPW may be seen as a desirable feedstock for the production of bioethanol, particularly in nations where millions of tonnes of sweet potatoes are produced annually (Modupe et al. 2018). Lactic acid has been successfully produced by fermenting SPPs with lactic acid bacteria like *Lacticaseibacillus rhamnosus*. According to Pagana et al. (2014), 1 L of sweet potato waste might yield 10 g of lactic acid.

Sweet potato food waste can be used to create biogas through anaerobic digestion. Ge et al. (2014) examined the methane production of tropical forestry (albizia) wastes with food wastes such as sweet potato, taro, and papaya and discovered that while the albizia waste had the largest volumetric production, food wastes such as the sweet potato had the highest yield.

Food processors are gradually switching over to using natural colors derived from fruits, flowers, and vegetables in place of artificial coloring additives. The color of the flesh and peel is one of the traits that set distinct sweet potato cultivars apart. The anthocyanins and carotenoids that give tubers their purple, orange, and yellow hues are responsible for this coloring. There have been investigations towards removing the pigments from the tubers because purple sweet potatoes are known to have a higher anthocyanin concentration than the other colors (Li et al. 2019). The colored peels of the sweet potato are just as good sources of the chemicals, even though the majority of study focuses on using the tuber to extract the colors. Anthocyanins are more concentrated in the peels than in the flesh or whole tubers, according to studies. According to Zhu et al. (2017) and Steed and Truong (2008), purple-fleshed sweet potato peels often contain twice as much anthocyanin as the flesh and may be a source for extracting the chemicals.

7.3.3 Utilization of Residues from Starch Production

Wastewater from the manufacturing of sweet potato starch has a high chemical oxygen demand and a high biological oxygen demand, which can cause severe pollution of nearby water systems (Akoetey et al. 2017). For every tonne of sweet potato roots processed, 6 m^3 of effluent is produced. The effluent from the manufacture of starch must be used in a way that eliminates nutrients and chemical oxygen demand to generate environmentally sustainable starch. The production of *Paenibacillus polymyxa* to be utilized as a bio-fertilizer for tea plants requires a medium with a chemical oxygen demand of roughly 1900 mg/L and significant levels of nitrogen, phosphorus, and other nutrients (Xu et al. 2014). After starch extraction, a solid fraction that is high in fiber can be recovered using milling and sifting to provide dietary fiber (Takamine et al. 2000). Antibiotics can also be made from the leftovers from the manufacturing of starch (Yang and Ling 1989).

The water from sweet potato processing contains protein (sporamin) that serves as a nutritious drink (Vithu et al. 2019). Foam separation is a method for removing protein from the waste stream from the manufacturing of sweet potato starch. When sweet potato protein recovery is optimized, it can reach 84.1% and lower both the biochemical and chemical oxygen demands (Mu et al. 2014). It has been documented that wastewater from the processing of sweet potatoes contains polyphenol oxidase, a phenolic compound cleaner (Cheng et al. 2015). Heavy metal hazards can be reduced by using stable environmentally acceptable adsorbents made from waste products from the processing of sweet potatoes (Hao et al. 2014; Asuquo and Martin 2016).

The leftover sweet potato material that results from the starch-making process can also be fermented to create single-cell protein. When the substrate was pre-treated with HCl and gamma irradiation, fermentation using a mixed culture of *Fusarium moniliforme* and *Saccharomyces cerevisiae* produced nitrogen utilization of 95.81% vs. 78.20% for the control (Aziz and Mohsen 2002). The resulting protein's characteristics make it effective both as a food ingredient and a food supplement (Yang and Ling 1989).

By cultivating the leftovers from the manufacturing of sweet potato starch with *Clostridium butyricum* and *Enterobacter aerogenes*, the leftovers have also been used to produce hydrogen (Yokoi et al. 2002). By using solid-state fermentation, amylolytic yeasts may increase the protein content of sweet potato waste from 6% to 21% in just 3 days (Yang 1988; Yang et al. 1993; El Sheikha and Ray 2022).

Dietary fiber is crucial for maintaining human health. The incidence of constipation, obesity, cardiovascular illnesses, colon cancer, and diabetes mellitus are all correlated with dietary fiber consumption, according to a significant quantity of research (Redondo-Cuenca et al. 2006). In addition to being utilized for nutrition, fiber can also be used in technology as a bulking agent or a fat substitute (de Escalada Pla et al. 2007). By using sieving and enzymatic techniques, Japanese researchers were able to extract the dietary fiber from sweet potatoes (Takamine et al. 2000; Yoshimoto et al. 2005; Mei Xin and Tai-Hua 2010), although there are no obvious changes in the content and characteristics of dietary fiber among cultivars.

For the creation of bio-composites based on polymer (3-hydroxybutyrate-co-3-hydroxyvalerate), the solid residue left over after the starch extraction from Chinese sweet potatoes can be employed as an addition (Vannini et al. 2021). The resulting bio-composites can be utilized as bio-alternatives to the materials often used in the packaging sector because they are entirely made from biological sources and may be biodegradable.

Tetracycline was produced by *Streptomyces viridifaciens* ATCC 11989 using solid-state fermentation to conserve energy in the synthesis of antibiotics and reduce the amount of agricultural waste (Yang and Ling 1989).

A common dish in several Asian nations, such as China, South Korea, Vietnam, and Indonesia, starch noodles are made in China from 28% of sweet potato products (Tan et al. 2009; Ariani et al. 2017). According to Menon et al. (2015), gluten-free starch noodles manufactured from tuber starch (such as sweet potato or cassava) are a great option for persons with celiac disease or gluten intolerance. However, this noodle performs poorly in terms of its total nutritional value because it lacks protein, vitamins, minerals, polyphenols, cellulose, and other essential ingredients. To increase the starch noodles' nutritional content, it is required to improve their quality. Natural antioxidants like curcumin, tea polyphenols, and lycopene can be added to sweet potato starch noodles to raise their broken rates, iodine blue values, hardness, and chewiness, according to Guo et al. (2022).

The superior quality of sweet potato starch can be precisely predicted by the thorough quality evaluation system created by Ma et al. in 2022. The creation of such a system is crucial for the breeding of high-quality, high-starch sweet potato cultivars with characteristics for post-harvest processing. The findings offer a fresh perspective on how to assess the quality of sweet potato starch and its byproducts.

7.3.4 Utilization of Waste from Alcoholic Fermentations

In Japan, sweet potatoes are fermented to create shochu, an alcoholic beverage. Making fibrous paper pots for growing decorative and veggie seedlings is a unique application for part of this waste. As the pots break down after being inserted into the ground, the seedlings are additionally fertilized by the pots (Yamauchi et al. 2006).

If the dissolved phosphorus from the wastewater from the alcoholic distillation of sweet potatoes isn't eliminated, it will cause eutrophication of the waterways. When cultured with shochu waste, the yeast *Hansenula anomala* successfully flocculates 70–80% of the dissolved phosphorus (Watanabe et al. 2009).

In a full-scale, fixed-bed reactor, biogas has also been created by digesting the waste from the distillation of shochu with a varied microbial population, including methanogens and bacteria from the phyla, *Firmicutes* and *Bacteroidetes* (Kobayashi et al. 2014).

For growing Shiitake mushrooms (*Lentinula edodes*), Azman et al. (2019) developed an alternate culture medium created from Shochu distillation lees. In comparison to mushrooms harvested from the standard medium and other traditional substrates, the average weight and count of mushrooms gathered from 20% shochu distillation lees medium were 1.4 times higher. Additionally, supplementing the medium with shochu distillation lees shortens the cultivation period compared to using the conventional medium and gives mushrooms 1.5 times more umami components.

According to Ray and Naskar (2008), the sweet potato waste biorefinery is a sustainable approach that helps with the production of ethanol and distilled beverages. To profitably valorize SPW within the circular economy concept and reduce greenhouse emissions by producing renewable resources and marketable products, Trevisan et al. (2020) evaluated integrated processing via a combination of different technologies to produce sustainable by-products.

7.4 Conclusion and Future Perspectives

The management of the waste generated during harvesting, post-harvest handling of any crop, and in the processing industry is challenging. The leftovers from sweet potatoes are made up of protein, dietary fiber, and carbohydrates. As a result, there are numerous ways to further valorize these wastes, including the creation of hydrogen, lipopeptides, poly-glutamic acid, cellulose nanocrystals, and food materials including leftover flour, soluble dietary fiber, and pectin, as well as solid emulsifiers. In terms of green processing, the transformation of leftovers and waste from sweet potatoes into more valuable goods is encouraging. The methods for separating products from garbage must be improved for better acceptance while remaining adaptive in the supply chain segments where root discarding is most prevalent.

References

Abano EE, Teye E, Amoah RS, Tetteh JP (2011) Design, construction and testing of an evaporative cooling barn for storing sweet potatoes in the tropics. Asian J Agric Res 5:1–12

Adeyosoye OI, Adesokan IA, Afolabi KD, Ekeocha AH (2010) Estimation of proximate composition and biogas production from *in vitro* gas fermentation of sweet potato (*Ipomoea batatas*) and wild cocoyam (*Colocasia esculenta*) peels. Afr J Environ Sci Technol 4(6):388–391

Ahmed M, Nigussie-Dechassa R, Bekele A (2012) Effect of planting methods and vine harvesting on shoot and tuberous root yields of sweet potato [*Ipomoea batatas* (L.) Lam.] in the Afar region of Ethiopia. Afr J Agric Res 7(7):1129–1141

Akoetey W, Britain MM, Morawicki RO (2017) Potential use of by-products from cultivation and processing of sweet potatoes. Ciência Rural 47(5):1–8

An LV, Frankow-Lindeberg BE, Lindeberg JE (2003) Effect of harvesting interval and defoliation on yield and chemical composition of leaves, stems, and tubers of sweet potato (*Ipomoea batatas* (L) Lam) cultivars. Field Crop Res 82(1):49–85

Anastacio A, Carvalho IS (2013) Phenolics extraction from sweet potato peels: key factors screening through a Placket-Burman design. Ind Crop Prod 43:99–105

Anastacio A, Silva R, Carvalho IS (2016) Phenolics extraction from sweet potato peels: modelling and optimization by response surface modelling and artificial neural network. J Food Sci Technol 53(12):4117–4125

Aregheore EM (2004) Nutritive value of sweet potato (*Ipomea batatas* (L) Lam) forage as goat feed: voluntary intake, growth and digestibility of mixed rations of sweet potato and batiki grass (*Ischaemum aristatum* var. *indicum*). Small Rumin Res 51:235–241

Ariani D, Herawati ERN, Nastiti A, Angwar M, Pranoto Y (2017) Effect of additional suji leaves and turmeric extract on physico-chemical characteristic and antioxidant activity of arenga-canna noodle. IOP Conf Ser Earth Environ Sci 101:012017. https://doi.org/10.1088/1755-1315/101/1/012017

Asuquo ED, Martin AD (2016) Sorption of cadmium (II) ion from aqueous solution onto sweet potato (*Ipomoea batatas* L.) peel adsorbent: characterisation, kinetic and isotherm studies. J Environ Chem Eng 4(4):4207–4228

Aziz NH, Mohsen GI (2002) Bioconversion of acid- and gamma-ray-treated sweet potato residue to microbial protein by mixed cultures. J Ind Microbiol Biotechnol 29:264–267

Azman NF, Yamauchi M, Yamada M et al (2019) Utilization of distillation waste of sweet potato Shochu lees for *Lentinula edodes* cultivation. J Mater Cycles Waste Manag 21:336–344. https://doi.org/10.1007/s10163-018-0795-5

Backer J, Ruiz ME, Munoz H, Pichnat AM (1980) The use of sweet potato (*Ipomoea batatas* (L) Lam) in animal feeding II. Beef production. Trop Anim Prod 5(2):152–160

Bechoff A, Tomlins K, Dhuique-Mayer C, Dove R, Westby A (2011) On-farm evaluation of the impact of drying and storage on the carotenoid content of orange-fleshed sweet potato (*Ipomoea batata* Lam.). Int J Food Sci Technol 46:52–60

Bovell-Benjamin A (2007) Sweet potato: a review of its past, present, and future role in human nutrition. Adv Food Nutr Res 52:1–59

Burri JB (2011) Evaluating sweet potato as an intervention food to prevent vitamin A deficiency. Compr Rev Food Sci Food Saf 10:118–130

Chakrabarti A, Buragohain SC, Baruah KK (2017) Carcass characteristics of broiler rabbit fed on sweet potato bsed ration. Int J Agric Sci Res 7(1):347–358

Chattopadhyay A, Chakraborty I, Sen H (2002a) Physico-chemical characteristics of sweet potato as a function of cultivars. Hortic J 15:45–50

Chattopadhyay A, Sen H, Chakraborty I, Satapathy M, Prakasa Rao LLT, Dutta A (2002b) Evaluation and characterization of sweet potato genetic resources in West Bengal. Indian J Plant Genet Resour 15:223–226

Chattopadhyay A, Chakraborty I, Mukhopadhyay SK, Kumar PR, Sen H (2006a) Compositional changes of sweet potato as influenced by cultivar, harvest date and cooking. Acta Hortic 703: 211–218

Chattopadhyay A, Chakraborty I, Kumar PR, Nanda MK, Sen H (2006b) Uncontrolled storage behaviour of sweet potato (*Ipomoea batatas* Lam). J Food Sci Technol 43:41–45

Cheng S, Zhang YF, Zeng ZQ, Lin J, Zhang YW, Ni H, Li HH (2015) Screening, separating, and completely recovering polyphenol oxidases and other biochemicals from sweet potato wastewater in starch production. Appl Microbiol Biotechnol 99(4):1745–1753

Choi JH, Hwang YP, Choi CY, Chung YC, Jeong HG (2010) Anti-fibrotic effects of the anthocyanins isolated from the purple-fleshed sweet potato on hepatic fibrosis induced by dimethyl nitrosamine administration in rats. Food Chem Toxicol 48:3137–3143

de Escalada Pla MF, Ponce NM, Stortz CA, Gerschensoon LN, Rojas AM (2007) Composition and functional properties of enriched fibre products obtained from pumpkin (*Cucurbita moschata* Duchesne ex Poiret). LWT Food Sci Technol 40:1176–1185

Dom MT, Ayalew WK, Glatz PC, Kirkwood RN, Hughes PE (2017) Nutrient utilisation in grower pigs fed a protein concentrate blended with sweet potato roots either boiled or ensiled with or without vines. Anim Prod Sci 57(8):1645–1652

Dongmeza E, Steinbronn S, Francis G, Focken U, Becker K (2009) Investigations on the nutrient and antinutrient content of typical plants used as fish feed in small scale aquaculture in the mountainous regions of Northern Vietnam. Anim Feed Sci Technol 149:162–178

El Sheikha AF, Ray RC (2017) Potential impacts of bioprocessing of sweet potato. Crit Rev Food Sci Nutr 57(3):455–471

El Sheikha AF, Ray RC (2022) Bioprocessing of horticultural wastes by solid-state fermentation into value-added/innovative bioproducts: a review. Food Rev Int 39:3009. https://doi.org/10.1080/87559129.2021.2004161

Eldiehy KSH, Niran D, Borah D, Sarmah D, Bora U, Mandal M, Deka D (2022) Towards biodiesel sustainability: waste sweet potato leaves as a green heterogeneous catalyst for biodiesel production using microalgal oil and waste cooking oil. Ind Crop Prod 187:115467

Etela I, Oji UI, Kalio GA, Tona GO (2008) Studies on sweet potato forage and dried brewers' grains as supplements to green panic for Bunaji cows. Trop Grassl 42:245–251

FAO (2009) FAO statistics. http://apps.fao.org

FAO (2016). http://faostat3.fao.org/home

Farrell DJ, Jibril H, Perez-Maldonado RA, Mannion PF (2000) A note on a comparison of the feeding value of sweet potato vines and lucerne meal for broiler chickens. Anim Feed Sci Technol 85:145–150

Fasuyi AO (2006) Nutritional potentials of some tropical vegetable leaf meals: chemical characterization and functional properties. Afr J Biotechnol 5(1):49–53

Ge X, Matsumoto T, Keith L, Li Y (2014) Biogas energy production from tropical biomass wastes by anaerobic digestion. Bioresour Technol 169:38–44

Giang HH, Ly LV, Ogle B (2004) Digestibility of dried and ensiled sweet potato roots and vines and their effect on the performance and economic efficiency of F_1 cross bred fattening pigs. Livest Res Rural Dev 16:6–7

Gonzalez C, Diaz I, Vecchionacce H, Ly J (2003) Performance traits of pigs fed sweet potato (*Ipomoea batatas* L.) foliage *ad libitum* and graded levels of protein. Livest Res Rural Dev 15(9):9–15

Guo W, Fan L, Wang Y, Li G, Gao X, Chen Z, Huang J (2022) Effect of three natural antioxidants on the structure and physicochemical properties of sweet potato starch noodles. Front Nutr 9: 1020281. https://doi.org/10.3389/fnut.2022.102028

Hao Z, Wang D, Chen H, Sun J, Xu Y (2014) Sweet potato starch residue as starting material to prepare polyacrylonitrile adsorbent via SI-SET-LRP. J Agric Food Chem 62(8):1765–1770

Huang M-H, Chu H-L, Juang L-J, Wang B-S (2010) Inhibitory effects of sweet potato leaves on nitric oxide production and protein nitration. Food Chem 121:480–486

Hue S-M, Boyce AN, Somasundram C (2014) Extraction from agricultural waste, *Ipomoea batatas* L. lam leaves as a cheap source of natural dye. Int J Environ Bioremediat Biodegrad 2:1–4

Isghiguro K, Yoshimoto M (2006) Content of the eye-protective nutrient lutein in sweetpotato leaves. In: Concise paper presented in 2nd International symposium on sweet potato and cassava, 14–17 June 2005, Kuala Lumpur, Malaysia, pp 213–214

Islam MS, Yoshimoto M, Yahara S, Okuno S, Ishiguro K, Yamakawa O (2002) Identification and characterization of foliar polyphenolic composition in sweet potato (*Ipomoea batatas* L.) genotypes. J Agric Food Chem 50:3718–3722

Katongole CB, Katongole C, Sabiiti E, Ledin I (2009) Intake, growth and carcass yield of indigenous goats fed market wastes of sweet potato (*Ipomoea batatas*) vines and scarlet eggplant (*Solanum aethiopicum*). Trop Anim Health Prod 41:1623–1631

Khan SH (2017) Sweet potato (*Ipomoea batatas* (L.) Lam) as feed ingredient in poultry diets. Worlds Poult Sci J 73(1):77–88

Kidmose U, Christensen LP, Agili SM, Thilsted SH (2007) Effect of home preparation practices on the content of provitamin A carotenoids in coloured sweet potato varieties (*Ipomoea batatas* Lam.) from Kenya. Innov Food Sci Emerg Technol 8:399–406

Kobayashi T, Tang Y, Urakami T, Morimura S, Kida K (2014) Digestion performance and microbial community in full-scale methane fermentation of stillage from sweet potato-shochu production. J Environ Sci China 26:423–431

Lebot V (2009) Tropical root and tuber crops: cassava, sweet potato, yams, and aroids. CABI, Oxfordshire, pp 91–274

Leon-Verlarde CU (2000) Using competing traits to select dual-purpose sweet potato in native germplasm. International Potato Center (CIP), Lima, Peru, pp 289–294

Li P, Ji S, Wang Q, Qin M, Hou C, Shen Y (2017) Adding sweet potato vines improve the quality of rice straw silage. Anim Sci J 88(4):625–632

Li A, Xiao R, He S, An X, He Y, Wang C, Yin S, Wang B, Shi X, He J (2019) Research advances of purple sweet potato anthocyanins: extraction, identification, stability, bioactivity, application, and biotransformation. Molecules 24(21):3816. https://doi.org/10.3390/molecules24213816

Lila MA (2004) Anthocyanins and human health: an in vitro investigative approach. J Biomed Biotechnol 5:306–313

Ma C, Zhang Y, Yue R, Zhang W, Sun J, Ma Z, Niu F, Zhu H, Liu Y (2022) Establishment of a quality evaluation system of sweet potato starch using multivariate statistics. Front Nutr 9: 1025061. https://doi.org/10.3389/fnut.2022.10250

Maloney Katherine P, Van-Den T, Allen Jonathan C (2012) Chemical optimization of protein extraction from sweet potato (*Ipomoea batatas*) peel. J Food Sci 77:E307–E312

Mei Xin M, Tai-Hua HJ-J (2010) Composition and physico-chemical properties of dietary fibre extracted from residues of 10 varieties of sweet potato by a sieving method. J Agric Food Chem 58:7305–7310

Menon R, Padmaja G, Sajeev MS (2015) Cooking behaviour and starch digestibility of NUTRIOSE R (resistant starch) enriched noodles from sweet potato flour and starch. Food Chem 182:217–223

Mitra S, Tarafdar J, Palaniswami MS (2010) Impacts of different maturity stages and storage on nutritional changes in raw and cooked tubers of orange-fleshed sweet potato (*Ipomoea batatas*) cultivars. Acta Hortic 858:205–212

Modupe EO, Akwayo IJ, Olugbenga ST, Oyinlola MO, Ayodeji AA, Emmanuel OO, Esther AO (2018) Bio-conversion of sweet potato peel waste to bio-ethanol using *Saccharomyces cerevisiae*. Int J Pharm Phytopharmacol Res 8(3):46–54

Mu T-H, Liu Y, Zhang M, Sun H-N (2014) Protein recovery from sweet potato starch wastewater by foam separation. Sep Sci Technol 49:2255–2260

Mukhopadhyay SK, Chattopadhyay A, Chakraborty I, Bhattacharya I (2011) Crops that feed the world 5. Sweet potato. Sweet potatoes for income and food security. Food Secur 3:283–230

Mussoline WA, Wilkie AC (2015) Anaerobic digestion potential of coproducts associated with ethanol production from sweet potato: a review. Indian Biotechnol 11(2):113–126

Nath R, Kundu CK, Majumder A, Islam Sk J, Gunri S, Chattopadhyay A (2005) Promotion of sweet potato in alleviating hunger and nutritional deficiencies in the poverty stricken and environmentally fragile red and laterite ecosystem of West Bengal, India. In: Concise paper presented at 2nd International symposium on sweet potato and cassava, held on 14–17 June 2005, Kuala Lumpur, Malaysia, pp 197–198

Nedunchezhiyan M, Ray RC (2010) Sweet potato growth, development, production and utilization: overview. In: Ray RC, Tomlins KI (eds) Sweet potato: post-harvest aspects in food, feed and industry. Nova Science Publishers Inc., Hauppauge, NY, pp 1–26

Nguyen THL, Ngoan LD, Verstegen MWA, Hendriks WH (2012) Ileal and total tract apparent protein and amino acid digestibility of ensilage and dry cassava leaves and sweet potato vines in growing pigs. Anim Feed Sci Technol 172:171–179

Ofori G, Oduro I, Ellis WO, Dapaah HK (2009) Assessment of vitamin A content and sensory attributes of new sweet potato (*Ipomoea batatas*) genotypes in Ghana. Afr J Food Sci 3:184–192

Oke M, Workneh T (2013) A review on sweet potato postharvest processing and preservation technology. Afr J Agric Res 8:4990–5003

Oluyori AP, Shaw AK, Preeti R, Reddy S, Atolani O, Olatunji GA, Fabiyi OA (2016) Natural antifungal compounds from the peels of *Ipomoea batatas* Lam. Nat Prod Res 30(18):2125–2129

Omojasola PF, Adeniran EA (2014) The production of itaconic acid from sweet potato peel using *Aspergillus niger* and *Aspergillus terreus*. Albanian J Agric Sci 13(4):72

Omoregie E, Igoche L, Ojobe TO, Absalom K, Onusiriuka BC (2009) Effect of varying levels of sweet potato (*Ipomoea batatas*) peels on growth, feed utilization and some biochemical responses of the cichlid (*Oreochromis niloticus*). Afr J Food Agric Nutr Dev 9:700–712

Pagana I, Morawicki R, Hager T (2014) Lactic acid production using waste generated from sweet potato processing. Int J Food Sci Technol 49:641–649

Panda SK, Swain MR, Singh S, Ray RC (2012) Proximate compositions of a purple sweet potato (*Ipomoea batatas* L.) herbal wine. J Food Process Preserv 37:596. https://doi.org/10.1111/j.1745-4549.2012.00681.x

Ray RC, Naskar SK (2008) Bio-ethanol production from sweet potato (*Ipomoea batatas* L.) by enzymatic liquefaction and simultaneous saccharification and fermentation. Dyn Biotechnol Process Biochem Mol Biol 2:47–49

Ray RC, Ravi V (2005) Post harvest spoilage of sweet potato in tropics and control measures. Crit Rev Food Sci Nutr 45:623–644

Ray RC, Tomlins KI (eds) (2010) Sweet potato: post-harvest aspects in food, feed, and industry. Nova Science Publishers, Inc, Hauppauge, NY. 316pp

Ray RC, Ravi V, Rao KR, Hegde V, Tomlins KI (2010) Post-harvest handling, storage methods, pests and diseases of sweet potato. In: Ray RC, Tomlins KI (eds) Sweet potato: post-harvest aspects in food, feed and industry. Nova Science Publishers Inc., Hauppauge, NY, pp 27–58

Redondo-Cuenca A, Villanueva-Suárez MJ, Rodríguez-Sevilla MD, Mateos-Aparicio I (2006) Chemical composition and dietary fibre of yellow and green commercial soybeans (*Glycine max*). Food Chem 101:1216–1222

Régnier C, Jaguelin Y, Noblet J, Renaudeau D (2012) Ileal digestibility of amino acids of cassava, sweet potato, cocoyam and erythrina foliages fed to growing pigs. Animal 6:586–593

Régnier C, Bocage B, Archimède H, Noblet J, Renaudeau D (2013) Digestive utilization of tropical foliages of cassava, sweet potatoes, wild cocoyam and erythrina in Creole growing pigs. Anim Feed Sci Technol 180:44–54

Schaub S, Leonard J (1996) Composting: an alternative waste management option for food processing industries. Trends Food Sci Technol 7:263–268

Smart J, Simmonds NW (1995) Evolution of crop plants, 2nd edn. Wiley, New York, pp 57–61

Solomon E (1999) Yield performance of sweet potato cultivars at Werer under irrigation. AgriTopia Quarterly Newsletter of Ethiopian Agricultural Research Organization (EARO), Addis Ababa, pp 2–14

Steed L, Truong V (2008) Anthocyanin content, antioxidant activity, and selected physical properties of flowable purple-fleshed sweet potato purees. J Food Sci 73:S215–S221

Sun H, Mu T, Xi L, Zhang M, Chen J (2014) Sweet potato (*Ipomoea batatas L.*) leaves as nutritional and functional foods. Food Chem 156:380–389

Takamine K, Abe J, Iwaya A, Maseda S, Hizukuri S (2000) A new manufacturing process for dietary fibre from sweet potato residue and its physical characteristics. J Appl Glycosci 47:67–72

Tan H-Z, Li Z-G, Tan B (2009) Starch noodles: history, classification, materials, processing, structure, nutrition, quality evaluating and improving. Food Res Int 42(5–6):551–576

Thankappan M (1989) Status of sweet potato pathology in India. In: Lead paper presented at International seminar on pest and diseases of sweet potato in Asia, 7–8 August, CTCRI, Trivandrum, India

Trevisan CW, Ferreira LT, Trierweiler JO (2020) Food waste biorefinery advocating circular economy: bioethanol and distilled beverage from sweet potato. J Clean Prod 268:121788. https://doi.org/10.1016/j.jclepro.2020.121788

Vannini M, Marchese P, Sisti L, Saccani A, Mu T, Sun H, Celli A (2021) Integrated efforts for the valorization of sweet potato by-products within a circular economy concept: biocomposites for packaging applications close the loop. Polymers 13:1048. https://doi.org/10.3390/polym13071048

Vithu P, Dash Sanjaya K, Rayaguru K (2019) Post-harvest processing and utilization of sweet potato: a review. Food Rev Int 35:726. https://doi.org/10.1080/87559129.2019.1600540

Watanabe T, Masaki K, Iwashita KO, Fujii T, Iefuji H (2009) Treatment and phosphorus removal from high-concentration organic wastewater by the yeast *Hansenula anomala* J224 PAWA. Bioresour Technol 100:1781–1785

Woolfe JA (1992) Sweet potato, an untapped food resource. Cambridge University Press, UK

Xu S, Bai Z, Jin B, Xiao R, Zhuang G (2014) Bioconversion of wastewater from sweet potato starch production to *Paenibacillus polymyxa* biofertilizer for tea plants. Sci Rep 4:4131. https://doi.org/10.1038/srep04131

Yamauchi M, Masuda S, Kihara M (2006) Recycled pots using sweet potato distillation lees. Resour Conserv Recycl 47:183–194

Yang S (1988) Protein enrichment of sweet potato residue with amylolytic yeasts by solid-state fermentation. Biotechnol Bioeng 32:886–890

Yang S, Ling M (1989) Tetracycline production with sweet-potato residue by solid-state fermentation. Biotechnol Bioeng 33:1021–1028

Yang SS, Jang HD, Liew CM, du Preez JC (1993) Protein enrichment of sweet-potato residue by solid-state cultivation with mono-cultures and cocultures of amylolytic fungi. World J Microbiol Biotechnol 9:258–264

Yang L, Xi Y, Luo XY, Ni H, Li HH (2019) Preparation of peroxidase and phenolics using discarded sweet potato old stems. Sci Rep 9:3769. https://doi.org/10.1038/s41598-019-40568-9

Yokoi H, Maki R, Hirose J, Hayashi S (2002) Microbial production of hydrogen from starch-manufacturing wastes. Biomass Bioenergy 22:389–395

Yoshimoto M, Yamakawa O, Tanoue H (2005) Petential chemopreventive properties and varietal difference of dietary fibre from sweet potato root. Jpn Agric Res Q 39(1):37–43

Zhu Z, Guan Q, Koubaa M, Barba FJ, Roohinejad S, Cravotto G, Yang X, Li S, He J (2017) HPLC-DAD-ESI-MS^2 analytical profile of extracts obtained from purple sweet potato after green ultrasound-assisted extraction. Food Chem 215:391–400

Yams and Aroid Crop Waste: Bio Valorization into Bioproducts and Platform Chemicals

8

Chintha Pradeepika, Deep J. Kalita, C. Visalakshi Chandra, Senthil and Sankar, and K. Hanume Gowda

Abstract

With the increase in the production of good products, there is an increase in crop wastes and byproducts that either end up in landfills or lead to environmental pollution due to poor compost management. This statement is a better fit for the horticultural tuber crops that usually have thick non-edible skin which is removed before their utilization to produce food products. To reduce the burden from these food wastes and increase their stability and applicability, their bio valorization into bioproducts and platform chemicals has become very crucial. Several research studies explored nano and biotechnological approaches as novel and green applications to valorize tuber crop waste to produce platform chemicals to produce biofuel, biopolymer, functional food additives, energy storage material, wastewater treatment, etc. In this chapter, recent research and their findings are covered in detail with interest in the bio valorization of yam and aroid tuber waste. In addition, the problems and limitations of waste management and valorization methods are investigated, considering future perspectives.

C. Pradeepika (✉) · Senthil and Sankar
Section of Crop Utilization, ICAR-Central Tuber Crop Research Institute, Thiruvananthapuram, Kerala, India
e-mail: pradeepika.chintha@icar.gov.in

D. J. Kalita
Renuvix LLC, Fargo, ND, USA

C. Visalakshi Chandra
Division of Crop Improvement, ICAR-Central Tuber Crop Research Institute, Thiruvananthapuram, Kerala, India

K. Hanume Gowda
Division of Crop Improvement, Regional Station, ICAR-Central Tuber Crop Research Institute, Bhubaneswar, Odisha, India

R. C. Ray (ed.), *Roots, Tubers, and Bulb Crop Wastes: Management by Biorefinery Approaches*, https://doi.org/10.1007/978-981-99-8266-0_8

Keywords

Yams · Aroids · Peel · Residue · Valorization · Bioproducts

Abbreviations

ABE	Acetone–butanol–ethanol
AD	Anaerobic digestion
ARA	Arachidonic acid
EFY	Elephant foot yam
F-SAF	Fed-batch simultaneous saccharification and fermentation
F-SSF	Fed-batch separate hydrolysis and fermentation
KGM	Konjac Glucomannan
SHF	Separate hydrolysis and fermentation
SSF	Simultaneous saccharification and fermentation
TPS	Taro peel starch
TPW	Taro peel waste
TW	Taro waste
VC	Volatile solids
YP-AC	Yam peel-based activated carbon
YPM	Yam peel waste

8.1 Introduction

With the increase in the production of agricultural commodities and population, there is an increase in crop and food waste. Due to a lack of general awareness and composting facilities, these untreated crops and food waste often end up in landfills and freshwater bodies thereby increasing environmental pollution (Rao et al. 2021). According to the Food and Agriculture Organization (FAO), food waste is defined as "the mass of wholesome food material that is lost, degraded, damaged, or discarded at any stage of the food supply chain system" (FAO 1981). In another comprehensive definition, food waste is defined as "the inedible parts and others of food discarded from the food supply chain such as which disposed to sewer, landfill or discarded to the sea, composted, or crops plowed in/not harvested to be recovered and treated by anaerobic digestion, co-generation and incineration for producing, for example, bioenergy" (Bos-Brouwers et al. 2014). The fact that these wastes are generated in large quantities and contain compounds that can benefit humans recently made them attractive low-cost raw materials with high economic value. The use of these biowastes such as agricultural residues, food processing waste, and food residuals to produce platform chemicals and fuel provides a means to efficiently valorize the waste by-products, thereby contributing to the global circular economy (Teigiserova et al. 2019).

Among various waste streams, horticultural crops contribute to almost 40–50% of global food waste per year (Corrado and Sala 2018; Bancal and Ray 2022). Also, developed countries in Europe and North America contribute to the waste generation that occurs at the later stages of the supply chain and the values are much higher (95–115 kg/year/consumer) than those in the developing countries (6–11 kg/year/consumer) and mostly occur at the harvesting and storage stages. Among various horticultural crops such as root crops, fruits, and vegetables, the waste produced from tuber crops is considered a major contributor as they usually have a thick peel that is removed before utilization in various food products. With the advances in science, new waste-managing techniques that require low capital investment and energy and result in improved product recovery, have been developed to turn the biowastes into value-added products/chemicals. This transformation of waste into value-added products is termed bio-valorization. Valorization of waste materials could lead to the production of energy, platform chemicals, fuels, and other value-added products, thereby reducing the dependency on diminishing nonrenewable petroleum feedstock (Nizami et al. 2017).

Recently, the biovalorization of agro-waste using fermentation techniques such as solid-state fermentation has been extensively studied to derive various platform chemicals (El Sheikha and Ray 2022). These platform chemicals such as bioactive compounds (Sooch et al. 2022), enzymes, organic acids (Kumari et al. 2022), microbial polysaccharides (Kotsanopoulos et al. 2022), bioethanol (Moodley et al. 2023), etc. serve as building blocks to produce high-value-added products like polymers, fuels, solvents, plastics, pharmaceuticals, etc. (El Sheikha and Ray 2022). Among various platform chemicals, the production of ethanol, lactic acid, succinic acid, levulinic acid, sorbitol, and 2,5-furan dicarboxylic acid is proven to be viable on an industrial scale. The production of these chemicals has proven to be currently economically viable, for example, the production of lactic acid from food waste has cost US$1066/million ton and was reported to generate an annual net profit of US$ 422,699 at a return rate of 18.98% (Kover et al. 2022). Among the biowaste raw materials, horticultural waste has tremendous potential to provide a platform/substrate to produce various added-value products such as antibiotics, antioxidant, anti-microbial, anti-aging and food additive products, edible fungi, enzymes, organic acids, flavor and aroma compounds due to their high content of polysaccharides, sugar, and phenolic compounds (Ray 2022; El Sheikha and Ray 2022). Horticultural waste from tuber crops has been explored as a valuable source to produce biofuel to alleviate the global petroleum-based fuel crisis (El Sheikha and Ray 2017). However, it is noteworthy to mention that utilizing horticultural food/crop waste in biorefineries requires meticulous pre-treatment due to their complex structure to break the cellulosic/lingo-cellulosic materials in monosaccharides, building blocks to produce value-added products.

This book chapter has been developed to provide the readers with extensive information about the recent development in the valorization of tuber crop waste, especially targeting yams and aroid crop waste. Taro and elephant foot yam (EFY) tubers are underutilized in the food industry due to their high content of oxalates despite their high nutritional values. In this chapter, we have discussed various technologies leading to the value-added products and valorization of their waste (mostly from their peels, and other parts of the plants) as a source to produce bioproducts and platform chemicals.

8.2 Yams and Aroids: Taxonomy, Botany, Global Production (FAO Data), and Waste Generation

8.2.1 Yams: Species, Botany (Morphology), and Edible Parts

Yams are starchy root vegetables belonging to the *Dioscorea* genus and Dioscoreaceae family. The four common edible species are *D. rotundata*, *D. alata*, *D. esculenta* and *D. hispida*. They are widely grown in tropical and subtropical regions of Africa, Asia, Latin America, and the Pacific Islands. They are a staple food in Africa, Asia, and the Caribbean. Yams are a good source of carbohydrates, fiber, vitamins, and minerals. Globally, Yams are cultivated over an area of 8.68 Mha with an annual production of 75.14 million tons (FAOSTAT 2021). They are particularly high in vitamin C, potassium, and manganese. Yams are also rich in antioxidants, which help to protect the body from damage caused by free radicals. Yams are typically larger, cylindrical, and have rough, bark-like skin, and higher starch content, and come in a variety of shapes, sizes, and colors. Yams also contain antioxidants that help protect the body against oxidative stress and may have anti-inflammatory properties (Lebot et al. 2023).

8.2.1.1 Taxonomy

Kingdom	Plantae
Sub-kingdom	Viridiplantae
Division	Tracheophyta
Sub-division	Spermatophytina
Class	Magnoliopsida
Order	Dioscoreales
Family	Dioscoreaceae
Genus	*Dioscorea*
Species	*alata, rotundata, esculenta, hispida*

8.2.1.2 Botany

Yams are a group of perennial herbaceous vines. They are climbing or trailing vines that can reach impressive lengths, often up to several meters. They have long, cylindrical, or tuberous underground storage roots (Fig. 8.1), which are the edible part of the plant. The leaves of yam plants are usually large, heart-shaped, and alternate along the vine. Flowers are small, unisexual, and arranged in clusters on a raceme. Tubers are the most distinguishing feature of yams, which can produce large, elongated tubers with rough, bark-like skin. The flesh of yams varies in color

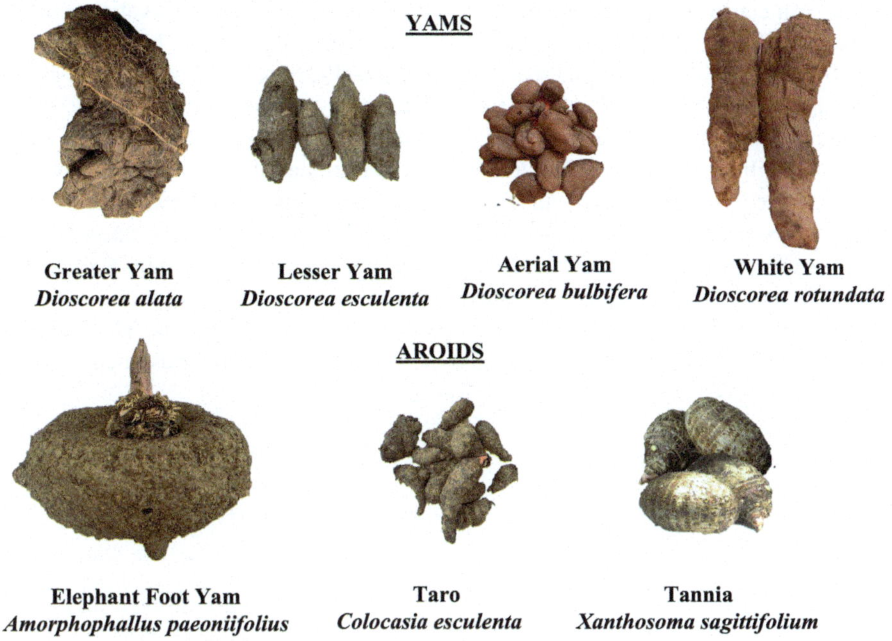

Fig. 8.1 Yams and aroid tubers

and can be white, yellow, purple, or pink, depending on the species. They are harvested after 6–12 months of growth (Obidiegwu et al. 2020).

8.2.2 Aroids: EFY—Taro, Xanthosoma, Etc.—Botany (Morphology), Edible Parts

8.2.2.1 Taro (*Colocasia esculenta* (L.) Schott)

Taro, scientifically known as *Colocasia esculenta* var. *esculenta* and *C. esculenta* var. *antiqurom*, is a root vegetable that has been cultivated and consumed for thousands of years. It belongs to the Araceae family and is widely grown in tropical and subtropical regions around the world. Globally, Taro is cultivated over an area of 1793 thousand hectares with 12.39 million ton with a major share of production coming from Africa (FAOSTAT 2021). It is known by various names in different cultures, such as "dasheen" in the Caribbean, "kalo" in Hawaii, and "elephant ear" or "cocoyam" in some English-speaking countries. Taro plants are herbaceous perennials with large, heart-shaped leaves that can reach impressive sizes. It has a rough, brownish outer skin and starchy, white, or purple flesh, depending on the variety. Taro has long been a staple crop in many cuisines, particularly in Asian and Pacific Island cultures. It is versatile and can be prepared in various ways, including boiling, steaming, roasting, or frying. Nutritionally, taro is a good source of dietary fiber, vitamins, and minerals. However, raw taro contains calcium oxalate crystals, which can cause skin irritation and a sensation of itchiness Taro cultivation requires a warm and moist climate, as it thrives in areas with temperatures between 20 and 30 °C (68–86 °F) (Chaïr et al. 2016). It prefers well-drained soil and can be grown both in flooded fields or upland areas. Taro is a root vegetable that is widely consumed in many parts of the world and belongs to the Araceae family. It has a starchy, slightly sweet flavor and a texture similar to that of a potato. It is grown for its large, heart-shaped leaves and its underground corms or tubers, which are an important source of carbohydrates and are also rich in fiber, vitamins, and minerals. Taro is used in many different culinary traditions, including in Asian, Polynesian, and Caribbean cuisine. It can be boiled, roasted, fried, or mashed, and is used in a wide range of dishes, from savory soups and stews to sweet desserts. The leaves of the taro plant are also edible and can be used as a wrap for other foods or cooked as a vegetable.

Kingdom	Plantae
Division	Magnoliophyta
Class	Liliopsida
Order	Arales
Family	Araceae
Genus	*Colocasia* Schott (colocasia)
Species	*Colocasia esculenta* (L.) Schott

8.2.2.2 Botany

The taro plant belongs to the Araceae family and is widely cultivated in tropical and subtropical regions worldwide for its edible corms and leaves. It has large, heart-shaped leaves that grow on long petioles or stems, and the edible part of the plant is the corm, which is an underground swollen stem. It is rich in carbohydrates, fiber, and various vitamins and minerals. Taro plants prefer moist, well-drained soil and thrive in warm, tropical climates. It is a staple food in several cuisines, and the leaves of the taro plant are also utilized in cooking. There are different cultivars and varieties of taro, each with its own unique characteristics and culinary uses.

8.2.3 Elephant Foot Yam (*Amorphophallus paeoniifolius* (Dennst.) Nicolson)

Kingdom	Plantae
Phylum	Tracheophyta
Class	Liliopsida
Order	Alismatales
Family	Araceae
Genus	*Amorphophallus*
Species	*Amorphophallus paeoniifolius*(Dennst.) Nicolson

EFY, also known as *A. paeoniifolius*, is a tropical perennial plant that belongs to the Araceae family. It is native to Southeast Asia and is widely cultivated in various parts of the world, including India, Sri Lanka, and other tropical regions. The name "elephant foot yam" refers to the large, tuberous underground corm of the plant, which can grow to a considerable size. The leaves are usually green and have several leaflets that radiate from a central stalk. The plant produces a single flower on a tall, erect stalk, which emits a strong odor to attract pollinators. The edible part of the plant is starchy and has a slightly fibrous texture, with a mild flavor similar to that of a potato. It is commonly used in various culinary preparations, has medicinal properties, and is used in traditional medicine systems. Elephant foot yam is believed to possess various health benefits and is used to treat digestive issues, inflammation, and respiratory ailments in certain traditional practices. It requires a warm and humid climate and requires well-drained soil and regular watering. It is propagated by planting small cormels, which are the small corm-like structures that grow around the main corm. It is cultivated for its edible corm, which is a significant food source in many tropical regions and is propagated by planting cormels or mature corms in prepared soil beds. It is not only valued for its culinary uses but also plays a role in traditional medicine systems due to its perceived medicinal properties. Its unique botany and striking appearance make it an intriguing plant to study and cultivate.

8.2.3.1 Botany

EFY is a tropical perennial plant characterized by its large, underground corm, which serves as the storage organ for the plant. Above the ground, the plant features large, palmate leaves with several leaflets attached to a central stalk. It produces a single inflorescence, known as an "inflorescent spathe," on a tall, erect stalk. The corm consists of several layers, with the outermost layer being the protective tunic and the central region containing the primary bud and other growing points. The inflorescence is a tall, erect structure that emerges from the ground and consists of a fleshy, tubular structure known as a spathe, which surrounds and protects the spadix. The spathe is typically maroon or purplish and has a strong odor to attract pollinators. EFY is an intriguing plant to study and cultivate due to its unique botany and striking appearance. It reproduces through both sexual and asexual means, with the flowers pollinated by insects attracted to the odor of the inflorescence. After successful pollination, the plant produces berries containing seeds. It thrives in warm and humid tropical climates and is usually cultivated in regions with temperatures above 20 °C (68 °F) and high rainfall (Ravi et al. 2009). It is cultivated for its edible corm, which is a significant food source in many tropical regions. It requires careful management to control pests, diseases, and weeds during cultivation. It is not only valued for its culinary uses but also plays a role in traditional medicine systems due to its perceived medicinal properties.

8.2.4 Tannia (*Xanthosoma sagittifolium* (L.) Schott)

Tannia is a tropical root vegetable that belongs to the Araceae family and is commonly referred to as "cocoyam" or "arrow leaf elephant ear." It is cultivated for its edible corms and leaves, which are widely consumed in various cuisines around the world. Tannia plants are herbaceous perennials that can reach a height of 1–1.5 m (3 ft) and have large, arrow-shaped leaves that are green and glossy. The edible part of the plant is the corm, which is an underground tuber. It is a good source of carbohydrates, dietary fiber, vitamins, and minerals. Tannia is cultivated in tropical and subtropical regions around the world and requires warm temperatures and high humidity for optimal growth. It is propagated by planting cormels or mature corms in prepared soil. Tannia has some cultural significance and is associated with traditional rituals and ceremonies in certain regions.

8.2.4.1 Taxonomy

Kingdom	Plantae
Phylum	Tracheophyta
Class	Liliopsida
Order	Alismatales
Family	Araceae
Genus	*Xanthosoma* Schott
Species	*Xanthosoma sagittifolium* (L.) Schott

8.3 Waste Generation

Rapid intensification of agriculture production has been generating huge quantities of agricultural waste which comprises crop waste, animal waste, food processing waste, and toxic agricultural waste. In general, agricultural waste can be divided into solid or liquid based on the nature of agricultural activities. According to Kee et al. (2023), solid waste is projected to increase up to 3.4 billion tons/year globally. This huge quantity of waste can lead to serious threats to human health due to improper waste disposal practices followed in many countries. Unsurprisingly, in many developing countries little is known about the risks caused by traditional waste disposal methods as well as hidden opportunities of agricultural waste as a resource. Longjan and Dehouche (2018) characterized the nutrient profile of nine different food wastes generated in Nigeria which includes yam, cassava, cocoyam, groundnut, beans, egusi seeds, corn, plantain, and Ugwu. As per the study, the specific waste index was reported to be 0.3–0.5 (yam peels), 0.2–0.3 (cassava peel), 0.3 (cocoyam peels), 0.5 (plantain peels), 1.4–1.5 (corn cob and husk), 0.2–0.3 (Egusi shells), 0.3 (bean skin and groundnut shell), and 0.5–0.6 (Ugwu stalk). The renewable energy potential of these nine food wastes revealed the yam waste (peel) as a good candidate among studied food wastes for the generation of methane and energy. Overall, this study indicated a great scope for solid food waste generated in Nigeria for the

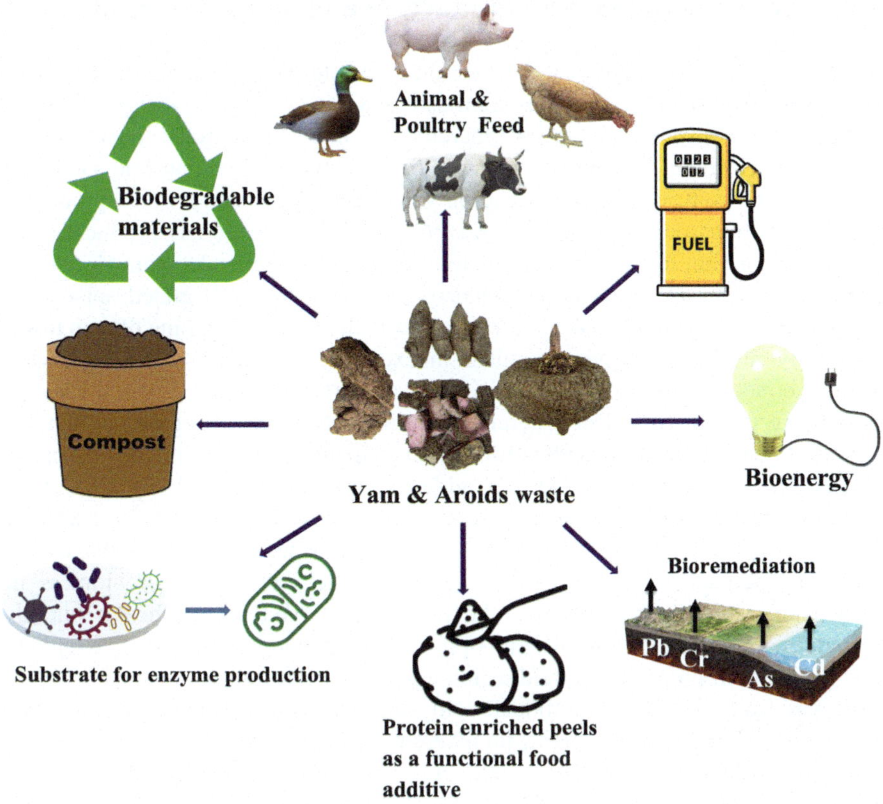

Fig. 8.2 Yam and aroids waste valorization into bioproducts and platform chemicals

generation of renewable energy to complement fossil fuel-based energy production. Similarly, another study reported generation of 62% of agricultural waste in Lago state and 78% of food waste in Benin City daily in Nigeria can be effectively managed through sustainable renewable energy production technologies (Orhorhoro and Oghoghorie 2019). Furthermore, Giwa et al. (2017) presented an excellent report on biomass resource potential and substantial bioenergy production in Nigeria. In addition, enormous amounts of agricultural residue and their respective energy equivalents produced yearly in different geopolitical zones between the years 2001 and 2006 were reported (Giwa et al. 2017). The major solid agricultural residues studied are yam peels, cassava peels, maize cobs, rice husks, and ground husks. Among different agricultural residues, yam peels waste generation ranked highest, 6372.83 (10^9 g) among different geopolitical zones studied as well as their energy potential (74.93 PJ). Therefore, valorization of yam and aroid tuber waste into various bioproducts and platform chemicals (Fig. 8.2) offers sustainable development by solving the pollution problem in all possible directions.

8.3.1 Yams Waste into Bioproducts and Platform Chemicals

Mass production of yam (*Dioscorea* spp.) tubers and their processing/utilization in various food products results in a significant amount of yam peel that goes through inappropriate disposal methods thereby leading to serious environmental consequences mainly pollution of land, air, and water (Longjan and Dehouche 2018). In this context, Nigeria, a leading yam-producing country with a production of 50.1 million tons accounts for 66.89% of the world's yam production and processing of yam tubers into various food products like pounded yam (*iyan*), yam flour (*pounded-yam*), yam gruel (*amala*), fried yam (*dundu*), grated spiced yam (*ojojo*) etc. (Amenaghawon et al. 2023; Aruna et al. 2017). Processing of yam tubers, especially manual processing resulted in excessive quantities of wastage which mainly contains brown, rough, and scaly skin (Ojolo et al. 2016). Moreover, processing steps involved in different food product preparations result in unavoidable peel generation followed by indiscriminate disposal in undesignated dumpsites (Nweke and Nwabanne 2021). To address these unsustainable and unreliable methods of yam peel disposal numerous studies have been reported on different bio-valorization processes to convert biomass into feedstock, biofuels, energy, platform chemicals, etc. for multidimensional use. Valorization of biomass not only addresses the challenges to the environment and human health but also generates income and employment opportunities among rural youth. In general bio-valorization processes, hybrid technologies are used for the conversion of yam peel waste into valuable bioproducts with minimal residue generation. Moreover, yam peels rich in lignocelluloses with inherent recalcitrant nature require extra steps/processes to breakdown complex links between lignin and hemicelluloses without deteriorating their basic chemical structure (Ginni et al. 2021). These extra steps/processes include chemical, biological, physical, biochemical, thermo chemical, etc. (Kumar et al. 2020). Yam peels are a storehouse of different active compounds such as diosgenin, flavonoids, polyphenols, saponins, etc. (Wu et al. 2011). In addition, yam peels contain fractions of nutrients like calcium, potassium, magnesium, protein, fat, and vitamins (Onu et al. 2022). Specifically, white yam tuber peel is described as a store house of good amounts of elemental carbon (45%), which can be effectively utilized for alternative adsorbents preparation for the removal of contaminants from aqueous solutions (Herrera-García et al. 2019; Wang et al. 2022). Hence yam peels becoming promising sources of materials for the development of various useful bioproducts and platform chemicals by protecting the environment as well as promoting sustainable development.

8.3.1.1 Bioethanol Production

Bioethanol production from abundantly available, renewable, and low-cost plant-based biomasses such as agriculture waste, wood chips, and grasses represent the second generation of biofuel technology (Waghmare et al. 2018; Ray 2020). On the other hand, the first generation of bioethanol primarily derives from food crops as feedstock (Ray and Ramachandran 2018) which increases the prices of food crops globally. Hence second generation bioethanol production from plant biomasses can be used as an economical alternative to first-generation bioethanol as well as to fossil fuels to run gasoline engines (Mohapatra et al. 2018). The demand for bioethanol is increasing globally as an alternative source of fossil fuels to achieve energy security, decrease the dependence on crude oil imports as well as reduce the rate of greenhouse gas emissions (Research and Markets 2020). According to 'Research and Markets' (2020), report bioethanol market size was US$ 43.2 billion in 2019 and is projected to reach US$ 64.8 billion by 2025. Especially starch-based feedstock in terms of volume and value to produce bioethanol is projected to grow at a faster rate at the highest CAGR from 2020 to 2025 (Research and Markets 2020). In this context, food and energy crops such as corn, cassava, yam, sorghum, etc. that are primary sources of starch and sugar are suitable crops for bioconversion of cellulosic biomass into fermentable sugars thereby bioethanol production using microorganisms (Ibeto et al. 2011). Hence, utilization of agro-waste from food and energy crops (cassava peels, yam peels, potato peels, corn cob waste, corn stalk waste, etc.) is gaining momentum in the production of bioethanol in recent times. In this context, a study on the characterization of yam peel for bioethanol production described yam peels as potential candidates for bioethanol production mainly due to the presence of higher holocellulose content and lower lignin content along with fractional amounts of nutrients (Bashir et al. 2021). Awoyale et al. (2021) characterized different Nigerian lignocellulosic biomasses (corn sobs, rice, husks, cassava peels, white yam peels, and sugarcane bagasse) for bioethanol production. The study demonstrated the presence of a substantial amount of cellulose content among the selected crops ranging from 39.8 to 25.8 wt.% which can be used as potential substrates for bioethanol and energy production (Awoyale et al. 2021). Banjo et al. (2019) compared bioethanol production between two yam species [bitter yam (*D. dumentorm*) and water yam (*D. alata*)] peels by a submerged fermentation process using strains of *Aspergillus* spp. and *Saccharomyces cerevisiae*. The results of the study revealed that bitter yam peels are the potential substrate for bioethanol production when compared to water yam peels. Similarly, a comparative study on the production of bioethanol from yam and banana peel indicated yam peels as potential candidates for higher bioethanol production (146.33 g/L) compared to banana peels (133.53 g/L) (Isah et al. 2019). In another study, Ezejiofor et al. (2018) reported the highest quantity of ethanol ($18.40 \pm 0.18\%$) and glucose ($38.7 \pm 0.90\%$) extraction from yam peels by employing acid hydrolysis, fermentation, and distillation when compared to other crop peels like potato, pineapple, and watermelon. On the contrary, Akponah and Akpomie (2011) reported higher bioethanol yield from cassava peels followed by potato while yam peels yielded the lowest bioethanol concentration. Villadiego-del Villar et al. (2021) reported bioethanol production with the best productivity of 0.19 g/L/h from 10% w/w yam peel sample concentration using simultaneous saccharification and fermentation method. Furthermore, Braide et al. (2018) reported yam peels in combination with cassava and potato peels as good combination substrates for the production of bioethanol using *Zymomonas mobilis* (8.36% v/v) and *S. cerevisiae* (6.64% v/v) at 72 h of the fermentation process. In conclusion, these research studies indicate that yam peels have the potential to serve as excellent and cheap sources of starch and sugars, which can be used as potential substrates for fermentation for bio-ethanol production especially in areas where they are in abundance.

8.3.1.2 Energy Generation

Agricultural wastes are potential sources of energy-producing materials; from which gasified or liquid fuel production is possible for the generation of heat energy. In this context, the anaerobic digestion (AD) of these agricultural wastes was found as an effective strategy for waste management in terms of eco-friendly decomposition and production of renewable energy to mitigate energy scarcity (Rahmat et al. 2019). In general, the AD process employs microbes for the decomposition of agricultural waste into byproducts for immediate use. Some of the major merits of the AD process are the transformation of agricultural waste into bioenergy, lower biomass generation, the potential to treat bigger loads of waste, and more importantly generation of organic manures after bioenergy production (Ziganshin et al. 2013). Nweke and Nwabanne (2021) studied different kinetic models to determine the feasibility of the production of biogas from yam peels; the results revealed that the first-order kinetic model explained the most acceptable biogas production (440 mL) when 400 g of shredded yam peels with 200 mL water fed into 1 L anaerobic digester. In another study co-digestion of different peels such as yam and plantain with cow dung was attended to observe the biogas production potential (Makinde and Odokuma 2015). Results from the study noted that yam peel treatment can produce higher biogas (428 mL) compared to plantain peel treatment, suggesting that waste generated from these crops should not limit to only animal feed and should be used for biogas production.

8.3.1.3 Bioremediation

The release of enormous quantities of industrial sewage polluted with heavy metals (lead, copper, cobalt, mercury, chromium, and arsenic) into water, land, and soil creates global environmental problems causing irreversible impacts on human, animal, and aquatic life. These toxic heavy metals are nondegradable and can accumulate in living bodies by contaminating the entire food chain (Roy Chowdhury et al. 2018). Hence remediation of these heavy metals is crucial to protect human and animal, plant, and soil health. Among different remediation methods selection of an appropriate method for the removal of heavy metals with a minimum effect on local ecosystems is important. As physical and chemical-based heavy metal remediation methods are proven to be very expensive and can lead to environmental deterioration (Roy Chowdhury et al. 2018), emerging green technology 'phytoremediation' can be a potential method. Phytoremediation technology uses green plants (living or dead) to remove the noxious pollutants in soil, water, and air cost-effectively and sustainably compared to conventional methods as mentioned above. More importantly, the biomass produced during the phytoremediation process can be economically valorized to bioenergy (Roy and Pandey 2020). In addition, agro-industrial waste can be treated as low-cost and sustainable 'green and clean' materials that can help in the removal of toxic pollutants. In this regard, research studies reported the effective utilization of agro-industrial wastes such as yam peels, banana peels, muskmelon peel, and *Cucumis melo* peel (Marimón-Bolívar et al. 2018; Vilardi et al. 2018) as effective biosorbents for toxic heavy metals. Marimón-Bolívar et al. (2018) reported excellent results in the adsorption of mercury heavy metal from an aqueous system using magnetic nanoparticles functionalized with yam peels-based biomass. In another study, Villabona-Ortíz et al. (2022) utilized yam peel and oil palm bagasse for single-component absorption of nickel (Ni) and chromium (Cr) in an aqueous system. In this study, a higher percentage (87%) removal of Cr with a maximum capacity of 11.45 mg/g and 50.12 mg/g was reported for oil pal bagasse and yam peel, while Ni values were recorded as 103.49 and 30.04 mg/g (Villabona-Ortíz et al. 2022). Wang et al. (2022) provide the details of alternative technology for the targeted exploitation of Chinese yam peel (*D. rhizoma*) for the removal of contaminants (Congo red dye) from water. The removal efficacy of Chinese yam peel-based adsorbent (CYP-A) on Congo red dye was found 86% at a contact time of 120 min with 100 mg/L polypyrrole-modified CYP-A. Later this removal efficacy was significantly increased to an acceptable 98.9%. In terms of thermodynamics, the adsorption process was described as endothermic for dye removal which was reported to have a maximum at 45 °C. In addition, this study provides the reusability capacity of polypyrrole-modified CYP-A up to three cycles of adsorption and desorption processes. This study concluded with the production of low-cost and versatile adsorbent materials from CYP for the removal of dyes. Similarly, Herrera-García et al. (2019) studied the adsorption capacity of yam (*D. rotundata*) peel-based activated carbon (YP-AC) and magnetically activated carbon adsorbent (YP-MAC) with Fe_3O_4 for the removal of 2,4-D from aqueous media. Thermodynamic studies showed efficient adsorption of 2,4-D by YP-AC and YP-MAC through an exothermic, entropic, and spontaneous process. Among YP-AC and YP-MAC, the latter showed greater results, with 60% of 2,4-D adsorption capacity up to five cycles, hence YP-MAC was described as a promising adsorbent of pesticides in aqueous solutions. In addition, Utsev and Onah (2017) investigated the chromium removal efficacy of yam peel-activated carbon from wastewater. From the results optimal carbonization process was found to be 200 °C for 40 min for efficient removal of chromium from wastewater.

8.3.1.4 Substrate for the Production of Enzymes and Organic Acids

Converting the readily available agricultural waste to bio-based formulations that support microbial cellular growth and maintenance is the best sustainable way for their utilization. In this context, the valorization of yam wastes as good substrates for the production of various useful enzymes and organic acids was studied. Yam waste is described as a storehouse of different enzymes (hydrolytic, proteolytic, amylase) which play an important role in pharmaceutical preparations and industries such as dairy, leather processing, textile, tanning, etc. (Okunwaye et al. 2021; Chaturvedi et al. 2019; Racheal et al. 2015; Godwin-Egein and Arinze 2000). Moreover, these enzymes find a wide application in food and wastewater treatment, metal leaching process, degumming of silk goods, bating hides in the biotechnology field, stain removal in detergent industries, etc. (Bhosale et al. 1995). In this context, Racheal et al. (2015) studied the extraction and purification of protease-isolated *Aspergillus niger* from yam peel waste. The results showed excellent enzyme activity (8.51 U/ mg) of purified protease isolated from yam peel than the crude samples of *A. niger*. The purified protease enzyme showed maximum activity on casein, which shows it can be effectively utilized in food, cosmetic, pharmaceutical, etc. industries. Moreover, this study revealed that yam peels can be an excellent source for the extraction of protease enzymes from *A. niger*. Similarly, the study conducted by Chaturvedi et al. (2019) showed yam peel waste can be an excellent substrate for oleaginous yeast production with high lipid content. Among different oleaginous yeast, *Rhodotorula mucilaginous* showed higher lipid content followed by *Saccharomyces pastorianus* on yam peel waste (25.48% and 19.57% dry biomass basis). Furthermore, higher enzyme activities of these oleaginous yeast such as *R. mucilaginous* with the highest FPase activity (1.63 IU/mL), *S. pastorianus* with higher amylase activity (0.63 IU/mL), *Trichosporon cutaneum* and *Lipomyces starkeyi* recorded showed higher xylanase activity (0.27 and 0.50 IU/mL), *Ganoderma wiiroense* showed higher inulinase activity (13.95 IU/mL) in yam peel waste as substrate. In addition, Okunwaye et al. (2021) identified amylolytic bacteria (*Corynebacterium* sp., *Bacillus* sp., *Pseudomonas* sp.) and fungi (*Penicillium* sp.) from yam tuber wastes with high α-amylase activity. Chaturvedi et al. (2019) and Godwin-Egein and Arinze (2000) extracted a few common hydrolytic enzymes in *Trichoderma harzianum* and *Fusarium oxysporam* using food wastes from yam, cassava, taro, sweet potato, plantain, and orange. Hydrolytic enzymes such as C1-cellulase, Cx-cellulase, and polygalacturonase with diverse activities were identified from different food wastes studied. Itelima et al. (2014) studied different food crop wastes for the development of formulations to cultivate soil fungi. In this study, food crop wastes such as yam, sweet potato, Irish potato, and carrot peels were utilized for the development of media formulations to support soil fungi growth. Similarly, Ugah and Nwoba (2018) reported the feasibility of yam peel, cassava peel, and rice husk for the development of formulations for soil fungi such as *Saccharomyces carlsbergensis* and *Penicillium* species. Findings from the study revealed that yam peel, cassava peel, and rice husk containing different nutrients could be effective alternative substrates for formulations of media for the cultivation of fungi.

8.3.1.5 Protein Enrichment as a Functional Food Additive

Proteinaceous food supply from conventional protein sources is inadequate and results in pollution from food and agricultural wastes. Such concerns have encouraged researchers to convert such food and crop waste into nutritious, safe, and valuable food or feed for humans and animals. Proper identification and management of waste are very important to ensure desirable environmental sanitation. One way to achieve this is by biotransformation of wastes into commercially valuable products like protein-enriched products. Efforts have been made to ensure that agricultural products, which are at times detrimental to man and animal health, are positively utilized by converting wastes from root tubers such as yam into value-added products (Iyayi and Losel 2001). Yam is an important food crop in developing countries like Africa, America, the Caribbean, South Pacific, and Asia. Among various types, white yam (*D. rotundata*) is the most cultivated yam and it is processed into a wide range of products in different parts of the world. The processing results in large amounts of yam peels as waste, and due to low protein content peels are either discarded or burnt causing environmental pollution. Usually, the price of conventional protein feed is high, thus increasing the cost of livestock production. Fermentation is an essential technology that is required for converting lignocellulosic materials such as yam peels into value-added products that have the potential to provide solutions to problems of animal nutrition such as inadequate intake of protein and calories. This will also reduce waste; pollution as well as increases the quality of animal by-products. According to Anupama and Ravindra (2000), fermented products contain proteins that can feed both humans and animals thereby replacing expensive conventional sources of protein like fishmeal and soy meal. Aruna et al. (2017) compared unfermented, non-supplemented, and supplemented *S. cerevisiae* (BY4743) fermented yam peels. Supplemented *S. cerevisiae* (BY4743) fermented yam peels had the highest composition for all the proximate parameters followed by non-supplemented biomass while unfermented yam peels had the lowest. An increase in protein, crude fat, and fiber contents was observed after fermentation. *S. cerevisiae*, enhanced the protein content of yam peels. Yafetto et al. (2020) demonstrated that protein contents of yam substrates can be improved with mono- and co-cultures of *A. niger* and *Trichoderma viride*. Similarly, Meng et al. (2019) investigated the Chinese yam peel (*D. opposite* Thunb) extract as a potential food additive for improving the immune responses of common carp (*Cyprinus carpio* L.) through modulations of intestinal microflora. The study revealed the positive impact of Chinese yam peel on common carp by inhibiting the proliferation of intestinal pathogens, maintaining homeostasis of gut microflora, and secretion of short-chain fatty acids and cytokines (Meng et al. 2019). Furthermore, Meng et al. (2019) suggest that Chinese yam peel could be used as an immunopotentiator in aquatic food products. Uchewa et al. (2014) evaluated the performance of rabbits fed with yam peel meal (YPM) as a replacement for maize meal at varying levels for weight gain, carcass characteristics, feed conversion ratio, and cost economics of production. The results of the study revealed significant differences for feed intake by rabbit; however non-significant differences were recorded for feed conversion ratio and body weight among treatments. Moreover, the replacement of maize meal with YPM showed a significant reduction in the cost of feed to body weight gain ratio of rabbits. On the other hand, Uchegbu et al. (2010) designed the finisher broiler feed using yam peels collected from different hotels and restaurants surrounding the Federal University of Technology, Owerri, Nigeria. Collected fresh peels of yam were sundried for 6–7 days and finely ground to peel meal. Finisher broiler fed with diets formulated using 0%, 10%, and 20% of yam peel meal and analyzed for hematological and serum biochemical tests. Results showed that there was an increase in serum creatinine, serum urea, and serum cholesterol in finisher broilers, therefore study suggested that proper attention is needed for recommending the yam peel meal for broilers.

8.3.2 Aroids (Taro, Elephant Foot Yam, Tannia) Waste into Bioproducts and Platform Chemicals

8.3.2.1 Taro

Taro is the fifth most important root crop harvested in the world with an estimated 11.80 million tons per annum (Liu et al. 2006). Taro tubers are the staple food in tropical and subtropical countries and gained importance as one of the major food sources due to their good sources of carbohydrates, resistant starch, mucilage, and energy. Moreover, taro starch has been a suitable food ingredient for the development of various food products mainly due to its gluten-free nature, smaller starch granule size, and higher digestibility (Liu et al. 2006). In Asia, taro has been playing an important role in food product diversification through the development of various popular extruded, weaning, fried, bakery, and confectionary products (Kreike et al. 2004). As a result of different processing steps large quantities of taro waste (corm residue and peels) are generated and usually disposed of traditionally. Chintagunta et al. (2016) observed the presence of good amounts of starch content (61.50% dw) in taro waste compared to potato (52.14% dw) and cassava peel (7.10% dw). In addition, higher volatile solids (VS) to total solids ratio (87.53%) in taro straw indicates the presence of high-quality organic matter suitable for anaerobic digestion thereby the production of biofuel (Liu et al. 2023). Lee and Jun (2011) investigated the enzyme activities associated with starch hydrolysis and the increase in glucose release from taro waste using ohmic heating and microwave pretreatment methods. Results of the study showed that ohmic heating contributed to an increase in sugar content through increased starch hydrolysis from taro waste. Also, added to the enhancement of enzyme activities associated with starch hydrolysis within a short period of incubation. Lee and Jun (2011) concluded that developed ohmic heating could be used with other feedstock as a pretreatment technology for the production of biofuels. Therefore, taro waste has vast potential to create avenues for the development of feedstock for various value-added products such as food additives, bioethanol, glucose, fertilizer, animal feed, platform chemicals, and media for microbial cultivation (Hsieh et al. 2016).

8.3.2.2 Biofuels (Bioethanol and Biogas)

Several research studies reported bioethanol production from taro tubers, however very few studies were carried out using taro root peel waste. Hence this section intends to overview the available literature on bioethanol production from taro root waste. Rosalina and Permana (2022) reported bioethanol production employing two-step of processes, first hydrolysis step of taro peel polysaccharides to glucose followed by a fermentation step using *S. cerevisiae* and tape yeast to convert glucose to ethanol. The results of bioethanol production using *S. cerevisiae* were noted as promising gas ethanol of 41.14% and glycerol of 29.08%. However, fermentation with tape yeast resulted in glycerol of 31.39%, cyclobutanol of 15.26%, pentane (12.93%), and methyl 2-oxo propanoate of 10.50% production. Wu et al. (2016) evaluated and optimized taro waste using simultaneous saccharification and fermentation (SSF) to produce bioethanol. Results of the study revealed optimal conditions for bioethanol production as SSF using 5% *Kluyveromyces marxianus* K21 inoculum at 40 °C after 22 h of cultivation which yielded maximal ethanol production of 48.98 g/L and productivity of 2.23 g/L/h. Moreover, a scaling-up study with a 5 L digester using the same K21 inoculum after 20 h of cultivation showed promising results (ethanol production of 43.78 g/L with productivity of 2.19 g/L/h) (Wu et al. 2016). On the other hand biogas production from taro peel waste was studied by Adeyosoye et al. (2010). Results of the study showed higher biogas production (72.0 mL) using taro peels compared to inoculum (30.0 mL) after 24 h of fermentation. Similarly, Ge et al. (2014) investigated biogas energy production using taro skin and cooked taro flesh with different combinations of papaya waste and sweet potato waste. The feedstock combinations for co-digestion on a fresh weight basis include taro skin and taro flesh (1:1), taro flesh and papaya (1:1), sweet potato and papaya (1:1), and taro skin, taro flesh, papaya, and sweet potato (1:1:1:1). In this study, tropical biowaste based feedstock were compared with woody plants' biowaste for the CH_4 yield. As concluded from the study, tropical biowaste showed higher CH_4 yield ranges between 345 and 411 L/kg VS compared to leaves and woody biowaste (161 and 113 L/kg VS), furthermore, the robustness of liquid anaerobic digestion was demonstrated (Ge et al. 2014). In addition, Liu et al. (2023) reported rapid production of methane from taro straw due to the presence of higher total soluble sugars compared to banana pseudostem. However, methane production steadily increased in a later stage with banana pseudostem treatment due to higher cellulose content. Rana et al. (2017) investigated the effect of metallic magnesium (Mg^0) catalysts in biofuel generation using different agricultural, kitchen, and aquatic waste materials-based feedstock. Among different feedstock, taro stems and leave slurry in the presence of artificial light without metallic Mg^0 catalyst produced a total volume of product gas of 6870 mL with the highest flame temperature of 740 °C and individual gas composition including H_2 of 12.27 mL, CH_4 of 4700.99 mL, and CO_2 of 2056.70 mL respectively. With 45 g Mg^0 metallic catalyst and artificial light showed total volume of product gas was 8598 mL with the highest flame temperature of 780 °C and individual gas composition include H_2 of 1290.91 mL, CH_4 of 6051.86 mL, and CO_2 of 1255.00 mL. Rana et al. (2017) concluded that the presence of artificial light and Mg^0 metallic catalyst helped in the higher production of calorific fuel gas via the activation of photosynthetic bacteria.

8.3.2.3 Compost and Vermicompost

With the help of microorganisms, waste (organic matter) can be converted to wealth (waste-free soil-like product) using biochemical processes which aid the increase in decomposition rate in its native state. Sembera et al. (2019) investigated the composting potential of wild taro waste as an environmentally friendly and potentially profitable means of disposing of wild taro. In addition, the pH of compost (between 8.3 and 8.4) prepared from the wild taro waste was found within the ideal pH range (between 5 and 8.5) described by US Composting Council (2002). Moreover, bioassay tests for seedling vigor (>95%) and emergence (>90%) noted 'very manure' and respirometry readings (1.9 mg CO_2/g organic matter/day) observed as 'stable' as per US Composting Council (2002). Furthermore, taro waste can be a candidate ingredient for specific quality vermicompost production. Hassoub et al. (2022) studied the effect of taro leaf, fish sludge, sawdust, and sugar beet on the physicochemical properties of vermicompost. During the vermicomposting, the addition of taro leaf increased amino acids and microbial activities with cow dung and fish sludge. Results from the study concluded great potential for taro leaf to produce good quality vermicompost.

8.3.2.4 Biopolymer

Renewable biodegradable alternatives are much needed to replace petroleum-based non-biodegradable plastic packing materials and to protect the environment from growing plastic pollution. Many research studies have reported the utilization of plant biomasses for the development of biodegradable bioplastics. Bidari et al. (2023) reported the novelty of taro peel starch (TPS) usage in the production of biodegradable films. The effect of TPS in the range of 2.5–3.5% w/w and glycerol as 25–35% w/w was investigated on the physicomechanical and water barrier properties of TPS-based biofilm. Overall, TPS-based biofilm exhibited transparent, homogenous, amorphous crystalline nature, and flexible surface. Moreover, the increased percentage of TPS, plasticized with glycerol showed a positive linear effect on thickness, opacity, and water vapor permeability with moisture content of biofilm developed. However, TPS-based film was found non-durable compared to other organic biomass-based biodegradable films.

8.3.2.5 Animal Feed

The cost of feed takes a significant portion (44.55–55%) in livestock business investment due to the high price of main feed ingredients such as soybeans and corn, also utilization of these ingredients competes with human food (Kueain et al. 2017). Therefore, efficient utilization of low-cost locally available agricultural waste with considerable amounts of nutrients plays a potential role in alternative animal feed without comprising the performance of livestock production. In this context, the utilization of taro waste has been reported in several studies. Rodríguez et al. (2016) studied physicochemical properties and the presence of biological indicators in taro-based silage for animal feed. Different silage formulations include T1: Chopped damaged taro tubers, yogurt, and water; T2: Chopped damaged taro tubers, and whey; T3: Chopped damaged taro tubers, 5% molasses B, and whey; T4: Chopped damaged taro tubers, 10% molasses B, and whey. Results of the study revealed that rapid stabilization of pH (3.8–4.2) among four silage formulations showed its stability as well as restricting the growth of proteolytic enzymes such as clostridium and enterobacteria. In addition, pH stabilization was well supported by the presence of a good amount of available carbohydrates in taro tuber waste. Therefore, even after 180 days of preservation, all four silage formulations showed excellent physicochemical parameters with good microbial indicators, which showcase the potential of taro waste as a candidate for the development of animal feed. In another study, Caisedo et al. (2020) determined the biochemical composition of taro tuber waste and foliage for pig rearing. Buntha et al. (2008) investigated the effect of ensiled taro leaf at a rate of 25%, 50%, 75%, and 100% replacement with fish meal on digestibility and protein content of silage. Positive results were observed for dry matter content, digestibility, and N retention in 70–75% ensiled taro leaf replacing a fish meal. In addition, Khayyal et al. (2017) evaluated the effect of dried taro waste (TW) with or without yeast on rabbit growth performances, blood parameters, nutrient digestibility, cecum activity, carcass traits, and economic efficiency. The experiments were divided into two batches, in the first batch rabbits were fed with 0%, 7%, 15%, and 22.50% dried TW without yeast, in the second batch rabbits were fed with the same feed of dried TW with 0.5% yeast along with a control diet for 8-weeks post-weaning period. No significant difference was observed between the two diets for different parameters analyzed in the study. However economic efficacy and profitability were remarkably noted for the dried TW diet with or without yeast compared to the control diet. The results concluded that dried TW could be used for rabbit rearing up to 22.50% without any adverse effects on productive performance and physiological performance. Giang et al. (2010) observed taro leave and petiole-based silage can replace 60% of rice bran in feed for raising ducks without compromising on growth performances. These findings were supported by Caicedo et al. (2019) where the chemical composition of taro waste was reported to be a good feedstock for animal feed development.

8.3.2.6 Media for Beneficial Organisms

Reutilization of taro waste for the development of growth media for beneficial bacteria assures the effective utilization of waste as well as the cost-effectiveness of the products developed. In this context, Hsieh et al. (2016) optimized *Lactobacillus acidophilus* growth and biological activity using taro peel waste, the study was carried out with taro waste and different N sources for media preparation. The results demonstrated that taro peel can be used as a main carbon source for growth media which can support the production of non-probiotics with enhanced biological activities. Chen et al. (2022) studied the feasibility of taro peel waste (TPW) utilization on fungal *Mortierella alpine* for the production of arachidonic acid (ARA). ARA is described as essential polyunsaturated fatty acids which act as a precursor for prostaglandins, leukotrienes, and other important substances in the animal and human body. Therefore, acidic and enzymatic hydrolysates of TPW were utilized as *M. alpine* culture substrates to observe the production of lipid content and ARA. The results concluded that both TPW-based hydrolysates yielded similar ARA (34.58% and 33.05%) during fermentation, whereas higher lipid content (30.68%) and biomass concentration (16.79%) were observed in enzymatic hydrolysate fermentation after 5 days. Hence enzymatic hydrolysate was selected for scale-up bioreactor experiments, among batch and fed-batch, the latter one yielded higher lipid content of 36.97% and ARA content of 46.04%. Overall, this experiment demonstrated the practicability of TPW hydrolysates as a low-cost alternative substrate for *M. alpine* cultivation.

8.3.3 Elephant Foot Yam (EFY)

EFY is one of the nutritious crops native to South Asia and a fine source of starch, sugars, proteins, minerals, and riboflavin. The normal nutritive profile comprises starch (11–28%), sugar (0.7–1.7%), protein (0.8–2.60%), fat (0.07–0.40%) and means energy value of 236–566.70 kJ/100 g (Singh et al. 2013; Ray 2015). However, due to the presence of about 0.78–6.24% calcium oxalate and an anti-nutritional factor that causes acridity, the use of EFY in high-value-added products is limited (Ray 2015). The peel of EFY accounts for nearly 20–25% of the corm weight and 1–4 cm in thickness is used as a supplement in animal feed or waste disposed into the environment causing biological litter (Ravi et al. 2011). Recently, a significant amount of research has been carried out to increase the industrial utilization of EFY and its peel into value-added products which are discussed below.

8.3.3.1 EFY Waste Utilization as Animal Feed

Fermentation of EFY peel using *Rhizopus oligosporus* has been reported to reduce the oxalate content and increase the nutrient content making it suitable for broiler feed up to 5%. Reduction in the oxalate content was reported to increase calcium absorption and higher growth rate (Koni et al. 2015). Similarly, the suitability of EFY peel as animal feed was also reported when fermented using *Bacillus subtilis* which reduces the oxalate content from feed stuff, as this microorganism can produce the oxalate decarboxylase enzyme that breaks down oxalate salts. When incubated for different lengths (7, 14, and 21 days), fermented EFY peel starch showed an increase in nutritional value and a decrease in oxalate content. Fermentation with *B. subtilis* resulted in a maximum decrease in the oxalate content of EFY (by 53.2%) and an increase in the fat content making the EFY starch suitable for poultry feed (Koni et al. 2017).

8.3.3.2 EFY Peel Utilization to Extract Phytochemicals and Antioxidants

EFY peels were used to extract phytochemicals for their potential utilization in therapeutics. Extraction of phytochemicals from the peel of EFY was carried out by dissolving dried peel powder in solvents like water, methanol, petroleum ether, and chloroform and investigated for the presence of alkaloids, tannins, phenols, carbohydrates, and fat in the peel. Total phytochemical content was reported to be 28.8% in methanolic extract followed by chloroform, aqueous, and petroleum extract which were 25%, 18.74%, and 11.86%, respectively. Based on the phytochemical constitution of the various extracts it was observed that the peel contains a substantial amount of tannins and phenols, alkaloids, and carbohydrates (Singh et al. 2013).

8.3.3.3 Utilization of EFY to Produce Biofuel

The industrial utilization of EFY results in a large amount of waste (mostly from its thick peel) which contains a substantial amount of carbohydrates. Unlike other lignocellulosic materials, that require pentose fermentation and pretreatment methods, Konjac waste results in all hexose sugars after hydrolysis that are relatively easy to utilize.Recently, the potential to use EFY waste as a cheap raw material to extract bio-ethanol has been reported by Anbuselvi and Kumar (2016). In Japan and China, Konjac waste is a byproduct of the Konjac Glucomannan (KGM) processing industry (Behera and Ray 2016) and with growing demand for KGM worldwide, waste has increased to approximately 2000 tons per year in China alone. Recently, Shao and Chen (2015) investigated the feasibility of using konjac waste as a source of acetone–butanol–ethanol (ABE) fermentation by *Clostridium acetobutylicum* ATCC 824 via two distinct hydrolysis and fermentation combination strategies namely simultaneous saccharification and fermentation (SSF) and separate hydrolysis and fermentation (SHF). The obtained ABE concentration from SHF was 10.95 g/L with 7.10 g/L of butanol at a productivity rate of 0.30 g/L/h, which was higher than the SSF (6.64 g/L) and that from the control experiment (4.34 g/L). The study also confirmed that the presence of bioactive alkaloids affected the ABE fermentation using *C. acetobutylicum*. Higher production of butanol with reduced fermentation time was expected with effective extraction of alkaloids before fermentation from the Konjac waste. A similar impact in the ethanol production from elephant foot yam peel waste due to the presence of phenolic compounds was reported by Mithra and Padmaja (2016). When steam and dilute sulphuric acid pretreated EFY peel substrates were utilized with detoxification chemical mix, both fed-batch separate hydrolysis and fermentation (F-SHF) and simultaneous saccharification and fermentation (F-SSF) using *S. cerevisiae* resulted in higher ethanol production at lower enzyme dosage. The study confirmed that the use of Tween 20 in combination with polyethylene glycol (PEG 4000) could remove 70–82% of phenolic compounds thereby increasing the ethanol production from the stem and dilute sulphuric acid pretreated EFY peel substrates (Mithra et al. 2018a). In a similar study, ethanol production from F-SHF and F-SSF was compared where in F-SHF, enzyme cocktail (Ecozyme RT80 + Stargen + Ecozyme XY50) was applied only one time compared to multiple time dosage in F-SSF along with detoxification mix comprising Tween 20, polyethylene glycol and sodium borohydride. Results concluded that fed-batch saccharification of ST and SSF enhanced the reducing sugar yield, ethanol productivity, and ethanol yield during the first 24 h fermentation under F-SHF mode, while continuous production and utilization of RS occurred up to 72 h in F-SSF. Again, dilute sulphuric acid pretreated EFY residues under F-SHF gave higher ethanol yield (35 g/L) and productivity (282 mL/kg dry biomass) than stem (29 g/L and 231 mL/kg respectively), while F-SSF resulted in higher ethanol yield for stem EFY residues 302 mL/kg than dilute sulphuric acid treated EFY 294 mL/kg. Compared to F-SHF, a higher accumulation of inhibitors in F-SSF was estimated despite the use of a detoxification mix. F-SHF of dilute sulphuric acid pretreated biomass concluded as the advantageous process to produce ethanol with only one-time feeding of enzyme cocktail compared to multiple feeds of enzymes along with detoxification mix or nutrient solution in F-SSF (Mithra et al. 2018b). Direct conversion of industrial konjac waste residues as a feasible feedstock to produce biobutanol was reported by Ujor et al. (2014). The study revealed the direct conversion of konjac waste to ABE production using batch fermentation with *Clostridium beijerinckii* NCIMB 8052 without the addition of exogenous enzymes yielded 15.1 g/L of ABE. Similarly, Shao and Chen (2015) evaluated *Amorphophallus* konjac waste as a suitable substrate for biobutanol production. Under separate hydrolysis and fermentation processes, 7.1 g/L of biobutanol was produced with a productivity of 0.3 g/L/h. Similarly, the utilization of Konjac flying powder (Tobiko) generated in the manufacturing process of Konjac-derived products to produce bioethanol, a form of biomass energy was reported by Kazuki et al. (2017), from Konjac flying powder to establish an effective application of this waste product. First, polysaccharides contained in Konjac flying powder were decomposed to monosaccharides (e.g., glucose) by sulfuric acid, hydrochloric acid, or nitric acid. Second, ethanol was generated from the monosaccharides by fermentation and refined by single distillation. Components analysis was performed with high-performance liquid chromatography and gas chromatography at each step: saccharification, fermentation, and distillation. The results revealed that ethanol (9.1 g/L, 600 mL, before distillation) could be successfully produced from Konjac flying powder (30 g) similar to other food industry waste products. In conclusion, we have demonstrated bioethanol production as a new manner of utilization of Konjac flying powder, in addition to its use as fertilizer and livestock feed.

8.3.3.4 Utilization of EFY Waste in Supercapacitor

To produce flexible electrochemical double-layer capacitors porous carbon materials as the electrode is highly recommendable due to their high chemical stability and large surface area but they lack higher specific capacitance and energy density. To solve such an issue nitrogen-doped porous carbon materials using polyacrylonitrile has been synthesized, but such materials lead to higher production cost. To reduce the cost of resources, Bai et al. (2020) synthesized nitrogen-doped porous carbon material via a simple thermally induced phase separation method with subsequent carbonization and activation of biomass konjac/polyacrylonitrile composites as the raw materials. As synthesized, composite nitrogen doped-carbon material with a hierarchical porosity and large specific surface areas showed a high specific capacitance of 390 F/g and 70% capacitance retention at 20 A/g in a three-electrode system. Also, the excellent capacitance decay of fewer than 10,000 cycles at 5 A/g was only 4.5%, which means excellent stability. When utilized as composite porous carbon material on a flexible symmetrical supercapacitor a maximum energy density of 9.0 W h/kg with a power density of 250.2 W/kg was obtained. The study concludes that Konjac waste can be used as a cheap carbon source for the production of nitrogen-doped carbon materials applicable to flexible energy storage devices.

8.3.3.5 EFY Waste Utilization in Wastewater Treatment

Removal of heavy metals from wastewater before being discharged into the environment is very crucial for human and aquatic animal safety. Use of agro-industrial waste-based activated carbon for removal of heavy metals via absorption has become one of the best promising solutions to reduce wastewater problems mostly from the textile industry. Recently, Lestari et al. (2018) reported the production of cost-effective adsorbent material from *A. paeoniifollus* skin via acid treatment and utilized it to remove calcium and magnesium ions present in synthetic hard water.

Compared to dried raw EFY skin with an adsorption capacity of 10.85 mg/g Ca and 1.79 mg/g Mg in 100 ppm solution, acid-modified EFY skin adsorbents resulted in adsorption of 27.64 mg/g Ca and 20.1 mg/g Mg. The study confirms the potential of using modified *A. paeoniifolius* skin as a low-cost adsorbent for the removal of metal ions mostly present in groundwater. In another consecutive study, the adsorption ability of raw starch, porous starch, and citric acid-modified porous starch toward Cd and Ni ions in an aqueous solution was investigated (Lestari and Dewi 2018). The study showed that the acid modification caused starch to become more porous and the absorption behavior is dependent on the initial concentration of Ni and Cd ions. Adsorption capacity Cd and Ni on of raw starch, porous starch, and citric acid-modified porous starch in 500 ppm aqueous solution was 68 mg/g and 25 mg/g, 199 mg/g and 61.88 mg/g, 256 mg/g and 87.45 mg/g confirming citric acid-modified porous starch to be the promising adsorbent than the raw starch and porous starch.

8.3.3.6 Use of EFY in Syrup

EFY waste peels have proven to be a high source of phytochemicals such as tannins, flavonoids, terols, polyphenols, and coumarins known as powerful natural antioxidants (Nataraj et al. 2012). Putri et al. (2018) utilized EFY ethanol extract in syrup to provide antioxidant health benefits flavonoids. The amount of flavonoid in EFY was reported to be 46.33 mg/g. EFY extract was added at three different concentrations (0.6%, 2%, and 4% extract). All syrup formulas showed significant antioxidant activity and the values were higher with higher levels of EFY extract (91.86 ppm for 0.6%, 83.05 ppm for 2%, and 71.39 ppm for 4% extract). However, based on the sensory properties test, syrup with an extract concentration of 0.6% was reported to be most preferred. EFY has the potential to be developed as a healthy local beverage for its qualified supplementation of antioxidants (Putri et al. 2018).

8.3.3.7 EFY Waste Utilization for Enzyme Production

Waste peels of EFY were utilized as substrates in solid-state fermentation to produce alkaline xylanase, an enzyme that contributes to the breakdown of xylan, and is used in an enormous range of biotechnological applications. *Aspergillus terreus* Thom isolated from elephant dung was used in the solid-state fermentation process. When compared to other sun-dried agricultural substrates such as apple peel, banana peel, orange peel, pea pod, citrus peel, corn husk, EFY peel, yam peel, wheat bran, sugar cane bagasse, EFY displayed the highest 83.10 U/mL activity followed by yam peel 64.85 U/mL, and wheat bran 53.24 U/mL; while other solid wastes showed moderate to low xylanase activity (<20 U/mL). EFY peel gave the maximum production of xylanase when fermented at 60 °C maintaining pH 8.0 for 96 h (Nenava et al. 2022). The use of EFY as a low-cost substrate in the production of xylanase media would be beneficial to the pulp and paper industry, recycling by deinking in the paper industry, bioethanol production, bakery, fruit juice clarification, and many more.

8.3.4 Tannia

Tannia is a tropical root crop grown widely in West Africa, tropical America, and Asia. Tannia is rich in carbohydrates, digestible crude protein, vitamins, and minerals such as Ca, Mg, and P (Opara 2000). The processed cocoyams (non-edible) waste accounts for a part of the entire cocoyam production resulting in the generation of vast quantities of waste (Ugwuanyi 2008). The peel waste includes skin, a thin outer cortex, and a small proportion of tuber constituting about 10–13% of the tuber with high carbohydrate content (Yahaya et al. 2013). Yahaya et al. (2013) investigated the nutritive attributes of tannia peel and the effect of the sampling period on the production of animal feed. The study reveals the presence of higher macronutrients (Ca, Na, P, K) and micronutrients (Fe, Zn, Cu, Mg) in the dry season compared to wet season sampling. Moreover, protein, fiber, carbohydrates, and lipid content ranged between 6.30–17.60%, 10.7–19.70%, 41.20–46.00%, and 0.70–2.14% respectively. Interestingly, lower concentration of antinutritional factors was observed in dry season sampling such as phytate, cyanide, soluble oxalates, and tannin content ranged between 1.26–1.43%, 3.17–3.20%, 1.18–1.69%, and 1.43–8.24%, respectively. Moreover, tannia peels are rich in lignocellulosic biomass with higher cellulose composition (48%), which is well suited for bio valorization into bioproducts and platform chemicals.

8.3.4.1 Bioethanol Production

Although many reports are available on Taro bio-ethanol production, the work available on Tannia is scarce. A preliminary analysis of bio-ethanol production using Tannia as feedstock with *S. cerevisiae* showed satisfactory yield. The simulation-based studies showed cocoyam can serve as a potential alternative for ethanol production with better economic performance and low environmental impact (Serna-Loaiza et al. 2018a, b). Cocoyam peels on fermentation using co-culture of *A. niger* and *S. cerevisiae* showed a maximum ethanol yield of 7.15% v/v with a pH value of 7.3 for 4 days (Nnaji et al. 2021). In addition, Maxwell et al. (2021) carried out a feasibility study on tannia peel via acid and enzyme hydrolysis for the production of fermentable sugars. HCl and H_2SO_4 acidic hydrolysis of tannia peels resulted in maximum glucose content of 27.30% and 26.70% at 70 °C for 180 min on the other hand, enzymatic hydrolysis yielded 44% maximum glucose at 30 °C and pH 5 in 3 days. The results of this study revealed the suitability of tannia peel for the production of fermented sugars that can be further utilized for biofuel production.

8.3.4.2 Protein Enrichment of Peel

Cormels of cocoyam are rich sources of starch but a large quantity of carbohydrate-rich solid waste is generated during cormel processing causing a waste disposal problem. Duru and Uma (2003) reported that the fermentation of cocoyam cormel solid process using *A. oryzae* produced increased quantities (over 50%) of crude proteins thereby effective conversion of solid waste into a proteinaceous product. The protein-enriched waste could be used as animal feed.

Table 8.1 summarizes the bioproducts and platform chemicals produced from yams and steroid wastes.

Table 8.1 Yams and aroids waste bio valorization into bioproducts and platform chemicals

S. No.	Crop	Waste bio valorization into bioproducts and platform chemicals	References
1	Yams	Biofuels (bioethanol and biogas)	Banjo et al. (2019), Isah et al. (2019), Ezejiofor et al. (2018), Villadiego-del Villar et al. (2021)
		Adsorbents for bioremediation	Utsev and Onah (2017), Marimón-Bolívar et al. (2018), Vilardi et al. (2018), Herrera-García et al. (2019), Villabona-Ortíz et al. (2022), Wang et al. (2022)
		Bioenergy	Nweke and Nwabanne (2021), Makinde and Odokuma (2015)
		Protein enrichment as a functional food additive	Uchewa et al. (2014), Uchegbu et al. (2010), Aruna et al. (2017), Meng et al. (2019), Yafetto et al. (2020)
		Substrate for the production of enzymes and organic acids	Racheal et al. (2015), Ugah and Nwoba (2018), Chaturvedi et al. (2019), Okunwaye et al. (2021)
2	Taro	Biofuels (bioethanol and biogas)	Adeyosoye et al. (2010), Ge et al. (2014), Wu et al. (2016), Rana et al. (2017), Rosalina and Permana (2022), Liu et al. (2023)
		Compost and vermicompost	Sembera et al. (2019), Hassoub et al. (2022)
		Biopolymer	Bidari et al. (2023)
		Animal feed	Buntha et al. (2008), Giang et al. (2010), Rodríguez et al. (2016), Caicedo et al. (2019), Caisedo et al. (2020)
		Media for beneficial organisms	Hsieh et al. (2016), Chen et al. (2022)
3	Elephant foot yam	Animal feed	Koni et al. (2015, 2017)
		Extract phytochemicals and antioxidants	Singh et al. (2013)
		Bio-fuel	Anbuselvi and Kumar (2016), Behera and Ray (2016), Shao and Chen (2015), Mithra and Padmaja (2016), Mithra et al. (2018a, b), Ujor et al. (2014), Kazuki et al. (2017)
		Supercapacitor	Bai et al. (2020)
		Wastewater treatment	Lestari et al. (2018), Lestari and Dewi (2018)
		Syrup preparation	Putri et al. (2018)
		Enzyme production	Nenava et al. (2022)
4	Tannia	Bioethanol	Serna-Loaiza et al. (2018a, b), Maxwell et al. (2021)
		Protein enrichment of peel	Duru and Uma (2003)

8.4 Future Prospectus

Tuber crops are an excellent source of starch and serve as a major source of industrial raw materials in tropical and subtropical countries. The presented chapter summarizes (Table 8.1) that the tuber crop waste can be utilized to generate bio-beneficial products like biofuel, animal feed, vermicompost, bioinoculants, adsorbents, etc. The starch-rich tubers, tuber wastes, and tuber industrial wastes can be processed for producing promising value-added components including biofertilizers, bioethanol, methane, poultry feed, etc. (Nedunchezhiyan et al. 2011). However, there are still many challenges that need to be addressed to exploit tuber crops' potential completely. Increased starch content tuber crop varieties for industrial purposes with reduced conventional traits should be developed using breeding and molecular methods are crucial. The animal feed and bio-ethanol industry has an increasing demand for alternative raw materials besides plants and fossil fuels. The use of agro-industrial by-products is a viable option to reduce solid-waste generation, add value to the agricultural by-products, and an alternative to dependence on depleting conventional raw materials. Waste utilization for bio valorization focuses on sustainable processes and reducing environmental impact. Bioconversion of low-nutrient industrial tuber wastes into protein-enriched products reduces environmental threats of waste disposal and is used as animal feed (Pandey and Soccol 2000; Soccol and Vandenberghe 2003). In view of ethanol fermentation, microorganisms are generally involved in producing enzymes that aid conversion into simple sugars and convert fermentable substrates into ethanol (Skoog and Hahn-Hägerdal 1988). Microorganisms and associated enzymes could play a chief role in the hydrolysis of starch and bioethanol production. However, to meet the global requirement, bioethanol production, and waste conversion level needs to be elevated. Biotechnological approaches should be utilized to develop efficient enzyme systems, and microbial strains in better utilizing the raw materials, sustaining fermentation conditions, and maximizing yield with reduced costs. Nanotechnological and biotechnological approaches can be utilized to valorize tuber crop wastes to generate bioactive compounds and increase their applicability. Adopting integrated processing methods in tuber crop waste could result in more bioactive products and thereby achieving zero waste without any impact on the environment.

8.5 Conclusion

In conclusion, the utilization of waste from horticultural tuber crops, such as yam and aroid waste presents a promising but underexplored area for resource utilization. While there is potential for transforming these waste materials into valuable resources, significant challenges, and limitations must be addressed to ensure their practicality and effectiveness. The chapter highlighted the diverse forms of waste generated from horticultural tuber crops and the associated environmental concerns. It acknowledged the potential benefits of waste utilization, such as bioenergy, biopolymer production, extraction of valuable functional compounds, etc. However, it also emphasized the need for careful consideration of the economic viability, technological feasibility, and scalability of waste utilization methods. One of the critical challenges in utilizing horticultural tuber crop waste is the variability in its quality, and waste conversion processes, such as enzymatic digestion. Addressing these variations requires advanced technologies and tailored approaches to ensure consistent and efficient waste utilization. In conclusion, while the utilization of waste from horticultural tuber crops holds potential, careful consideration of the variability, economic viability, and scalability of waste utilization methods is crucial. Only through comprehensive and holistic approaches can the utilization of horticultural tuber crop waste become a viable and effective strategy for resource optimization and environmental sustainability.

References

Adeyosoye OI, Adesokan IA, Afolabi KD, Ekeocha AH (2010) Estimation of proximate composition and biogas production from in vitro gas fermentation of sweet potato (*Ipomoea batatas*) and wild cocoyam (*Colocasia esculenta*) peels. Afr J Environ Sci Technol 4(6):388–391

Akponah E, Akpomie OO (2011) Analysis of the suitability of yam, potato and cassava root peels for bioethanol production using *Saccharomyces cerevisiae*. Int Res J Microbiol 2(10):393–398. http://www.interestjournals.org/IRJM

Amenaghawon AN, Omoruyi BO, Kenneth I, Okedi MO, Esenogho GO, Oyefolu PK, Muojama OE, Otuya IC, Eshiemogie SO, Okoh R, Anyalewechi CL (2023) Biotechnological conversion of yam peels for enhanced citric acid production: data-driven machine learning modeling and global sensitivity analysis of the impact of metabolic stimulants. Ind Crop Prod 191:116022

Anbuselvi S, Kumar J (2016) Production of ethanol from elephant yam and its optimization. Pharm Lett 8(8):55–57

Anupama RP, Ravindra P (2000) Value-added food: single cell protein. Biotechnol Adv 18(6): 459–479

Aruna TE, Aworh OC, Raji AO, Olagunju AI (2017) Protein enrichment of yam peels by fermentation with *Saccharomyces cerevisiae* (BY4743). Ann Agric Sci 62(1):33–37

Awoyale AA, Lokhat D, Eloka-Eboka AC (2021) Experimental characterization of selected Nigerian lignocellulosic biomasses in bioethanol production. Int J Ambient Energy 42(12): 1343–1351

Bai Q, Li H, Zhang L, Li C, Shen Y, Uyama H (2020) Flexible solid-state supercapacitors derived from biomass konjac/polyacrylonitrile-based nitrogen-doped porous carbon. ACS Appl Mater Interfaces 12(50):55913–55925. https://doi.org/10.1021/acsami.0c16752

Bancal V, Ray RC (2022) Overview of food loss and wastes in fruits and vegetables: from issues to resources. In: Ray RC (ed) Fruits and vegetable wastes: valorization to bioproducts and platform chemicals. Springer, Singapore, pp 3–32

Banjo TT, Ogbonna CB, Banjo TO, EZE, O. (2019) Bioethanol production from bitter yam (*Dioscorea dumetorum*) and water yam (*Dioscorea alata*) peels. Nig J Microbiol 33(2): 4687–4696

Bashir A, Baba NM, Akpomie TM (2021) Characterization of biomass residue (yam peels) for bioethanol production. Int J Adv Chem 9(3):146–149

Behera SS, Ray RC (2016) Konjac Glucomannan, a promising polysaccharide of *Amorphophallus konjac* K. Koch in healthcare. Int J Biol Macromol 92:942–956

Bhosale SH, Rao MB, Deshpande VV, Srinivasan MC (1995) Thermostability of high-activity alkaline protease from *Conidiobolus coronatus* (NCL 86.8. 20). Enzym Microb Technol 17(2): 136–139. https://doi.org/10.1016/0141-0229(94)00045-S

Bidari R, Abdillah AA, Ponce RAB, Charles AL (2023) Characterization of biodegradable films made from taro peel (*Colocasia esculenta*) starch. Polymers 15(2):338

Bos-Brouwers HEJ, Timmermans AJM, Soethoudt JM, Ostergren K, Gustavsson J, Hansen OJ, Möller H, Anderson G, O'Connor C, Quested T (2014) FUSIONS definitional framework for food waste. FUSIONS. FP7-KBBE-2012-6-311972

Braide W, Oji IO, Adeleye SA, Korie MC (2018) Comparative study of bioethanol production from agricultural wastes by *Zymomonas mobilis* and *Saccharomyces cerevisiae*. Int J Appl Microbiol Biotechnol Res 6:50–60

Buntha P, Borin K, Preston TR, Ogle B (2008) Digestibility and nitrogen balance studies in pigs fed diets with ensiled taro (*Colocasia esculenta*) leaves as a replacement for fish meal. Livest Res Rural Dev 20(supplement)

Caicedo W, Ferreira FNA, Barrionuevo J, Oña L (2019) Chemical composition of the taro tuber (*Colocasia esculenta* L. Schott) of waste with and without processing for use in pigs. In: MOL2NET 2019, International conference on multidisciplinary sciences. https://doi.org/10.3390/mol2net-05-xxxx

Caisedo W, Viáfara D, Pérez M, Ferreira FNA, Asitimbay M, Gavilanes Z, Valle S, Ferreira WM (2020) Chemical composition and productive performance of silage of taro tubers and foliage (*Colocasia esculenta* L. Schott) in rearing pigs. Cuban J Agric Sci 54(3):1–9. https://www.redalyc.org/journal/6537/653767640006/html/

Chaïr H, Traore RE, Duval MF, Rivallan R, Mukherjee A, Aboagye LM et al (2016) Genetic diversification and dispersal of taro (*Colocasia esculenta* (L.) Schott). PLoS One 11(6): e0157712. https://doi.org/10.1371/journal.pone.0157712

Chaturvedi S, Bhattacharya A, Nain L, Prasanna R, Khare SK (2019) Valorization of agro-starchy wastes as substrates for oleaginous microbes. Biomass Bioenergy 127:105294. https://doi.org/10.1016/j.biombioe.2019.105294

Chen YH, Ong CC, Lin TY (2022) Effect of sea salt and taro waste on fungal *Mortierella alpina* cultivation for arachidonic acid-rich lipid production. Fermentation 8(2):81. https://doi.org/10.3390/fermentation8020081

Chintagunta AD, Jacob S, Banerjee R (2016) Integrated bioethanol and biomanure production from potato waste. Waste Manag 49:320–325

Corrado S, Sala S (2018) Food waste accounting along global and European food supply chains: state of the art and outlook. Waste Manag 79:120–131. https://doi.org/10.1016/j.wasman.2018.07.032

Duru CC, Uma NU (2003) Protein enrichment of solid waste from cocoyam (*Xanthosoma sagittifolium* (L.) Schott) cormel processing using *Aspergillus oryzae* obtained from cormel flour. Afr J Biotechnol 2(8):228–232

El Sheikha AF, Ray RC (2022) Bioprocessing of horticultural wastes by solid-state fermentation into value-added/innovative bioproducts: a review. Food Rev Intl 39:1–57. https://doi.org/10.1080/87559129.2021.2004161

Ezejiofor TIN, Enenebeaku UE, Enenebeaku CK, Nwankwo MU, Ogbonnaya CIA (2018) Comparative study of bioethanol yield from yam, potato, watermelon, and pineapple peels using different concentrations of hydrochloric acid. World News Nat Sci 16:18–32

FAOSTAT (2021). https://www.fao.org/faostat/en/#data/QCL/visualize

Food and Agriculture Organization of the United Nations (FAO) (1981) Food loss prevention in perishable crops. FAO Agricultural Service Bulletin, No. 43. FAO Statistics Division, Rome. http://www.fao.org/3/s8620e/S8620E00.htm. Accessed 15 Mar 2021

Ge X, Matsumoto T, Keith L, Li Y (2014) Biogas energy production from tropical biomass wastes by anaerobic digestion. Bioresour Technol 169:38–44

Giang NT, Preston TR, Ogle B (2010) Effect on the performance of common ducks of supplementing rice polishings with taro (*Colocacia esculenta*) foliage. Livest Res Rural Dev 22(10):194

Ginni G, Kavitha S, Kannah Y, Bhatia SK, Kumar A, Rajkumar M, Kumar G, Pugazhendhi A, Chi NTL (2021) Valorization of agricultural residues: different biorefinery routes. J Environ Chem Eng 9(4):105435

Giwa A, Alabi A, Yusuf A, Olukan T (2017) A comprehensive review on biomass and solar energy for sustainable energy generation in Nigeria. Renew Sust Energ Rev 69:620–641

Godwin-Egein MI, Arinze AE (2000) The induction of some common hydrolytic enzymes and antibiotics in *Trichoderma harzianum* and *Fusarium oxysparum* using some food wastes. Global J Pure Appl Sci 6(1):31–36

Hassoub MA, Zhang L, Zuhair R, Ghonim SMMES, El-Dahshouri MF, Mustafa NS, Moustafa YTA (2022) Influence of wastes of taro leaf, sugar beet and saw dust on physiochemical parameters of produced vermicompost. Asian J Soil Sci Plant Nutr 8(3):29–40

Herrera-García U, Castillo J, Patiño-Ruiz D, Solano R, Herrera A (2019) Activated carbon from yam peels modified with Fe_3O_4 for removal of 2,4-dichloro phenoxy acetic acid in aqueous solution. Water 11(11):2342. https://doi.org/10.3390/w11112342

Hsieh SC, Liu JM, Pua XH, Ting Y, Hsu RJ, Cheng KC (2016) Optimization of *Lactobacillus acidophilus* cultivation using taro waste and evaluation of its biological activity. Appl Microbiol Biotechnol 100:2629–2639

Ibeto CN, Ofoefule AU, Agbo KE (2011) A global overview of biomass potentials for bioethanol production: a renewable alternative fuel. Trends Appl Sci Res 6(5):410

Isah SI, Abbas MI, Rahman M (2019) Comparative study on the production of bioethanol from banana and yam peel. Pharmacol Online 1:100–106

Itelima J, Onyimba I, Ogbonna A, Nyam M, Nwaukwu I (2014) Utilization of food crop wastes for the formulation of laboratory media used for cultivating soil fungi. Glob J Microbial Res 2:10–14

Iyayi EA, Losel DM (2001) Protein enrichment of cassava by-products through solid state fermentation by fungi. J Food Technol Afr 6(4):116–118

Kazuki K, Saito E, Hirokami L, Hiroi T, Onoda A (2017) Bioethanol production from waste flying powder (Tobiko), generated in the manufacturing process of konjac-derived products. J Jpn Inst Energy 96:199–205

Kee PE, Cheng YS, Chang JS, Yim HS, Tan JCY, Lam SS, Lan JCW, Ng HS, Khoo KS (2023) Insect biorefinery: a circular economy concept for biowaste conversion to value-added products. Environ Res 221:115284

Khayyal AA, Bakr EOA, Phillip YL, Hussein AM, Khir AA (2017) Effect of diets containing dried taro (*Colocasia esculanta*) waste and dried yeast (*Saccharomyces cerevisiae*) on the performance of growing rabbits. J Anim Poultry Prod 8(6):109–117. https://doi.org/10.21608/jappmu.2017.45794

Koni TNI, Paga A, Wea R, Foenay TA (2015) Nutritive values and metabolizable energy of *Amorphophallus campanulate* fermented by *Rhizopus oligosporus* as poultry feed. Pak J Nutr 14(6):322–324

Koni TNI, Zuprisal R, Hanim C (2017) Nutrient evaluation of fermented *Amorphophallus campanulatus* as poultry feed. Int J Pollut Sci 16(12):511–514

Kotsanopoulos K, Behera SS, Ray RC (2022) Microbial production of polysaccharides, oligosaccharides and sugar alcohols from vegetables and fruit wastes. In: Ray RC (ed) Fruits and vegetable wastes: valorization to bioproducts and platform chemicals. Springer, Singapore, pp 343–364

Kover A, Kraljić D, Marinaro R, Rene ER (2022) Processes for the valorization of food and agricultural wastes to value-added products: recent practices and perspectives. Syst Microbiol Biomanuf 2:50–66. https://doi.org/10.1007/s43393-021-00042-y

Kreike CM, Van Eck HJ, Lebot V (2004) Genetic diversity of taro, *Colocasia esculenta* (L.) Schott, in Southeast Asia and the Pacific. Theor Appl Genet 109:761–768

Kueain YA, Suamba IK, Wijayanti PU (2017) Business financial analysis on piggery (a case study of piggery of UD Karang at Jagapati village, Sub-District of Abiansemal, Badung). Jurnal Agribisnis dan Agrowisata 6:96–104

Kumar B, Bhardwaj N, Agrawal K, Chaturvedi V, Verma P (2020) Current perspective on pretreatment technologies using lignocellulosic biomass: an emerging biorefinery concept. Fuel Process Technol 199:106244

Kumari P, Mehta A, Sehgal R, Ray RC, Gupta R (2022) Microbial enzymes and organic acids production from vegetables and fruit wastes and their applications. In: Ray RC (ed) Fruits and vegetable wastes: valorization to bioproducts and platform chemicals. Springer, Singapore, pp 237–257

Lebot V, Lawac F, Legendre L (2023) The greater yam (*Dioscorea alata* L.): a review of its phytochemical content and potential for processed products and biofortification. J Food Compos Anal 115:1–18. https://doi.org/10.1016/j.jfca.2022.104987

Lee SH, Jun S (2011) Enhancement of sugar release from taro waste using ohmic heating and microwave heating techniques. Trans ASABE 54(3):1041–1047

Lestari AYD, Dewi LK (2018) Characteristics of *Amorphophallus campanulatus* modified starch as novel adsorbent for nickel and cadmium removal from aqueous solution. In: IOP conference series: materials science and engineering, vol 358(1). IOP Publishing, p 012021. https://doi.org/10.1088/1757-899X/358/1/012021

Lestari AYD, Malik A, Ilmi MI, Sidiq M (2018) Removal of calcium and magnesium ions from hard water using modified *Amorphophallus campanulatus* skin as a low-cost adsorbent. In: MATEC Web of conferences, vol 154. EDP Sciences, p 01020. https://doi.org/10.1051/matecconf/201815401020

Liu Q, Donner E, Yin Y, Huang RL, Fan MZ (2006) The physicochemical properties and in vitro digestibility of selected cereals, tubers and legumes grown in China. Food Chem 99(3):470–477

Liu X, Zou J, Xu S, Li X, Xie J, Yu Q (2023) Hydrothermal pretreatment based on biogas slurry recycling to improve methane yield of typical plant wastes from South China. Ind Crop Prod 200:116686

Longjan GG, Dehouche Z (2018) Nutrient characterization and bioenergy potential of common Nigerian food wastes. Waste Manag Res 36(5):426–435

Makinde OA, Odokuma LO (2015) Comparative study of the biogas potential of plantain and yam peels. Braz J Sci Technol 9:354–359

Marimón-Bolívar W, Tejeda-Benítez L, Herrera AP (2018) Removal of mercury (II) from water using magnetic nanoparticles coated with amino organic ligands and yam peel biomass. Environ Nanotechnol Monit Manag 10:486–493

Maxwell OI, Oghenerivwe AL, Onyekachi E (2021) Study of the effect of process parameters on the yield of fermentable sugar from tuber peels via acid and enzyme hydrolysis. J Energy Res Rev 9(2):24–32. https://doi.org/10.53294/ijfetr.2021.1.1.0047

Meng X, Hu W, Wu S, Zhu Z, Lu R, Yang G, Qin C, Yang L, Nie G (2019) Chinese yam peel enhances the immunity of the common carp (*Cyprinus carpio* L.) by improving the gut defence barrier and modulating the intestinal microflora. Fish Shellfish Immunol 95:528–537

Mithra MG, Padmaja G (2016) Lime pretreatment associated compositional and ultrastructural changes in selected root and vegetable processing residues. Int J Environ Agric Biotechnol 2(1): 238675

Mithra MG, Sreekumar J, Padmaja G (2018a) Binary-and triple-enzyme cocktails and their application mode affect fermentable sugar release from pretreated lignocellulo-starch biomass. Biomass Convers Biorefinery 8:97–111. https://doi.org/10.1007/s13399-017-0237-y

Mithra MG, Jeeva ML, Sajeev MS, Padmaja G (2018b) Comparison of ethanol yield from pretreated lignocellulo-starch biomass under fed-batch SHF or SSF modes. Heliyon 4(10): e00885. https://doi.org/10.1016/j.heliyon.2018.e00885

Mohapatra S, Ramachandran S, Ray RC (2018) Bioethanol from biorenewable feedstocks: technology, economics and challenges. In: Ray RC, Ramachandran S (eds) Bioethanol production from food crops. Academic, London, pp 3–27

Moodley P, Ray RC, Kana EBG (2023) Advances in lignocellulosic biofuel production systems. Elsevier. 436p

Nataraj HN, Murthy RLN, Setty R (2012) Evaluation of gastro-protective ability of *Amorphophallus paeoniifolius* corms against indomethacin-induced gastric ulcers. RGUHS J Pharm Sci 2:67–73

Nedunchezhiyan M, Jata SK, Byju G, Veena SS (2011) Effect of tuber crop wastes/byproducts on nutritional and microbial composition of vermicomposts and duration of the vermicomposting process. J Bot. https://doi.org/10.1155/2011/801703

Nenava R, Nighojkar S, Kumar A, Nighojkar A (2022) Fungal xylanase produced using *Amorphophallus paeoniifolius* peels as substrate which aided in the making of bread dough fluffy. J Sci 5(3):4. https://doi.org/10.21786/bbrc/15.2.12

Nizami AS, Rehan M, Waqas M, Naqvi M, Ouda OK, Shahzad K, Miandad R, Khan MZ, Syamsiro M, Ismail IM, Pant D (2017) Waste biorefineries: enabling circular economies in developing countries. Bioresour Technol 241:1101–1117. https://doi.org/10.1016/j.biortech.2017.05.097

Nnaji AC, Mbah GO, Onoh MI, Okorie O, Udeh BC (2021) Bio-ethanol production from cocoyam peels using enzymatic hydrolysis. J Mater Sci Res Rev 8(2):13–28

Nweke CN, Nwabanne JT (2021) Anaerobic digestion of yam peel for biogas production: a kinetic study. J Eng Appl Sci 18(1):275–286

Obidiegwu JE, Lyons JB, Chilaka CA (2020) The *Dioscorea* genus (yam)—an appraisal of nutritional and therapeutic potentials. Foods 9(9):1304. https://doi.org/10.3390/foods9091304

Ojolo SJ, Orisaleye JI, Badiru N (2016) Design and fabrication of a yam peeling machine. J Eng Res 21(2):50–59

Okunwaye T, Uadia PO, Okogbenin BO, Okogbenin EA, Onyia DC, Obibuzor JU (2021) Amylase-producing fungi and bacteria associated with some food processing wastes. Nig J Biotechnol 38(1):74–82. https://doi.org/10.4314/njb.v38i1.9

Onu CE, Nweke CN, Nwabanne JT (2022) Modeling of thermo-chemical pretreatment of yam peel substrate for biogas energy production: RSM, ANN, and ANFIS comparative approach. Appl Surf Sci Adv 11:100299

Opara LU (2000) Edible aroids—post operation. FAO, Rome. https://www.fao.org/fileadmin/user_upload/inpho/docs/Post_Harvest_Compendium_-_Edible_aroids.pdf

Orhorhoro EK, Oghoghorie O (2019) Review on solid waste generation and management in sub-Saharan Africa: a case study of Nigeria. J Appl Sci Environ Manag 23(9):1729–1737

Pandey A, Soccol CR (2000) Economic utilization of crop residues for value addition: a futuristic approach. J Sci Ind Res 59(1):12–22

Putri I, Hintono A, Susanti S (2018) Optimization of elephant foot yam (*Amorphophallus paeoniifolius*) extracts syrup formula as a nutritive antioxidant drink. Int J Sustain Agric Res 5(1):1–9

Racheal OO, Ahmed ATF, Ndigwe EV, Morakinyo SD (2015) Extraction, purification and characterization of protease from *Aspergillus niger* isolated from yam peels. Int J Nutr Food Sci 4(2):125–131. https://doi.org/10.11648/j.ijnfs.20150402.11

Rahmat B, Hodiyah I, Supriadi A, Hikmat M, Purnama G (2019) Design of biogas digester with thermophilic pretreatment for reducing fruit wastes. Int J Recycl Org Waste Agric 8:291–297

Rana G, Mandal T, Mandal NK, Mandal S (2017) Hydrogen-containing fuel gas generation from organic wastes using photon-activated magnesium metal catalyst. S Afr J Chem Eng 23(1):124–131

Rao M, Bast A, De Boer A (2021) Valorized food processing by-products in the EU: finding the balance between safety, nutrition, and sustainability. Sustainability 13(8):4428. https://doi.org/10.3390/su13084428

Ravi V, Ravindran CS, Suja G (2009) Growth and productivity of elephant foot yam (*Amorphophallus paeoniifolius* (Dennst.) Nicolson): an overview. J Root Crops 35(2):131–142

Ravi V, Ravindran CS, Suja G, George J, Nedunchezhiyan M, Byju G, Naskar SK (2011) Crop physiology of elephant foot yam [*Amorphophallus paeoniifolius* (Dennst. Nicolson)]. Adv Hort Sci 25(1):51–63

Ray RC (2015) Post-harvest handling, processing and value addition of elephant foot yam—an overview. Int J Innov Hortic 4(1):1–10

Ray RC (2020) Sustainable biofuels: challenges and opportunities. Academic. 502p

Ray RC (2022) Fruits and vegetable wastes: valorization into bioproducts and platform chemicals. Springer, Singapore. 448p

Ray RC, Ramachandran S (2018) Bioethanol production from food crops: sustainable sources, interventions, and challenges. Academic. 460p

Research and Markets (2020) Global bioethanol market (2020 to 2025)—rising demand from the pharmaceutical industry presents opportunities. https://www.globenewswire.com/news-release/2020/06/11/2046814/0/en/Global-Bioethanol-Market-2020-to-2025-Rising-Demand-From-the-Pharmaceutical-Industry-Presents-Opportunities.html

Rodríguez R, Valle S, Ly J, Lezcano P, Flores L, Ferreira FNA, Caicedo W (2016) Physicochemical and biological indicators in silages of taro (*Colocasia esculenta* (L.) Schott) tubers for animal feeding. Cuban J Agric Sci 50(1):121–129

Rosalina AM, Permana AH (2022) Bioethanol production using taro roots waste (*Colocasia esculenta*) from Bogor, Indonesia and analysis of chemical compounds. Rasayan J Chem 15(2):870–874

Roy Chowdhury A, Datta R, Sarkar D (2018) Heavy metal pollution and remediation. In: Green chemistry. Elsevier, pp 359–373

Roy M, Pandey VC (2020) Role of microbes in grass-based phytoremediation. In: Phytoremediation potential of perennial grasses, pp 303–336

Sembera JA, Waliczek TM, Meier EJ (2019) Composting as an alternative management strategy for wild taro waste. HortTechnology 29(2):205–209

Serna-Loaiza S, Martínez A, Pisarenko Y, Cardona-Alzate CA (2018a) Integral use of plants and their residues: the case of cocoyam (*Xanthosoma sagittifolium*) conversion through biorefineries at small scale. Environ Sci Pollut Res 25:35949–35959. https://doi.org/10.1007/s11356-018-2313-7

Serna-Loaiza S, Pisarenko YA, Cardona CA (2018b) Ethanol production from cocoyam (*Xanthosoma sagittifolium*): application of thermodynamic-topological analysis. Fine Chem Technol 13(2):40–50

Shao M, Chen H (2015) Feasibility of acetone–butanol–ethanol (ABE) fermentation from *Amorphophallus* konjac waste by Clostridium acetobutylicum ATCC 824. Process Biochem 50(8):1301–1307. https://doi.org/10.1016/j.procbio.2015.05.009

El Sheikha AF and Ray RC 2017 Potential impacts of bioprocessing of sweet potato. Crit Rev Food Sci Nutr 57(3): 455-471.

Singh A, Srivastava KC, Banerjee A, Wadhwa N (2013) Phytochemical analysis of peel of *Amorphophallus paeoniifolius*. Int J Pharma Bio Sci 4(3):810–815

Skoog K, Hahn-Hägerdal B (1988) Xylose fermentation. Enzym Microb Technol 10(2):66–80

Soccol CR, Vandenberghe LP (2003) Overview of applied solid-state fermentation in Brazil. Biochem Eng J 13(2–3):205–218

Sooch BS, Mann MK, Sharma P, Ray RC (2022) Phenolics and other bioactive compounds from vegetables and fruit wastes: extraction methods and their possible utilizations. In: Ray RC (ed) Fruits and vegetable wastes: valorization to bioproducts and platform chemicals. Springer, Singapore, pp 81–114

Teigiserova DA, Hamelin L, Thomsen M (2019) Review of high-value food waste and food residues biorefineries with focus on unavoidable wastes from processing. Resour Conserv Recycl 149:413–426. https://doi.org/10.1016/j.resconrec.2019.05.003

U.S. Composting Council (2002) Test methods for the examination of composting and composts. (CDROM only). Composting Council Res. Educ. Foundation, Holbrook, NY

Uchegbu MC, Ezuma CC, Ogbuewu IP, Opara MN (2010) Physiological responses of finisher broilers to yam peel meal: haematology and serum biochemistry. Electron J Environ Agric Food Chem 9(10):1657–1664

Uchewa EN, Orogwu CE, Nwakpu PE (2014) Effect of yam peel meal (YPM) replacement for maize on the growth performance and carcass traits of weaner rabbits. Int J Agric Innov Res 2(4):536–541

Ugah UI, Nwoba ST (2018) Microbial media formulation using rice husks, yam and cassava peels on some selected microorganisms. Sci Res 26(1):122–131

Ugwuanyi JO (2008) Moisture sorption isotherm and xerophilic moulds associated with dried cocoyam chips in storage in Nigeria. Int J Food Sci Technol 43(5):846–852

Ujor V, Bharathidasan AK, Cornish K, Ezeji TC (2014) Feasibility of producing butanol from industrial starchy food wastes. Appl Energy 136:590–598. https://doi.org/10.1016/j.apenergy.2014.09.040

Utsev TJ, Onah MO (2017) Removal of chromium from wastewater using yam peel activated carbon. Eur J Eng Technol Res 2(1):9–14. https://ej-eng.org/index.php/ejeng/article/view/240

Vilardi G, Di Palma L, Verdone N (2018) Heavy metals adsorption by banana peels micro-powder: equilibrium modeling by non-linear models. Chin J Chem Eng 26(3):455–464

Villabona-Ortíz A, Tejada-Tovar C, González-Delgado ÁD (2022) Elimination of chromium (VI) and nickel (II) ions in a packed column using oil palm bagasse and yam peels. Water 14(8):1240

Villadiego-del Villar AE, Sarmiento-Zea N, León-Pulido J, Rojas-Pérez LC (2021) Bioethanol production from yam (*Dioscorea rotundata*) using simultaneous saccharification and fermentation (SSF). TecnoLógicas 24(50):116–125

Waghmare PR, Khandare RV, Jeon BH, Govindwar SP (2018) Enzymatic hydrolysis of biologically pretreated sorghum husk for bioethanol production. Biofuel Res J 5(3):846–853

Wang Y, Chen R, Dai Z, Yu Q, Miao Y, Xu R (2022) Facile preparation of a polypyrrole modified Chinese yam peel-based adsorbent: characterization, performance, and application in removal of Congo red dye. RSC Adv 12(15):9424–9434. https://pubs.rsc.org/en/content/articlehtml/2022/ra/d1ra08280a

Wu XT, Zhu T, Zheng QM, Sheng C (2011) Association, study on extraction of sweet Chinese yam skin saponins by using response surface methodology. J Chin Cereals Oils Assoc 26:91–96

Wu WH, Hung WC, Lo KY, Chen YH, Wan HP, Cheng KC (2016) Bioethanol production from taro waste using thermo-tolerant yeast *Kluyveromyces marxianus* K21. Bioresour Technol 201: 27–32

Yafetto L, Odamtten GT, Biikorang E, Adu S (2020) Protein enhancement of yam (*Dioscorea rotundata*) peels with single-or co-inoculation of *Aspergillus niger* van Tieghem and *Trichoderma viride* Pers ex Fr. under solid-state fermentation. Ghana J Sci 61(2):27–37. https://doi.org/10.4314/gjs.v61i2.3

Yahaya IA, Nok AJ, Bonire JJ (2013) Chemical studies of the peel of *Xanthosoma sagittifolium* (Tannia Cocoyam). Pak J Nutr 12(1):40–44

Ziganshin AM, Liebetrau J, Pröter J, Kleinsteuber S (2013) Microbial community structure and dynamics during anaerobic digestion of various agricultural waste materials. Appl Microbiol Biotechnol 97:5161–5174

Valorization of Carrot and Turnip Processing Wastes and By-Products

9

Akashdeep Kaur and Gargi Ghoshal

Abstract

From the beginning of the food production chain, around 15% of all fruits and 25% of all vegetables are lost, according to the Food Wastage Footprint and Climate Change Report. Mostly due to the effects of microbial deterioration, the large losses and wastes in the fresh and processing industries are developing into a serious environmental problem. To solve these issues, there has been a recent increase in research and invention in the fields of food, packaging, and pharmaceuticals. The underused wastes (seed, peel, rind, and pomace) have the potential to be excellent sources of beneficial bioactive substances, such as functional nutrients, amylopectin, phytochemicals, vitamins, enzymes, dietary fibers, and oils. Fruit and vegetable wastes are among the most common agro-industrial wastes. This discarded biomass can be used to create biofuels such as bioethanol, biomethane, biohydrogen, and biobutanol. This article intends to highlight the applications of carrot and turnip processing wastes and their usage in various value-added products.

Keywords

Carrot and turnip waste · Waste valorization · By-products · Nutrient composition · Morphology · Applications

A. Kaur
Energy Research Centre, Panjab University, Chandigarh, India

G. Ghoshal (✉)
Dr. S. S. Bhatnagar University Institute of Chemical Engineering and Technology, Panjab University, Chandigarh, India

R. C. Ray (ed.), *Roots, Tubers, and Bulb Crop Wastes: Management by Biorefinery Approaches*, https://doi.org/10.1007/978-981-99-8266-0_9

Abbreviations

°C	Degree Celsius
°F	Degree Fahrenheit
AD	Anno Domini
BC	Before Christ
BOD	Biochemical oxygen demand
Bt	Billion tonnes
Ca	Calcium
CE	Common era
CF	Carrot fibers
CH	Chitosan
CP	Carrot pomace
CR	Congo Red
CWP	Cell wall particles
DPPH	α,α-Diphenyl-β-picrylhydrazyl
EtOH	Ethyl alcohol
FAO	Food and Agriculture Organization
Fe	Iron
FTIR	Fourier transform infrared
g	Grams
GAE	Gallic acid equivalent
GC	Gas chromatography
Ha	Hectares
HPP	High-pressure processing
IR spectroscopy	Infrared spectroscopy
kg	Kilograms
MCC	Microcrystalline cellulose
Mg	Magnesium
mg	Milligrams
Mha	Million hectares
mL	Milliliters
MMT	Million metrics tonnes
MPUMH	Multiple-pass ultrasonication with mechanical homogenization
MT	Million tonnes
NaCl	Sodium chloride
NCIM	National Collection of Industrial Microorganisms
P	Phosphorus
PDCs	Processed carrot discards
pH	Potential of hydrogen
RW	Refractance window
SEM	Scanner electron microscopy
TP	Turnip peroxidase

TPC	Total phenolic contents
TPE	Turnip peel extract
UAE	Ultrasound-assisted extraction
USDA	United States Department of Agriculture
UV-Vis spectroscopy	Ultraviolet and visible regions of the electromagnetic spectrum
W	Watts
WVP	Water vapor permeability

9.1 Introduction

9.1.1 History of Carrot and Turnip

9.1.1.1 Carrots

Carrots are one of the most important root vegetables in the world. In the wild, it is a small, bitter root with little food appeal, but over the years of cultivation and domestication by humans through nature, it has become a very versatile vegetable that comes in many colors, shapes, and sizes. The domestic carrot (*Daucus carota* var. *sativus*), Family Apiaceae, has been cultivated for over 2000 years and was originally used only as a medicinal plant, but remains an important global crop with increasing production in Asia. According to FAOSTAT (2022), the major producing countries of carrot and turnip are China, Russia, Uzbekistan, USA (Table 9.1). The swollen roots are eaten raw, cooked, and in sweet/savory dishes. It contains a high amount of β-carotene which is converted into vitamin A in the body. Also, as a main ingredient in the food industry and an important ingredient in cosmetics, it has long been used as an image representing healthy eating habits. The leaves are also used in salads and the seeds are used to make herbal tea.

Table 9.1 Turnip- and carrot-producing countries (data taken from FAOSTAT 2022)

Name of country rank wise	Production (tonnes)	Yield (100 g/ha)	Area harvested (ha)
China	18,088,583	459,532	395,525
Uzbekistan	3,155,745	702,411	44,927
USA	1,432,740	497,237	28,814
Russian Federation	1,303,288	294,548	44,247
Germany	961,970	644,752	14,920
UK	888,851	623,974	14,245
Ukraine	863,290	199,836	43,200
Pakistan	728,137	253,636	28,708
Indonesia	720,090	194,064	37,106
France	706,750	425,497	16,610

9.1.1.2 Turnips

For centuries, turnips (*Brassica rapa* L. ssp. *rapa*) have been one of the most important crops in the Alps. As a food and fodder source, it was primarily used. A very important winter food staple in the past was fermented turnips (Ruebenkraut), along with sauerkraut. There are relatively few farmers remaining who are capable of producing fermented turnips, including soil management, sowing, crop management, harvesting, storing, processing, and propagating seeds (Vogl-Lukasser et al. 2007).

9.1.2 Nutritional Value

9.1.2.1 Carrots

Throughout the world, carrots are an edible crop that belongs to the Apiaceae family (formerly Umbelliferae). A tropical or sub-tropical region typically cultivates this crop between September and November, while temperate regions cultivate it all year round. Cold temperatures are necessary for seed production in the crop (Raees-ul and Prasad 2015). The carrot contains carotenoids and flavonoids which have good antioxidant properties. The bioactive components are primarily concentrated on the outside of the root (cortex), and the presence of significant amounts of vitamins, bioactive components, and minerals has caused it to be ranked among the top ten fruits and vegetables in terms of nutrition. The distinctive, carotenoids rich carrot roots have a distinctive flavor brought on by the presence of terpenoids and polyacetylenes. Because of the slow production of sugars in the root during growth, the bitter carrot flavor is subdued to an acceptable extent. The different pigments cause the development of red, orange, yellow, purple, black, and white roots (Raees-ul and Prasad 2014). The pigments found in various colored roots have a variety of health benefits for people, such as lutein, which is contained in yellow carrots and helps to create the pigment that prevents macular degeneration which is needed for appropriate eye function (Chrong et al. 2007). Lycopene, which is found in tomatoes and red carrot cultivars, has a potent singlet oxygen scavenging activity and high blood plasma levels have an impact on reducing the incidence of many malignancies. It is understood that provitamin A carotenoids found in the human body are routinely converted by the body into vitamin A. According to studies, carrots account for 14–17% of all vitamin A consumed (Scarano et al. 2018).

9.1.2.2 Turnips

These are biennial herbaceous plants of the Cruciferae Brassica family that matures in 2 months and can be sown in the spring, late summer, or fall for roots or greens (Paul et al. 2019). One of the earliest crops to be produced, turnips are frequently consumed in high-altitude regions of Asian nations. According to one study on the plant's chemical makeup, it is a rich source of glucosinolates (Baenas et al. 2012), organic acids, phenolic compounds, volatiles, and sulfur compounds, all of which have been linked to a variety of healthy benefits (Cao et al. 2021). Turnip roots are abundant in a variety of nutritional profiles, including carbohydrates, protein, dietary fiber, vitamin C, vital amino acids, and mineral elements, although they contain less fat than other vegetables. Compared to turnip roots, turnip leaves have more nutritional values. In addition to the vitamins and minerals listed above, this study claimed that turnip leaves are also a rich source of carotenoids, lutein, vitamin A, vitamin B, and vitamin K (Javed et al. 2019).

9.1.3 Fruit and Vegetable-Based Industries

About 40% of the food produced in India is wasted, according to the Food and Agricultural Organization (FAO) (Plazzotta et al. 2017). Food waste generated during food production, preparation, and consumption totals around 1.3 billion metric tonnes annually, making its reduction a top priority for the food and agriculture industries. By 2030, food loss and waste should be cut by 50% from present levels, according to a national objective set by the US Department of Agriculture (USDA 2015). According to the Food and Agriculture Organization of the United Nations (FAO), food loss is the term used to describe any food that is lost along the supply chain between the producer and the retailer, typically as a result of production and processing inefficiencies. Food waste is defined as any food that is discarded but is still safe and edible for human consumption. According to the FAO, fruits, and vegetables account for over 60% of all food losses and waste, the greatest percentage of any other type of food. At 180 kg per person per year in North America, the majority of food loss and waste occurs between the production and retail sectors (Gustavsson et al. 2011; Bansal and Ray 2022). The amount of edible product that is wasted due to inefficient fruit and vegetable processing might range from 25% to 30% (Ray 2022). Consumer food waste in North America totals close to 110 kg per person per year, which is the greatest amount globally. Food waste at the production and retail levels has been significantly impacted from in-field harvest to processing and packaging facilities to supermarket retail stores; production to retail encompasses all of these activities. Since the 1930s, consumers have had access to fresh-cut or minimally processed fruit and vegetable products, but it wasn't until the last two decades that they started to acquire popularity as people started to be interested in eating convenient and healthful meals. Value-added (processed, ready-to-eat, convenient) veggies were purchased by 49% of US households in 2016 (Sagar et al. 2018).

9.1.4 Waste Generation

One-third of the food produced for human use worldwide is wasted each year from agricultural production to ultimate household consumption (Gustavsson et al. 2011). Food waste in the European Union currently totals over 89 million tonnes, and it is predicted that this number will rise 40-fold in the upcoming years (Coman et al. 2020). Fruits and vegetables are crucial for human nutrition because they provide a significant amount of the vitamins, minerals, and fiber we need each day.

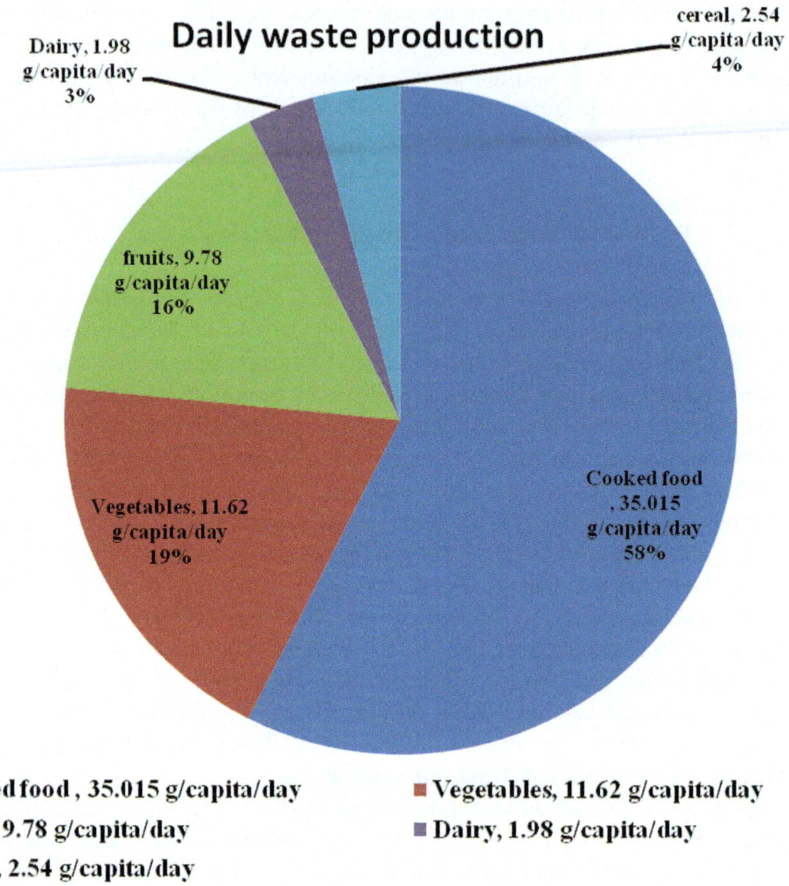

Fig. 9.1 Daily production of food waste (based on Garcia and Raghavan 2022)

Unfortunately, half of all fruits and vegetables produced worldwide are wasted, leading to environmental problems mostly brought on by microbial decay. The majority of waste, or so-called by-products, is produced during industrial operations. Even after processing, these byproducts still include a variety of bioactive substances, including macronutrients (proteins and carbs) and phytochemicals (polyphenols and carotenoids) (Sooch et al. 2022). Recent years have seen a major increase in interest in recovering these bioactive substances from industrial waste, largely because of the potential advantages they may have for human health (Coman et al. 2020). According to the data, 3% of the waste produced from dairy, 4% from cereals, 16% from fruits, 58% from cooked food and 19% from vegetables (Fig. 9.1). Similar issues with the majority of food waste produced from all municipal solid waste were noted in developing countries including China (52.6%), the USA (25%), Bangladesh (75%), and France (32%). A recent poll done in Scandinavia in 2019 predicted that households there lost more than 50% of their food supply as a result of the difficulties in deciding between "best before" and "use by" (Filimonau and De Coteau 2019).

9.1.5 Conversion of Carrot and Turnip Wastes into Value-Added Products

The by-products of the food processing industries that cannot be reused are referred to as food processing wastes. These are the raw material flows that are not products and are discarded as waste because their economic value is lower than the expense of recovery and collection for reuse. These wastes result in a loss of valuable biomass and nutrients and pose growing disposal and environmental challenges. The need for natural pigment synthesis at the industrial level for prospective use in food, medicine, and cosmetics can be met by the valorization of wastes (fruits and vegetables) and their by-products (e.g., peels, seeds, or pomace) (Ashfaq et al. 2022). These by-products and wastes are a great source of natural pigments such as anthocyanins, betalains, carotenoids, and chlorophyll. It is anticipated that these natural pigments will have a substantial impact on the creation of functional foods and offer a wealth of bio-therapeutic possibilities. With a focus on sustainability, the key research findings and advancements made in the fields of natural pigments derived from plant waste, environmentally friendly extraction and processing methods, encapsulation strategies, and potential bioactivities are critically examined (Sharma et al. 2021). Of all the agro-industrial wastes, fruit and vegetable wastes (FVWs) are among the most prevalent. Bio-fuels including bio-ethanol, bio-methane, bio-hydrogen, and bio-butanol can be produced from this leftover biomass. FVWs can also be used as a raw material to recover bioactive substances like antioxidants, enzymes, or antibiotics as well as to create industrially useful organic acids like acetic acid, lactic acid, and citric acid (Sanchez et al. 2021; Kumari et al. 2022; Ray 2022).

9.1.5.1 Carrots

Carrots are relatively affordable and very nourishing since they are abundant in β-carotene, vitamins B1, B2, B6, and B12. Many malignancies, including prostate cancer, are linked to high carotenoid intake. Moreover, vitamin A is an antioxidant that is essential for tissue growth and repair, aids in infection control, maintains eye health, and nourishes epithelial tissues in the lungs as well as the skin. Consequently, it is crucial to retain as much β-carotene as possible to maintain the product's appealing appearance and nutritional value. Carrots are high in dietary fibre in addition to being strong in carotenoids. Moreover, a large amount of calcium, potassium, and phosphorus can be found in carrots. Due to their high sugar content, carrots are also a strong source of energy. If handled appropriately, this low-cost crop could be turned into a value-added product (Singh et al. 2006). Along with being used normally in salads and the production of curries in India, carrot roots can also be processed into nutrient-rich processed foods like beverages, extract, powder, canned, candy, pickles, and sweets. When used to enrich foods like cake, bread, and

Fig. 9.2 Illustrative diagram of percentage of wastes generated from carrot processing (Seregelj et al. 2021)

biscuits as well as to create a variety of functional items, carrot pomace, which contains roughly 50% β-carotene, can be profitable (Sharma et al. 2012; Seregelj et al. 2021). The pomace and peels include 30–50% and 14%, respectively of waste generated from carrot processing (Fig. 9.2) (Seregelj et al. 2020).

9.1.5.2 Turnips

Due to its nutritional makeup, the purple turnip is one of the family's most significant members (Aysun 2021). Turnips are the oldest cultivated vegetables and have been consumed by people since the Stone Age (Kaveh and Chayjan 2017). A few beneficial ingredients in turnips not only support health advantages but also have therapeutic properties. Bioactive chemicals like kaempferol, glucosinolates, vitamin K, peroxidase, sulforaphane, phenolics, and organic acids are highlighted in this article (Table 9.2). Several minerals, such as copper, manganese, and calcium as well as organic acids such as sinapic and ferulic acids and their derivatives which are present in fresh greens and turnip roots in variable amounts are also briefly discussed. Minerals like copper, manganese, and calcium are briefly discussed as well as organic acids like sinapic and ferulic acids and their derivatives, which are found in different amounts in fresh greens and turnip roots. Turnips have significant

Table 9.2 Turnip peel extract composition, conversion to alcohol, sugar, and kinetics of biomass production (sources: Javed et al. 2019; Kandari and Gupta 2012; Taveira et al. 2009)

Composition of turnip extract							
Parameters	**Total sugars (%)**	**Fermentable sugars (%)**	**Non-fermentable sugars (%)**	**Protein (%)**	**Lipid (%)**	**Vitamins (mg/100mL)**	**Dry weight (%)**
Turnip	6.6	4.6	2.0	0.70	0.08	392	3.70
Alcohol production and sugar conversion of extra after 48 h of fermentation in turnip peel extract							
Alcohol (%)		**Residual fermentable[a] sugar (%)**		**Sugar conversion (%)**		**Biomass (g/L)**	
1.55		0.69		85.00		2.20	
Kinetics of biomass production in peel extract of turnip							
Time (h)	**Biomass (g/L)**	**Productivity (g/L/h)**	**Residual fermentable sugars (%)**			**Sugar conversion (%)**	
12	1.8	0.15	2.6			43.48	
24	3.5	0.14	1.8			60.87	
36	4.2	0.12	0.75			83.60	
48	6.8	0.14	0.69			85.00	
60	7.2	0.12	0.25			94.57	
72	7.0	0.09	0.20			95.66	

[a] Parameters for fermentation: temperature of 30 °C, pH 5.8, and inoculum age of 24 h

antioxidant activity, an excellent hepato-protective role, and a favorable effect on the liver damage brought on by diabetes due to the presence of bioactive elements (Javed et al. 2019). Some are harmful for human and animal consumption (perhaps goitrogenic), while some have been demonstrated to be advantageous (such as the chemoprotective action against specific malignancies in humans) (Taveira et al. 2009). Progoitrin and gluconapin, which are found in turnip greens and tops, give turnips a lingering bitter flavor (Francisco et al. 2011). The creation and delivery of therapeutic proteins have recently attracted the attention of researchers. According to this viewpoint, both α- and β-interferons, which are therapeutic proteins created under specific circumstances, such as fermentation, may play a key role in the treatment of a variety of diseases, including anemia, cancer, hepatitis (A and B), diabetes, and heart attacks (Auger et al. 2010). These proteins are produced by microorganisms through fermentation and are used in modified turnip roots, cell cultures, and other products. Parenteral pathways account for the majority of therapeutic protein production. Many unique approaches have been investigated and developed to get these therapeutic proteins to achieve maximum efficacy with the lowest risk (Muheem et al. 2016).

9.2 Total World Production, Harvested Areas, and Yield of Carrot and Turnip (Recent FAO Data)

9.2.1 Geographical Distribution of Carrot and Turnip

9.2.1.1 Carrots

The vegetable carrot (*Daucus carota* L. subsp. *sativus* L.) has a well-known and widespread weedy ancestor known as the wild carrot or Queen Anne's lace (*D. carota* subsp. *carota*). According to FAO (2011), the most extensively planted crop in the Apiaceae family is the carrot, which is grown on 1.2 million ha worldwide. The US produces 34,000 ha of carrots per year, with an estimated crop value of $758 million (USDA, National Agricultural Statistics Service 2011). In most temperate regions of the world, wild carrot is present. In India, 414 lakh tonnes of carrots are produced annually from a total area under cultivation of 22,538 ha. The major producing States are Uttar Pradesh, Assam, Karnataka, Andhra Pradesh, Punjab, and Haryana (Thamburaj and Singh 2005).

9.2.1.2 Turnips

These were first domesticated very long ago. Since 1800 BC, the Assyrian language has used the word *laptu*. Across the entire Middle East, turnips are a particularly popular vegetable (Arab: lift, Persian: salgham). A significant domesticated crop plant in ancient Greece and Rome was the turnip. *B. rapa* has undergone extensive phenotypic variation as a result of breeding and selection in both East Asia and Europe. Certain green vegetables, including Chinese cabbage (ssp. *pekinensis*), may have been naturally bred in central China from pak choi (ssp. *chinensis*) and turnip rape (ssp. *oleifera*), and then adapted to various locations and heading kinds. Turnips were even shown on the coats of arms of prominent families, demonstrating their significance in Alpine nations. Turnips were first domesticated very long ago. Laptu first appeared in ancient Assyrians around 1800 BC. Across the entire Middle East, turnips are a popular vegetable (Arab: lift, Persian: salgham). A significant domesticated crop plant in ancient Greece and Rome was the turnip. It was known as gongylis by the Greek philanthropist Theophrast (371–285 BC), and the Roman sociologist Columella (35–65 AD) distinguished between the cultivation and preservation of *napus* and *rapa* by lactic acid Fermentation (Shiri et al. 2020).

9.2.2 Production, Processing, and Storage

Fruits and vegetables have an intriguing nutritional profile that is high in bioactive metabolites, has a high antioxidant potential, and has health advantages. Yet, their usefulness, the shorter shelf life resulting from their high water content, and their seasonality nature result in significant food losses and waste. To reduce food losses and waste, it is crucial to value-add goods and apply preservation techniques to vegetables and fruit by-products. Almost 21.6% of the world's fruits and vegetables were lost in 2016, according to the Food and Agricultural Organization of the United Nations' 2019 report (Jenkins et al. 2016). These enormous losses and waste are caused in part by the seasonality of fruits and vegetables and their short shelf lives (Kamiloglu et al. 2016). According to the Food and Agricultural Organization of the United Nations (FAO), 40 million tonnes of carrots and turnips were produced globally in 2018, China accounted for 45% of this amount. Many cuisines frequently serve cooked or raw carrots. The physical and chemical composition of plant food is influenced by several factors, including plant genetics, the growing environment, the ripening stage, and post-harvest circumstances. Being one of the most important factors in the food's bioactive composition and concentration, food-processing methods may also include several adjustments to the nutritional composition of the meal. Processing may encourage the chemical alteration or oxidation of natural phytochemicals, which would reduce their antioxidant activity (Neri et al. 2020). Because these wastes release powerful greenhouse gases like methane and carbon dioxide into landfills, in addition to suffering economic losses, this practice also poses major environmental risks (Clementz 2019). Yet, the by-products or waste from the processing of carrots could be turned into foods that can be consumed or into value-added products that have significant functional and nutritional advantages (Kaur et al. 2021). Because they are edible and contain valuable bioactive ingredients such as dietary fibers, carotenoids, polyphenols, and antioxidants, by-products of the processing of carrots are distinctive (Kaur et al. 2019). By creating sustainable bio-processing methods that adhere to the zero-waste circular economy principles, high-value bioactive components like colorants, antioxidants, and phytochemicals can be extracted from these by-products and sequentially processed into biofuels (Casazza et al. 2020; Elkhalifa et al. 2022).

9.2.3 Yield of Carrot and Turnip Wastes

After potatoes, tomatoes are the second most popular vegetable in the world, and among the top ten most popular vegetables are carrots and cucumbers. Global production of tomatoes, gherkins, and cucumbers, as well as the combined production of carrots and turnips, is estimated by FAOSTAT to be 182 million tonnes (MT), 75 MT, and roughly 40 MT, respectively (FAO 2021). Every year, almost 1.3 billion tonnes of the world's entire food production, or more than one-third of it is wasted somewhere along the food supply chain (Gustavsson et al. 2011). The majority of solid waste is comprised of fruit and vegetable waste, which accounts for between 40% and 50% of the total waste (Sagar et al. 2018). With a share of 44% of the country's total production, Ontario is the leading producer of fruits and vegetables. With a peak farm gate value of CAD 129.3 million, carrots are Ontario's second-largest field vegetable behind tomatoes (Kaur et al. 2022a).

9.2.3.1 Carrots

These carrots must pass strict quality grading standards for size and shape before they can be sold to consumers. If the carrots do not meet the standards, they are discarded and are known as carrot waste. As a result, the enterprises that process carrots produce a lot of waste, amounting to around 25–30% of the overall harvest (Kee 2021). Along with suffering financial losses, the dumping of these wastes into landfills has major negative effects on the environment because it releases strong greenhouse gases like methane and carbon dioxide (Clementz 2019). The possibility exists, however, to further transform carrot wastes, or by-products into consumable foods or products with significant functional and nutraceutical benefits (Kaur et al. 2021).

9.2.3.2 Turnips

According to Stahl and Berghel (2011), the mechanical strength and density of the resulting granulate, and the energy consumption of the agglomeration process, was reduced as the turnip waste content in the compacted mixture with sawdust increased. Only four of the five tested fungal isolates could grow on turnip waste and produce lovastatin at concentrations ranging from 140 to 410 mg/L. The other three isolates could only thrive on carrot waste (Zohri et al. 2007).

9.3 Botany, Morphology, and Composition of the Crops

9.3.1 Botanical Origin

9.3.1.1 Carrots

Native to Eurasia, the wild carrot (subspecies *D. carota carota*, also known as *Queen Anne's lace*) is believed to have been domesticated in Central Asia about the year 1000 CE (Iorizzo et al. 2013). Archaeological digs have uncovered prehistoric seeds, indicating that the plant was utilized medicinally before the domestication of its edible root. During the thirteenth century, carrots were grown in China and northwest Europe, and during European colonial rule, the wild carrot was mistakenly introduced as a weed in the United States. There is currently widespread cultivation of domesticated carrots (ssp. *D. carota sativus*) in temperate zones (Iorizzo et al. 2020).

9.3.1.2 Turnips

The white turnip, also known as a turnip, is a root vegetable that is frequently grown in temperate regions of the world for its white, meaty taproot. The name "turnip" is a combination of the Latin word "napus," which refers to the plant, and "turn," as in turned or rounded on a lathe. Larger types are planted as cattle feed, while smaller, more delicate forms are grown for human use. In regions of Canada (Quebec, Newfoundland, Manitoba, and the Maritimes), Scotland, Ireland, Cornwall, and Northern England, the term "turnip" (or "neep") frequently refers to rutabaga, commonly known as "swede," a bigger, yellow root vegetable belonging to the same genus (*Brassica*).

9.3.2 Morphology

9.3.2.1 Carrots

In the Apiaceae family, the herbaceous carrot produces an edible taproot. Common forms of roots come in a variety of morphologies, from spherical to lengthy, with blunt to pointy bottom ends. White, yellow, and purple flesh variations are also known in addition to the orange-colored roots (Iorizzo et al. 2013). Sub-types of flowers were categorized based on the stamen's and petal's level of alteration. Light microscopy was used to examine flower anatomy to explain organ alteration and pinpoint the moment when morphology starts to change. According to model plants like Antirrhinum and Arabidopsis, the early stages of floral development in the fertile male flowers of cultivated carrots were characterized, and the comparable stages in the unique cytoplasmic male-sterile flower types were compared.

9.3.2.2 Turnips

These can be kept for months after harvest and do well in chilly areas. The leaves are typically sparsely downy, thin, and light green. The huge, sphere-shaped root of the turnip plant forms beneath the flowering stems and leaf petioles and has white-fleshed edible parts. The flowers are grouped at the apex of the raceme, and they are typically elevated above the terminal buds. Turnip plants begin to bolt in the late. This then leads to the growth of flower buds which are similarly consumed when still green before opening. On the other hand, turnip tops are recognizable for their color, fibrosity, moisture content, and bitter flavor and turnip greens stand out for their potent aroma, bright leaf color, and salty flavor. Little, tender turnips and huge turnips are the two varieties farmed, with the smaller turnips being grown for human use and the larger ones for use as cattle feed (Javed et al. 2019).

9.3.3 Composition of Carrot and Turnip Crop

9.3.3.1 Carrots

It ranges in moisture content from 86% to 89% and minerals including Ca, P, Fe, and Mg, as well as sugars, are abundant in carrots. Calcium (80 mg/100 g), iron (2.2 mg/100 g), and phosphorus (53 mg/100 g) are the other chemical components of carrots. They also contain protein (0.9%), fat (0.2%), carbohydrates (10.6%), crude fiber (1.2%), and total ash (1.1%). Six varieties of carrots were found to contain 1.67–3.35% reducing sugars, 1.02–1.18% non-reducing sugars, and 2.71–4.53% total sugars (Sharma et al. 2012). According to researchers, four hybrid types of carrots contained reducing sugars that made up 6–32% of the free sugars. In carrot roots, cellulose, hemicellulose, and lignin makeup, respectively, 71.7%, 13.0%, and 15.2% of the crude fiber. Carrot flavor is mostly brought about by glutamic acid's presence and the buffering effects of free amino acids. Glycolic acid, succinic acid, lactic acid, and ketoglutaric acid have all been reported to exist in trace amounts. Caffeine is the major phenolic acid present in carrots. In significant levels, carrot roots include thiamin, riboflavin, niacin, folic acid, and vitamin C. Black carrots have an anthocyanin level of 1750 mg/kg, while pink cultivars have trace quantities. The three primary anthocyanins, cyanidin 3-xylosylglucosylgalactoside, cyanidin 3-ferulylxyloglucosyl galactoside, have been found (Raees-ul and Prasad 2014).

9.3.3.2 Turnips

It contains a lot of glycosinolates, which are tri-substituted by glucose, amino acids, and sulfate. According to the different types of amino acids they include, glucosinolates are divided into three categories: aliphatic, aromatic, and indole. Comparative analysis of 59 kinds of turnip tubers from 4 agro-climatic regions in China yielded a total of 10 glucosinolates. The two primary glucosinolates among them are gluconasturtiin and glucobrassicanapin (Paul et al. 2019). It has been observed that phenolic compounds and organic acids, two widely distributed bioactive molecules in plants, have a variety of positive health effects, including antioxidation, anti-inflammation, anti-diabetes, and neuroprotection. Turnip flower buds, leaves, and stems all displayed high levels of total organic acids (between 36 and 51 g/kg, dry basis), which was much greater than that of roots. Malic acid predominates in all three of the turnip's edible sections, although its ratio varies greatly depending on which part of the plant is consumed: the roots had the highest level (81%), followed by the leaves and stems (65%), and flower buds had the lowest level (41%) (Fernandes et al. 2007). Turnip has a total soluble sugar content of roughly 345 g/kg (including leaf and bulb), which is greater than kale and rape but lower than swede. Glucose, fructose, and sucrose are the three main sugars that are soluble (Keim et al. 2020).

9.4 Harvesting and Storage of the Crops

9.4.1 Carrots

In carrots, under ideal conditions, the majority of cultivars need between 70 and 80 days to reach maturity, carrots can reach maturity in as short as 4 months (120 days) when grown from seeds. Although they prefer full sun for growth, they can withstand some shadow. The ideal temperature ranges from 16 to 21 °C (61 to 70 °F). The best type of soil is loamy or sandalwood soil with a pH of 6.3–6.8. The crop requires low nitrogen, moderate phosphate, and high amounts of potash, so fertilizer should be used in accordance with the type of soil. Plants shouldn't be planted in very rich or rocky soil because doing so will cause the roots to grow hair or twist (Abbott 2012). Irrigation is used as needed to maintain soil moisture. After sprouting, the crop is eventually weeded and cut to an 8–10 cm spacing to prevent competition below the soil. Carrots can be stored over the winter in the refrigerator for a few months or in a cool, damp environment. Unwashed carrots can be kept for a long time in soil, a sand-and-wood shavings mixture, or a bucket filled with sand layers. Carrots should be stored in an environment with a humidity of 98% and a temperature of 0–4 °C (32–40 °F). Browning, white flushing, and bitterness that might arise during storage can all lead to losses in carrot production. Bitterness can be decreased by storage in tight containers with little ethylene. White blush and browning can be avoided by applying edible coatings, heating the product, adding hydrogen sulfide, and using ultraviolet light (Papoutsis and Edelenbos 2021).

9.4.2 Turnips

Cold conditions are ideal for growing turnips as a root crop; warmer climates cause the roots to become woody and disagreeable. They are typically planted in the spring in places with frigid climates (like the northern US and Canada), where the vegetation only lasts 3–4 months. For a second fall crop, turnips can also be sown in temperate climates in the late summer (those having a growth season of 5–6 months). In areas with warm winters and growth seasons that span seven or more months, they are sown in the fall. It typically takes 55–60 days from planting to harvest. Turnips can be preserved for 2 weeks in the refrigerator or 8–10 months when frozen. To prevent cross-contamination, turnips are kept away from liquids and raw meat (Tanguler et al. 2021).

9.5 Processing and Production of Wastes of the Crops and Waste Composition

9.5.1 Carrots

In 2014, the world's total production of carrots and turnips was estimated at 38.84 million metric tonnes (MMT). With 1.44 MMT, the United States contributed less than 4% to world carrot production. Up to 50% of the raw resources are still in the form of pomace due to the low yields associated with the manufacturing of carrot juice, which is primarily disposed of as feed or manure. Nonetheless, this pomace is rich in beneficial substances such as carotenoids, dietary fiber, uronic acids, and neutral sugars. New solutions are required to lessen the problem because the pomace has occasionally caused environmental issues. Fruit and vegetable processing facilities are having trouble managing solid wastes with high BOD (Biochemical Oxygen Demand) since they are situated in crowded places with little room and a poor water supply. These wastes constitute a loss of priceless biomass and nutrients and face growing disposal challenges (Sharma et al. 2012). Up to 50% of the carrot's carotene is lost when it is processed into commercial juice, leaving 30–50% of the carrot as pomace. Depending on the processing conditions, the pomace's total carotene level could reach 2 g/kg dry matter (Singh et al. 2006). Adding value to the waste lowers the cost of the primary product, which directly benefits the processors and customers. During the main growing season, carrot shreds can be dried with or without juice extraction to create year-round availability for carrot products. Carrot pomace has been used to make functional drinks and meals such as bread, high-fiber biscuits, cake, sauces, pickles, and fortified wheat bread. Consumer acceptance of such products still needs to be demonstrated, especially in light of negative effects on sensory quality. To make high-fiber sweet or salty biscuits, carrot pomace powder has been examined for its approximate composition and total dietary fiber and added to wheat flour at levels of 10%, 20%, and 30%. Since powder had a sizable amount of dietary fiber and ash, it increased the concentrations of minerals and fiber in both types of biscuits (Kumari and Grewal 2007). According to a study on the kinetics of carrot pomace dehydration, 65 °C was the ideal drying temperature for retaining carotene and ascorbic acid (Upadhyay et al. 2008). The researchers claimed that high-quality toffees can be made from carrot pomace with a higher concentration of total reducing or non-reducing sugars, total soluble solids, ascorbic acid, and acidity (Jagtap et al. 2000). Statistical analysis of the data for several characteristics found significant variation between Asian and European carrot pomace toffees, despite the yield and chemical composition of the two types of toffees suggesting little difference between them. The effect of adding dry carrot pomace at 5%, 7.5%, and 10% was examined in wheat bread, and it became clear that the pomace added carotenoids, fiber, and mineral components to the bread. The addition of 5% carrot pomace was the optimum treatment from a rheological and organoleptic perspective (Tanska et al. 2007). Carotenoids are easily isomerized at high storage temperatures (45 °C) or during prolonged exposure to light, according to storage tests on dried carotenoid powder isolated from discarded carrots. Wheat is a poor source of micronutrients (1.4 mg/100 g of iron), so dried carrot pomace can improve the mineral composition of wheat bread. Furthermore, potassium, zinc, iron, and manganese make up 5.5% of the minerals in dried carrot pomace (Sharma et al. 2012).

9.5.2 Turnips

One of the fermented products is a reddish, sour, and hazy beverage known as (salgam, shalgam, or fermented turnip beverage). Both lactic acid and ethyl alcohol fermentation are the main processes that produce it. The manufacturing of this beverage as well as its consumption have been rising quickly in recent years in Turkey and other major European cities. As a result, more leftover turnips and bulgur flour (bulgur bran or setik) are produced as byproducts. Usually, these byproducts of the manufacture of shalgam are fed to animals or otherwise disposed of in the environment (especially in small businesses whose numbers are growing rapidly). A novel method of evaluation might be the utilization of these residuals to enhance certain quality aspects of tarhana (a dried food ingredient, based on a fermented mixture of grain and yogurt or fermented milk). Traditional Turkish alcoholic beverage shalgam is produced through lactic acid fermentation. Some European towns also have markets where shalgam is sold. Bulgur flour (It is the portion of the boiling dry wheat that remains under the screen after the outer shells have been broken during the crushing process), salt, water, purple carrot, turnip, and occasionally red beetroot are used in the preparation of shalgam. The standard production process can take 10–12 days. Production for commercial use takes 4–5 days (Tanguler et al. 2021). The turnip is regarded as a well-known, healthy root vegetable throughout Eastern America, Asia, and Europe. It was cultivated as a main food source during the ancient Roman and Greek eras (Lo Scalzo et al. 2008). Young turnips, commonly referred to as baby turnips, are consumed raw in salads. Compared to young turnips, which have a sweeter flavor and a more delicate texture, mature turnips have a stronger flavor and a harder, woodier texture.

9.6 Aspects of Waste Valorization

The agri-food industry generates a significant quantity of waste, primarily as a result of inefficiencies caused by product loss and damage across the supply chain (Paini et al. 2022).

9.6.1 Nutrients and Bioactive Components

9.6.1.1 Carrots

On a dry weight basis, carrot pomace includes 4–5% protein, 8–9% reducing sugar, 5–6% minerals, and 37–48% total dietary fiber, making carrot products a well-known source of dietary fiber. The amount of total β-carotenes in fresh, unblanched, and blanched carrots, respectively, is found in carrot pomace at 17% and 31–35%. Moreover, it has been stated that the nutritional fiber components of carrot pomace are pectin (3.88%), hemicellulose (12.3%), cellulose (51.6%), and lignin (32.1%) (on a dry weight basis) (Nawirska and Kwasniewska 2005).

Hence, after carrot juice has been extracted, the remains may contain substances with bioactive qualities that could be studied for use in the creation of food ingredients and dietary supplements. Phytonutrients like phenolics and carotenoids are abundant in carrots. Carrots are categorized as vitaminized foods since they are high in β-carotene, ascorbic acid, and tocopherol. Carrots are regarded as a functional food with substantial health-promoting benefits because of the high variety of various chemicals they contain. The physiological properties of phenols, also known as polyphenols, such as their antioxidant, antimutagenic, and anticancer activity, have drawn a lot of attention. According to reports, they could be a contender in the fight against free radicals, which injure our bodies and food systems (Sharma et al. 2012).

9.6.1.2 Turnips

Comparatively speaking, turnip leaves contain more nutrients than turnip roots. In addition to the elements listed above, some researchers claimed that turnip leaves are also a rich source of lutein, vitamin A, vitamin B, and vitamin K, as well as carotenoids (Javed et al. 2019). Various amino acid components, glucosinolates have been divided into aliphatic, aromatic, and indole glucosinolates (Paul et al. 2019). The breakdown product of glucosinolates is isothiocyanate. Turnips have been discovered to successfully lower the risk of cancer due to their high glucosinolate content. Many naturally occurring foods and their constituent parts are regarded as antioxidants and can be taken to protect cells from oxidative stress-induced damage. Turnips are a rich source of organic acids and phenolic compounds, which are natural antioxidants. The DPPH (2,2-diphenylpicryl hydrazyl) radical-scavenging assay was used to evaluate the antioxidant properties of turnip extracts from various sections (Fernandes et al. 2007).

9.6.2 Extraction by Using Conventional and Non-conventional Methods

Natural pigments and other bioactive substances can be found in waste or byproducts produced during food processing. Contrary to synthetic coloring agents, which have several detrimental impacts on human health, natural pigments, and coloring agents have several health benefits. As a result, the extraction of pigments from natural sources has attracted significant interest on a global scale. Traditional extraction methods including soxhlet, solvent extraction, and water distillation have several drawbacks including extended extraction times, high solvent consumption, and poor extraction yields. To improve the pigment extraction process, modern methods like high pressure, ultrasound, negative pressure, and electric field have been used. The development of new extraction techniques is primarily focused on increasing pigment output, using less solvent, minimizing environmental impact, and making extraction convenient (Wani et al. 2021).

9.6.2.1 Carrots

The bio-wastes from the food sector (such as peels, seeds, etc.) represent a rich source of bioactive ingredients but are not properly managed. Carotenoids were recovered from carrot pomace using high-shear dispersion and ultrasonication techniques, along with flaxseed oil as a sustainable solvent (green biorefinery approach). Maximum carotenoids were produced by high shear dispersers (94.8 ± 0.08% recovery) (Tiwari et al. 2019). The objectives of this study were to evaluate the high-pressure processing (HPP) method's capacity to extract bioactive compounds from black carrot pomace and to optimize the treatment environment (pressure, temperature, and processing time) to achieve the highest bioactive chemical yield possible. Pilot-scale high-pressure food processors with water as the extraction solvent were used in the experiments. For maximum extraction of bioactive chemicals, combined pressure-thermal treatment at 266.6 MPa, 78.3 °C for 13.6 min was shown to be the optimal process (Agcam et al. 2021).

The effect of microwave power, extraction duration, and oil-to-waste ratio on carotenoid recovery percentage was optimized using response surface methods. The optimal carotenoid recovery of 77.48% was achieved with the following conditions: 165 W of microwave power, 9.39 min of extraction time, and 8.06:1 g/g of oil to the waste ratio (Elik et al. 2020). Sunflower oil was used to extract carotenoids from carrot processing waste. The spray-dried encapsulate had the lowest water activity, moisture content, and particle size, while the freeze-dried encapsulate displayed the highest hygroscopicity, oxidative stability, and color characteristics (Seregelj et al. 2021). The optimal extraction conditions, which included a treatment temperature of 50 °C for 5 min and a solvent mixture of hexane and ethanol (1:3) with a solid-to-solvent ratio of 1:40 (w/v), resulted in a much greater yield of total carotenoids (192.81 ± 0.32 mg carotenoids/100 g DW) (Kaur et al. 2022b). Using two sources of ultrasonic irradiations, ultrasound-assisted extraction (UAE) of -carotene was explored to better understand the effects of key operational parameters and to investigate the usage of vegetable oils as novel solvents. The maximum extraction yield of β-carotene from an ultrasonic horn was 83.32% with 50 °C temperature, 100 W power, 60% duty cycle, and a solid-to-solvent ratio of 0.3:20 (g/mL of solvent), whereas the maximum extraction yield from an ultrasonic bath was 64.66% with optimal operating conditions. Higher β-carotene extraction yields were obtained by operating UAE (Purohit and Parag 2015). The use of multiple-pass ultrasonication with mechanical homogenization (MPUMH), optimized at 9 min of ultrasonication followed by 1 min of mechanical homogenization, was suggested as a sustainable valorization procedure for puree processing from processed carrot discards (PDCs) (Kaur et al. 2023a).

9.6.2.2 Turnips

Copper has numerous uses including construction, agriculture, shipbuilding, and the chemical industry. In a recent study, turnip peel extract (TPE) as an affordable, eco-friendly vegetable waste and green bio-inhibitor for copper corrosion in a 3.5 NaCl (%)solution was described. The TPE was prepared via ultrasound-assisted water extraction, which enables it to be employed as a green inhibitor in corrosive conditions naturally and directly (Fekri et al. 2022). Turnip peel extract has a low amount of sugar compared to the vegetable's sugar content. The compositions imply the viability of using biomass and alcohol produced through fermentation. Alcohol generated from turnip peel extract was 1.5% (w/v) (Kandari and Gupta 2012).

In another study, the TPC (total phenolic contents) of the polyphenolic extract of turnip produced with ultrasound assistance was 169.29 ± 6.89 mg GAE (gallic acid equivalent)/g dry weight (Yucetepe 2021). It was determined whether certain agri-horticultural wastes, such as leftover residues from turnip and cabbage had any synergistic effects when combined and were potentially useful as nutraceuticals (Gupta et al. 2017).

The ability of the enzyme peroxidase to remove phenolic compounds and aromatic amines from aqueous solutions as well as to decolorize textile effluents is well recognized. Turnip peroxidase extract was used in biocatalysis, and the results show that it is useful in facilitating the biotransformation of simple phenols like guaiacol (Rosa et al. 2020). In another study, a turnip peroxidase (TP) preparation's ability to discolor textile azo colors and effluents was evaluated. As a model pollutant for the enzyme's purification capabilities, Congo Red (CR), an azo dye, was used. To remove the most color, the best conditions were pH 2.0, 40 °C, 50 mM hydrogen peroxide, 50 mg/L CR dye, and 0.45 U/mL of TP activity during 10 min of incubation. By-products of the enzymatic treatment were examined using UV-Vis and IR spectroscopy, which revealed a polymeric precipitate and no remaining chemicals in the aqueous phase (Ahmedi et al. 2012; Ragab et al. 2022).

9.6.3 Applications of Carrot and Turnip Wastes in the Food Industry

9.6.3.1 Carrots

Carrots are a large source of bioactive chemicals with strong health-promoting qualities. A stronger immune system and a lower risk of cancer and degenerative diseases are linked to higher intakes of carotenoids. Moreover, carrots' flavor, fragrance, and bitterness—as well as their high antioxidant properties—are caused by the presence of phenolic chemicals. Wastes produced by the processing of carrots, such as peels and pomaces, can pose significant nutritional, financial, and environmental issues (Seregelj et al. 2020). After extraction juice pulp was utilized to make carrot tomato blended sauce (Ghoshal et al. 2009). For 6 months, the dried carrot pomace was kept at room temperature (28–42 °C) and used to make condensed milk with carrots as the main ingredient. A new technique was used to make condensed milk products using carrots after conducting early trials. The finished product made from osmo-convective dehydrated pomace was generally well-tolerated (Singh et al. 2006). Carrot pulp is an everyday agricultural and food waste in Romania. There are significant amounts of bioactive substances and fibers in this kind of food waste. The principal ingredients responsible for the color, aroma, and bitterness of carrots are anthocyanins (in the case of purple carrots) and carotenoids (in the other cases). Also, it has been shown that polyphenolic acids are capable of acting as antioxidants and that anthocyanins play a role in the decline of cardiovascular disease. Also, it has been shown that polyphenolic acids are capable of acting as antioxidants and that anthocyanins play a role in the decline of cardiovascular disease. If there is little demand for novel products made from vegetable industry waste, it is quite likely that they will not affect food waste or human health. Carrot powders, which are used in Indian foods like *chapati* and *halwa*, serve as an illustration of such a product. Another illustration is the use of carrot pomace, a by-product of the extraction of carrot juice, to boost the nutritional content of ready-to-eat snacks (Turturica and Bahrim 2021). This type of snack may be an intriguing suggestion for customers hoping to boost their daily diet of veggies due to the high concentration of carotenoids in it. Sliced carrots can be dried using hot air drying or a microwave-vacuum combination, or they can be fried in oil in a low-pressure medium. Both methods can be used to transform carrots into crispy treats (Carpentier et al. 2009). In this investigation, varying amounts of microcrystalline cellulose (MCC) and carrot fibers (CF) were added to chitosan (CH)-based films (0–5%, w/w). Scanning electron microscopy (SEM), Fourier transform infrared (FTIR) spectroscopy, water vapor permeability (WVP), and mechanical, thermal, and optical properties were used to characterize the produced films. With increasing filler concentrations, CH films' tensile characteristics and thermal stability were enhanced (Sogut and Cakmak 2020). Using carrot pomace-derived cell wall particles (CWP), a low-fat yogurt product was made (McCann et al. 2011). To create more nutrient-dense and fiber-rich extrudates using rice and gram pulse flour, the use of carrot pomace was examined (Upadhyay et al. 2010). The physicochemical characteristics of the developed products were enhanced by substituting carrot pomace flour for pseudocereal flour. Due to their high-quality protein, high fiber content, and antioxidant activity, pseudocereals and CP have the potential to fortify bakery goods for the development of functional foods to reduce the risk of health disorders like diabetes, obesity, CVDs, and many types of cancer (Saini et al. 2022). The study was focused on manufacturing whey-carrot-based functional beverages and examine how the 28-day storage period affected certain physicochemical characteristics. Given the advantages of the bioactive components found in both fruits and whey, it is conceivable that the whey-carrot beverage could be a potential product for the beverage industry (Naik et al. 2023). In another study, the objective was to design and standardize a bioprocess for the shelf-stable Kanji mix that contains freeze-dried lactic acid bacteria along with spices and Refractance-Window (RW)-dried black carrot powder (Kaur et al. 2023b). Jain et al. (2015) studied the microencapsulation of β- carotene by complex coacervation method and its release properties when whey protein isolates and gum acacia were used as coacervates. The same group made a similar type of study using casein/gum tragacanth (Jain et al.

2016) and casein/guar gum (Thakur et al. 2017). Kaur et al. (2019) studied fermented bio-color production from carrot waste in solid-state fermentation. Jain et al. (2019) studied β-carotene encapsulation using solid lipid nanoparticles.

9.6.3.2 Turnips

Brassica species are very effective for nitrogen and nitrogenous fertilizers which is one of the main expenses of crop production (Nawaz et al. 2020). In addition to their obvious potential for usage as biomaterials and energy sources in the future, it is also clear that these bio-wastes might be exploited as unconventional food sources in light of the current food shortage. These compounds can be used not only for direct consumption but also after processing to create new functional foods and as a source of natural ingredients (Chichoub et al. 2019).

The removal of arsenic(III) from drinking water using low-cost biosorbents made from orange peels, turnip peels, and peanut shells has been studied. In 360 min, orange peels adsorb the most arsenic (87.77%) and peanut shells adsorb the most arsenic (86.38%) from water at the ideal temperature (30 °C), pH, and adsorbent dosage of (1 g). In 330 min, turnip peels exhibit the highest rate of adsorption (81.73%) (Sidhu et al. 2014). Shalgam juice plant waste is among the most prevalent wastes in Turkey and its neighboring nations. In a recent work, Shalgam juice plant waste was used as a growth medium for the production of carotenoids from *Rhodotorula glutinis* (Keskin et al. 2023).

9.6.4 Utilization of Carrot and Turnip Waste in Pharmaceuticals/Nutraceuticals (Table 9.3)

9.6.4.1 Carrot

The waste from the food industry typically consists of dietary fibers, proteins and peptides, lipids, fatty acids, and phenolic compounds, which depend on the nature of the product produced (Sooch et al. 2022). It is essential for the commercialization of these bioactive chemicals so that they can be used in pharmaceutical and nutraceutical goods. Studies are conducted to recover these fibers from the waste residues of carrots since they have been suggested to have cholesterol-lowering properties that can prevent coronary heart ailments. Additionally, numerous initiatives are being made to use the beneficial chemicals from carrot wastes in the creation of functional foods and beverages (Anal 2018). Recovery of bioactive substances has a great deal of potential for their use as functional additives and nutraceuticals in the food sector as well as in pharmaceutical and cosmetic applications. Therefore, converting plant fruit processing byproducts into high-value-added compounds represents a promising alternative for managing fruit residue issues as well as for producing functional food products with high nutritional value and several potential positive health effects (Charalampia et al. 2019). Waste (wild carrot seeds, *Daucus carota* L.) created during the commercial extraction of 96% ethanol was re-extracted with 40% and 70% ethanol to recover important compounds for prospective application as safe food additives and functional food ingredients or nutraceuticals. According to the

Table 9.3 Studies on applications of carrot and turnip waste in the food industry, pharmaceuticals/nutraceuticals

Uses	Major findings	References
Carrot waste		
Bioactive components	Higher intakes of carotenoids are linked to improved immunological function, lowered risk of cancer, and fewer degenerative illnesses.	Seregelj et al. (2020) Ghoshal (2022)
Food applications	Carrot-tomato sauce: Different concentrations of xanthan and guar gum were used to make the blended carrot-tomato sauce. The most acceptable sauce, in terms of sensory qualities, was a blended carrot-tomato sauce with 0.1% xanthan gum. Condensed milk: After conducting preliminary tests, a novel method was developed to produce condensed milk products utilizing carrots. Osmo-convective dehydrated pomace was used to create a finished product that was typically well tolerated. Ready-to-eat snacks: Such a product is exampled by carrot powders, which are used in Indian dishes like *chapati* and *halwa*. Snacks: Sliced carrots can be dried using hot air-drying or a microwave-vacuum combination, or they can be fried in oil in a low-pressure medium to transform carrots into crispy treats. Yogurt: Using carrot pomace-derived cell wall particles (CWP), a low-fat yogurt product was developed. Extruded products: To create more nutrient-dense and fiber-rich extrudates using rice and gram pulse flour, the use of carrot pomace was examined. Bakery products: The physicochemical characteristics of the gluten-free bakery-developed products were enhanced by substituting carrot pomace flour for pseudocereal flour. Buckwheat flour and quinoa flour scored at levels of 72% overall acceptability, although amaranth flour performed best at levels of 66%. Beverage: The study was focused on manufacturing whey-carrot-based functional beverages. It was discovered that beverages with a lower percentage of carrot juice have better quality. Kanji: The objective was to design and standardize a bioprocess for the shelf-stable Kanji mix that contains freeze-dried lactic acid bacteria along with spices and Refractance-Window (RW)-dried black carrot powder.	Ghoshal et al. (2009) Singh et al. (2006) Turturica and Bahrim (2021) Carpentier et al. (2009) McCann et al. (2011) Upadhyay et al. (2010) Saini et al. (2022) Naik et al. (2023) Kaur et al. (2023b)
Chitosan films	In this investigation, varying amounts of microcrystalline cellulose (MCC) and carrot fibers (CF) were added to chitosan (CH)-based films (0–5%, w/w).	Sogut and Cakmak (2020)

(continued)

Table 9.3 (continued)

Uses	Major findings	References
Encapsulation	Microencapsulation of β-carotene by complex coacervation method and its release properties when whey protein isolates and gum acacia were used as coacervates. β-Carotene encapsulation using solid lipid nanoparticles was studied.	Jain et al. (2015, 2019)
Pharmaceutical/ nutraceutical industries	Sunscreen can be made by mixing wild carrot extract into a gel formulation. Carrot pomace has the potential to be used as a functional element for men with hyperlipidemia.	Pavlyuk et al. (2015)
Bioethanol production	Enzymatic hydrolysis of carrot pomace and subsequent fermentation were conducted to produce bioethanol for use as a biofuel. Carrot pomace including *S. cerevisiae* and *P. stipitis* produced the highest amounts of bioethanol, 6.91 and 2.66 g/L, respectively. To obtain ethanol, the study suggested using both waste products (beer yeast and carrot wastes) in a fermentation process. In this study using two inexpensive raw materials from local agro-industrial waste, it is possible to produce ethanol.	Barzee et al. (2019) Demiray et al. (2016) Aimaretti and Ybalo (2012)
Turnip waste		
Fertilizer industry	*Brassica* species is very effective for nitrogen and nitrogenous fertilizers, which is one of the main expenses of crop production.	Nawaz et al. (2020)
Biosorbents	The removal of arsenic(III) from drinking water using low-cost biosorbents made from orange peels, turnip peels, and peanut shells has been studied. In 330 min, turnip peels exhibit the highest rate of adsorption (81.73%).	Sidhu et al. (2014)
Growth medium	Shalgam juice plant waste was used as a growth medium for the production of carotenoids from *Rhodotorula glutinis*.	Keskin et al. (2023)
Pharmaceutical/ nutraceutical industry	The findings demonstrated that turnip root ethanol extract protects against early liver damage in diabetic rats. According to the findings, turnips are a simple dietary supply of physiologically active substances. Turnip juice dramatically reduced the amount of time people slept after taking phenobarbital.	Daryoush et al. (2011) Fernandes et al. (2007) Javed et al. (2019)
Bioethanol production	This study was the first to use discarded turnip juice as a raw material for the synthesis of bioethanol. *S. cerevisiae*, *Kluyveromyces marxianus*, and *P. stipitis* had the highest bioethanol concentrations and Qp values, which were 7.32, 5.34, and 1.55 g/L and 0.61, 0.44, and 0.05 g/L, respectively.	Demiray et al. (2018)

findings, the 70% EtOH extract contains more total phenol and flavonoids than the 40% EtOH extract. The 40% EtOH extract had a scavenging activity of 78.72% ± 3.276 while the 70% EtOH extract had an antioxidant activity of 86.88% ± 3.118. The findings of the tests on photo-protective activity demonstrated that some forms of sunscreen can be made by mixing wild carrot extract into a gel formulation (Pavlyuk et al. 2015). The residue left over after extracting carrot juice, known as carrot pomace, is frequently thrown away despite being nutrient-rich and having potential use as a functional element for men with hyperlipidemia.

9.6.4.2 Turnip

Aspergillus niger NCIM 595 was evaluated on two potentially useful substrates for the synthesis of citric acid: turnip whey and beetroot (Yadav et al. 2019). The impact of turnip root ethanolic extract was studied against early liver damage in rats with diabetes induced by alloxan (Daryoush et al. 2011). Moreover, turnips have several medicinal benefits, including a preventive effect against diabetic liver damage and other pharmacological properties (Fernandes et al. 2007). These advantageous qualities have been partly attributed to physiologically active turnip chemicals, which exhibit strong antioxidant activity. In the literature, it was noted that giving turnip juice dramatically reduced the amount of time people slept after taking phenobarbital (Javed et al. 2019).

9.6.5 Bioethanol Production

9.6.5.1 Carrots

Investigations have been conducted on the enzymatic hydrolysis of carrot pomace and subsequent fermentation to produce bioethanol for use as a biofuel. Utilizing xylanase, pectinase, and cellulase for enzymatic hydrolysis, the sugars generated were fermented to produce 28 g/L of ethanol (70% of the theoretical yield) (Barzee et al. 2019).

Another study claimed to have used dilute acid pretreatment without enzymatic hydrolysis before fermentation with yeasts, i.e., *Saccharomyces cerevisiae* or *Pichia stipites* but was only successful in obtaining a maximum ethanol titer of 6.91 g/L with *S. cerevisiae* (Demiray et al. 2016). To obtain ethanol, the study suggested using both waste products (beer yeast and carrot wastes) in a fermentation process. In this study using two inexpensive raw materials from local agro-industrial waste, it is possible to produce ethanol (Aimaretti and Ybalo 2012).

9.6.5.2 Turnips

This study was the first to use discarded turnip juice as a raw material for the synthesis of bioethanol. Methods for pretreatment, starting biomass loading, and fermentation duration were examined. Different types of nitrogen sources were supplemented, together with mineral salts and hydrolysates of agricultural waste, in the growth medium. *S. cerevisiae*, *Kluyveromyces marxianus*, and *P. stipitis* had the highest bioethanol concentrations and Qp values, which were 7.32, 5.34, and 1.55 g/L and 0.61, 0.44, and 0.05 g/L, respectively (Demiray et al. 2018).

9.6.6 Technical Difficulties

Fruit and vegetable byproducts are frequently abundant in both potentially food-grade and non-food-grade constituents. The potential economic benefits of utilizing these "high value" components from these by-products remain under consideration by scientists (Gupta et al. 2015). However, several significant technological and scientific challenges contribute to undermining the economic case for by-product upgrading. The following are two significant technological challenges:

9.6.6.1 Extraction Process

These phytochemical extracts must be prepared using extraction techniques that belong to certain standards. Some of the factors should be considered, such as the preference for using fresh raw materials for extraction. Using solvents like water, ethanol, or mixtures of those two that are compatible with food is also important. Enzymes that could destroy the phytochemicals during the extraction process need to be inactivated using thermal treatments in most cases. Depending on the cost of the acquired extract on the market, spray drying or freeze drying are viable methods that might be used to concentrate the extracts (Gupta et al. 2015). Before drying, extract purification using non-ionic polymeric resins may be utilized in some circumstances to concentrate the phytochemicals. In these situations, the phytochemicals are maintained in the stationary phase while the water extract (or a water solution of the ethanol-water extract) is filtered through the resin column. These substances are then eluted with ethanol, and the resulting extract is concentrated. The manufactured extracts are often in dry form, though liquid extracts are another option. These extracts can be utilized to make useful juices or dry extracts for use in tablets (Larrosa et al. 2002).

9.6.6.2 Microbiological Instability

Many byproducts, especially those made from fruit and vegetable trimmings, have high water content. The residues offer fungi and bacteria a top-notch breeding ground, and as they begin to degrade, they quickly lose their promise as a source of ingredients fit for use in food. This is the case when byproducts are kept at room temperature over a prolonged period. As a result, stabilization requires immediate attention, perhaps by the use of drying, cooling, or freezing techniques (Gupta et al. 2015).

9.7 Future Research and Conclusion

The scalable procedures are required for the preservation or enhancement of functional characteristics and bioactive substances from fruits and vegetable wastes which are necessary for the development of functional food ingredients. The wastes of agricultural fruits and vegetables are prospective sources of dietary fiber and other useful substances. The qualities and functionality of the developed products are determined by the method used to process these wastes. Carrot and turnip pomace may be converted into an ingredient for use in food production or may be extracted and purified as a source of antioxidant and phenolic components that can be commercialized as nutritional supplements. It is crucial to ensure that the extracts do not include high concentrations of pesticides and other agrochemicals. The regular analysis of pesticides in all of these items is consequently necessary. Recovery and purification of high-value bioactives from carrot and turnip pomace could be accomplished using effective, low-investment techniques on a wide scale. This study demonstrated that the waste part of these vegetables can be used as an essential source of valuable nutrients and new bioactive components that are helpful for human health.

Acknowledgments Ms. Akashdeep is thankful to Panjab University, Chandigarh for her fellowship.

References

Abbott C (2012) The year-round harvest: a seasonal guide to growing, eating, and preserving the fruits and vegetables of your labor. Adams Media, pp 54–55

Agcam E, Akyildiz A, Kamat S, Balasubramaniam VM (2021) Bioactive compounds extraction from the black carrot pomace with the assistance of high pressure processing: an optimization study. Waste Biomass Valoriz 12:5959–5977

Ahmedi A, Abouseoud M, Couvert A, Amrane A (2012) Enzymatic degradation of Congo red by turnip (*Brassica rapa*) peroxidase. Z Nat C 67:429–436

Aimaretti N, Ybalo C (2012) Valorization of carrot and yeast discards for the obtention of ethanol. Biomass Bioenergy 42:18–23

Anal AK (ed) (2018) Food processing by-products and their utilization: introduction. Food processing by-products and their utilization. John Wiley & Sons, pp 1–10

Ashfaq I, Alvi T, Hussain S, Adnan M, Syed A (2022) Nutritional profiling, management strategies, and valorization of vegetable wastes. RADS J Food Biosci 2:42–49

Auger B, Marnet N, Gautier V, Maia-Grondard A, Leprince F, Renard M, Guyot S, Nesi N, Routaboul JM (2010) A detailed survey of seed coat flavonoids in developing seeds of *Brassica napus* L. J Agric Food Chem 58:6246–6256

Aysun Y (2021) Valorization of peel wastes of purple turnip (*Brassica rapa* L.): extraction of polyphenolics through ultrasonic-assisted extraction and investigation of changes in total phenolic content, total monomeric anthocyanin content and total antioxidant capacity during in vitro gastro-intestinal digestion. Eur J Sci Technol 27:152–157

Baenas N, Moreno DA, Garcia-Viguera C (2012) Selecting sprouts of Brassicaceae for optimum phytochemical composition. J Agric Food Chem 60:11409–11114

Bansal V, Ray RC (2022) Overview of food loss and wastes in fruits and vegetables: from issues to resources. In: Ray RC (ed) Fruits and vegetable wastes: valorization to bioproducts and platform chemicals. Springer, Singapore, pp 3–32

Barzee TJ, El-Mashad HM, Zhang R, Pan Z (2019) sChapter 12 – Carrots. In: Pan Z, Zhang R, Zicari S (eds) Integrated processing technologies for food and agricultural by-products. Academic Press, pp 297–330. ISBN 9780128141380

Cao Q, Wang G, Peng Y (2021) A critical review on phytochemical profile and biological effects of turnip (*Brassica rapa* L.). Front Nutr 8:721733

Carpentier S, Knaus M, Suh M (2009) Associations between lutein, zeaxanthin, and age-related macular degeneration: an overview. Crit Rev Food Sci Nutr 49:313–326

Casazza AA, Pettinato M, Perego P (2020) Polyphenols from apple skins: a study on microwave-assisted extraction optimization and exhausted solid characterization. Sep Purif Technol 240: 116640

Charalampia D, Haralabos KC, Dimitrios S, Koutelidakis AE (2019) In vitro antioxidant, antithrombotic, antiatherogenic and antidiabetic activities of *Urtica dioica*, *Sideritis euboea* and *Cistus creticus* water extracts and investigation of pasta fortification with the most bioactive one. Curr Pharm Biotechnol 20:874–880

Chichoub W, Dias MI, Barros L, Calhelha RC, Alves MJ, Harzallah-Skhiri F, Ferreira ICFR (2019) Valorisation of the green waste parts from turnip, radish and wild cardoon: nutritional value, phenolic profile and bioactivity evaluation. Food Res Int 126:108651

Chrong EWI, Wong TY, Kreis AJ, Simpson JA, Guymer RH (2007) Dietary antioxidants and primary prevention of age related macular degeneration: systemic review and meta-analysis. Br Med J 335:755–759

Clementz A (2019) Novel method for valorization of by-products from carrot discards. LWT Food Sci Technol 100:374–380

Coman V, Teleky BE, Mitrea L, Martau GA, Szabo K, Calinoiu LF, Vodnar DC (2020) Bioactive potential of fruit and vegetable wastes. Adv Food Nutr Res 91:157–225

Daryoush M, Bahram AT, Yousef D, Mehrdad N (2011) Protective effect of turnip root (*Brassica rapa* L) ethanolic extract on early hepatic injury in alloxanized diabetic rats. Aust J Basic Appl Sci 5:748–756

Demiray E, Karatay SE, Donmez S, Donmez G (2016) The usage of carrot pomace for bioethanol production. J Chilean Chem Soc 61:2996–2998

Demiray E, Karatay SE, Ekici H, Dönmez G (2018) Introducing a new salty waste for second-generation bioethanol production. Energy Sources A Recov Util Environ Effects 40:2070–2078

Elik A, Yanik DK, Gogus F (2020) Microwave-assisted extraction of carotenoids from carrot juice processing waste using flaxseed oil as a solvent. LWT Food Sci Technol 123:109100

Elkhalifa S, Mariyam S, Mackey HR, Al-Ansari T, McKay G, Parthasarathy P (2022) Pyrolysis valorization of vegetable wastes: thermal, kinetic, thermodynamics, and pyrogas analyses. Energies 15:6277

FAO (2021) Fruit and vegetables-your dietary essentials. The International Year of Fruits and Vegetables

FAO, Food and Agricultural Organization of the United Nations (2011) The statistics division of FAO (FAOSTAT)

FAO, Food and Agricultural Organization of the United Nations (2022) The statistics division of FAO (FAOSTAT)

Fekri MH, Omidali F, Alemnezhad MM, Ghaffarinejad A (2022) Turnip peel extract as green corrosion bio-inhibitor for copper in 3.5% NaCl solution. Mater Chem Phys 286:126150

Fernandes F, Valentao P, Sousa C, Pereira J, Seabra R, Andrade P (2007) Chemical and antioxidative assessment of dietary turnip (*Brassica rapa* var. *rapa* L.). Food Chem 105: 1003–1010

Filimonau V, De Coteau DA (2019) Food waste management in hospitality operations: a critical review. Tour Manag 71:234–245

Francisco M, Cartea ME, Soengas P, Velasco P (2011) Effect of genotype and environmental conditions on health promoting compounds in *Brassica rapa*. J Agric Food Chem 59:2421–2431

Garcia SLR, Raghavan V (2022) Green extraction techniques from fruit and vegetable waste to obtain bioactive compounds—a review. Crit Rev Food Sci Nutr 62:6446–6466

Ghoshal G (2022) Vegetable and fruit waste: utilization in novel industrial applications. In: Roy RC (ed) Fruits and vegetable waste, pp 207–235. ISBN: 978-981-16-9527-8. https://doi.org/10.1007/978-981-16-9527-8

Ghoshal G, Basu S, US, Shivhare (2009) Effect of hydrocolloids on rheological, physico-chemical and sensory properties of carrot-tomato blended sauce. J Food Sci Technol 46(3): 202–206. ISSN: 0022-1155

Gupta P, Ray J, Aggarwal BK, Goyal P (2015) Food processing residue analysis and its functional components as related to human health: recent developments. Austin J Nutr Food Sci 3:1068

Gupta C, Prakash D, Gupta S (2017) Antioxidant potential of some agri-horticultural wastes. Int Food Res J 24:1156–1159

Gustavsson J, Cederberg C, Sonesson U (2011) Global food losses and food waste, vol 38. The Food and Agriculture Organization of the United Nations, pp 28–34

Iorizzo M, Senalik DA, Ellison SL, Grzebelus D, Cavagnaro PF, Allender C, Brunet J, Spooner DM, Van DA, Simon PW (2013) Genetic structure and domestication of carrot (*Daucus carota* subsp. *sativus*) (*Apiaceae*). Am J Bot 100:930–938

Iorizzo M, Curaba J, Pottorff M, Ferruzzi MG, Simon P, Cavagnaro PF (2020) Carrot anthocyanins genetics and genomics: Status and perspectives to improve its application for the food colorant industry. Genes 11:906

Jagtap SA, Adsule RN, Chavan UD (2000) Utilization of carrot pomace for preparation of toffees. Beverage Food World 27:30–32

Jain A, Sharma G, Ghoshal G, Singh B, Shivhare US, Katare OP (2015) Microencapsulation by complex coacervation using whey protein isolates and gum acacia: an approach to preserve the functionality and controlled release of β- carotene. Food Bioprocess Technol 8(8):1635–1644. https://doi.org/10.1007/s11947-015-1521-0

Jain A, Thakur D, ghoshal G, Katare OP, Shivhare US (2016) Characterization of microcapsulated β-carotene formed by complex coacervation using casein and gum tragacanth. Int J Biol Macromol 87:101–113

Jain A, Sharma G, Ghoshal G, Singh B, Shivhare US, Katare OP (2019) Beta carotene-encapsulated solid lipid nanoparticle (BC-SLNs) as a promising vehicle for cancer: an investigative assessment. AAPS PharmaSciTech 20(3):100

Javed A, Ahmad A, Nouman M, Hameed A, Tahir A, Shabbir U (2019) Turnip (*Brassica rapus* L.): a natural health tonic. Brazil J Food Technol 22:e2018253

Jenkins W, Tucker ME, Grim J (2016) Routledge handbook of religion and ecology. Routledge, London

Kamiloglu S, Toydemir G, Boyacioglu D, Beekwilder J, Hall R, Capanoglu E (2016) A review on the effect of drying on antioxidant potential of fruits and vegetables. Crit Rev Food Sci Nutr 56: 110–129

Kandari V, Gupta S (2012) Bioconversion of vegetable and fruit peel wastes in viable product. J Microbiol Biotechnol Res 2:308–312

Kaur P, Ghoshal G, Jain A (2019) Bio-utilization of fruits and vegetables waste to produce β-carotene in solid-state fermentation: characterization and antioxidant activity. Process Biochem 76:155–164

Kaur P, Elsayed A, Subramanian J, Singh A (2021) Encapsulation of carotenoids with sucrose by co-crystallization: physicochemical properties, characterization and thermal stability of pigments. LWT Food Sci Technol 140:110810

Kaur P, Singh S, Ghoshal G, Ramamurthy PC, Parihar P, Singh J, Singh A (2022a) Valorization of agri-food industry waste for the production of microbial pigments: an eco-friendly approach. In:

Nayak K et al (eds) Advances in agricultural and industrial microbiology. https://doi.org/10.1007/978-981-16-8918-5_8

Kaur P, Subramanian J, Singh A (2022b) Green extraction of bioactive components from carrot industry waste and evaluation of spent residue as an energy source. Sci Rep 12:16607

Kaur GJ, Kumar D, Orsat V, Singh A (2023a) Techno-economic analysis of the multiple-pass ultrasonication with mechanical homogenization (MPUMH) processing of processed carrot discards to puree. Foods 12:157

Kaur P, Zalpouri R, Modi R, Sahota PP, Dhillon TS, Kaur A (2023b) Development and standardization of processing technique for ready-to-use lab fermented Kanji mix using refractance window dried black carrot powder. Sci Rep 13:185

Kaveh M, Chayjan RA (2017) Modeling thin-layer drying of turnip slices under semi-industrial continuous band dryer. J Food Process Preserv 41:e12778

Kee SH (2021) Bioconversion of agro-industry sourced biowaste into biomaterials via microbial factories—a viable domain of circular economy. Environ Pollut 271:116311

Keim JP, Gandarillas M, Benavides D, Cabanilla J, Pulido RG, Balocchi OA (2020) Nutrient concentrations and profile of non-structural carbohydrates vary among different Brassica forages. Anim Prod Sci 60:1503–1513

Keskin A, Fırat M, Ünlü Büyüktopcu AE (2023) Valorization of shalgam juice plant waste for the production of carotenoids by *Rhodotorula glutinis*. Int J Agric Environ Food Sci 7:79–87

Kumari S, Grewal RB (2007) Nutritional evaluation and utilization of carrot pomace for preparation of high-fiber biscuits. J Food Sci Technol 44:56–58

Kumari P, Mehta A, Sehgal R, Ray RC, Gupta R (2022) Microbial enzymes and organic acids production from vegetables and fruit wastes and their applications. In: Ray RC (ed) Fruits and vegetable wastes: valorization to bioproducts and platform chemicals. Springer, Singapore, pp 237–257

Larrosa M, Llorach R, Espín JC, Tomás-Barberán FA (2002) Increase of antioxidant activity of tomato juice upon functionalisation with vegetable byproducts extracts. LWT- Food Sci Technol 35:532–542

Lo Scalzo R, Genna A, Branca F, Chedin M, Chassaigne H (2008) Anthocyanin composition of cauliflower (*Brassica oleracea* L. var. *botrytis*) and cabbage (*B. oleracea* L. *var. capitata*) and its stability in relation to thermal treatments. Food Chem 107:136–144

McCann TH, Fabre F, Day L (2011) Microstructure, rheology and storage stability of low-fat yoghurt structured by carrot cell wall particles. Food Res Int 44:884–892

Muheem A, Shakeel F, Jahangir MA, Anwar M, Mallick N, Jain GK, Warsi MH, Ahmad FJ (2016) A review on the strategies for oral delivery of proteins and peptides and their clinical perspectives. Saudi Pharm J 24:413–428

Naik B, Walter N, Kohli D, Gupta AK, Mishra S, Khan JM, Saris PEJ, Irfan M, Rustagi S, Kumar V (2023) Whey-carrot based functional beverage: development and storage study. J King Saud Univ Sci 35:102775

Nawaz MQ, Ahmed K, Qadir G, Rizwan M, Nawaz MF, Sarfraz M (2020) Growth and yield of turnip (*Brassica rapa* L.) in response to different sowing methods and nitrogen levels in salt-affected soils. Pak J Agric Res 33:126–134

Nawirska A, Kwasniewska M (2005) Dietary fiber fractions from fruit and vegetable processing waste. Food Chem 91:221–225

Neri L, Faieta M, Di Mattia C, Sacchetti G, Mastrocola D, Pittia P (2020) Antioxidant activity in frozen plant foods: effect of cryoprotectants, freezing process and frozen storage. Foods 9:1886

Paini J, Benedetti V, Ail SS, Castaldi MJ, Baratieri M, Patuzzi F (2022) Valorization of wastes from the food production industry: a review towards an integrated agri-food processing biorefinery. Waste Biomass Valoriz 13:31–50

Papoutsis K, Edelenbos M (2021) Postharvest environmentally and human-friendly pre-treatments to minimize carrot waste in the supply chain caused by physiological disorders and fungi. Trends Food Sci Technol 112:88–98

Paul S, Geng CA, Yang TH, Yang YP, Chen JJ (2019) Phytochemical and health beneficial progress of turnip (*Brassica rapa*). J Food Sci 84:19–30

Pavlyuk I, Stadnytska N, Jasicka-Misiak I, Górka B, Wieczorek PP, Novikov V (2015) A study of the chemical composition and biological activity of extracts from wild carrot (*Daucus Carota.* L) Seeds waste. Res J Pharm Biol Chem Sci 6:603–611

Plazzotta S, Manzocco L, Nicoli MC (2017) Fruit and vegetable waste management and the challenge of fresh-cut salad. Trends Food Sci Technol 63:51–59

Purohit AJ, Parag RG (2015) Ultrasound-assisted extraction of β-carotene from waste carrot residue: effect of operating parameters and type of ultrasonic irradiation. Sep Sci Technol 50: 1507–1517

Raees-ul H, Prasad K (2014) Carrot is one of the most nutritious root crops. In: Ingredients South Asia. FFC Information Solution Pvt. Ltd., pp 94–95

Raees-ul H, Prasad K (2015) Nutritional and processing aspects of carrot (*Daucus carota*)—a review. South Asian J Food Technol Environ 1:1–14

Ragab MM, Othman HA, Hassabo AG (2022) Various extraction methods of different enzymes and their potential applications in various industrial sectors (a review). Egypt J Chem 65:495–508

Ray RC (2022) Fruits and vegetable wastes: valorization into bioproducts and platform chemicals. Springer, Singapore, 448p

Rosa GP, Barreto MDC, Pinto Diana CGA, Seca Ana ML (2020) A green and simple protocol for extraction and application of a peroxidase-rich enzymatic extraction. Methods Protocols 3:25

Sagar NA, Pareek S, Sharma S, Yahia EM, Lobo MG (2018) Fruit and vegetable waste: bioactive compounds, their extraction, and possible utilization. Compr Rev Food Sci Food Saf 17:512–531

Saini HK, Sharma KD, Anand A, Sharma R, Gautam A, Soni P (2022) Development of gluten-free bakery products enriched with bio-waste from carrot juice processing. Pharma Innov J 11:2310–2316

Sanchez M, Laca A, Laca A, Diaz M (2021) Value-added products from fruit and vegetable wastes: a review. Clean Soil Air Water 49:2000376

Scarano A, Gerardi C, D'Amico L, Accogli R, Santino A (2018) Phytochemical analysis and antioxidant properties in colored tiggiano carrots. Agriculture 8:102

Seregelj V, Vulic J, Cetkovic G, Canadanovc-Brunet J, Saponjac VT, Stajcic S (2020) Natural bioactive compounds in carrot waste for food applications and health benefits. Stud Nat Prod Chem 67:307–344

Seregelj V, Cetkovic G, Canadanovc-Brunet J, Saponjac VT, Vulic J, Levic S, Nedovic V, Brandolini A, Hidalgo A (2021) Encapsulation of carrot waste extract by freeze and spray drying techniques: an optimization study. LWT Food Sci Technol 138:110696

Sharma KD, Karki S, Thakur NS, Attri S (2012) Chemical composition, functional properties and processing of carrot—a review. J Food Sci Technol 49:22–32

Sharma M, Usmani Z, Gupta VK, Bhat R (2021) Valorization of fruits and vegetable wastes and by-products to produce natural pigments. Crit Rev Biotechnol 41:535–563

Shiri T, Kumar A, Sharma S, Kumar A, Singh G, Rashmi. (2020) Effects of liquid amla curd organic biomanure (LACOB) on growth and yield of turnip (*Brassica rapa*). Pharma Innov J 9: 200–207

Sidhu M, Sama P, Parmar J, Bhatt SM (2014) Biosorption of arsenic (III) from drinking water by using low cost biosorbents derived from peels of oranges, turnip and peanut shells. Int J Pharm Res Drug Dev 1:66–69

Singh B, Panesar PS, Nanda V (2006) Utilization of carrot pomace for the preparation of a value added product. World J Dairy Food Sci 1:22–27

Sogut E, Cakmak H (2020) Utilization of carrot (*Daucus carota* L.) fiber as a filler for chitosan based films. Food Hydrocoll 106:105861

Sooch BS, Mann MK, Sharma P, Ray RC (2022) Phenolics and other bioactive compounds from vegetables and fruit wastes: extraction methods and their possible utilizations. In: Ray RC (ed) Fruits and vegetable wastes: valorization to bioproducts and platform chemicals. Springer, Singapore, pp 81–114

Stahl M, Berghel J (2011) Energy efficient pilot-scale production of wood fuel pellets made from a raw material mix including sawdust and rapeseed cake. Biomass Bioenergy 35:4849–4854

Tanguler H, Cankaya A, Agcam E, Uslu H (2021) Effect of temperature and production method on some quality parameters of fermented carrot juice (Shalgam). Food Biosci 41:100973

Tanska M, Zadernowski R, Konopka I (2007) The quality of wheat bread supplemented with dried carrot pomace. Pol J Nat Sci 22:126–136

Taveira M, Fernandes F, Pinho PG, Andrade PB, Pereira JA, Valentao P (2009) Evolution of *Brassica rapa* var. *rapa* L. volatile composition by HS-SPME and GC/IT-MS. Microchem J 93: 140–146

Thakur D, Ghoshal G, Jain A, Shivhare US, Katare OP (2017) Microencapsulation of β-carotene based on Casein/Guar gum blend using zeta potential-yield stress phenomenon: an approach to enhance photo-stability and retention of functionality. AAPS PharmSciTech 18(5:1447

Thamburaj S, Singh N (2005) Textbook of vegetables, tuber crops and spices. Indian Council of Agriculture Research, New Delhi

Tiwari S, Upadhyay N, Singh AK, Meena GS, Arora S (2019) Organic solvent-free extraction of carotenoids from carrot bio-waste and its physicochemical properties. J Food Sci Technol 56: 4678–4687

Turturica M, Bahrim GE (2021) Carrot- application in food industry and health benefits. Innov Romanian Food Biotechnol 21:1–21

Upadhyay A, Sharma HK, Sarkar BC (2008) Characterization and dehydration kinetics of carrot pomace. Int Agric Eng J 10:1–9

Upadhyay A, Sharma HK, Sarkar BC (2010) Optimization of carrot pomace powder incorporation on extruded product quality by response surface methodology. J Food Qual 33:350–369

USDA (2015) https://www.usda.gov/foodwaste/faqs#:~:text=What%20is%20the%20U.S.%20Food,percent%20by%20the%20year%202030

USDA, National Agricultural Statistics Service (2011) Data and statistics (vegetables)

Vogl-Lukasser B, Vogl CR, Reiner H (2007) The turnip (*Brassica rapa* L. subsp. *rapa*) in Eastern Tyrol (Lienz district; Austria). Ethnobot Res Appl 5:305–317

Wani FA, Rashid R, Jabeen A, Brochier B, Yadav S, Aijaz T, Makroo HA, Dar BN (2021) Valorisation of food wastes to produce natural pigments using non-thermal novel extraction methods: a review. Int J Food Sci Technol 56:4823–4833

Yadav Y, Sunny S, Sharma KC, Peter H, Adhana KR (2019) Review: a novel ayurvedic techniques approach for value addition of waste materials. Res J Life Sci Bioinformatics Pharm Chem Sci 5:473–481

Yucetepe A (2021) Valorization of peel wastes of purple turnip (Brassica rapa L.): extraction of phenolics through ultrasonic-assisted extraction and investigation of changes in total phenolic content, total monomeric anthocyanin content and total antioxidant capacity during in vitro gastro-intestinal digestion. Eur J Sci Technol 27:152–157

Zohri A, El-Kady IA, Ragab S (2007) Bioproduction of lovastatin (mevacor; a hypocholesterolemic drug) by some fungi using agro-industrial by-products. In: International conference on: world perspectives for sugar beet and cane as a food and energy crop, Sharm El Sheikh, Egypt

Sugar Beet Waste as Substrate for Microbial Production of Food Ingredients

10

Negar Rostami and Kianoush Khosravi-Darani

Abstract

This chapter explored the potential of sugar beet waste as a valuable resource for biorefinery processes through fermentation. Sugar beet is a major crop used for large-scale sugar production. The chapter emphasized the types of liquid and solid waste generated during sugar beet processing, including molasses and sugar beet pulp, which contain valuable compounds like cellulose, hemicellulose, and pectin. These waste materials can be utilized in various applications, including animal feed, bioethanol production, and the generation of biochemicals and pharmaceutical intermediates. Fermentation is presented as a powerful tool for valorization, converting organic compounds into valuable products through microbial action. The biorefinery approach, inspired by traditional petroleum refineries, aims to extract maximum value from renewable resources and waste streams. Sugar beet waste, including pulp and residual components, can be effectively utilized in biorefinery processes, providing a long-term and comprehensive approach to value development.

Keywords

Sugar beet · Fermentation · Molasses · Sugar beet pulp · Single-cell protein

N. Rostami · K. Khosravi-Darani (✉)
Department of Food Technology Research, National Nutrition and Food Technology Research Institute, Faculty of Nutrition Science and Food Technology, Shahid Beheshti University of Medical Sciences, Tehran, Iran
e-mail: k.khosravi@sbmu.ac.ir

R. C. Ray (ed.), *Roots, Tubers, and Bulb Crop Wastes: Management by Biorefinery Approaches*, https://doi.org/10.1007/978-981-99-8266-0_10

Abbreviations

DSA	Dilute sulfuric acid
HMF	Hydroxyl-methyl-furfural
LHW	Liquid hot water
PHAs	Polyhydroxyalkanoates
SBM	Sugar beet molasses
SBP	Sugar beet pulp
SCP	Single-cell protein
SSF	Solid-state fermentation

10.1 Introduction

In 1747, a German scientist named Andreas Marggraf demonstrated that the crystals formed from powdery beetroots during crude extraction bore a resemblance to sucrose crystals. Franz Carl, motivated by the desire to explore sugar extraction from beets, embarked on experimental endeavors that ultimately sparked the emergence and progression of the sugar beet industry (Cosyn et al. 2011; Sadras and Calderini 2020). As a result, during the late 1800s, new sugar factories were established throughout nearly the entire European continent (Sadras and Calderini 2020). Sugar beet (*Beta vulgaris*) and sugar cane (*Saccharum officinarum*) are two plants that are used for large-scale sugar production in many countries. Sugar cane is the primary crop, while sugar beet is the second most widely grown crop in the world, accounting for around one-quarter of production (Gonzalez and Björnsson 2022; Bradshaw 2010). Despite being more expensive than sugar cane production, sugar beet cultivation is on the rise in several countries due to financial support for the sugar industry and certain agricultural and social advantages. Moreover, the recent advancement of using sugar beets for producing ethanol fuel is another advantage that has garnered significant attention (Erdal et al. 2007). Initially, Sugar beet (*Beta vulgaris* ssp. *vulgaris*) was grown mainly for animal feed, but later it replaced sugar cane as the primary source of sucrose in continental Europe (Finkenstadt 2014). It is now one of the most crucial crops in Europe and accounts for approximately 20% of global sugar production (2018). It should be mentioned that the farming methods utilized for cane and beet differ due to differences in the anatomy and physiology of the plants as well as the geographic and other conditions in which they are commonly grown (Cheesman 2004; Zheng et al. 2013). Sugar canes are well-suited for growing in the tropics and subtropics region including India, Australia, Cuba, Brazil, etc., whereas sugar beets are typically considered to be temperate region crops and are grown in countries like France, Germany, and the northern United States. However, new resistant varieties have made it a viable cash crop in the tropics and subtropics as well (Erdal et al. 2007; Saulnier and Thibault 1999; Bradshaw 2010). In 52 countries' agricultural systems, sugar beet is a crucial component that thrives in regions with a temperate climate (i.e., central and south Europe, USA, etc.) (Stevanato et al. 2019). For instance, in the USA, sugar beet accounts for approximately 50% of the edible sugar market (Webster et al. 2016).

Sugar beet is a very consistent plant. It can be resistant to water stress (Mall et al. 2018), resistant to frost (Webster et al. 2016), or even salinity tolerant (Mall et al. 2021). Sugar beet is a major agricultural crop and nutrient-rich commodity. It is the raw material for numerous industrial products and is used primarily in sucrose production (Usmani et al. 2022). The root is robust and conical, traversed by two sacchariferous furrows, and has a flat neck; it is almost completely buried in the ground (Clapham et al. 1990; Zicari et al. 2019). The plant is a glabrous or slightly hairy annual, biennial, or perennial with a wide range of heights ranging from 30 to 120 cm (or even 200 cm). Furthermore, stems are more or less branching, decumbent, rising, or upright. The size, shape, and color of leaves are extremely variable; they are often dark green or reddish and somewhat glossy, frequently creating a radicle rosette (Clapham et al. 1990; Højland and Pedersen 1994). According to studies, differing varietals, genetics, agronomy, ecological factors, harvesting, and post-harvesting conditions all affect the nutritional content of fresh beetroot (Chawla et al. 2016; Zicari et al. 2019). But commonly, it contains carbohydrates (9.96 g/100 g), including starch, fructose, sucrose, glucose, and dietary fiber; protein (1.68 g/100 g), fat (0.18 g/100 g), and leaves also contain carbohydrates (5 g/100 g), starch (4.5 g/100 g), and protein (14.8 mg/100 g) (Agarwal and Varma 2014; Richardson 2014). Vitamins A, B1, B2, B3, B6, C, and folate, as well as minerals including sodium, calcium, iron, phosphorus, potassium, magnesium, and zinc, are all present in beets. Beet leaves are also higher in nutrients than beetroot (Agarwal and Varma 2014). Nowadays, sugar has a wide range of uses in human nutrition (Sanchez and Cardona 2008) as a source of energy, sweeteners, and preservatives (Asadi 2006). Additionally, a sugar market is emerging recently as both academics and manufacturers are trying to expand the use of sucrose beyond food products (Van Putten et al. 2013).

10.2 World Production of Sugar Beet (FAO Data)

In 2021, around 270.16 MT (million metric tons) of sugar beet was produced globally, an increase from roughly 255.5 MT produced in the prior year (Shahbandeh 2022). Sugar beet production may reach 284 MT by 2031(FAO 2023). Research has focused on drought-resistant varieties 80% of the sugar crops will be utilized to produce sugar (FAO 2023). Sugar is increasingly becoming more and more readily affordable and available in the future, which may lead to more risk for nonchronic disease due to excessive consumption of sugar (Shahbandeh 2022). Agricultural waste is a byproduct derived from the processing of crops and other agricultural products (Tsang et al. 2019). According to the Food and Agriculture Organization of the United Nations, nearly one-third of all food produced globally for human nourishment is lost or wasted, or 1.6 billion tons annually (Bancal and Ray 2022). In addition, fruits, vegetables, roots, and tubers account for around

40–50% of all the food that is wasted (Searle and Malins 2013; Ray 2022). For instance, from the processing of sugar beet to sugar, large amounts of various types of waste, comprising leaves, sugar beet pulp, bioethanol, and vinasse, are obtained (Cárdenas-Fernández et al. 2017), which can be used as food/chemical additives and/or as nutraceuticals (Poel et al. 1998; Barnabas et al. 2015), and can be used as valuable substrates in biotechnology. Such biomass may be used for microbial cultivation to generate cellular proteins, organic acids, biologically important secondary metabolites, enzymes, prebiotic oligosaccharides, and other invaluable products (Ptak et al. 2022). Sugar beet pulp can be applied in the production of animal feed, while molasses is usually used for the biosynthesis of biomass and alcohol (Duraisam et al. 2017; Ray and Ramachandran 2018). Recently, sugar beet waste are applied for the production of biochemical, biofuels (Olmos and Hansen 2012), and pharmaceutical intermediates (Cárdenas-Fernández et al. 2017) as well as lactic fermentation (Tomaszewska et al. 2018), and ethanol production (Arshad et al. 2017).

10.3 Taxonomy and Botany

In ancient times, different varieties of the beet plant, in both white and red versions (Székely and Máte 2022), were already being cultivated on the shores of the Mediterranean (Cooke and Scott 2012). The sugar beet presently cultivated is very remote from the garden plant (Draycott 2008). The sea beet (*Beta vulgaris* subsp. *maritime*) is the forerunner of all the beet cultivars cultivated today (Székely and Máte 2022). The *Beta* genus is divided into the sections *Beta*, *Nanae*, *Procumbentes*, and *Corollinae* (Abu-Ellail et al. 2021), wherein cultivated beets are implemented in the subspecies *vulgaris* and include leafy beets (chard), garden beets (red), fodder beets (forage), and sugar beets (Biancardi et al. 2010). Sugar beet (*B. vulgaris* L. ssp. *vulgaris*) is a member of the family Chenopodiaceae (known today as Amaranthaceae) and the order Caryophyllales (Abu-Ellail et al. 2021). Cultivated beets (*B. vulgaris* ssp. *vulgaris*), comprising sugar beet, rank among the highly influential crops (Wascher et al. 2022). The cultivated beets are biennial plants (Elliott and Weston 1993); with a two-year vegetative cycle (Kaffka and Grantz 2014). They need to have a vernalization period (overwintering) to start the reproductive phase (Letschert 1993), so they generate a rosette and a carrot-shaped body in their first year. The next year, following a period of winter dormancy, the seed stalk and bloom appear (Smit 1983). Bare, simple, and glossy leaves vary in color from dark green to dark red. The most frequently consumed part is the body of the beet, which can be divided into cylindrical, round, and, flat groups according to its form (Barnóczki et al. 2010) (Fig. 10.1).

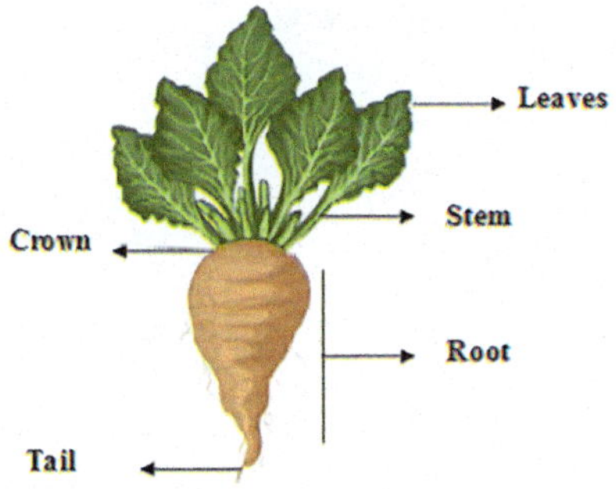

Fig. 10.1 Main parts of sugar beet

10.4 Types of Liquid and Solid Waste (Source of Cellulose, Hemicellulose, and Pectin)

Sugar beet contains 75% water, 20% sugar, and 5% pulp and accounted for 27% of the world's sugar production (Bhat et al. 2015). The principal objective of growing sugar beet is the generation of refined sugar. However, any additional by-products that are left over or produced during diffusion, juice purification, evaporation, crystallization, dried pulp manufacturing, and sugar recovery (Kopitzky 2021), may be worth much more commercially than the original commodity (Caliceti et al. 2022). Sugar beet waste includes molasses, sugar beet pulp, and sugar beet mud (Bhat et al. 2015). The sheer volume of extracted waste from sugar production is molasses and pulp (Paroha et al. 2020; Sauthoff et al. 2016). Sugar beet pulp (SBP), the water-insoluble residue which arises after the extraction of beet chips (Mcginnis 1982), is mostly composed of cell-wall polysaccharides (Sun and Hughes 1999): cellulose, hemicelluloses, and pectin (75–85%), a little amount of lignin (<9%), protein (<7%), lipids (<2%), saponins (<2%) and ash. Approximately 45–50 kg of SBP (dry matter) can be derived from one ton of fresh sugar beet (Berlowska et al. 2018; Kopitzky 2021; Rouilly et al. 2006; Van Der Poel 2000). Sugar beet pulp is mostly utilized as feed for livestock, and to a smaller extent in chemical processing in the ethanol production process (Rouilly et al. 2009). Moreover, the rich nutrient and mineral content of sugar beet pulp presents them as efficient raw substrate for industrial generation ofavariety of high-value products including bioethanol, biogas, biopolymers, and bio plastics (Usmani et al. 2022). sugar beet molasses (SBM) is the by-product that is left over after the crystallizable during the downstream process (Tondro et al. 2020), sugar has been separated from the concentrated beet juice (Handbook 2009), so it encompasses approximately 50% of residual sugars which sucrose is the main fermentable sugar, some other non-sugar molecules includes 10–15% minerals and trace elements (Mustafa et al. 2020; Mladenović et al. 2016), 15–20% non-sugar organic substances and about 20% water (Mischopoulou et al. 2016). As a result, molasses is used as animal feed, for yeast cultivation, and the generation of ethanol and organic acids (Sindhu et al. 2016). There are several ways for applications of sugar beet, which are summarized

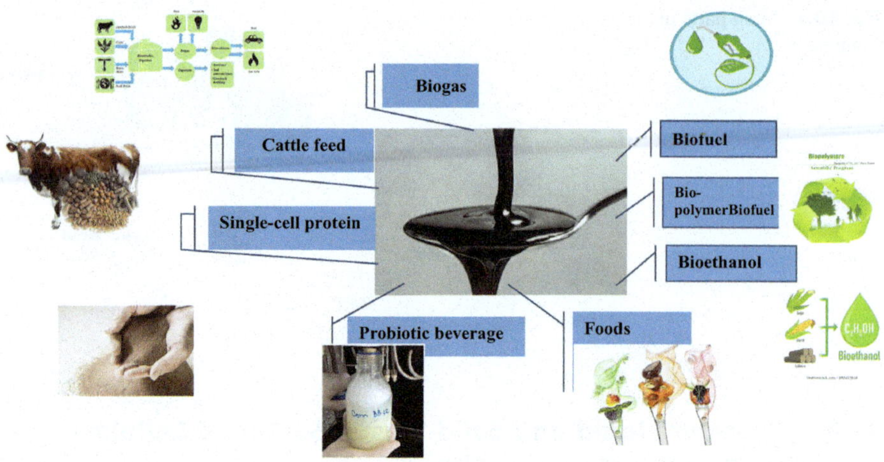

Fig. 10.2 Different alternative applications of sugar beet waste in bioprocess engineering

in Fig. 10.2. Table 10.1 has been summarized the application of this magic plant in food. In the next section will be focused on valorization of sugar beet waste.

10.5 Fermentation as a Valorization Tool and Biorefinery Approach

Fermentation has been an integral part of human civilization for thousands of years, used for the production of food and beverages (Behera et al. 2019; Tamang et al. 2020). However, in recent times, fermentation has gained increased attention as a powerful tool for valorization and a key component of biorefinery approaches (Subramanian et al. 2022). Fermentation, broadly defined, is a metabolic process that converts organic compounds into simpler compounds through the action of microorganisms, such as bacteria, yeasts, or fungi. It is a versatile tool that enables the conversion of various substrates into valuable products, including biofuels, chemicals, enzymes, and pharmaceuticals (Teng et al. 2021; Lima et al. 2022). Valorization refers to the process of extracting maximum value from raw materials or waste streams.

Fermentation plays a crucial role in valorization by utilizing renewable resources, waste materials, and by-products to produce high-value compounds (Negro et al. 2016; Capanoglu et al. 2022). It offers several advantages, such as high selectivity, mild reaction conditions, and compatibility with a wide range of feedstock (Kubičková and Kubička 2010). The biorefinery concept takes inspiration from traditional petroleum refineries, but instead of relying on fossil fuels, it aims to convert biomass into a spectrum of valuable products through integrated and sustainable processes (Cherubini 2010). One of the key strategies is the utilization of sugarbeet as a feedstock for biorefinery processes. Sugarbeet offers more than just its

Table 10.1 Different applications of beetroot in food

Substrate	Product/results	M.O.	Reference
Red beets	Probiotic beet juice	Lactic acid bacteria	Yoon et al. (2005)
Fermented beetroot juice	Modified metabolic activity and stabilized intestinal microbial ecosystem of rats	*Lactobacillus*	Klewicka et al. (2009)
Beetroot and fermented beetroot juices	Metabolomics, antioxidant levels, and anticancer activity	Lactic acid bacteria	Kazimierczak et al. (2014)
Fermented red beetroots and red beetroot juice	Bioactive compounds	Lactic acid bacteria	Czyzowska et al. (2020)
Fermented beetroot juice	Spray dried powder	Lactic acid bacteria	Janiszewska-Turak et al. (2022b)
Dried fermented beetroot	A potential source of lactic acid bacteria/health-promoting pigments, e.g., in the form of chips/shelf life is extended	Lactic acid bacteria	Janiszewska-Turak et al. (2022a)
Sugar beet's byproduct	Biogas	–	Sauthoff et al. (2016)
Sugar beet	Molasses	–	Adbhai et al. (2022)
Beet molasses	Erythritol	–	Hijosa-Valsero et al. (2021)
Beetroot waste powder	Production of gluten-free and high-fiber cookies		Chauhan and Rajput (2018)
Minimally processed beetroot waste	Functional ingredients		Costa et al. (2017)
Beetroot pomace	Betalains and phytochemicals	–	Kushwaha et al. (2018)

sucrose content. The remaining biomass, consisting of pulp and other residual components, can be a valuable resource for biorefinery processes (Kolfschoten et al. 2014; Glaser et al. 2022).

10.5.1 Solid-State Fermentation

Solid-state fermentation (SSF) is a microbial fermentation method in which selected microorganisms (bacteria, fungi, and yeasts) are cultivated on an enough moist (to support cell growth), solid, non-soluble organic material that provides an excellent substratum for the growth of the microorganisms and produces metabolites (Pandey et al. 1996, 2000; Gaitán-Hernández et al. 2006; Mienda et al. 2011).

Utilizing bio waste, such as sugarcane bagasse, cassava bagasse, and sugar beet pulp, as a solid substrate in solid-state fermentation offers a promising approach to effectively managing solid waste and introduces a novel method in waste management (Idris et al. 2017). Operation in SSF is easy with low capital investment (De Oliveira Felipe et al. 2017). Initially, mostly fungi were used in such fermentation; later, many thermotolerant bacterial species and yeasts were used to carry out such fermentation also (Singhania et al. 2009). Most of the microorganisms used in fermentation are generally regarded as safe and free from toxins (Yafetto et al. 2019; Suman et al. 2015; El Sheikha and Ray 2022). SSF is used by several industries like pharmaceuticals, food, energy, textile, etc. (Ghosh 2016). Agro-industrial residues serve as raw materials for the production of numerous valuable bio-based products, including enzymes, animal feed, organic acids, antibiotics, bioethanol, mushrooms, antioxidants, single-cell proteins (SCP), secondary metabolites, and biofuels such as ethanol, butanol, and hydrogen (Diaz et al. 2018). Additionally, a recent development includes the utilization of agro-industrial wastes for the production of biosurfactants. These biosurfactants play a crucial role in the bioremediation of environmental pollution caused by the improper disposal of agro-industrial residues (Mitchell et al. 2002; Lizardi-Jiménez and Hernández-Martínez 2017; Yafetto et al. 2019). Simultaneously, these fermentation processes enable the reduction of waste residues and ensure a secure and environmentally friendly final disposal method (Gaitán-Hernández et al. 2006; Pandey et al. 2000; Couto and Sanromán 2006).

10.5.2 Submerged Fermentation

Submerged fermentation is a commonly employed fermentation technique where microorganisms are grown in a liquid medium (Sánchez et al. 2015). There are several advantages for the solid state in comparison to submerged fermentation including less water, energy, and cheap agro-industrial substrate but the disadvantage is an important dependency of gas phase to particle size and moisture. Also, thermal conductivity is low. All mentioned variables have a great impact on process control and design. Oxygen transfer as the main problem in submerged fermentation depends upon the scale, shape, as well as agitation, and aeration system (Durand 2003). Through submerged fermentation, sugarbeet pulp can be transformed into bioethanol, organic acids (such as lactic acid and succinic acid), enzymes, and other bio-products (Lu et al. 2020).

10.6 Pre-treatment Strategies: Acid Hydrolysis, Enzymatic Hydrolysis, Acid-Enzyme Hydrolysis, and Hydrothermal Hydrolysis

The low dry-matter content of SBP makes it suitable for combustion. However, its high sugar content and low lignin make it particularly attractive for the production of bioethanol and biorefinery processes. The yeast's ability to efficiently ferment pentose sugars further facilitates the effective utilization of biomass for bioethanol production (Bettiga et al. 2009). Lignocellulose materials are pretreated (Alvira et al. 2010). Hydrolyzing lignocellulose into fermentable monosaccharides is a challenging task due to various physicochemical, structural, and compositional factors that hinder the digestibility of cellulose. Because of these structural characteristics, pretreatment is a crucial step in obtaining potentially fermentable sugars during the hydrolysis process. The purpose of pretreatment is to dismantle the lignin structure and disturb the crystalline structure of cellulose, thereby improving enzyme accessibility to cellulose during the hydrolysis step (Mosier et al. 2005). There are two approaches to carrying out this operation: enzymatic hydrolysis and chemical hydrolysis. In the enzymatic route, a pretreatment of lignocellulose biomass is necessary to ensure that cellulose becomes accessible for enzymatic breakdown (Balat 2011; Nguyen et al. 2020). On the other hand, in the chemical route, the initial stage of pretreatment involves hydrolysis. In this case, it is crucial to prevent the degradation of sugars to maximize the yield. It is important to note that the pretreatment step is the most costly and least technologically advanced stage in the overall process of converting cellulosic biomass into fermentable sugars (Ortiz and de Oliveira 2014). Due to the varying physicochemical characteristics of different lignocellulose materials, it is important to select appropriate pretreatment technologies based on the properties of each specific raw material. Moreover, the choice of a particular pretreatment method significantly influences all subsequent steps in the overall conversion process. These impacts include cellulose digestibility, the production of toxic compounds that could potentially inhibit yeast, the power requirements for stirring, the energy demand in downstream processes, and the demands of wastewater treatment. Therefore, careful consideration of the pretreatment approach is crucial to optimize the overall conversion scheme and address these important factors (Galbe and Zacchi 2007).

10.6.1 Acid Hydrolysis

The primary goal of acid pretreatments is to dissolve the hemicellulose fraction of biomass and enhance the accessibility of cellulose to enzymes. This type of pretreatment can be carried out using concentrated or diluted acid. However, the use of concentrated acid is less desirable for ethanol production due to the formation of inhibitory compounds. Additionally, concentrated acid pretreatments pose challenges such as equipment corrosion and the need for acid recovery. These drawbacks result in high operational and maintenance costs, diminishing the attractiveness of applying concentrated acid pretreatment on a commercial scale (Wyman 1996). Out of the various methods proposed, pretreatment with dilute sulfuric acid (DSA) is considered the most appealing for industrial applications and is widely used globally (Vasconcelos et al. 2020). DSA pretreatment can be conducted at high temperatures, such as 180 °C, for a short duration or at lower temperatures, like 120 °C, for a longer retention time of 30–90 min. This method offers the advantage of solubilizing hemicellulose, particularly xylan, and converting the solubilized hemicellulose into fermentable sugars. However, depending on the process temperature, certain sugar degradation compounds like furfural, HMF (hydroxyl-methyl-furfural), and aromatic lignin degradation compounds may be produced, which can negatively impact microbial metabolism during the fermentation step. Nevertheless, this pretreatment approach generates fewer degradation products compared to concentrated acid pretreatments (Saha et al. 2005). Under specific pretreatment conditions, acid pretreatment has been shown to eliminate small portions of cellulose and modify the lignin structure by removing acid-soluble lignin. Typically, acid pretreatments are followed by an enzyme hydrolysis step to recover the remaining cellulose sugars. Although these pretreatments have been successful in removing hemicellulose and making the cellulose accessible for enzymatic breakdown, they present challenges. Harsh pretreatment conditions (high severity factor) can lead to the formation of inhibitors, and there is a requirement to neutralize the effluent and remaining solids before proceeding with downstream processing (Rajendran et al. 2018).

10.6.2 Enzymatic Hydrolysis

Lignocellulose can be broken down into simple sugars through hydrolysis, either enzymatically using (hemi)cellulolytic enzymes or chemically using sulfuric acid or other acids (Zhang et al. 2012). However, enzymatic hydrolysis is gaining popularity as it requires less energy and occurs under mild environmental conditions. Moreover, enzymatic hydrolysis results in the generation of fewer fermentation inhibitor products compared to chemical methods (such as acid hydrolysis) (Brummer et al. 2014). Enzymatic deconstruction of lignocellulose is a challenging process due to its complex structure, which makes it highly resistant to breakdown. The intricate network formed by cellulose, hemicellulose, and lignin contributes to this recalcitrance. Additionally, some enzymes can be absorbed by condensed lignin, leading to decreased hydrolysis yield due to non-specific interactions between these enzymes and lignin, forming undesired linkages. These factors further contribute to the complexity of the enzymatic deconstruction of lignocellulose (Poovaiah et al. 2014). The degradation of lignocellulose biomass can be achieved through various methods, including the use of acids or bases, specific enzymes, or hydrothermal treatment (Alvira et al. 2010; Zheng et al. 2013). However, the efficiency of enzymatic saccharification of lignocellulose is influenced by multiple factors, and the results of acidic hydrolysis are unsatisfactory (Larsson et al. 1999).

10.6.3 Hydrothermal Hydrolysis

As a result, there is a potential for these approaches to be replaced by hydrothermal processing, specifically Liquid Hot Water (LHW) treatment. In LHW treatment, plant biomass is subjected to elevated pressure for a short duration of time in an aqueous system, typically with an average dry matter concentration of 1–8%. The treatment is carried out at temperatures ranging from 120 to 260 °C (Zhang and Shahbazi 2011). At the elevated temperatures used in LHW treatment, the dissociation of water molecules is enhanced, leading to water acting as an acidic catalyst. This unique property of water eliminates the need for additional reagents in the process, making it a self-sufficient method (Mood et al. 2013). During LHW treatment, a significant amount of biomass is dissolved, with approximately 40–60% of the biomass being solubilized. This includes the dissolution of nearly all hemicelluloses, 30–60% of the lignin, and 4–22% of the cellulose present in the biomass (Hu et al. 2008). The LHW treatment offers several advantages over enzymatic and chemical methods. Firstly, it is a cost-effective technique compared to other approaches. Additionally, water, which is the solvent used in the process, is readily available and inexpensive. The hydrolysate obtained from LHW treatment is characterized by high concentrations of monosaccharides and low levels of inhibitors that can hinder further enzymatic digestion. This increases the efficiency of subsequent enzymatic processes (Weil et al. 1998). Considering these benefits, the LHW method is considered highly attractive and promising for the de-polymerization of plant biomass prior to anaerobic fermentation (Ziemiński et al. 2014).

10.7 Sugar Beet Bagasse as Substrate

Sugar beet bagasse, also referred to as SBP, is the primary leftover material from the beet sugar industry (Cosyn et al. 2011; Finkenstadt 2014). It contains approximately 20% cellulose and is primarily composed of pectin, galacturonic acid, rhamnose, arabinose, galactose, methanol, and acetic acid (Duraisam et al. 2017). SBP offers numerous advantages for both domestic and industrial applications. It can be utilized as animal feed, particularly for the production of fiber-based products like pulp and paper. Additionally, it can be used in the synthesis of activated carbon, which is employed for treating environmental waste. Moreover, SBP can serve as a source for generating fuel, such as bio-ethanol (Vaccari et al. 2005). Given its abundant availability as a by-product of the beet sugar industry, sugar beet pulp shows promise for producing various value-added products (Chamy et al. 1994).

10.7.1 Cattle Feed

Sugar beet pulp is a residue obtained during the processing of sugar beet, and it serves as a valuable source of animal feed for cattle and other livestock. It is available in two forms: dried flakes or compressed pellets. Sugar beet pulp is known for its appealing taste and provides ample energy levels for animals. It has a low sugar content and contains minimal non-structural carbohydrates. The fiber present in sugar beet pulp is highly digestible, making it suitable for ruminant animals as it supports the health of the rumen and promotes the production of acetate. Additionally, sugar beet pulp has a high capacity to absorb liquids, which makes it useful as an additive in silage, enabling the retention of the nutritional value of effluent (Habeeb et al. 2017).

10.7.2 Single-Cell Protein

An ongoing challenge in today's world is to ensure the production of proteins at maximum capacity to meet the nutritional needs of a growing global population. This challenge becomes even more crucial when considering the importance of deriving proteins from renewable resources. One potential solution to address this issue is the utilization of SCP as a source of protein in both human food and animal feed (Adedayo et al. 2011). SCP is obtained from cultivated microbial biomass and offers a cost-effective protein alternative that promotes animal growth, thus playing a significant role in addressing protein deficiencies (Ravindra 2000). SCP exhibits a high protein content, typically ranging from 60% to 82% of dry matter, and it also contains fats, carbohydrates, nucleic acids, vitamins, and minerals (Asad et al. 2000; Jamal et al. 2008). Furthermore, SCP stands out for its rich concentration of essential amino acids, including lysine and methionine, which are often limited in many plant and animal food sources. Incorporating SCP as an additive in the main diet provides a more affordable alternative to costly protein sources like soybeans and fish (Mondal et al. 2012). Agricultural waste represents a renewable resource with vast potential for various biotechnological applications. Recently, a range of waste materials such as bagasse, rice straw, rice hulls, citrus wastes, sulfite waste liquor, molasses, animal manure, whey, starch, and sewage have been used as substrates for microbial growth (Patelski et al. 2015). Sugar beet pulp, currently used primarily for animal feed production, holds great promise due to its composition rich in polysaccharides (approximately 75%), such as cellulose, hemicellulose, lignin, and pectin. This composition makes sugar beet pulp an appealing carbon source for the production of microbial biomass protein (Pessoa et al. 1996). Different types of microorganisms, including bacteria, fungi, mold, algae, and yeasts, can serve as sources for SCP production (Mondal et al. 2012).

10.7.3 Foods with Improved Aroma

Casciano et al. (2022) reported spontaneous or controlled fermentation of red beetroot by three different starter cultures. The results showed that *Leuconostoc mesenteroides* caused the lowest diversity of molecules. Kefir starter produced metabolites associated mainly with yeast, e.g. phenylethyl alcohol and isoamyl alcohol. Finally, the *Leu. mesenteroides* mix led to more complex volatile materials. Spontaneous micro-flora could be characterized by molecules with a positive aroma. Pasteurization could be recommended (Casciano et al. 2022).

10.7.4 Microbial Growth Enhancer

Couvreur et al. reported a formulation study to develop a growth medium based on wheat processing products and sugar for the cultivation of *Limosilactobacillus reuteri*. Also, MRS (De Man, Rogosa, and Sharpe) medium was used as the control medium for lactic bacteria cultivation. The results showed that in optimized formulation 3-hydroxypropanoic acid production was increased (Couvreur et al. 2017).

10.7.5 Agro-industrial Waste like Beet Pulp as a Potential Source of Polyhydroxyalkanoates (PHA) Production

Tripathi et al. studied the hydrothermal treatment of bagasse and other lignocellulose waste for the production of PHA copolymer as an eco-friendly biopolymer with many applications in the food and pharmaceutical industries (Tripathi et al. 2022). However, the high production cost of PHAs limits their worldwide application. This can be minimized by screening of PHA producers and raw materials (Khosravi-Darani et al. 2003, 2013; Bozorg et al. 2015; Mokhtari-Hosseini et al. 2009; Tripathi et al. 2021; Brigham and RiedeL 2018).

10.7.6 Bioethanol Production Using Sugar Beet

As the world becomes more advanced, more energy is needed to maintain these changes. Using bioethanol as one of the most important biofuels is a suitable response to these changes (Behera et al. 2022). Also, an important crop in the pandemic of COVID 19 was sugar beet due to its high potential for ethanol production in less growth span. Bioethanol, which is produced from biomass, is a suitable disinfectant and fuel.

Berlowska et al. (2017) depolymerized waste of sugar beet pulp biomass enzymatically to use hydrolyzates as a carbon source of fermentation for the production of biomethane, biohydrogen, and bioethanol. The results showed that the highest hydrogen yield was achieved with sugar beet stillage. However, dark fermentation of biohydrogen was more efficient when the carbon source was hydrolyzate of sugar beet pulp. Anaerobic digestion methods for whole beets for the production of biomethane require higher energy (137%) in comparison to bio-ethanol production. In each hectare crop, 5250 L of ethanol may be produced. Lactic acid may be produced by the fermentation of sugar beet. Dried beet pulp may be applied as a source of vitamin C, due to the content of 30% galacturonic acid. Pectin as a dietary fiber could be applicable in the baking and meat industries. The vinasse produced as an industrial by-product is useful as a fertilizer. Material sciences and pharmaceutical industries are coming soon industry for the application of sugar beet waste (Mall et al. 2021).

10.8 Future Perspectives and Conclusion

Bioprocessing of sugar beet waste for the production of value-added products has contributed to the solution of dwindling soil, food, energy resources, and water. Also, such a process reduces the total costs of products. Not only cost but also nutritional composition are reasons for gaining the attention of beet waste in the biotechnological process. The high sugar content of sugar beet wastes leads to convert it to an interesting carbon source in microbial production of several metabolites. Both solid-state and submerged fermentation are successful strategies for the bioconversion of waste to value-added products. As it can be concluded from this chapter, there are lots of reports about the production of oligosaccharides, acids, and enzymes from sugar beet waste but still relatively few studies on the production of enzymes, bioplastics, and biofuels, and this area is open to research. The application of food wastes such as sugar beet waste for the production of bio-products is promising and requires new industrial breakthroughs. Handling of sugar beet waste is considered a challenging strategy in future years, however, integration and proper management of these wastes may play a key role to produce commercial as well as sustainable products.

References

Abu-Ellail FF, Salem KF, Saleh MM, Alnaddaf LM, Al-Khayri JM (2021) Molecular breeding strategies of beetroot (*Beta vulgaris* ssp. *vulgaris* var. *conditiva* Alefeld). In: Advances in plant breeding strategies: vegetable crops, Bulbs, roots and tubers, vol 8. Springer, pp 157–212

Adbhai AR, Dewanjee S, Patel KG, Karmakar N (2022) Sugar beet molasses production and utilization. In: Sugar beet cultivation, management, and processing. Springer

Adedayo M, Ajiboye E, Akintunde J, Odaibo A (2011) Single cell proteins: as nutritional enhancer. Adv Appl Sci Res 2:396–409

Agarwal K, Varma R (2014) Biochemical screening of beetroot leaves. Int J Pharm Sci Rev Res 1: 127–134

Alvira P, Tomás-Pejó E, Ballesteros M, Negro MJ (2010) Pretreatment technologies for an efficient bioethanol production process based on enzymatic hydrolysis: a review. Bioresour Technol 101: 4851–4861

Arshad M, Hussain T, Iqbal M, Abbas M (2017) Enhanced ethanol production at commercial scale from molasses using high gravity technology by mutant *S. cerevisiae*. Braz J Microbiol 48:403–409

Asad MJ, Asghar M, Yaqub M, Shahzad K (2000) Production of single cell protein from delignified corn cob by *Arachniotus* species. Pak J Agric Sci 37:3–4

Asadi M (2006) Beet-sugar handbook. John Wiley & Sons

Balat M (2011) Production of bioethanol from lignocellulosic materials via the biochemical pathway: a review. Energy Convers Manag 52:858–875

Bancal V, Ray RC (2022) Overview of food loss and wastes in fruits and vegetables: from issues to resources. In: Ray RC (ed) Fruits and vegetable wastes: valorization to bioproducts and platform chemicals. Springer, Singapore, pp 3–32

Barnabas L, Ramadass A, Amalraj RS, Palaniyandi M, Rasappa V (2015) Sugarcane proteomics: an update on current status, challenges, and future prospects. Proteomics 15:1658–1670

Barnóczki A, Csontos G, Deme P, Feher B, Glits M, Gólya E, Gyúrós J, Hájos M, Hodossi S, Hraskó I (2010) Zöldségtermesztés szabadföldön. Mezőgazda Kiadó 309:1–39

Behera SS, Ray RC, Das U, Panda SK, Saranraj P (2019) Microorganisms in fermentation. In: Berenjian A (ed) Essentials in fermentation technology. Springer, Cham, pp 1–39

Behera SS, Saranraj P, Ray RC (2022) Microbial bioethanol fermentation technologies—recent trends and future prospects. In: Castro FIG, Guti-Errez-Antonio C (eds) Biofuels and biorefining. Elsevier, pp 75–107

Berlowska J, Pielech-Przybylska K, Balcerck M, Cieciura W, Borowski S, Kregiel D (2017) Integrated Bioethanol Fermentation/Anaerobic Digestion for Valorization of Sugar Beet Pulp. Energies 10(9):1255. https://doi.org/10.3390/en10091255

Berlowska J, Cieciura-Włoch W, Kalinowska H, Kregiel D, Borowski S, Pawlikowska E, Binczarski M, Witonska I (2018) Enzymatic conversion of sugar beet pulp: a comparison of simultaneous saccharification and fermentation and separate hydrolysis and fermentation for lactic acid production. Food Technol Biotechnol 56:188

Bettiga M, Bengtsson O, Hahn-Hägerdal B, Gorwa-Grauslund MF (2009) Arabinose and xylose fermentation by recombinant *Saccharomyces cerevisiae* expressing a fungal pentose utilization pathway. Microb Cell Factories 8:1–12

Bhat SA, Singh J, Vig AP (2015) Vermistabilization of sugar beet (*Beta vulgaris* L) waste produced from sugar factory using earthworm *Eisenia fetida*: genotoxic assessment by *Allium cepa* test. Environ Sci Pollut Res 22:11236–11254

Biancardi E, Mcgrath JM, Panella LW, Lewellen RT, Stevanato P (2010) Sugar beet. In: Root and tuber crops. Springer Science and Business Media

Bozorg A, Vossoughi M, Kazemi A, Alemzadeh I (2015) Optimal medium composition to enhance poly-β-hydroxybutyrate production by *Ralstonia eutropha* using cane molasses as sole carbon source. Appl Food Biotechnol 2:39–47

Bradshaw JE (2010) Root and tuber crops. Springer Science and Business Media

Brigham CJ, Riedel SL (2018) The potential of polyhydroxyalkanoate production from food wastes. Appl Food Biotechnol 6:7–18

Brummer V, Skryja P, Jurena T, Hlavacek V, Stehlik P (2014) Suitable technological conditions for enzymatic hydrolysis of waste paper by Novozymes® enzymes NS50013 and NS50010. Appl Biochem Biotechnol 174:1299–1308

Caliceti C, Malaguti M, Marracino L, Barbalace MC, Rizzo P, Hrelia S (2022) Agri-food waste from apple, pear, and sugar beet as a source of protective bioactive molecules for endothelial dysfunction and its major complications. Antioxidants 11:1786

Capanoglu E, Nemli E, Tomas-Barberan F (2022) Novel approaches in the valorization of agricultural wastes and their applications. J Agric Food Chem 70:6787–6804

Cárdenas-Fernández M, Bawn M, Hamley-Bennett C, Bharat PK, Subrizi F, Suhaili N, Ward DP, Bourdin S, Dalby PA, Hailes HC (2017) An integrated biorefinery concept for conversion of sugar beet pulp into value-added chemicals and pharmaceutical intermediates. Faraday Discuss 202:415–431

Casciano F, Mayr H, Nissen L, Putti A, Zoli F, Gianotti A, Conterno L (2022) Red beetroot fermentation with different microbial consortia to develop foods with improved aromatic features. Foods 11:3055

Chamy R, Illanes A, Aroca G, Nuñez L (1994) Acid hydrolysis of sugar beet pulp as pretreatment for fermentation. Bioresour Technol 50:149–152

Chauhan S, Rajput H (2018) Production of gluten free and high fiber cookies using beet root waste powder and wheat flour husk. Pharma Innov J 7:556

Chawla H, Parle M, Sharma K, Yadav M (2016) Beetroot: a health promoting functional food. Inventi Rapid Nutraceuticals 1:976–3872

Cheesman OD (2004) Environmental impacts of sugar production the cultivation and processing of sugarcane and sugar beet. CABI

Cherubini F (2010) The biorefinery concept: using biomass instead of oil for producing energy and chemicals. Energy Convers Manag 51:1412–1421

Clapham AR, Tutin TG, Moore DM (1990) Flora of the British isles. CUP Archive

Cooke DA, Scott J (2012) The sugar beet crop. Springer Science and Business Media

Costa APD, Hermes VS, Rios ADO, Flores SH (2017) Minimally processed beetroot waste as an alternative source to obtain functional ingredients. J Food Sci Technol 54:2050–2058

Cosyn S, Woude KVD, Sauvenier X, Evrard J-N (2011) Sugar beet: a complement to sugar cane for sugar and ethanol production in tropical and subtropical areas. Int Sugar J 113:120–123

Couto SR, Sanromán MA (2006) Application of solid-state fermentation to food industry—a review. J Food Eng 76:291–302

Couvreur J, Teixeira AR, Allais F, Spinnler H-E, Saulou-Berion C, Clement T (2017) Wheat and sugar beet coproducts for the bioproduction of 3-hydroxypropionic acid by *Lactobacillus reuteri* DSM17938. Fermentation 3:32. https://doi.org/10.3390/fermentation3030032

Czyzowska A, Siemianowska K, Sniadowska M, Nowak A (2020) Bioactive compounds and microbial quality of stored fermented red beetroots and red beetroot juice. Pol J Food Nutr Sci 70:35–44

De Oliveira Felipe L, De Oliveira AM, Bicas JL (2017) Bioaromas—perspectives for sustainable development. Trends Food Sci Technol 62:141–153

Diaz AB, Blandino A, Caro I (2018) Value added products from fermentation of sugars derived from agro-food residues. Trends Food Sci Technol 71:52–64

Draycott AP (2008) Sugar beet. Wiley, 496p

Duraisam R, Salelgn K, Berekete AK (2017) Production of beet sugar and bio-ethanol from sugar beet and it bagasse: a review. Int J Eng Trends Technol 43:222–233

Durand A (2003) Bioreactor designs for solid state fermentation. Biochem Eng J 13:113–125

El Sheikha AF, Ray RC (2022) Bioprocessing of horticultural wastes by solid-state fermentation into value-added/innovative bioproducts: a review. Food Rev Int 39:3009. https://doi.org/10.1080/87559129.2021.2004161

Elliott M, Weston G (1993) Biology and physiology of the sugar-beet plant. In: The sugar beet crop. Springer, pp 37–66

Erdal G, Esengun K, Erdal H, Gunduz O (2007) Energy use and economical analysis of sugar beet production in Tokat province of Turkey. Energy 32:35–41

FAO (2023). Fao.org. Sugarbeet and land and water. Food and Agriculture Organization of the United Nations. Retrieved 24 Jan 2023, from https://www.fao.org/land-water/databases-and-software/crop-information/sugarbeet/en/

Finkenstadt VL (2014) A review on the complete utilization of the sugarbeet. Sugar Tech 16:339–346

Gaitán-Hernández R, Esqueda M, Gutierrez A, Sánchez A, Beltrán-García M, Mata G (2006) Bioconversion of agrowastes by *Lentinula edodes*: the high potential of viticulture residues. Appl Microbiol Biotechnol 71:432–439

Galbe M, Zacchi G (2007) Pretreatment of lignocellulosic materials for efficient bioethanol production. Biofuels 108:41–65

Ghosh JS (2016) Solid state fermentation and food processing: a short review. J Nutr Food Sci 6:1–7

Glaser SJ, Abdelaziz OY, Demoitie C, Galbe M, Pyo S-H, Jensen JP, Hatti-Kaul R (2022) Fractionation of sugar beet pulp polysaccharides into component sugars and pre-feasibility analysis for further valorisation. Biomass Convers Biorefin:1–14. https://doi.org/10.1007/s13399-022-02699-4

Gonzalez MNG, Björnsson L (2022) Life cycle assessment of the production of beet sugar and its by-products. J Clean Prod 346:131211

Habeeb A, Gad A, El-Tarabany A, Mustafa M, Atta M (2017) Using of sugar beet pulp by-product in farm animals feeding. Int J Sci Res Sci Technol 3:107–120

Handbook FFA (2009) White sugar. FAO, Rome

Hijosa-Valsero M, Garita-Cambronero J, Paniagua-García AI, Díez-Antolínez R (2021) By-products of sugar factories and wineries as feedstocks for erythritol generation. Food Bioprod Process 126:345–355

Højland J, Pedersen S (1994) Sugar beet, beetroot and fodder beet (*Beta vulgaris* L. subsp. *vulgaris*): dispersal, establishment and interactions with the environment. The National Forest and Nature Agency, Copenhagen, 73p

Hu G, Heitmann JA, Rojas OJ (2008) Feedstock pretreatment strategies for producing ethanol from wood, bark, and forest residues. Bioresources 3:270–294

Idris ASO, Pandey A, Rao S, Sukumaran RK (2017) Cellulase production through solid-state tray fermentation, and its use for bioethanol from sorghum stover. Bioresour Technol 242:265–271

Jamal P, Alam M, Fairoza N (2008) Process optimization for bioprotein production from cheaper carbon source. In: 4th Kuala Lumpur international conference on biomedical engineering 2008: BIOMED 2008, 25–28 June 2008, Kuala Lumpur, Malaysia. Springer, pp 889–892

Janiszewska-Turak E, Rybak K, Pobiega K, Nikodem A, Gramza-Michałowska A (2022a) Sustainable production and characteristics of dried fermented vegetables. Fermentation 8:659. https://doi.org/10.3390/fermentation8110659

Janiszewska-Turak E, Walczak M, Rybak K, Pobiega K, Gniewosz M, Woźniak Ł, Witrowa-Rajchert D (2022b) Influence of fermentation beetroot juice process on the physico-chemical properties of spray dried powder. Molecules 27:1008. https://doi.org/10.3390/molecules27031008

Kaffka SR, Grantz DA (2014) Sugar crops. In: Alfen NV (ed) Encyclopedia of agriculture and food systems, vol 5. Elsevier, Amsterdam, pp 240–260

Kazimierczak R, Hallmann E, Lipowski J, Drela N, Kowalik A, Pussa T, Matt D, Luik A, Gozdowski D, Rembiałkowska E (2014) Beetroot (*Beta vulgaris* L.) and naturally fermented beetroot juices from organic and conventional production: metabolomics, antioxidant levels and anticancer activity. J Sci Food Agric 94:2618–2629

Khosravi-Darani K, Vasheghani Farahani E, Shojaosadati SA (2003) Application of the plackett-burman statistical design to optimize poly ($\hat{I}^2$-hydroxybutyrate) production by ralstonia eutropha in batch culture. Iran J Biotechnol 1:155–161

Khosravi-Darani K, Mokhtari Z-B, Amai T, Tanaka K (2013) Microbial production of poly (hydroxybutyrate) from C 1 carbon sources. Appl Microbiol Biotechnol 97:1407–1424

Klewicka E, Zduńczyk Z, Juśkiewicz J (2009) Effect of lactobacillus fermented beetroot juice on composition and activity of cecal microflora of rats. Eur Food Res Technol 229:153–157

Kolfschoten RC, Bruins ME, Sanders JP (2014) Opportunities for small-scale biorefinery for production of sugar and ethanol in the Netherlands. Biofuels Bioprod Biorefin 8:475–486

Kopitzky R (2021) Poly (lactic acid)–poly (butylene succinate)–sugar beet pulp composites; part II: water absorption characteristics with fine and coarse sugar beet pulp particles; a phenomenological investigation. Polymers 13:3558. https://doi.org/10.3390/polym13203558

Kubičková I, Kubička D (2010) Utilization of triglycerides and related feedstocks for production of clean hydrocarbon fuels and petrochemicals: a review. Waste Biomass Valoriz 1:293–308

Kushwaha R, Kumar V, Vyas G, Kaur J (2018) Optimization of different variable for eco-friendly extraction of betalains and phytochemicals from beetroot pomace. Waste Biomass Valoriz 9: 1485–1494

Larsson S, Palmqvist E, Hahn-Hägerdal B, Tengborg C, Stenberg K, Zacchi G, Nilvebrant N-O (1999) The generation of fermentation inhibitors during dilute acid hydrolysis of softwood. Enzym Microb Technol 24:151–159

Letschert JPW (1993) Beta section beta: biogeographical patterns of variation, and taxonomy. Wageningen University and Research

Lima PJM, Da Silva RM, Neto CACG, Silva NCGE, da Silva Souza JE, Nunes YL, Sousa Dos Santos JC (2022) An overview on the conversion of glycerol to value-added industrial products via chemical and biochemical routes. Biotechnol Appl Biochem 69:2794–2818

Lizardi-Jiménez M, Hernández-Martínez R (2017) Solid state fermentation (SSF): diversity of applications to valorize waste and biomass. 3 Biotech 7:44. https://doi.org/10.1007/s13205-017-0692-y

Lu J, Lv Y, Qian X, Jiang Y, Wu M, Zhang W, Zhou J, Dong W, Xin F, Jiang M (2020) Current advances in organic acid production from organic wastes by using microbial co-cultivation systems. Biofuels Bioprod Biorefin 14:481–492

Mall A, Misra V, Kumar M, Pathak A (2018) Assessment of ethanol recovery and juice quality parameters in sugarbeet in drought conditions of sub-tropical India. In: Singh J, Nigam R, Hasan W, Kumar A, Singh R, Naz H, Kapoor N, Bairwa SL, Rani A, Kumar M (eds) Second international conference on advances in agricultural, biological and applied sciences for sustainable future (ABAS-2018)-Theme I. Swami Vivekanand Subharti University, Meerut, UP, pp 20–22

Mall AK, Misra V, Pathak A, Srivastava S (2021) Sugar beet cultivation in India: prospects for bio-ethanol production and value-added co-products. Sugar Tech 23:1218–1234

Mcginnis RA (1982) Beet-sugar technology

Mienda BS, Idi A, Umar A (2011) Microbiological features of solid state fermentation and its applications—an overview. Res Biotechnol 2:21–26

Mischopoulou M, Naidis P, Kalamaras S, Kotsopoulos T, Samaras P (2016) Effect of ultrasonic and ozonation pretreatment on methane production potential of raw molasses wastewater. Renew Energy 96:1078–1085

Mitchell D, Berovic M, Krieger N (2002) Overview of solid state fermentation. Biotechnol Annu Rev 2002(8):183–122. https://doi.org/10.1016/s1387-2656(02)08009-2

Mladenović DD, Djukić-Vuković AP, Kocić-Tanackov SD, Pejin JD, Mojović LV (2016) Lactic acid production on a combined distillery stillage and sugar beet molasses substrate. J Chem Technol Biotechnol 91:2474–2479

Mokhtari-Hosseini ZB, Vasheghani-Farahani E, Heidarzadeh-Vazifekhoran A, Shojaosadati SA, Karimzadeh R, Darani KK (2009) Statistical media optimization for growth and PHB production from methanol by a methylotrophic bacterium. Bioresour Technol 100:2436–2443

Mondal AK, Sengupta S, Bhowal J, Bhattacharya D (2012) Utilization of fruit wastes in producing single cell protein. Int J Sci Environ Technol 1:430–438

Mood SH, Golfeshan AH, Tabatabaei M, Jouzani GS, Najafi GH, Gholami M, Ardjmand M (2013) Lignocellulosic biomass to bioethanol, a comprehensive review with a focus on pretreatment. Renew Sust Energ Rev 27:77–93

Mosier N, Wyman C, Dale B, Elander R, Lee Y, Holtzapple M, Ladisch M (2005) Features of promising technologies for pretreatment of lignocellulosic biomass. Bioresour Technol 96:673–686

Mustafa, G., Arshad, M., Bano, I. And Abbas, M. 2020. Biotechnological applications of sugarcane bagasse and sugar beet molasses. Biomass Convers Biorefin, 1–13, 13

Negro V, Mancini G, Ruggeri B, Fino D (2016) Citrus waste as feedstock for bio-based products recovery: review on limonene case study and energy valorization. Bioresour Technol 214:806–815

Nguyen TVT, Unpaprom Y, Manmai N, Whangchai K, Ramaraj R (2020) Impact and significance of pretreatment on the fermentable sugar production from low-grade longan fruit wastes for bioethanol production. Biomass Convers Biorefin 12:1–13

Olmos JC, Hansen MZ (2012) Enzymatic depolymerization of sugar beet pulp: production and characterization of pectin and pectic-oligosaccharides as a potential source for functional carbohydrates. Chem Eng J 192:29–36

Ortiz PS, de Oliveira S Jr (2014) Exergy analysis of pretreatment processes of bioethanol production based on sugarcane bagasse. Energy 76:130–138

Pandey A, Selvakumar P, Ashakumary L (1996) Performance of a column bioreactor for glucoamylase synthesis by *Aspergillus niger* in SSF. Process Biochem 31:43–46

Pandey A, Soccol CR, Mitchell D (2000) New developments in solid state fermentation: I-bioprocesses and products. Process Biochem 35:1153–1169

Paroha S, Singh S, Gupta A (2020) Sugar beet-the potential feedstocks for alcohol production. Int J Agric Sci 12:9778–9780

Patelski P, Berlowska J, Dziugan P, Pielech-Przybylska K, Balcerek M, Dziekonska U, Kalinowska H (2015) Utilisation of sugar beet bagasse for the biosynthesis of yeast SCP. J Food Eng 167: 32–37
Pessoa A Jr, Mancilha I, Sato S (1996) Cultivation of *Candida tropicalis* in sugar cane hemicellulosic hydrolyzate for microbial protein production. J Biotechnol 51:83–88
Poel PW, Schiweck H, Schwartz TK (1998) Sugar technology: beet and cane sugar manufacture. Verlag Dr Albert Bartens KG
Poovaiah CR, Nageswara-Rao M, Soneji JR, Baxter HL, Stewart CN Jr (2014) Altered lignin biosynthesis using biotechnology to improve lignocellulosic biofuel feedstocks. Plant Biotechnol J 12:1163–1173
Ptak M, Skowrońska A, Pińkowska H, Krzywonos M (2022) Sugar beet pulp in the context of developing the concept of circular bioeconomy. Energies 15:175
Rajendran K, Drielak E, Sudarshan Varma V, Muthusamy S, Kumar G (2018) Updates on the pretreatment of lignocellulosic feedstocks for bioenergy production–a review. Biomass Convers Biorefin 8:471–483
Ravindra P (2000) Value-added food: single cell protein. Biotechnol Adv 18:459–479
Ray RC (2022) Fruits and vegetable wastes: valorization into bioproducts and platform chemicals. Springer, Singapore, 448p
Ray RC, Ramachandran S (2018) Bioethanol production from food crops: sustainable sources, interventions, and challenges. Academic Press, 460p
Richardson K (2014) Preliminary evaluation of the leaf and root nutrient composition of a fresh market beetroot variety. J Plant Nutr 20:408–420
Rouilly A, Jorda J, Rigal L (2006) Thermo-mechanical processing of sugar beet pulp. I. Twin-screw extrusion process. Carbohydr Polym 66:81–87
Rouilly A, Geneau-Sbartai C, Rigal L (2009) Thermo-mechanical processing of sugar beet pulp. III. Study of extruded films improvement with various plasticizers and cross-linkers. Bioresour Technol 100:3076–3081
Sadras VO, Calderini D (2020) Crop physiology case histories for major crops. Academic Press
Saha BC, Iten LB, Cotta MA, Wu YV (2005) Dilute acid pretreatment, enzymatic saccharification and fermentation of wheat straw to ethanol. Process Biochem 40:3693–3700
Sanchez OJ, Cardona CA (2008) Trends in biotechnological production of fuel ethanol from different feedstocks. Bioresour Technol 99:5270–5295
Sánchez OJ, Montoya S, Vargas LM (2015) Polysaccharide production by submerged fermentation. In: Polysaccharides. Springer
Saulnier L, Thibault JF (1999) Ferulic acid and diferulic acids as components of sugar-beet pectins and maize bran heteroxylans. J Sci Food Agric 79:396–402
Sauthoff S, Musshoff O, Danne M, Anastassiadis F (2016) Sugar beet as a biogas substrate? A discrete choice experiment for the design of substrate supply contracts for German farmers. Biomass Bioenergy 90:163–172
Searle S, Malins C (2013) Availability of cellulosic residues and wastes in the EU. International Council on Clean Transportation, Washington, DC
Shahbandeh M (2022) Global sugar consumption 2021/22. Statista. Retrieved 18 Feb 2023, from https://www.statista.com/statistics/249681/total-consumption-of-sugar-worldwide/
Sindhu R, Gnansounou E, Binod P, Pandey A (2016) Bioconversion of sugarcane crop residue for value added products—an overview. Renew Energy 98:203–215
Singhania RR, Patel AK, Soccol CR, Pandey A (2009) Recent advances in solid-state fermentation. Biochem Eng J 44:13–18
Smit AL (1983) Influence of external factors on growth and development of sugar-beet (*Beta vulgaris* L.). Wageningen University and Research
Stevanato P, Chiodi C, Broccanello C, Concheri G, Biancardi E, Pavli O, Skaracis G (2019) Sustainability of the sugar beet crop. Sugar Tech 21:703–716
Subramanian K, Sarkar MK, Wang H, Qin Z-H, Chopra SS, Jin M, Kumar V, Chen C, Tsang C-W, Lin CSK (2022) An overview of cotton and polyester, and their blended waste textile

valorisation to value-added products: a circular economy approach–research trends, opportunities and challenges. Crit Rev Environ Sci Technol 52:3921–3942
Suman G, Nupur M, Anuradha S, Pradeep B (2015) Single cell protein production: a review. Int J Curr Microbiol App Sci 4:251–262
Sun R, Hughes S (1999) Fractional isolation and physico-chemical characterization of alkali-soluble polysaccharides from sugar beet pulp. Carbohydr Polym 38:273–281
Székely D, Máte M (2022) Red beetroot (Beta vulgaris L.). In: Advances in root vegetables research. IntechOpen. https://doi.org/10.5772/intechopen.106692
Tamang JP, Cotter PD, Endo A, Han NS, Kort R, Liu SQ, Mayo B, Westerik N, Hutkins R (2020) Fermented foods in a global age: east meets west. Compr Rev Food Sci Food Saf 19:184–217
Teng TS, Chin YL, Chai KF, Chen WN (2021) Fermentation for future food systems: precision fermentation can complement the scope and applications of traditional fermentation. EMBO Rep 22:E52680
Tomaszewska J, Bieliński D, Binczarski M, Berlowska J, Dziugan P, Piotrowski J, Stanishevsky A, Witońska I (2018) Products of sugar beet processing as raw materials for chemicals and biodegradable polymers. RSC Adv 8:3161–3177
Tondro H, Musivand S, Zilouei H, Bazarganipour M, Zargoosh K (2020) Biological production of hydrogen and acetone-butanol-ethanol from sugarcane bagasse and rice straw using co-culture of *Enterobacter aerogenes* and *Clostridium acetobutylicum*. Biomass Bioenergy 142:105818
Tripathi AD, Paul V, Agarwal A, Sharma R, Hashempour-Baltork F, Rashidi L, Darani KK (2021) Production of polyhydroxyalkanoates using dairy processing waste—a review. Bioresour Technol 326:124735. https://doi.org/10.1016/j.biortech.2021.124735
Tripathi AD, Mishra PK, Darani KK, Agarwal A, Paul V (2022) Hydrothermal treatment of lignocellulose waste for the production of polyhydroxyalkanoates copolymer with potential application in food packaging. Trends Food Sci Technol 123(1):233. https://doi.org/10.1016/j.tifs.2022.03.018
Tsang YF, Kumar V, Samadar P, Yang Y, Lee J, Ok YS, Song H, Kim K-H, Kwon EE, Jeon YJ (2019) Production of bioplastic through food waste valorization. Environ Int 127:625–644
Usmani Z, Sharma M, Diwan D, Tripathi M, Whale E, Jayakody LN, Moreau B, Thakur VK, Tuohy M, Gupta VK (2022) Valorization of sugar beet pulp to value-added products: a review. Bioresour Technol 346:126580. https://doi.org/10.1016/j.biortech.2021.126580
Vaccari G, Tamburini E, Sgualdino G, Urbaniec K, Klemeš J (2005) Overview of the environmental problems in beet sugar processing: possible solutions. J Clean Prod 13:499–507
Van Der Poel PW (2000) Zuckertechnologie: Rüben-und Rohrzuckergewinnung. Bartens
Van Putten R-J, Van Der Waal JC, De Jong E, Rasrendra CB, Heeres HJ, De Vries JG (2013) Hydroxymethylfurfural, a versatile platform chemical made from renewable resources. Chem Rev 113:1499–1597
Vasconcelos MH, Mendes FM, Ramos L, Dias MOS, Bonomi A, Jesus CDF, Watanabe MDB, Junqueira TL, Milagres AMF, Ferraz A (2020) Techno-economic assessment of bioenergy and biofuel production in integrated sugarcane biorefinery: identification of technological bottlenecks and economic feasibility of dilute acid pretreatment. Energy 199:117422
Wascher FL, Stralis-Pavese N, Mcgrath JM, Schulz B, Himmelbauer H, Dohm JC (2022) Genomic distances reveal relationships of wild and cultivated beets. Nat Commun 13:2021
Webster TM, Grey TL, Scully BT, Johnson Iii WC, Davis RF, Brenneman TB (2016) Yield potential of spring-harvested sugar beet (*Beta vulgaris*) depends on autumn planting time. Ind Crop Prod 83:55–60
Weil J, Brewer M, Hendrickson R, Sarikaya A, Ladisch MR (1998) Continuous pH monitoring during pretreatment of yellow poplar wood sawdust by pressure cooking in water. In: Biotechnology for fuels and chemicals: proceedings of the nineteenth symposium on biotechnology for fuels and chemicals held, May 4–8, 1997, at Colorado Springs, Colorado. Springer, pp 99–111
Wyman C (1996) Handbook on bioethanol: production and utilization. CRC Press, 444p
Yafetto L, Odamtten G, Adu S, Ekloh E (2019) The biotechnological potential of using mono- and co-cultures of *Aspergillus niger* van Tieghem and *Trichoderma viride* Pers ex Fr. to enhance the

protein content of cassava (*Manihot esculenta* Crantz) peels by solid substrate fermentation. Stud Fungi 4:111–122

Yoon KY, Woodams EE, Hang YD (2005) Fermentation of beet juice by beneficial lactic acid bacteria. LWT Food Sci Technol 38:73–75

Zhang B, Shahbazi A (2011) Recent developments in pretreatment technologies for production of lignocellulosic biofuels. J Pet Environ Biotechnol 2:108. https://doi.org/10.4172/2157-7463.1000108

Zhang Z, Liu B, Zhao ZK (2012) Efficient acid-catalyzed hydrolysis of cellulose in organic electrolyte solutions. Polym Degrad Stab 97:573–577

Zheng Y, Lee C, Yu C, Cheng Y-S, Zhang R, Jenkins BM, Vandergheynst JS (2013) Dilute acid pretreatment and fermentation of sugar beet pulp to ethanol. Appl Energy 105:1–7

Zicari S, Zhang R, Kaffka S (2019) Sugar beet. In: Integrated processing technologies for food and agricultural by-products. Elsevier

Ziemiński K, Romanowska I, Kowalska-Wentel M, Cyran M (2014) Effects of hydrothermal pretreatment of sugar beet pulp for methane production. Bioresour Technol 166:187–193

Valorization of Beetroot Waste for Extraction of Natural Dye for Textile and Food Applications

11

Balwinder Singh Sooch, Navdeep Sandhu, Manpreet Kaur Mann, and Ramesh C. Ray

Abstract

The utilization of natural dyes for the coloration of various products has gained significant attention due to growing concerns about the adverse health and environmental impact of synthetic dyes. Beetroot, a natural source of color, holds immense potential as a sustainable source of natural dyes. Betalains, present in abundance in beetroot, are the major pigments responsible for their color. In addition, beetroot is an excellent dietary supplement, rich in minerals, vitamins, and many phytoconstituents with different medicinal properties. The beetroot waste, which includes pulp, peel, and stalks, is produced by industrial juice, jams, drinks, and sugar beet manufacturers. The valorization of this beetroot waste for the extraction of natural colors can contribute to its valuable use in addition to the reduction of this waste. The present chapter focuses on the extraction of natural dyes from beetroot waste for their applications in the food and textile sectors. Various extraction methods, such as maceration, solvent extraction, enzymatic extraction ultrasonication, and microwave-assisted methods, are discussed in detail, considering their efficiency and sustainability aspects. The potential challenges and future research directions in utilizing beetroot waste as a natural dye are also explored.

Keywords

Beetroot waste · Betalains · Natural dye · Betacyanins · Betaxanthins · Valorization

B. S. Sooch (✉) · N. Sandhu · M. K. Mann
Enzyme Biotechnology Laboratory, Department of Biotechnology and Food Technology, Patiala, India

R. C. Ray
ICAR-Central Tuber Crops Research Institute (Regional Centre), Bhubaneswar, India

R. C. Ray (ed.), *Roots, Tubers, and Bulb Crop Wastes: Management by Biorefinery Approaches*, https://doi.org/10.1007/978-981-99-8266-0_11

Abbreviations

ABS	Aqueous biphasic systems
BEW	Betalain water extracts
BRE	Betalain rich extract
DES	Deep eutectic solvents
DOPA	L-L-3,4-Dihydroxyphenylalanine
HPP	High-pressure processing
HTST	High-temperature short-time
Ils	Ionic liquids
MAE	Microwave-assisted extraction
NMWCO	Nominal molecular weight cut-offs
PEG	Polyethylene glycol
PPG	Polypropylene glycol
UAE	Ultrasound-assisted extraction
UV	Ultraviolet

11.1 Introduction

Fruits and vegetables are considered vital to health because they contain several essential constituents required for a healthy body, such as vitamins, zinc, phosphorus, and antioxidants. This is the reason that fruits and vegetables are utilized in large quantities worldwide. According to the United Nations Food and Agricultural Organization, global vegetable production will be around 14.4 billion in 2025, with a projected growth of 8.2% annually (Markets and Markets 2021). However, 60% of the vegetable produce is being wasted during harvest, processing, and packaging operations (Mirabella et al. 2014; Mann and Sooch 2020; Ray 2022; Sooch et al. 2022). Hence, fruits and vegetables have been identified as a commodity that generates massive food waste. Various reasons for perishable food waste are overproduction, mechanical damage during post-harvest handling, damage by microorganisms, insects, or pests, lack of proper storage facilities, product dumped due to inferior acceptable quality, and many more like these (Ray and Ravi 2005; Sooch et al. 2017; Mann and Sooch 2022). Different types of by-products and leftovers from food manufacturing and industrial processing units are some parts of vegetables such as the skin, stalk, pulp, peel, and seed, which generate 10–75% of waste materials and these are often high in nutrients (Bancal and Ray 2022; Sooch et al. 2023). Out of various vegetables grown the world over, Beet (*Beta vulgaris* L.), commonly known as beetroot, is a popular vegetable crop grown worldwide for various purposes. The European Union is the largest beetroot global producer (~70%), with the beetroot juice production in the United Kingdom alone, generating waste corresponding to ~35–40% (w/w) of the initial biomass. Beetroot is a rich source of many water-soluble nitrogenous pigments, including betalains (red and yellow tyrosine-derived pigments), which possess strong functional properties and these pigments are increasingly popular for different applications in food and other industries. These pigments, which are the derivatives of betalamic acid, are divided into two main groups, first the red–violet pigments called betacyanins and second the yellow pigments called betaxanthins primarily used as natural colorants. Betacyanins occur with color differences from purple to violet, whereas betaxanthins vary from yellow to orange. The average content of betalains in beetroot was around 1000 mg/100 g of total solids and 120 mg/100 g fresh weight (Marmion 1991; Cai et al. 2022). The pigment content in the roots depends on the degree of maturity, variety, and climatic conditions of the particular geographic area. Betanin constitutes 75–78% of betacyanin pigments' total content, and iso-betanin represents 95% of all red pigments found in beet juice. This beetroot waste includes pulp and peel and is produced by industrial juice, jams, drinks, and sugar beet manufacturers (del Amo-Mateos et al. 2023). This waste is rich in pigments and other valuable products. The valorization of this beetroot waste for the extraction of natural colors can contribute to the valuable use and reduction of this waste. The present chapter focuses on the extraction of natural dyes from beetroot waste for their applications in the food and textile sectors.

11.2 Taxonomy and Morphology of Beetroot Plant

The beetroot, commonly called beet, is a biennial plant that grows annually for the fleshy root and young leaves. It is a member of the Magnoliophyta division and class Mangnoliopsida. Beetroot, which belongs to the Chenopodiaceae family, ranks amongst the tenth most potent vegetables. It is a biennial or perennial plant and grows primarily in the temperate biome. The beetroot plant consists of three major parts: the taproot, stems, and leaves (Chhikara et al. 2019). Beetroot is a root vegetable with purple-green leaves, and it has a round taproot with minimal secondary roots. They produce seeds during the second year of growth. They are wind pollinated and can be dug starting in early October and harvested early for their greens. The seed propagates the beet, which germinate within 5–10 days. *Beta vulgaris* has three basic varieties such as chard (grown specifically for its leaves), beets (grown for its bulbous root and with edible leaves), and sugar beets (for making sugar from long, thick roots). Beetroot is grown worldwide mainly for food applications such as salads, pickles, and juices.

11.3 Chemical Composition of Beetroot

Beetroot makes excellent dietary supplements, rich in minerals and vitamins, and also has various phytoconstituents with different medicinal properties. The main compounds found in beetroot are carbohydrates, proteins, lipids, minerals, trace elements, vitamins, phenolic compounds, pigments, organic and inorganic acids, saponins, and the compound geosmin (Dhiman et al. 2021). Sucrose is the main

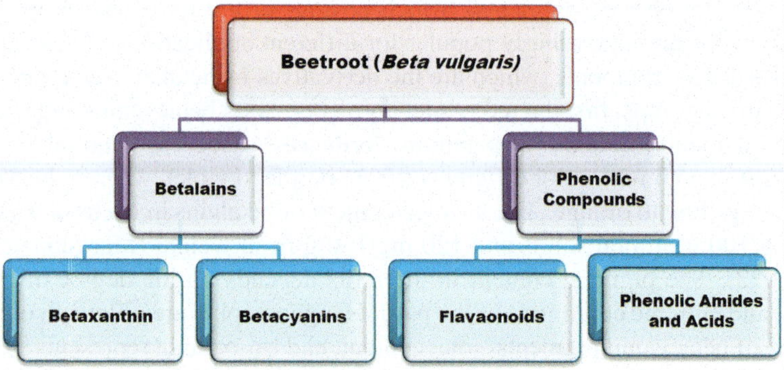

Fig. 11.1 Overview of bioactive compounds present in beetroot waste

sugar in beetroot, including small traces of glucose and fructose with around 87.4% of moisture content, 1.35% of proteins, 7.59% of fats, 1.9% of dietary fibers, and 7.59% of carbohydrates (Nemzer et al. 2011). A study showed that beetroot contains nine essential amino acids (histidine, isoleucine, leucine, lysine, methionine, phenylalanine, threonine, tryptophan, and valine) and nine non-essential amino acids (aspartic acids, serine, glutaminic acid, proline, glycine, alanine, tyrosine, arginine, and cysteine). Beetroot consists of high amounts of nitrates. Several traces of minerals have also been identified, such as iron (0.75 mg/100 g), copper (0.8 mg/100 g), potassium, and the highest amount of sodium (72.58 mg/100 g). The most predominant organic acid in beetroot is oxalic acid (Ninfali and Angelino 2013). Geosmin is the compound that gives beetroot its characteristic “earthy” taste and aroma. The beetroot has a significant amount of vitamin C and vitamins B1, B2, niacin, B6, and the leafy part comprises vitamin A. Betacyanins and betaxanthins are the main betalains found in beetroot, which impart several functional properties. The biosynthesis pathway of betalains involves the enzymatic conversion of L-tyrosine into L-L-3,4-dihydroxyphenylalanine (DOPA) with a hydroxylation enzyme. In the next step, the conversion of L-DOPA to betalamic acid and cyclo-DOPA is carried out through a catalysis reaction by two enzymes, namely DODA (4,5-dioxygenase) and CYP76AD1, respectively. Further, betalamic acid conjugates with amino acids and amines to produce betaxanthins, and betalamic acid combines automatically with cyclo-DOPA to give betalacyanin. An overview of the bioactive compounds present in the beetroot waste has been presented in Fig. 11.1.

The waste generated from beetroot processing includes essential components such as minerals, vitamins, polyphenols, and bioactive compounds, which can be recovered and converted to produce high-quality, sustainable products such as natural colorants and dyes (del Amo-Mateos et al. 2023). The food and beverage industry produces considerable amounts of waste that represent a promising source of natural dyes, which may be used because of their favorable technological or nutritional properties. The current trend in the textile materials heading is to investigate possible applications of bioactive agents for producing highly functional value-added textiles and natural dyes using beetroot waste as raw material. The use of natural dyes from this plant waste can be a sustainable alternative to synthetic dyes.

11.4 Waste Generation from Beetroot Processing and Its Use for Extraction of Natural Dyes

Beetroot waste is a potential source of colorants and pigments used in textile and food industries to develop novel and useful products. Various chemical or synthetic colors are generally used in different industries for providing an aesthetic aspect to various products. However, the replacement of these harmful chemical colors must be done using natural colorants because of the many harmful effects of these synthetic or chemical coloring materials. In contrast, natural colors such as beetroot pigments also enhance the physical and nutritional properties of food and food products, by fortification of vitamins, minerals, and antioxidants in addition to imparting color in the respective product, especially in food and textile industries. Additionally, natural dyes have functional properties such as UV (ultraviolet) protection and deodorizing properties. Hence, because of the different interactions of dye and fabric, many other functional properties with fastness can be achieved in addition to therapeutic properties. A new trend is evolving toward the use of natural colorants for food and textile applications. Betalains, a natural coloring pigment, is found throughout the root parts, including peels and pulp. The pulp and peel parts are the main waste parts produced by beetroot processing industries, such as industrial juice, jams, drinks, and sugar beet manufacturers. The waste material in the form of peels generated through the processing of beetroot contains high concentrations of phenolic compounds, and it is a rich source of phytochemicals and antioxidants (Baião et al. 2020). The valorization process of beetroot waste into value-added industrial-based products is shown in Fig. 11.2. Vulgaxanthin I, Vulgaxanthin II, indicaxanthin, betanin, prebetanin, isobutane, and neo betanin are among the betalains discovered in beetroot peel extract, and cyclo-dopa glucoside, V-formylcyclodopa glucoside, dihydroxy-indol carboxylic acid glucoside, betalamic acid, L-tryptophan, *p*-coumaric acid, ferulic acid, and quantities of unidentified flavonoids are also present (Zade et al. 2011). This waste has functionality as antioxidative (Azeredo 2009), antilipidemic and antimicrobial, antitumor (Amjadi et al. 2019), and anti-inflammatory (Tan et al. 2015) properties. The extractability of betalains from beetroot peels was carried out through microwave-assisted extraction (MAE) by Zin and Bánvölgyi (2021).

The ultrasound-assisted extraction (UAE) process for the recovery of betalains from waste red beetroot stalks was optimized by Maran and Priya (2016). Red beet wastes (stalks) were also used to extract betalain by the application of high-pressure processing (HPP) and high-temperature short-time (HTST) thermal treatment (Dos Santos et al. 2018). An eco-friendly method for extraction of betalains and phytochemicals from beetroot pomace, a waste from the beetroot processing industry, was developed after optimizing different process variables (Kushwaha et al.

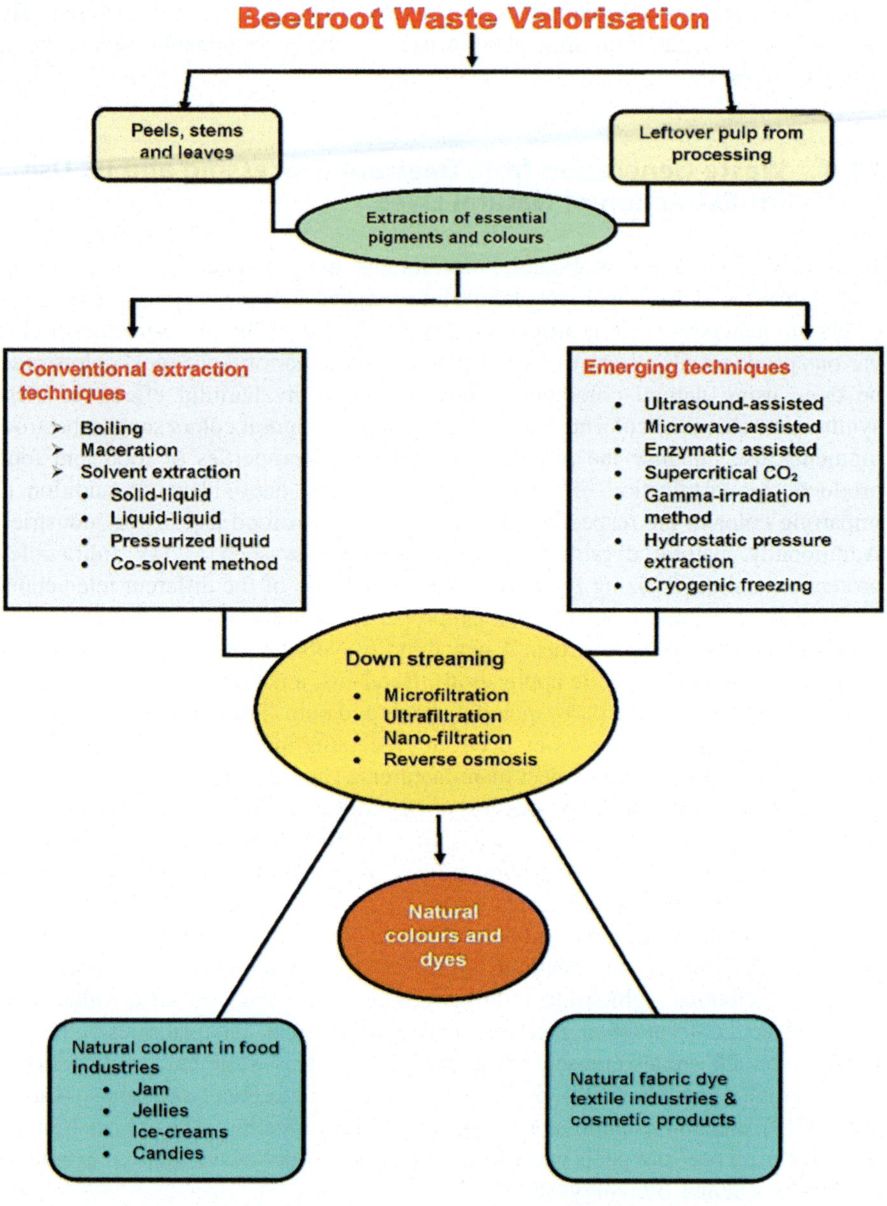

Fig. 11.2 Valorization of beetroot waste into value-added industrial-based products

2018). The beetroot processing waste material (peels, parings, and stalks) was used to obtain flour and colorant by drying it at different temperatures (Costa et al. 2017). The betalains were successfully recovered from wet and dried beet pulp waste generated from the juice industry (Fernando et al. 2021). The ultrasonication-assisted extraction process was successfully optimized to extract betalains from beetroot waste stalks by Singh et al. (2021). The red beetroot peel was also used to extract betalain and other nutraceutical compounds for food applications (El-Beltagi et al. 2022). An environmentally friendly process was adopted for the extraction of betalains and phytochemicals from beetroot waste (beetroot pomace) by optimizing several factors such as beetroot weight, pH, temperature, and time (Al-Amir et al. 2022). A new integrated process for one-step extraction and separation of betalains and chlorophylls from red beet waste (stems and leaves) was developed using thermoreversible aqueous biphasic systems (ABS) composed of quaternary ammonium-based ionic liquids (ILs) and polypropylene glycol (Rosa et al. 2023).

11.5 Methods of Extraction of Natural Dyes from Beetroot Waste

Synthetic dyes are made from different processes, including mixing different chemical constituents and involving several harsh treatments. Dye extraction from natural sources is carefully and meticulously controlled without using any chemical process in dyeing and processing. Different methods used for the extraction of colors from beetroot are illustrated in Fig. 11.3.

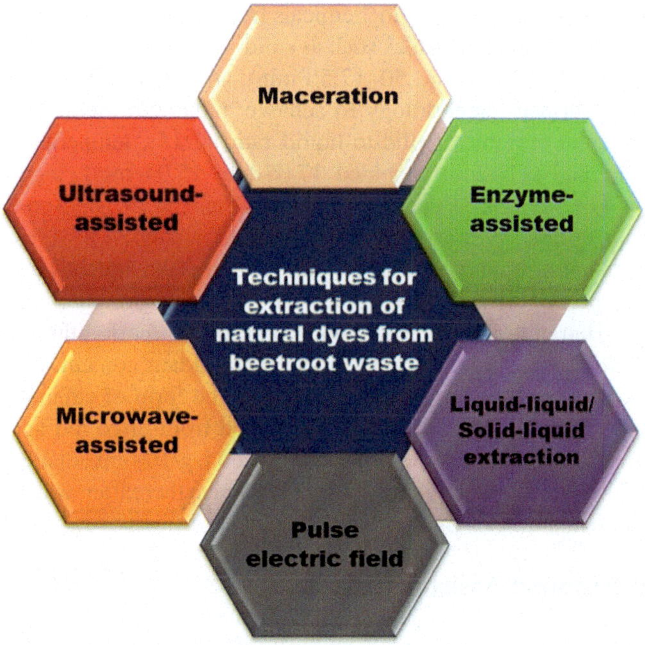

Fig. 11.3 Extraction methods for valorization of beetroot waste into natural dyes

Any extraction method should result in maximum recovery and minimum amount of residual. The selection of the appropriate technique plays a crucial part in extraction. A list of various methods used for the extraction of coloring compounds from beetroot and its waste material has been given in Table 11.1.

11.5.1 Physical Extraction Method

Physical extraction involves two types of methods, which are classified as conventional and nonconventional. The conventional approaches are maceration, blanching, and Soxhlet extraction, and they use some organic solvents with or without heat treatment (Castro-Enríquez et al. 2020). Nonconventional methods used to improve the extraction process are ultrasound-assisted, microwave-assisted, pulse electric field, cryogenic freezing, and membrane processing, which are employed to extract natural dyes from beetroot waste. These extractions have advantages such as less time extraction, eco-friendly, solvent energy process, and post-treatment waste (Fernando et al. 2021).

11.5.1.1 Maceration

Maceration involves the crushing of beetroot samples in a mortar and pestle with some solvent. Since beetroot pigments are water soluble, they can be extracted easily in water. Adding ethanol or methanol leads to a better yield of pigment in the presence or absence of acidification or heat treatment (Azeredo 2009). A method for extraction of betalains and phytochemicals from beetroot pomace was developed and different experimental variables such as solid-to-liquid (water) ratio (1:15–1:45), temperature (30–70 °C), time (2.50–12.50 min), and pH (1.50–5.50) were tailored using response surface methodology to achieve maximum yield. The use of optimum experimental variables (solid-to-liquid ratio (1:15), temperature (50.04 °C), time (10 min), and pH (2.50)) yielded 17.07 and 15.04 mg/L pigments, namely betacyanin and betaxanthin, respectively (Kushwaha et al. 2018). Some studies have shown that an optimal extraction time of 43 min yields 31.03% of betacyanins, 45.01% of betaxanthins, and 35.03% of total betalains. An aqueous ethanol medium is more appropriate than an individual solvent for maximal pigment recovery (Iahtisham-Ul-Haq et al. 2020). Paramita et al. (2022) reported that the maceration procedure yielded the optimal water content, antioxidant content, and color in the dye from beetroot waste using a single solvent in the form of ethanol. The betalain and other nutraceutical compounds were extracted from dried red beetroot peel by aqueous extraction through maceration technique (5 g/100 mL distilled water) and the dried red beetroot peel aqueous extract obtained thereafter was found to have 535 mg/100 g betalains (El-Beltagi et al. 2022).

11.5.1.2 Ultrasound-Assisted Technique

The ultrasound-assisted technique enhances color extraction efficiency through cavitation mechanism, solid matrix cell walls disruption mass transfer, and deep penetration of solvent to isolate the betalains. The cavitation process causes

Table 11.1 Various methods used for the extraction of colors from beetroot waste

Natural dye	Source	Method of extraction	Yield	Reference
Betalains	Beetroot waste (stalks)	Ultrasound-assisted extraction (UAE): • Temperature: 53 °C • Ultrasonic power: 89 W • Extraction time: 35 min • Solid/liquid ratio: 1: 19 g/mL	1.28 ± 0.02 mg/g (betacyanin) and 5.31 ± 0.09 mg/g (betaxanthin)	Maran and Priya (2016)
Betalains	Beetroot peels	Microwave irradiation method • Citric acid • Ethanol	229.264 mg/L 472.113 mg/L	Singh et al. (2017)
Betalains	Peels, parings, and stalks	Solid–liquid extraction (70% ethanol with solid–liquid ratio 1 g:5 mL)	68–81 mg/g dry matter	Costa et al. (2017)
Betalains	Beetroot stalks	High-pressure processing (HPP) and high-temperature short-time (HTST) thermal treatment	0.21–0.25 mg of betanin/g of stalks	Dos Santos et al. (2018)
Betalains	Beetroot pomace	Solid-to-liquid (water) ratio (1:15), temperature (50.04 °C), time (10 min), and pH (2.50)	Betacyanin (17.07) + betaxanthin (15.04) = 32.11 mg/L	Kushwaha et al. (2018)
Betalains	Peels	Microwave-assisted technique	180.38 mg/100 g	Zin et al. (2020)
Betacyanin	Beet stalks	Ultrasound method Power—89 W	1.28 mg/g of betacyanin	Singhee (2020)
Betaxanthin	Peel	Solvent extraction with thermostat water bath	839.12 mg/L	Zin et al. (2020)
		Solvent extraction (solvent ratio 0.8 w/v)	1512.14 mg/L	
Betalains	Beetroot waste (peels and pulp)	• Deep eutectic solvents (DES) • Betalain water extracts (BEW)	3.65–3.99 mg/g (in DES) 3.49–3.55 mg/g (in BEW)	Hernández-Aguirre et al. (2021)
Betalains	Beetroot waste (pulp)	The ultrasound-assisted technique (30% ethanol)	3 mg/g (approx.)	Fernando et al. (2021)
Betalains	Beetroot waste (stalks)	Ultrasonication approach: • Power intensity: 79.801 W/cm^2 • Solid/solvent ratio: 22.4 g/mL • Sonication time: 26.7 min	3 mg/g (betacyanin) and 3.36 mg/g (betaxanthin)	Singh et al. (2021)

(continued)

Table 11.1 (continued)

Natural dye	Source	Method of extraction	Yield	Reference
Betalains	Beetroot waste (peels)	Microwave-assisted extraction (MAE): • 150 s of MAE at 800 W • Pure water solvent	202.08 ± 2.23 mg/ 100 g	Zin and Bánvölgyi (2021)
Betacyanin	Beetroot skin	Solvent extraction (ethanol with aquades)	73.21%	Paramita et al. (2022)
Betalains	Beetroot pomace	Solid–liquid extraction (0.5 g/10 mL water)	55 mg/L	Al-Amir et al. (2022)
Betalains	Beetroot peel extract	Aqueous extraction by maceration technique (5 g/ 100 mL distilled water)	535 mg/100 g	El-Beltagi et al. (2022)
Betalains	Beetroot waste (stems and leaves)	• Thermoreversible aqueous biphasic systems (ABS) • Temperature 20 °C, time 70 min, and solid–liquid ratio of 0.12	6.67% (w/w)	Rosa et al. (2023)

compression and refraction, allowing bubbles formed in the medium to collapse. The collapse allows temperatures and pressures of 5000 K and 50 MPa to be reached at the molecular level, increasing the permeability of plant cell walls and the diffusivity of the solute in the solvent, which positively influences the release and extraction of compounds. The UAE process for the recovery of betacyanins and betaxanthins from waste red beetroot stalks was optimized and four factors, namely extraction temperature, ultrasonic power, extraction time, and solid–liquid ratio were adjusted using Box-Behnken response surface design. It was observed during studies that extraction temperature, ultrasonic power, and solid–liquid ratio is the significant factor that largely influence the extraction yield of pigments. A maximum yield of 1.28 ± 0.02 and 5.31 ± 0.09 mg/g was achieved for betacyanin and betaxanthin, respectively, at optimized extraction temperature of 53 °C, ultrasonic power of 89 W, extraction time of 35 min and the solid–liquid ratio of 1:19 g/mL (Maran and Priya 2016). Another study has shown that solvent β-cyclodextrin not only improves the extraction process but also maintains the stability of the pigment (Singh and Sengupta 2020). The ethanol/water mixtures used in ultrasound-assisted extraction were found to be more effective for the recovery of both betalains and polyphenols from dried beet pulp generated from the juice industry (Fernando et al. 2021). The multiple regression analysis was used to investigate the individual and interaction effects of different factors such as power intensity, solid-solvent ratio, and sonication period on the ultrasonication approach used to extract pigments from beetroot waste stalks. The maximum betacyanin and betaxanthin content of 3 and 4.36 mg/g, respectively were obtained under the optimized conditions used, such as power intensity of 79.801 W/cm^2, solid-solvent ratio of 22.4 g/mL, and sonication time of 26.7 min. All process variables used have shown a significant ($P < 0.05$) influence on the betalains yield from beetroot waste stalks (Singh et al. 2021).

11.5.1.3 Microwave-Assisted Extraction (MAE)

The microwave-assisted technique is a promising alternative to conventional harsh techniques. This technique reduces solvent consumption, energy demand, and processing time for the extraction process (Cardoso-Ugarte et al. 2014). The cells become thermally stressed during microwave irradiation treatment and as a result, the temperature and pressure inside the cell increase to levels that result in the rupture of cell walls releasing intracellular compounds. It was recorded during a study that the combination of a power of 400 W with a 100% duty cycle for 90–-120 seconds (s) resulted in the maximum extraction of betacyanins, whereas 150 s of irradiation increased the extraction of betaxanthins content. It has also been observed that the addition of ascorbic acid during the process resulted in an enhancement of extraction efficiency. In this strategy, ascorbic acid is used as the second solvent, developing a two-step microwave-assisted technique between cooling periods. Betalain yields obtained by MAE were twice as high as those obtained via conventional extraction at 80 °C. This two-step microwave-assisted technique leads to 50% more recovery of the pigments present in beetroot. Treating 400 W and 100% duty cycle for 1.5–2.0 min together yields the highest amount of betanins (52%), while betaxanthins are best obtained for 140–150 s. Ascorbic acid can protect from further degradation of pigments (Cardoso-Ugarte et al. 2014). The effect of MAE using citric acid and ethanol as solvents in different sets of beetroot peels was studied by Singh et al. (2017). The Box-Behnken design was employed to optimize the process and a maximum amount of 229.264 mg/L betanins were recovered using citric acid as solvent under optimized conditions (pH 5.20, microwave power of 224.61 MW and time 57.06 s). On the other hand, the enhanced betanins recovery of 472.113 mg/L was achieved using ethanol as solvent under optimized conditions of pH 4.74, microwave power 384.25 MW, and time 74.91 s. The extractability of betalains from beetroot peels was carried out through MAE at 100–800 W for 30–150 s with four different solvents. The maximum amounts of total betalains of 202.08 ± 2.23 mg/100 g were recovered using pure water solvent after 150 s of MAE at 800 W (Zin and Bánvölgyi 2021). Pectin was recovered from discarded red beetroot pomace using microwave extraction and a surfactant (polyethylene glycol, PEG4000) (del Amo-Mateos et al. 2023).

11.5.1.4 Pulse Electric Field

An electric pulse field implies giving short high-voltage pulses to a product between two electrodes. This method increases the membrane permeability and enhances the extraction yield (Loginova et al. 2011). It also causes less destruction to cell tissues and less energy consumption. Voltage energy of 1 kV/cm improves the seepage efficiency of betalains from red beetroot with a power consumption of 7 kJ/kg. It is an effective method to preserve the sensory and nutritive values of the extracted product. Different parameters influence extraction efficiency, including sample geometry, number of pulses, electric field strength, and electric conductivity of the sample. Pulsed electric field technology has advantages, such as the preservation of the nutritional and sensory value of products. The major drawback of this technique is the corrosive nature of the electrode, which can migrate into the pigments extracted and cause contamination (Fincan et al. 2004).

11.5.1.5 Other Methods

Some other methods like supercritical fluid extraction, pressurized liquid extraction, hydrostatic pressure extraction, cryogenic freezing, and gamma irradiation are other emerging nonconventional techniques for the extraction of pigments and dyes. Red beet stalks were used to extract betalain by the application of HPP and HTST thermal treatment. Different combinations of HPP (6000 bar for 10, 20, and 30 min) and HTST (75.7 °C for 80 s, 81.1 °C for 100 s, and 85.7 °C for 120 s) treatments were applied to extract betalain from the beetroot waste material. It was observed that there is no benefit to using the HPP treatment because lower betalain concentrations were obtained in HPP-treated samples than in the untreated control sample (Dos Santos et al. 2018). In supercritical extraction, the solvent can be subjected to high temperature and pressure beyond its critical point. This leads to high solvation power of the solvent and high diffusivity. In pressurized liquid extraction, the solvent is elevated at high pressures, reducing the extraction duration and less solvent usage. Deep eutectic solvents (DES) using magnesium chloride hexahydrate and urea were used in variable proportions for extraction and stabilization of red and violet betalains from beetroot waste (peels and pulp) generated from juice industry (Hernández-Aguirre et al. 2021).

11.5.2 Chemical Methods of Extraction of Natural Dyes from Beetroot Waste

Chemical methods include solvent extraction, which can be carried out in two ways: liquid and solid. It is an effective method for heat-sensitive pigments/dyes and has an advantage over other methods as it is easy to handle and requires less investment. The most commonly used solvents are methanol and ethanol in the case of betalains present in beetroot. In most scenarios, 20–50% (v/v) methanol or ethanol is used for complete extraction (Fu et al. 2020).

11.5.2.1 Liquid-Liquid Extraction

Liquid-liquid extraction is a common method for extracting natural compounds from various sources, including beetroot residue. In the case of natural dyes, liquid-liquid extraction is also used to separate the pigments from the beetroot waste and further isolate them in an appropriate organic solvent. Liquid-liquid extraction comprises of extraction of liquid obtained from waste with a liquid solvent. The aforementioned extraction technique leads to the transfer of solutes from one liquid phase to another liquid phase based on the polarity difference. The total phenolic content was estimated from discarded red beetroot using the gallic acid equivalents method. The supernatant obtained from centrifugation with double distilled water was mixed with the Folin-Ciocalteu reagent. After incubation sodium carbonate was added to the above solution and UV-VIS spectral analysis was carried out. The total content was expressed as Gallic acid equivalents. The total flavonoid content in discarded red beetroot was measured using catchein as standard for the colorimetric method, which was 471 mg CE (catechin equivalent)/L (del Amo-Mateos et al. 2023).

11.5.2.2 Solid–Liquid Extraction

Solid–liquid extraction involves the extraction of solid waste with a liquid solvent. It is one of the traditional methods used for extraction. The main aim of this process is to target a certain compound of bulk solid to be extracted into the liquid solvent (Patel et al. 2019). Beetroot pigments can be extracted by water, methanol, ethanol, and propan-2-ol (Paramita et al. 2022). However, methanol and propan-2-ol are permitted for all uses, with a maximum residue limit of 10 mg/kg in the final foodstuff or food ingredient. The beetroot waste (peels, parings, and stalks) dried at 70 °C yields flour with significant antioxidant activity and higher betalain content, whereas chlorination had no impact on the process yield. The betalain content of 68–81 mg/g dry matter was obtained after extraction using a solid–liquid ratio of 1 g:5 mL with 70% ethanol (Costa et al. 2017). Solid–liquid extraction carried out by continuous soxhlet extraction or pressurized liquid extraction, supercritical liquid method, ionic liquid extraction, and high-pressure extraction are the main techniques used under solid–liquid extraction (Aires 2017; Renard 2018). The maximum extraction of the betalains of 55 mg/L was obtained from beetroot pomace in water used in the ratio of 0.5 g/10 mL at 50 °C for 3 min at pH 5.0. The waste was used for effective aqueous extraction of natural color and the presence of the betalian and phenolic group was confirmed by FTIR, UV-VIS, and mass spectroscopy analysis (Al-Amir et al. 2022).

11.5.3 Enzymatic-Assisted Extraction

Enzyme-assisted extraction involves certain enzymes that speed up the process of recovery of pigments (Nadar et al. 2018). Enzymes such as cellulase, hemicellulase, pectinase, and ligninase are used to degrade the cell wall components. These enzymes are also used in combination for maximal recovery of pigments. The enzymatic extraction can be employed by enzymatic hydrolysis and extraction, and can also be performed by adding solvent before the enzymatic hydrolysis (Obama et al. 2012). The main parameters essential during extraction are the concentration of the enzyme, optimum pH, and temperature where the enzyme shows its best activity and the hydrolysis time. The enzymatic-aided extraction may offer several advantages in comparison to conventional extraction techniques, such as higher extraction yields, lower extraction times, lower energy consumption, and efficient recoveries with the use of simpler procedures, without, the use of toxic or hazardous solvents. The various conventional techniques can be combined with enzymatic extraction for better recovery. Enzymatic extraction has some limitations, such as it not being a cost-effective method and upscaling at the industrial is difficult (Puri et al. 2012). Betalains can be extracted from the pulp of red beetroot for the food industry in the presence of water, with an acid/alcohol medium (citric acid, ascorbic acid, citric acid and ascorbic acid mixture, ethyl alcohol, citric acid, and ethyl alcohol mixture, HCl mixture, and ethyl alcohol), or enzyme/microorganism-assisted processes (Yusuf et al. 2017; Chhikara et al. 2019).

11.5.4 Other Methods Involving the Extraction of Natural Dyes from Beetroot Waste

Other methods include supercritical extraction with CO_2, which ensures the selectivity toward a specific target compound with a certain temperature and pressure. It is also considered a novel green method and the solvent can be reused in the processing, thus reducing the total cost of the process. Co-solvents ethanol and methanol can be used to improve the extraction process (Khaw et al. 2017). Application of high pressures and temperatures can improve the extraction of betalains by improving the penetration of solvents into the cellular matrix and efflux of cellular components to the cell exterior. Another method enhances the pigment by subjecting the extraction method to gamma irradiation. Gamma-irradiation treatment improves extraction efficiency mainly due to cell wall permeabilization by altering the interior tissue structure and loss of turgor pressure (Rastogi and Raghavarao 2004). When 100% water is used as a solvent, this technique is called subcritical water extraction or superheated water extraction (Pronyk and Mazza 2009; Mustafa and Turner 2011). Betalain consists of a nitrogenous core structure, betalamic acid, which combines with an amino acid to produce betacyanin and betaxanthin (vulgaxanthin). Because betalain compounds are hydrophilic, they can be extracted with water, aqueous methanol, or ethanol. However, the extraction method and choice of solvents affect the extract's constituent compounds significantly. Moreover, extracted beetroots are sensitive to pH, water activity, temperature, light, oxygen, metal ions, and enzyme activities (Zin et al. 2020). An integrated process for one-step extraction and separation of betalains and chlorophylls from red root beet waste (stems and leaves) developed with the use of thermoreversible ABS composed of quaternary ammonium-based Ils and polypropylene glycol. A central composite design was used to optimize the extraction parameters (time, temperature, and solid–liquid ratio) for both pigment extraction yields and finally, the maximum extraction yield of 6.67% (w/w) for betalains and 1.82% (w/w) for chlorophylls was obtained with 20 °C temperature, 70 min time and a solid–liquid ratio of 0.12 (Rosa et al. 2023).

11.6 Post-extraction Processing

Post-extraction processing entails the down-streaming of the target compound. It involves the purification of the target compound from the undesired extract constituents. Membrane processing is one of the fundamental down-streaming processes used for the isolation, concentration, fractionation, and purification of liquids. It is based on the principle that a liquid sample is passed through a semi-permeable membrane based on the specific pore size of the target constituent. The target constituents pass the membrane, and undesired compounds are left behind. The membrane's selectivity is based on the specific compound to be purified. The sample can be force driven through the membrane called pressurized membrane processing. It is generally called size sieving. Membrane processing is done in two ways: cross-flow filtration and dead-end filtration. In cross-flow filtration, the feed is horizontal to the membrane, whereas in dead-end filtration, the feed flow is vertical. Membrane filtration is adequate for industrial utilization for heat-sensitive products as no heat is involved in the processing, and it can be done at room temperature (Drioli and Fontananova 2004). The membrane processing can be different depending on the membrane size and mode, such as microfiltration, ultrafiltration, nanofiltration, and reverse osmosis. Microfiltration and ultrafiltration work on the pore size mechanism and hydrostatic pressure. The only difference with reverse osmosis is that the hydrostatic pressure is greater than osmotic pressure because solvent movement is in the opposite direction. Membrane processing is usually used in the juice industry to clarify and purify juices.

An initial laboratory-scale evaluation of the separation characteristics of membranes with nominal molecular weight cut-offs spanning from 30 to 0.5 kD revealed that betalains are effectively separated in the 0.5 kD region. Subsequent pilot-scale tests utilizing 1 kD, loose reverse osmosis, and reverse osmosis spiral-wound membranes revealed loose reverse osmosis membrane to be very efficient, removing up to 96% of salt and 47% of other dissolved particles while retaining the majority of pigment (98%) in the betalain rich extract (BRE). The total amount of betalain in the BRE increased to 46%, the highest recovery reported at the pilot scale level to date. Interestingly, after three diafiltrations, more than 95% of the nitrates were removed from the BRE. These studies demonstrate that membrane technology is the most effective method for producing BRE with extremely reduced salt and nitrate levels (Mereddy et al. 2017).

11.7 Stability Issue of Natural Dye Extracted from Beetroot

The dye extracted from beetroot can compete with synthetic dyes, but some important parameters that affect the stability are the pH of the dye bath, salt concentration, time of dyeing, and temperature of dyeing. The color strength of betanin pigment decreases at pH 4 and is maximum absorbed at pH 5 (Guesmi et al. 2012). It is also observed that the stability and strength of the dye decreases as the salt concentration increases. Dyeing strength is maximum studied without salt. Betanin works best at

50 °C, and the coloration power decreases above and below the optimum range. A time duration of 30–45 min is considered best for dyeing with betanin pigment. Betanin starts to leave the fabric because of its thermal stability. Betalains degrade when exposed to light and oxygen (Azeredo 2009). The dyeing with betalains is usually done in the dark as exposure to light reduces the dye's strength (Azeredo 2009). The use of natural red colorants extracted from beetroot was tested in some processed foods (jelly and ice sherbets) and it was also recorded that the best carriers for betalains are dextrin followed by soluble starch, lactose, and glucose, respectively. The color and higher pigment stability of betalain pigments extracted from red beet were observed at a pH range of 3.0–7.0 and a temperature range of 40–50 °C (Attia et al. 2013). Betaxanthin and betacyanin pigments show the best activity at 60 °C. No significant effect of pH was observed on the stability of betacyanin and betaxanthin (Kushwaha et al. 2018). Red beet stalks were used to extract betalain and to improve its stability by the application of HPP and HTST thermal treatment. It was observed that there is no benefit of using the HPP treatment because lower betalain concentrations were obtained in HPP-treated samples than in the untreated control sample. This may be due to the reason that HPP treatment would have caused more exposure of the sample to oxygen and/or light. Whereas HTST was found to be the most suitable to maintain betalain stability (Dos Santos et al. 2018). In another study, a powdered red beetroot dye was produced, which was then used as a food coloring in the preparation of candies. Eight-week stability tests were conducted on both the prepared red beet dye and the confectionery at 4, 25, and 45 °C. In one study, the betanin content (%) was measured to evaluate the color's stability, and the highest betanin retention was observed in red beet dye, while the lowest betanin retention was observed at 45 °C, indicating the color degradation at higher temperatures. The pH stability of red beetroot dye was also measured at 4, 25, and 45 °C. An increase in pH was observed as storage temperatures rose. Maximum color stability was observed at 4 °C with the most acidic pH, while color degradation was observed with a progressive pH increase at 45 °C. Candies and beetroot powder are safe for consumption at both 4 and 25 °C storage temperatures, as determined by microbiological testing (Nisa et al. 2021). It was observed from the kinetic curves of betalain degradation that the exposure of water samples to light and atmospheric oxygen causes the alteration of the violet coloration from beetroot pigments. Betalain obtained in water extracts from the waste of the juice industry was found to be degraded in 5–7 days, whereas betalains obtained from DES extracts were stable under visible light for 150 days and 340 days in amber vessels (Hernández-Aguirre et al. 2021). The thermal susceptibility of beetroot peel extracts to processing conditions was investigated at different temperatures (30–70 °C) and the thermal instability of betalains was evaluated in terms of half-life ($t_{1/2}$), temperature quotient (Q_{10}), and activation energy (E_a) using the Arrhenius equation. It was concluded from the observations that the resulted in retention percentage ($R\%$) of total betalains was increased from 22% to 51% and of betacyanins from 3% to 29% by the addition of ascorbic acid but in contrast $R\%$ of betaxanthins was decreased from 56% to 40% after the heat treatment at 70 °C for 5 h (Zin and Bánvölgyi 2021).

11.8 Applications of Beetroot Color

Red beetroot is a rich source of pigments with a wide variety of applications in food and non-food industries, such as textile industries (Choo 2019). Betalains, a major component of beetroot, have attracted the wide attention of researchers and the market not only because of the orientation of people toward using natural food colors but also due to a wide range of their biological activities, including anti-inflammatory, hepatoprotective, antioxidant, antimicrobial, and anticarcinogenic properties and these can be exploited for providing solutions for human health improvement. Natural colors extracted from beetroot have many advantages over synthetic colors and are used in food and other industries. The extraction of natural dye for the preparation of herbal gulal (dye) from beetroot was also explored in the past (Tiwari 2020). The bioactive compounds extracted from beetroot waste are used as nutraceuticals and in medical fields. The applications of pigments extracted from the beetroot in the food and textile industries are discussed below.

11.8.1 Applications of Beetroot Color in the Food Industry

The vibrant color produced from beetroot waste has gained much attention in food industries. It is used as a colorant in fermented and heat-processed sausages. Beetroot is recognized as a health-promoting food due to the presence of components such as vitamins, minerals, phenolics, carotenoids, nitrate, ascorbic acids, and betalains. Betalains exist in two forms, betacyanin (red-violet pigment) and betaxanthin (yellow-orange pigment), and are commercially recognized as a food dye due to their non-hazardous, non-toxic, non-carcinogenic, and non-poisonous characteristics. As an anticipated boon for the food industry, beetroot is used as a food coloring or additive in ice cream, yogurt, and other food products. The beetroot extract is also added to tomato pastes, stews, sauces, desserts, jams, jellies, candies, and breakfast cereals to increase their redness (Chhikara et al. 2019).

Pomace is a by-product of the beetroot processing industry because it is rich in betalains and phytochemicals that can be utilized in edible applications. A large quantity of pomace is an inexpensive source of dietary fiber. The pomace of beetroot contains multiple bioactive compounds and antioxidant activity. Due to its natural, non-toxic pigments and coloring properties as well as its lack of toxicity, it is widely used in the food industry as an additive. As a result of its many functional properties, such as its water-holding capacity, water-retention capacity, swelling capacity, and oil-absorption capacity, beetroot powder is extensively used in a variety of food products, including cookies, cakes, snacks, candies, etc. The utilization of beetroot pomace is economical, and it can be utilized as a functional ingredient. Beetroot extract and powder developed from waste with high antioxidative properties prevent lipid oxidation in sausages with a significant amount of betalains and phenolic compounds (Theba et al. 2021).

Beetroot extract used instead of carmine shows better color development in beef sausages (Aykln-Dinçer et al. 2021). The color extracted from beetroot waste is stable for up to 20 days of storage and can be utilized in products with short to medium shelf life. Since the colorants are beetroot extract's residual parts, they show high antioxidative properties and are high in phenolic compounds, adding functionality to food products (Cardoso-Ugarte et al. 2014). Nowadays, jellies, jams, and marmalades are being prepared from beetroot peel showing a great amount of phytochemicals. They are prepared with or without sugar (Vishnupriya et al. 2022). Beetroot crisps are being developed with the help of multi-grains, spices, and different types of seeds from beetroot waste. Beetroot acts as a rich source of dietary nitrates and strengthens endothelial functions. These antioxidative-rich products scavenge free radicals and are anti-inflammatory (Singh and Sengupta 2020). Beetroot peels and extracts are used in dressings in salads, desserts, and sauces offering visually appealing products with a natural essence. Kumar and Brooks (2018) reported the antimicrobial potential of beetroot extract and reported the antimicrobial potential of beetroot extract and capitalized on these antimicrobial properties to create novel red beetroot products with added value.

The colorant obtained from this beetroot waste (peels, parings, and stalks) showed color stability for 20 days and was found as a suitable functional additive to improve the nutritional characteristics and appearance of food products (Costa et al. 2017). It was also observed in a study that the pressurized liquid extraction technique is a good alternative for extracting phenolic compounds with high biological potential from beetroot resides, leaves, and stems, and these can be used in formulations for nutraceuticals, functional foods or pharmaceutical industry (Battistella Lasta et al. 2019). The betalain and other nutraceutical compounds extracted from dried red beetroot peel were used for the preservation of Nile Tilapia fish and showed an excellent preservative effect on the fish fillet (El-Beltagi et al. 2022).

11.8.2 Applications of Beetroot Color in the Textile Industry

Besides the food industry, beetroot waste colors have also found their place in the textile industry as natural dyes. Many dyes from this waste show various colors, such as red, violet, yellow, and purple. Utilizing beetroot waste colors creates eco-friendly dyes and this is a step toward zero waste. Natural dyes show extensive properties like insect repellent, deodorizing, flame retardant, UV protection, fluorescence, antimicrobial activity, biodegradability, and non-toxicity (Widiawati 2009; Rather et al. 2016). Dyeing with beetroot pigments extracted from its peel was also used to dye silk fabrics using an ultra-sonic technique with or without mordant. The dye extracted from silk fabrics was used with eco-friendly mordants like citric acid, tartaric acid, acetic acid, and tannic acid (Yaqub et al. 2018). Betanin has also shown dyeing properties on wool fabrics and threads. Besides food and textile dyes extracted from beetroot, it is a fluorescent dye for tissue staining. Acidic betalain pigments are also used as hematological staining (Udonkang et al. 2019). The dye extracted from beetroot pomace is used to stain fungal species and replace the cotton blue synthetic dye. Regarding the dying process, protein materials (wool and silk) are not dyed with fresh red beetroot extract; if a mordant is added to the dye solution, only light colors such as "milk rose" or "pink baby" are produced (Kumar and Prabha 2018).

In another study, the beetroot peels can serve as a sustainable source of betalains, which can be used to dye wool using water- and energy-efficient procedures. The extraction of betalains from colored food waste/peels of red beetroot required the use of water as the only solvent, with no additional additives. It was necessary to functionalize betalains or the wool for the extract to be able to dye the wool using acetic acid, ethanol, and arginine, three varieties of sustainable functionalization were performed. The mechanism that justifies intensely colored wool dyeing was elucidated for each functionalization. With the information provided by UV-VIS and HPLC-MS analyses, the extract was characterized. Due to the information obtained from the FTIR and CIELab analyses, it was feasible to characterize the wool that had been dyed with an extract derived from red beetroot peels. In acidic environments, the functionalization of betalains and wool produces the most intense red hues. The hue differs according to the pH and betalain concentration (Popescu et al. 2021). Applications of naturally dyed textiles can be extended to a field such as sportswear and medicinal field due to their various functional properties such as antimicrobial as well as UV protection (Verma et al. 2021).

The pigments (betalains) extracted from pulp and crimson beetroot peels have no affinity for textiles. Betalains (betacyanins, betaxanthins) derived from beetroot are unstable pigments that change color in response to changes in pH or enzymatic hydrolysis when exposed to light. Long-term exposure of some beetroot segments to light in open areas causes their discoloration and browning, a phenomenon common to all betalain-containing products. However, under certain conditions, betalains could dye all protein-based textiles intensely, profoundly, and sustainably (Antigo et al. 2018; Popescu et al. 2021). The dried waste samples from the beetroot juicing industry demonstrate good betalain and polyphenol extractability and it was reported that betalains from dried beetroot waste can serve as a good source of color for many technological sectors (Fernando et al. 2021). The betalains were extracted from beetroot pomace by optimizing several factors and extracted color was used for dying cotton, wool, and silk fabrics. The maximum dyed fabric's color strength (K/S) values of 7.5, 11.03, and 15.6 were observed for treated cotton, wool, and silk fiber, respectively. All samples were also tested for color fastness tests by washing, perspiration, and exposure to light (Al-Amir et al. 2022).

11.9 Conclusion and Future Prospects

The valorization of beetroot waste for extracting pigments and colorants presents significant benefits and opportunities for the textile and food industries. These industries can lead the production of sustainable, eco-friendly, and safe quality products with natural colors by processing waste obtained from beetroot processing. A natural pigment can replace synthetic colors in food industries for processing products like candies, jams, jellies, and ice cream. These pigments also offer various benefits in textile industries for developing natural and nontoxic shades of different colors. Some studies have shown that natural dyes are insect-repellent and have antimicrobial activity. Various research and development efforts have focused on improving extraction techniques for getting maximum yields from waste in the recent past. More advancements in extraction techniques can lead to the development of cost-effective processes and scale up the technology on a larger scale for obtaining color from fruit and vegetable waste materials. The recent focus of research is also on the pigment's stability and shelf life to assure that pigments' color can be restored longer. The increasing demand for natural colorants has led to the development of standards and protocols for delivering the best quality products. Research for developing innovative and novel techniques for utilizing beetroot waste to design novel technologies and processes is continued in many laboratories. The future will witness the increased popularity of sustainable sourcing of pigments from beetroot to develop colors and dyes.

References

Aires A (2017) Phenolics in foods: extraction, analysis, and measurements. In: Phenolic compounds. IntechOpen, pp 61–88

Al-Amir A, Gomaa EH, El-Azabawy RE, El-Bayaa AA (2022) Valorization beetroot waste for eco-friendly extraction of natural dye for textile and food applications. Egypt J Chem 65:725–736. https://doi.org/10.21608/ejchem.2022.121319.5441

Amjadi S, Emaminia S, Heyat Davudian S et al (2019) Preparation and characterization of gelatin-based nanocomposite containing chitosan nanofiber and ZnO nanoparticles. Carbohydr Polym 216:376–384. https://doi.org/10.1016/j.carbpol.2019.03.062

Antigo JLD, Bergamesco R, Madrona GS (2018) Effect of pH on the stability of red beet extract (*Beta vulgaris* l.) microcapsules produced by spray drying or freeze drying. Food Sci Technol 38:72–77. https://doi.org/10.1590/1678-457X.34316

Attia GY, Moussa MM, Sheashea E (2013) Characterization of red pigments extracted from red beet (*Beta vulgaris* l.) and its potential uses as antioxidant and natural food colorants. Egypt J Agric Res 91:10.21608/ejar.2013.167086

Aykln-Dinçer E, Güngör KK, Çağlar E, Erbaş M (2021) The use of beetroot extract and extract powder in sausages as natural food colorant. Int J Food Eng 17:75–82. https://doi.org/10.1515/ijfe-2019-0052

Azeredo HMC (2009) Betalains: properties, sources, applications, and stability—a review. Int J Food Sci Technol 44:2365–2376

Baião DDS, da Silva DVT, Paschoalin VMF (2020) Beetroot, a remarkable vegetable: its nitrate and phytochemical contents can be adjusted in novel formulations to benefit health and support cardiovascular disease therapies. Antioxidants (Basel, Switzerland) 9. https://doi.org/10.3390/antiox9100960

Bancal V, Ray RC (2022) Overview of food loss and wastes in fruits and vegetables: from issues to resources. In: Ray RC (ed) Fruits and vegetable wastes: valorization to bioproducts and platform chemicals. Springer, Singapore, pp 3–32

Battistella Lasta HF, Lentz L, Gonçalves Rodrigues LG et al (2019) Pressurized liquid extraction applied for the recovery of phenolic compounds from beetroot waste. Biocatal Agric Biotechnol 21:101353. https://doi.org/10.1016/j.bcab.2019.101353

Cai M, Zhong H, Ma Q et al (2022) Physicochemical and microbial quality of *Agaricus bisporus* packaged in nano-SiO_2/TiO_2 loaded polyvinyl alcohol films. Food Control 131:108452. https://doi.org/10.1016/j.foodcont.2021.108452

Cardoso-Ugarte GA, Sosa-Morales ME, Ballard T et al (2014) Microwave-assisted extraction of betalains from red beet (*Beta vulgaris*). LWT 59:276–282. https://doi.org/10.1016/J.LWT.2014.05.025

Castro-Enríquez DD, Montaño-Leyva B, Del Toro-Sánchez CL et al (2020) Stabilization of betalains by encapsulation—a review. J Food Sci Technol 57:1587–1600. https://doi.org/10.1007/s13197-019-04120-x

Chhikara N, Kushwaha K, Sharma P et al (2019) Bioactive compounds of beetroot and utilization in food processing industry: a critical review. Food Chem 272:192–200. https://doi.org/10.1016/J.FOODCHEM.2018.08.022

Choo WS (2019) Betalains: application in functional foods. In: Bioactive molecules in food. Springer, pp 1471–1498

Costa APD, Hermes VS, Rios AO, Flôres SH (2017) Minimally processed beetroot waste as an alternative source to obtain functional ingredients. J Food Sci Technol 54:2050–2058. https://doi.org/10.1007/s13197-017-2642-4

del Amo-Mateos E, Fernández-Delgado M, Lucas S et al (2023) Valorization of discarded red beetroot through the recovery of bioactive compounds and the production of pectin by surfactant-assisted microwave extraction. J Clean Prod 389:135995. https://doi.org/10.1016/j.jclepro.2023.135995

Dhiman A, Suhag R, Chauhan DS et al (2021) Status of beetroot processing and processed products: thermal and emerging technologies intervention. Trends Food Sci Technol 114:443–458. https://doi.org/10.1016/J.TIFS.2021.05.042

Dos Santos CD, Ismail M, Cassini AS et al (2018) Effect of thermal and high pressure processing on stability of betalain extracted from red beet stalks. J Food Sci Technol 55:568–577. https://doi.org/10.1007/s13197-017-2966-0

Drioli E, Fontananova E (2004) Membrane technology and sustainable growth. Chem Eng Res Des 82:1557–1562

El-Beltagi HS, El-Mogy MM, Parmar A et al (2022) Phytochemical characterization and utilization of dried red beetroot (*Beta vulgaris*) peel extract in maintaining the quality of Nile tilapia fish fillet. Antioxidants 11. https://doi.org/10.3390/antiox11050906

Fernando GSN, Wood K, Papaioannou EH et al (2021) Application of an ultrasound-assisted extraction method to recover betalains and polyphenols from red beetroot waste. ACS Sustain Chem Eng 9:8736–8747. https://doi.org/10.1021/acssuschemeng.1c01203

Fincan M, DeVito F, Dejmek P (2004) Pulsed electric field treatment for solid–liquid extraction of red beetroot pigment. J Food Eng 64:381–388. https://doi.org/10.1016/j.jfoodeng.2003.11.006

Fu Y, Shi J, Xie SY et al (2020) Red beetroot betalains: perspectives on extraction, processing, and potential health benefits. J Agric Food Chem 68:11595–11611. https://doi.org/10.1021/ACS.JAFC.0C04241

Guesmi A, Ladhari N, Ben HN, Sakli F (2012) Isolation, identification and dyeing studies of betanin on modified acrylic fabrics. Ind Crop Prod 37:342–346

Hernández-Aguirre OA, Muro C, Hernández-Acosta E et al (2021) Extraction and stabilization of betalains from beetroot (*Beta vulgaris*) wastes using deep eutectic solvents. Molecules 26. https://doi.org/10.3390/molecules26216342

Iahtisham-Ul-Haq BMS, Randhawa MA, Muhammad SA (2020) Extraction, characterization and optimization of betalains from red beetroot using response surface methodology. Pak J Agric Sci 57:535–543

Khaw K-Y, Parat M-O, Shaw PN, Falconer JR (2017) Solvent supercritical fluid technologies to extract bioactive compounds from natural sources: a review. Molecules 22:1186

Kumar S, Brooks MSL (2018) Use of red beet (Beta vulgaris L.) for antimicrobial applications-a critical review. Food Bioprocess Technol 11:17–42. https://doi.org/10.1007/S11947-017-1942-Z

Kumar V, Prabha R (2018) Extraction and analysis of natural dye. J Nat Prod Plant Resour 8:32–38

Kushwaha R, Kumar V, Vyas G, Kaur J (2018) Optimization of different variable for eco-friendly extraction of betalains and phytochemicals from beetroot pomace. Waste Biomass Valoriz 9: 1485–1494. https://doi.org/10.1007/s12649-017-9953-6

Loginova KV, Lebovka NI, Vorobiev E (2011) Pulsed electric field assisted aqueous extraction of colorants from red beet. J Food Eng 106:127–133. https://doi.org/10.1016/J.JFOODENG.2011.04.019

Mann MK, Sooch BS (2020) Emerging trends in food industry waste valorization for bioethanol production. In: Verma P (ed) Biorefineries: a step towards renewable and clean energy. Clean energy production technologies. Springer, Singapore, pp 57–92

Mann MK, Sooch BS (2022) Utilization of fruit and vegetable wastes for the cultivation of edible mushrooms. In: Ray RC (ed) Fruits and vegetable wastes : valorization to bioproducts and platform chemicals. Springer Nature, Singapore, pp 117–138

Maran JP, Priya B (2016) Multivariate statistical analysis and optimization of ultrasound-assisted extraction of natural pigments from waste red beet stalks. J Food Sci Technol 53:792–799. https://doi.org/10.1007/s13197-015-1988-8

Markets and Markets (2021) Fruit and vegetable seeds market size, revenue, challenges—2025. In: Mark. Res. Rep. https://www.marketsandmarkets.com/Market-Reports/fruit-and-vegetable-seed-market-1257.html. Accessed 10 Jun 2023

Marmion DM (1991) Handbook of US colorants: foods, drugs, cosmetics and medical devices. John Wiley & Sons

Mereddy R, Chan A, Fanning K et al (2017) Betalain rich functional extract with reduced salts and nitrate content from red beetroot (*Beta vulgaris* L.) using membrane separation technology. Food Chem 215:311–317. https://doi.org/10.1016/J.FOODCHEM.2016.07.132

Mirabella N, Castellani V, Sala S (2014) Current options for the valorization of food manufacturing waste: a review. J Clean Prod 65:28–41. https://doi.org/10.1016/J.JCLEPRO.2013.10.051

Mustafa A, Turner C (2011) Pressurized liquid extraction as a green approach in food and herbal plants extraction: a review. Anal Chim Acta 703:8–18

Nadar SS, Rao P, Rathod VK (2018) Enzyme assisted extraction of biomolecules as an approach to novel extraction technology: a review. Food Res Int 108:309–330

Nemzer B, Pietrzkowski Z, Spórna A et al (2011) Betalainic and nutritional profiles of pigment-enriched red beet root (*Beta vulgaris* L.) dried extracts. Food Chem 127:42–53. https://doi.org/10.1016/J.FOODCHEM.2010.12.081

Ninfali P, Angelino D (2013) Nutritional and functional potential of *Beta vulgaris* cicla and rubra. Fitoterapia 89:188–199. https://doi.org/10.1016/j.fitote.2013.06.004

Nisa A, Hina S, Kalim I et al (2021) Quality assessment and application of red natural dye from beetroot (*Beta vulgaris*). Pak J Agric Res 34:10.17582/journal.pjar/2021/34.3.552.558

Obama P, Ricochon G, Muniglia L, Brosse N (2012) Combination of enzymatic hydrolysis and ethanol organosolv pretreatments: effect on lignin structures, delignification yields and cellulose-to-glucose conversion. Bioresour Technol 112:156–163

Paramita O, Ansori M, Kusumastuti A, Fatimah N (2022) Natural food colourant of beetroot skin (*Beta vulgaris* L): characterisation study. IOP Conf Ser Earth Environ Sci 969:012047. https://doi.org/10.1088/1755-1315/969/1/012047

Patel K, Panchal N, Ingle P (2019) Review of extraction techniques. Int J Adv Res Chem Sci 6:6–21

Popescu V, Blaga AC, Pruneanu M et al (2021) Green chemistry in the extraction of natural dyes from colored food waste, for dyeing protein textile materials. Polymers (Basel) 13. https://doi.org/10.3390/polym13223867

Pronyk C, Mazza G (2009) Design and scale-up of pressurized fluid extractors for food and bioproducts. J Food Eng 95:215–226

Puri M, Sharma D, Barrow CJ, Tiwary AK (2012) Optimisation of novel method for the extraction of steviosides from Stevia rebaudiana leaves. Food Chem 132:1113–1120

Rastogi NK, Raghavarao K (2004) Increased mass transfer during osmotic dehydration of γ-irradiated potatoes. J Food Sci 69:E259–E263

Rather LJ, Shabbir M, Bukhari MN et al (2016) Ecological dyeing of woolen yarn with *Adhatoda vasica* natural dye in the presence of bio mordants as an alternative copartner to metal mordants. J Environ Chem Eng 4:3041–3049

Ray RC (2022) Fruits and vegetable wastes: valorization into bioproducts and platform chemicals. Springer, Singapore, 448p

Ray RC, Ravi V (2005) Spoilage of sweet potato in tropics and control measures (review). Crit Rev Food Sci Nutr 45:623–644

Renard CMGC (2018) Extraction of bioactives from fruit and vegetables: state of the art and perspectives. LWT 93:390–395

Rosa ME, Ferreira AM, Neves CMSS et al (2023) Valorisation of red beet waste: one-step extraction and separation of betalains and chlorophylls using thermoreversible aqueous biphasic systems. Green Chem 25:1852–1864. https://doi.org/10.1039/D2GC04480C

Singh N, Sengupta R (2020) Multigrain beetroot crisps: a healthy snacking option. Asian J Med Health 18:19–25. https://doi.org/10.9734/ajmah/2020/v18i330189

Singh A, Ganesapillai M, Gnanasundaram N (2017) Optimizaton of extraction of betalain pigments from *Beta vulgaris* peels by microwave pretreatment. IOP Conf Ser Earth Environ Sci 263: 032004

Singh S, Omre PK, Chand K et al (2021) Process optimization of ultrasonic assisted extraction of betalains from red beet, *Beta vulgaris* L. waste stalks. Indian J Exp Biol 59:858–866. https://doi.org/10.56042/ijeb.v59i12.57831

Singhee D (2020) Review on natural dyes for textiles from wastes. In: Chemistry and technology of natural and synthetic dyes and pigments. IntechOpen, pp 1–24. https://doi.org/10.5772/intechopen.93178

Sooch B, Mann M, Singh R (2017) Current insights into proteomics of biofuel crops and cyanobacteria. In: Ram SS, Ashok P, Edgard G (eds) Biofuels: production and future perspectives. CRC Press/Taylor & Francis, pp 511–539

Sooch BS, Mann MK, Sharma P, Ray RC (2022) Phenolic and other bioactive compounds from vegetable and fruit waste: extraction methods and their possible utilization. In: Ray RC (ed) Fruits and vegetable wastes: valorization to bioproducts and platform chemicals. Springer Nature Singapore, Singapore, pp 81–114

Sooch BS, Mann MK, Kaur S (2023) 2—Lignocellulosic biomass: a feedstock to support the circular economy. In: Moodley P, Ray RC, Gueguim Kana EB (eds) Advances in lignocellulosic biofuel production systems. Woodhead Publishing, pp 23–46

Tan YM, Lim SH, Tay BY et al (2015) Functional chitosan-based grapefruit seed extract composite films for applications in food packaging technology. Mater Res Bull 69:142–146. https://doi.org/10.1016/j.materresbull.2014.11.041

Theba T, Ravani A, Bhatt H (2021) Utilization of beetroot pomace for food fortification. Int J Chem Stud 9:2653–2657. https://doi.org/10.22271/chemi.2021.v9.i1ak.11628

Tiwari SK (2020) Extraction of natural dye and preparation of herbal gulal from beetroot (*Beta vulgaris*). J Pharmacogn Phytochem 9:3206–3211

Udonkang MI, Inyang IJ, Ukorebi AN et al (2019) Spectrophotometry, physiochemical properties, and histological staining potential of aqueous and ethanol extracts of beetroot on various tissues of an albino rat. Biomed Hub 3:1–10

Verma M, Gahlot N, Singh SSJ, Rose NM (2021) UV protection and antibacterial treatment of cellulosic fibre (cotton) using chitosan and onion skin dye. Carbohydr Polym 257:117612. https://doi.org/10.1016/j.carbpol.2020.117612

Vishnupriya B, Manjudevi M, Agalya V (2022) Development of natural jelly from beetroot peel: analyze its stability and sensory properties. J Univ Shanghai Sci Technol 24:335–351

Widiawati D (2009) The revival of the usage of natural fibers and natural dyes in Indonesian textile. IBB J Vis Art Des 3:115–128

Yaqub A, Chaudhary N, Bhatti RAA et al (2018) Green extraction and dyeing of silk from *Beta vulgaris* peel dye with ecofriendly acid mordants. J Biol Environ Sci 13:308–321

Yusuf M, Shabbir M, Mohammad F (2017) Natural colorants: historical, processing and sustainable prospects. Nat Products Bioprospect 7:123–145. https://doi.org/10.1007/s13659-017-0119-9

Zade S, Charde R, Charde M (2011) Antioxidant and anti-inflammatory activity of ethanolic extract of *Beta vulgaris* Linn. roots. Int J Biomed Adv Res 2. https://doi.org/10.7439/ijbar.v2i4.28

Zin MM, Bánvölgyi S (2021) Emerging technology approach for extractability and stability of betalains from the peel of beetroot (*Beta vulgaris* L.). Biomass Convers Biorefin 13:10759. https://doi.org/10.1007/s13399-021-01975-z

Zin MM, Majrxi E, Banvolgyi S (2020) Conventional extraction of betalain compounds from beetroot peels with aqueous ethanol solvent. Acta Aliment 49:163–169. https://doi.org/10.1556/066.2020.49.2.5

Valorization of Jerusalem Artichoke and Its Crop Residues Using Green Technologies 12

Kurappalli Rohil Kumar and Kiran Babu Uppuluri

Abstract

The valorization of Jerusalem artichoke (JA) and its crop residues using environment-friendly green technologies, including enzymatic hydrolysis, fermentation, extraction, and biorefinery processes, is sustainable and efficient in converting into valuable products. Enzymatic hydrolysis of residual biomass yields sugars, serving as feedstock for producing biofuels, organic acids, and other biochemicals. Microbial strains can transform the extracted compounds into products such as single-cell oil, single-cell proteins, and 2,3-butanediol through fermentation. Additionally, extraction techniques enable the recovery of bioactive compounds with applications in the food, pharmaceutical, and cosmetic industries. Adopting green technologies for JA valorization offers several advantages, including reduced environmental impact, minimized waste generation, and enhanced resource efficiency. These approaches promote sustainable and economically viable processes, advancing a circular economy and reducing reliance on traditional fossil-based resources. The present chapter focuses on valorizing Jerusalem artichoke and its crop residues, promising feedstocks for sustainable and eco-friendly resource utilization, and advancing a circular bio-based economy.

Keywords

Jerusalem artichoke · Green technologies · Valorization · Biorefinery · Biofuels · Biopolymers

K. R. Kumar · K. B. Uppuluri (✉)
Bioprospecting Laboratory, Centre for Bioenergy, School of Chemical and Biotechnology, SASTRA Deemed University, Thanjavur, India

R. C. Ray (ed.), *Roots, Tubers, and Bulb Crop Wastes: Management by Biorefinery Approaches*, https://doi.org/10.1007/978-981-99-8266-0_12

Abbreviations

ABE Acetone-butanol-ethanol
JA Jerusalem artichoke
SHF Separate hydrolysis and fermentation
SSF Simultaneous saccharification and fermentation

12.1 Introduction

The Jerusalem artichoke (JA) (*Helianthus tuberosus* L.; Asteraceae family) was originally a North American perennial weed and was later introduced to Europe and Asia as a cultivated plant. It is now widely cultivated across Europe, Asia, and other world regions (Tesio et al. 2011). The JA can reach heights of up to 300 cm and features a branching stem with lanceolate, lanceolate-ovate, or, rarely, ovate leaves. Its root system gives rise to rhizomes, which develop into fleshy tubers. These tubers come in different colors, such as white, red, light brown, brown, violet-brown, or dark brown, and have primarily oval shapes, although they can also be rounder, oblong, or pear-shaped/slender (Slimestad et al. 2010; Puttha et al. 2013). The JA offers a wide range of applications due to its rich chemical composition in both the above and underground parts of the plant. These components include proteins, glucose, fructose, sucrose, and inulin (Gunnarsson et al. 2014). The aerial portions and tubers can produce bioethanol and contribute to the food sector (Gunnarsson et al. 2014; Johansson et al. 2015). Notably, the plant's high biomass yield has led to its recognition as a promising energy crop for bioethanol production, as stated by Matías et al. (2013). Moreover, the leaves and stems can serve as valuable animal feed, while the inulin-rich tubers possess uses in pharmacology, food production, and fodder utilization (Nabeshima et al. 2020). JA has numerous health benefits due to its anti-obesity, anti-diabetic, antifibrotic, and anti-inflammatory effects (Chang et al. 2014). JA tubers may be a possible prebiotic supplement because inulin and fructo-oligosaccharides affect intestinal morphology and improve blood metabolites (Kang et al. 2018). Kleessen et al. (2007) reported that adding JA to bakery items promotes the formation of fecal bifidobacteria after 1 week of consuming food (Kleessen et al. 2007). The review provides a thorough overview of JA cultivation, the production of various potential bioproducts, and relevant production technologies. Special focus is given to biofuel production (such as bio-ethanol, butanol, and biodiesel), chemicals, etc., from JA (Fig. 12.1).

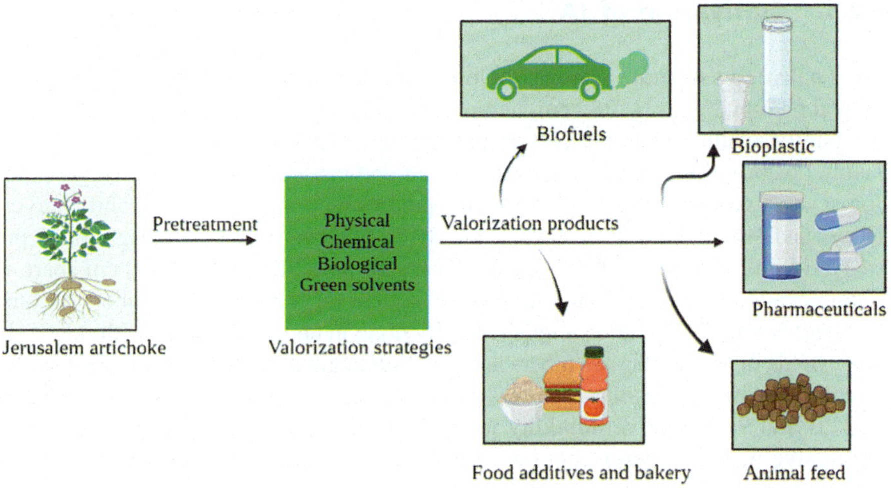

Fig. 12.1 Valorization of Jerusalem artichoke crop residues using green technologies

12.2 Jerusalem Artichoke Characteristics

The JA is a perennial plant belonging to the Asteraceae family, commonly called the sunflower or daisy family (Nguyen et al. 2021). This plant features a sturdy 1–3 m long stem with distinct ridges that may become woody as it matures. It bears small yellow flowers and has hairy oval-shaped leaves. The leaves are arranged alternately, with the lower ones being broader and larger, reaching lengths up to 30 cm, while the upper leaves are smaller and narrower (Kays and Nottingham 2007). Underground, the plant develops a rhizome system that produces small tubers. The flower heads of these plants measure 5–7.5 cm in width and consist of small, yellow, tubular disk flowers in the center, surrounded by florets that can be found individually or in groups. These flower clusters are at the end of alar branches and main stems. When it comes to tubers, they come in irregular shapes, ranging from knobby to round clusters. These tubers display various colors, including pale brown, white, red, and purple (Yang et al. 2015). Native Americans cultivated the JA roots long before the Europeans arrived. This versatile plant received various Latin and common names when introduced to Europe (Kays and Nottingham 2007). Nearly 100 common names were documented in different languages (Kays and Nottingham 2007). This plant's frequently used English names include Jerusalem artichoke, sunchoke, topinambur, woodland sunflower, or earth apple. Interestingly, the name "Jerusalem artichoke" can be misleading, as it belongs to the same genus as the garden sunflower but has no connection to Jerusalem and is unrelated to artichokes (Sawicka et al. 2020).

12.3 Cultivation of JA

JA is native to temperate regions of North America and can tolerate a wide range of annual precipitation (31–282 cm) and average temperatures (6.3–26.6 °C). It thrives in slightly alkaline soils but can adapt well to various soil types and pH levels in sunny positions. The tubers are cold-tolerant and can endure freezing temperatures for several months, allowing in-ground preservation during winter until harvest (Kays and Nottingham 2007; Qiu et al. 2018). Planting in early spring, at a depth of 10–15 cm, with seed tubers spaced 30–60 cm apart in rows 45–120 cm apart, is recommended. A 6–7 °C soil temperature is ideal for planting since tubers become dormant below 5 °C. Harvesting is typically done in about 125 days, with relatively high yields of 16–20 t/ha for tubers and 18–28 t/ha green weight for foliage. Careful handling during harvesting is necessary to prevent bruising. Storage conditions of 0–2 °C and 95% relative humidity for 4–5 months are recommended. The plant's aggressive nature requires cautious management, as missed tubers during harvest can grow aggressively in the next season. Still, this aggressiveness reduces the need for pest management due to the rapid growth rate (Swanton et al. 1992; Qiu et al. 2018).

12.4 JA Components

JA is a semi-wild plant with immense potential for development. During processing, the tubers of JA contain about 75.88% water, 12.46% total sugar, 1.58% total protein (w/w), 1.77% total fat, and various microelements and vitamins. Inulin is the predominant carbohydrate in these tubers, accounting for 80% of the total carbohydrates (Long et al. 2016). However, the composition of tubers can vary due to different cultivars, harvest periods, production conditions, postharvest storage, and processing methods. Inulin, a primary component of tubers, is a storage polysaccharide in many plants. It mainly consists of D-fructose molecules linked by (2→1) β-bonds, with D-glucose molecules bonded to fructose by (2→1) α-bonds at the ends. The degree of polymerization of standard inulin typically ranges from 2 to 60, with an average molecular weight of around 5500 Da (Szambelan et al. 2005). Such fermentable polysaccharides in JA make it highly valuable for various biorefinery processes. Apart from being rich in fermentable polysaccharides, JA also contains various amino acids, B vitamins, and minerals, making it advantageous for multiproduct biorefineries. Its diverse nutrient profile opens up possibilities for developing multiple valuable products. Table 12.1 provides the chemical composition of JA per 100 g, with data sourced from the US Department of Agriculture (USDA) website at https://www.usda.gov/.

Table 12.1 Chemical composition of Jerusalem artichoke (100 g) (data from the US Department of Agriculture (USDA), https://www.usda.gov/)

Name	Amount	Unit
Water	78	g
Energy	73	kcal
Energy	304	kJ
Protein	2	g
Total lipid (fat)	0.01	g
Ash	2.54	g
Carbohydrates, by difference	17.4	g
Fiber, total dietary	1.6	g
Sugars, total including NLEA	9.6	g
Calcium, Ca	14	mg
Iron, Fe	3.4	mg
Magnesium, Mg	17	mg
Phosphorus, P	78	mg
Potassium, K	429	mg
Sodium, Na	4	mg
Zinc, Zn	0.12	mg
Copper, Cu	0.14	mg
Manganese, Mn	0.06	mg
Selenium, Se	0.7	μg
Vitamin C, total ascorbic acid	4	mg
Thiamin	0.2	mg
Riboflavin	0.06	mg
Niacin	1.3	mg
Pantothenic acid	0.397	mg
Vitamin B6	0.077	mg
Folate, total	13	μg
Folate, food	13	μg
Folate, DFE	13	μg
Choline, total	30	mg
Vitamin A, RAE	1	μg
Carotene, beta	12	μg
Vitamin A, IU	20	IU
Vitamin E (alpha-tocopherol)	0.19	mg
Vitamin K (phylloquinone)	0.1	μg
Fatty acids, total polyunsaturated	0.001	g

12.5 JA Pretreatment

Efficient inulin extraction is critical for effectively using JA in various applications. Washing and drying fresh tubers is followed by grinding them into a coarse powder using a crushing machine. Because inulin is very soluble in hot water, hot water extraction is a simple method for getting inulin extract. Lingyun et al. (2007)

optimized the hot water extraction conditions, achieving an 83.6% inulin extraction yield after 20 min at 76.65 °C with a neutral pH and a solvent-to-solid ratio of 10.56: 1 (v/w) (Lingyun et al. 2007). Using hot water extraction, JA produced a greater yield of inulin 93% (Yi et al. 2010). The extracted inulin is subsequently transformed into fermentable sugars for microbial usage in industrial biorefinery applications via acid or enzyme hydrolysis. An early approach is acid hydrolysis, which uses sulfuric or hydrochloric acid (Qiu et al. 2018). The best results (>90% inulin hydrolysis) were obtained at a pH less than 2.0, a temperature greater than 90 °C, and 1 h. Using 5% (w/w) HCl at 60 °C for 1 h, Xu et al. (2016) achieved 92.3% of total sugars (20.2 0.1 g/L of glucose and 87.5 0.3 g/L of fructose). The glucose yield of 97.6% for the JA stalk was achieved through enzymatic hydrolysis after pretreating the biomass with a combination of 5% nitric acid and sodium hydroxide in the alkali-acid configuration (Dziekońska-Kubczak et al. 2019). Acid hydrolysis, on the other hand, produces colored byproducts and inhibitors that might limit microbial growth, making it less suitable for green production technologies. The most often employed enzyme, inulinase, has diverted researchers' attention to enzyme hydrolysis. The 94.5% inulin was converted in 48 h at 48 °C with an inulinase loading of 10 units g/ inulin (Sarchami and Rehmann 2014). Singh et al. (2017) employed immobilized yeast inulinase and found 84.5% inulin hydrolysis after 4 h at 125 rpm (Singh et al. 2017). Enzymatic hydrolysis has milder reactions, fewer byproducts, no coloring, and is easier to separate and refine than acid hydrolysis. Enzymatic procedures can also combine hydrolysis and fermentation, saving time and increasing productivity. Some microorganisms can produce inulinases directly, eliminating the need for further enzyme addition during refining and lowering expenses. Hot water extraction is a straightforward method, while enzymatic hydrolysis is preferred over acid hydrolysis due to its environmental friendliness and potential for combining hydrolysis and fermentation in biorefinery processes (Qiu et al. 2018).

12.6 Biorefinery Approaches on JA Tubers, Stalk, Leaves, etc.

12.6.1 Biofuels

Biofuels are valuable products that can be converted into liquid fuels for heating or transportation applications. Among the industrially significant biofuels, bioethanol, acetone, and butanol hold prominence. These biofuels can be derived from diverse substrates, and studies have documented their production from different inulin-rich feedstocks of JA.

12.6.1.1 Bioethanol

Bioethanol is a distilled biofuel typically produced from molasses or lignocellulosic materials. Various substrates have been utilized for the fermentative production of bioethanol, including inulin-rich feedstocks. These feedstocks, such as JA, are abundant in temperate and tropical regions and possess high solubility at elevated temperatures without forming viscous solutions. Single microorganisms are often employed in a simultaneous saccharification and fermentation (SSF) process for bioethanol production from inulin-rich feedstocks like JA. However, a mixed culture of two microorganisms has been utilized in certain cases. Several microorganisms, including *Kluyveromyces marxianus* (Gao et al. 2015), *Saccharomyces cerevisiae* (Singh et al. 2022), and *Zygosaccharomyces bailii* (Paixao et al. 2018), have been employed in batch systems for ethanol production from JA tubers. JA demonstrates ethanol yields equivalent to sugarcane and twice as much corn, making it a highly promising substrate for ethanol production. It has gained significant attention in New Zealand, China, and Europe. A fed-batch system can be employed to overcome the feedback inhibition issue caused by the accumulation of high ethanol concentrations in batch systems, utilizing a lower substrate concentration to control product formation. For instance, a fed-batch system employing a co-culture of *Aspergillus niger* and *Saccharomyces* sp. was used for bioethanol production from JA tubers (Ge and Zhang 2005). Only a few reports are available on the continuous system for bioethanol production from JA. Continuous fermentation of JA juice using whole cells of *K. marxianus* in a stirred tank reactor (Margaritis and Bajpai 1982a, b) and immobilized cells of *K. marxianus* in a packed-bed bioreactor yielded significant ethanol productivity (Margaritis and Bajpai 1982b). Recombinant microorganisms, particularly *S. cerevisiae*, have been extensively utilized as genetically engineered hosts for bioethanol production from JA. Recombinant *Saccharomyces* sp. carrying inulinase genes from various microbial strains, such as *Penicillium janthinellum* (Wang et al. 2015), *A. niger* (Wang et al. 2016), and *K. marxianus* (Khatun et al. 2017), have been used in SSF of JA tubers for ethanol production. In a series of studies conducted by Kim et al. (2013) and Kim and Kim (2014), they investigated the conversion of a mixture of JA stalk and tubers into ethanol using SSF with the aid of inulinase-producing yeast *K. marxianus*. However, this approach seemed to compromise the potential benefits of using JA tubers for the production of value-added inulin and other valuable products (Li et al. 2016). The enzymatic hydrolysis of JA stalk using alkaline hydrogen peroxide process yielded an impressive 93.8 g/L of glucose, along with 55.6 g/L of ethanol produced via the SSF process, which was 36.5% higher compared to the separate hydrolysis and fermentation (SHF) process (Khatun et al. 2015). Focused on lignocellulosic ethanol production exclusively from JA stalk by employing an engineered inulinase-producing yeast, *S. cerevisiae*. However, the heterologous expression of inulinases remains a primary challenge in implementing the recombinant strategy for efficient biomass conversion. The use of JA has opened up new opportunities for bioethanol production. After fermentation of JA, bioethanol is separated through distillation.

12.6.1.2 Acetone and Butanol

Traditionally, acetone-butanol production has relied on two primary substrates: corn and molasses. The cost of the substrate plays a significant role in the economics of the bioprocess, accounting for approximately 50% of the total production cost (Sarchami and Rehmann 2014). Hence, it is crucial to identify readily available and cost-effective raw materials for efficient acetone-butanol fermentation. Inulin-rich feedstocks, such as JA, present a promising alternative for acetone-butanol production. Under optimized conditions in a batch system, *Clostridium acetobutylicum* has demonstrated efficient production of approximately 23–24 g/L of acetone and butanol (Marchal et al. 1985), as well as 11.2 g/L of butanol (Chen et al. 2010), from JA tubers. In another consolidated bioprocess, *Clostridium saccharobutylicum* achieved the production of 0.25 g/L/h of acetone-butanol-ethanol (ABE) from JA tubers (Sarchami and Rehmann 2014). The production of biobutanol through NaOH-H_2O_2 pretreatment of JA stalk resulted in the best conditions, leading to a significant increase of 31.1% (19.7%) in butanol (ABE) production (Xue et al. 2017). The recovery of acetone and butanol is typically accomplished through distillation methods.

12.6.1.3 Biodiesel

Biodiesel, unlike bioethanol, boasts a higher heating value and exhibits lower water absorption, making it a favorable option for vehicle use without the need for engine modifications (Li et al. 2013). Researchers explored the potential of JA tubers as a promising feedstock for algal cultivation. They aimed to reduce the cost of microalgal cultivation in the context of bio-oil and biodiesel manufacturing (Cheng et al. 2009). Their study utilized JA hydrolysate as a carbon source to foster lipid accumulation in *Chlorella protothecoides*. Subsequently, after a 4-day scale cultivation, they successfully extracted a lipid concentration of 44% by dry mass, which was then converted into biodiesel through transesterification. The resulting biodiesel primarily contained oleic acid methyl ester, linoleic acid methyl ester, and cetane acid methyl ester as its main components. In a similar vein, Zhao et al. (2010b) also contributed to the research on lipid production from JA tubers, but this time utilizing the oleaginous yeast, *Rhodosporidium toruloides* Y4. Their efforts led to a lipid titer of 39.6 g/L and a cellular lipid content of 56.5% (w/w) when they employed JA hydrolysates as a nutrient source (Zhao et al. 2010c). In another study, a lipid productivity of 1.73 g/L/d was achieved (Sung et al. 2014). These studies collectively highlight the viability of an innovative method for biodiesel production using JA tubers and microalgal cultivation. Such an approach could lead to cost reductions in obtaining carbon sources for algal oil production, presenting a promising avenue for the biodiesel industry.

12.6.2 Organic Acids

Organic acids contain one or more functional carboxylic acid groups and have wide-ranging applications in the food and pharmaceutical sectors. The conventional methods for organic acid production involve chemical synthesis or fermentation of carbohydrates like glucose and sucrose. In some cases, a sequential approach combines chemical synthesis and fermentation. Additionally, organic acids can be derived from inulin-rich feedstocks such as Jerusalem artichoke.

Lactic acid, a significant naturally occurring organic acid, is commonly derived from raw materials such as molasses, lignocellulosic, barley, whey, and malt. The selection of economically viable raw materials is essential for cost-effective bioprocessing. Consequently, inulin-rich feedstocks like chicory and JA have gained attention as potential sources for lactic acid production. Several studies have reported lactic acid production from chicory root tuber extract using *Lactobacillus* sp. in batch systems (Baston and Constantin 2012). From JA tubers extract using *Lacticaseibacillus paracasei* (Choi et al. 2012) and *Lactobacillus bulgaricus* (Xu et al. 2016). To address the issue of feedback inhibition in batch systems, some studies have explored the use of fed-batch systems for lactic acid production from JA tubers (Wang et al. 2013b). In such systems, a mixed culture of *A. niger* and *Lactobacillus* sp. was employed for the solid-state fermentation of JA tubers (Ge et al. 2010).

Additionally, immobilized *Lactococcus lactis* has been utilized in a fed-batch system for lactic acid production from JA tubers (Shi et al. 2012). In recent years, there has been an increasing trend in using recombinant microorganisms with enhanced product yield. For example, a batch system employing recombinant *Kluyveromyces marxianus* has been successfully applied for lactic acid production from JA tubers (Bae et al. 2018).

Numerous studies have explored the fermentative production of various organic acids from inulin-rich feedstocks. For instance, in a packed-bed reactor, a bioprocess utilized co-immobilized inulinase from *A. niger* and whole cells of *Zymomonas mobilis* for continuous fermentation of JA tubers extract, resulting in gluconic acid production (Kim and Kim 1992). In another investigation, *Yarrowia lipolytica* was employed in a batch system to produce citric acid from JA tubers extract (Wang et al. 2013a). Additional research reports include the production of succinic acid (Gunnarsson et al. 2014) and glutamic acid (Qiu et al. 2017) from JA tubers extract using *Actinobacillus succinogenes* and *Bacillus amyloliquefaciens*, respectively, in batch systems. In a fed-batch system, malic acid production from JA tubers extract was achieved using *Aureobasidium pullulans* (Xia et al. 2017). Furthermore, butyric acid (Huang et al. 2011) and propionic acid (Liang et al. 2012) were produced from JA tubers extract through the use of immobilized *Clostridium tyrobutyricum* and *Propionibacterium acidipropionici*, respectively, in fed-batch systems. After the fermentation process, the purification of organic acids is commonly carried out using chromatography with strong anionic resins.

12.6.3 Wide-Ranging Applications

Apart from inulinases, fructooligosaccharides, biofuels, and organic acids, JA tubers have produced other important industrial metabolites. One example is the production of single-cell oils, lipids with specific functions and significant properties. Traditionally, various microorganisms, including yeasts, fungi, bacteria, and microalgae, have been known to accumulate more than 20% lipids in their biomass, making them valuable sources of single-cell oils. In a bioprocess involving JA tubers extract, *Rhodotorula mucilaginosa* produced single-cell oil in batch and fed-batch systems (Zhao et al. 2010b). Another approach utilized a co-culture of *R. mucilaginosa* and immobilized *Pichia guilliermondii* for single-cell oil production from JA tubers extract in a batch system (Zhao et al. 2011). Additionally, in a batch bioprocess, recombinant cells of *Y. lipolytica* were successfully employed for single-cell oil production using JA tubers extract (Zhao et al. 2010a). These studies demonstrate the potential of inulin-rich feedstocks for producing single-cell oils by utilizing various microorganisms.

Single-cell proteins are those obtained from dried cells of diverse microorganisms such as bacteria, filamentous fungi, yeast, algae, and more. These proteins serve as valuable dietary supplements in human nutrition and animal feed. Although limited, some reports on producing single-cell proteins using Jerusalem artichoke tubers are available. For instance, *Cryptococcus aureus* demonstrated efficient growth on JA tubers extract, facilitating the production of single-cell proteins in a batch system (Gao et al. 2007). Similarly, recombinant cells of *Y. lipolytica* were employed in a stirred tank reactor to produce single-cell proteins from JA tubers extract (Cui et al. 2011). These studies highlight the potential of JA tubers as a source for single-cell protein production using different microorganisms.

2,3-Butanediol is an organic metabolite widely used in the food, cosmetics, and pharmaceutical industries. Traditionally, it is derived from various sources such as molasses, whey, and lignocellulosic materials. Several microorganisms, including *Paenibacillus polymyxa* (Ge et al. 2010) and *Klebsiella pneumoniae* (Sun et al. 2009), have demonstrated efficient utilization of JA tubers extract for the production of 2,3-butanediol in batch systems. (Li et al. 2010). *Klebsiella pneumoniae* CICC 10011 efficiently utilized JA stalk and tuber for the production of 2,3-butanediol, which employed solid-state fermentation with stage-shift aeration, impressive yield of 901.2 mmol/L (80.5 g/L) of 2,3-butanediol. In a fed-batch system, both the stalks and tubers of JA have been employed for 2,3-butanediol production by *Klebsiella pneumoniae* (Sun et al. 2009). In a fed-batch system, 2,3-butanediol production from JA tubers extracts using *K. pneumoniae* was demonstrated. These studies highlight the potential of JA as a substrate for 2,3-butanediol production, offering an alternative and sustainable approach to obtaining this valuable organic compound.

Sorbitol, a sugar alcohol widely used in the chemical and food industry, can be synthesized through different methods. One common approach is the catalytic hydrogenation of glucose and sucrose at high temperatures using nickel as a catalyst. Another method involves the electrochemical reduction of glucose under alkaline conditions. Additionally, sorbitol can be produced from Jerusalem artichoke tubers extract. Researchers have successfully employed *S. cerevisiae* in a batch system to produce sorbitol from JA tubers extract (Duvnjak et al. 1991). In another study, the co-immobilization of inulinase from *A. niger* with whole cells of *Z. mobilis* in a packed-bed reactor enabled the continuous hydrolysis of JA tubers extract for sorbitol production (Kim and Kim 1992). Furthermore, the production of mannitol (Saha 2006) and pullulan (Xia et al. 2017) has been reported from chicory root tubers extract and JA tubers extract, respectively, using batch and fed-batch systems. These findings demonstrate the potential of JA tuber extract as a valuable source for producing various industrially important metabolites such as sorbitol, mannitol, and pullulan.

12.6.4 Applications of JA in Human Health

12.6.4.1 Antioxidant Capacity

Table 12.2 demonstrates that the phytochemicals identified in JA have been discussed (adopted from Oleszek et al. 2019).

JA is a valuable source of beneficial compounds in its tubers, leaves, and flowers. Researchers led by Oleszek et al. (2019) discovered 23 compounds in its flowers, with feradiol demonstrating antiproliferative effects against colon cancer cells and ent-kaur16-en-19-oic acid and β-sitostenone showing antimicrobial activity against *Enterococcus faecium*. The flower extract also contained 3,4,20-tri-hydroxychalcone-40-*O*-β-D-glucoside, which exhibited antioxidant properties. Tyrosol and quercetin-3-*O*-β-D-glucopyranoside, present in the flowers, showed potent free radical scavenging ability and inhibited the β-glucosidase enzyme, respectively. The ethyl acetate fraction from the flowers was found to enhance antioxidant enzyme activity in *Caenorhabditis elegans* (Kim et al. 2019). The leaves of JA displayed antioxidant potential, with specific compounds such as 5,8-diOH-6,7-diMeO-2-(3,4-diMeOPh)-4-benzopyrone and 5,8-diOH-6,7,40-triOMe showing higher antioxidant activity than the positive control (Kim et al. 2011). Comparative analysis of different JA organs revealed significant quantities of phenolic acids, including chlorogenic acid and its isomers, suggesting their potential as natural antioxidants with biocidal properties (Showkat et al. 2019).

12.6.4.2 Dermatological Treatments

JA extracts have been studied for their applications in dermatological treatments, relieving atopic dermatitis symptoms, reducing epidermal thickness, and alleviating skin irritation when used in cosmetic products (Kang et al. 2018).

12.6.4.3 Digestive System

The main beneficial compound found in JA is inulin, a fructan-type compound with prebiotic and probiotic effects on the human body. Inulin aids in probiotic adhesion to intestinal epithelial cells, supporting their survival under gastrointestinal conditions and promoting the growth of beneficial gut bacteria (Iraporda et al. 2019). Incorporating inulin derived from JA into foods like yogurt has positively affected blood glucose, cholesterol, and total lipid levels. Studies have also linked inulin intake to increased levels of beneficial gut bacteria in humans and animals (Méndez-Yáñez et al. 2022).

12.6.4.4 Improvement of Biochemical Parameters

JA consumption has been associated with improved biochemical parameters in rats, including gene expression related to fatty acid synthesis, fibrosis, and inflammation. This suggests potential prevention against Type 2 diabetes and nonalcoholic fatty liver disease. Additionally, it has demonstrated positive effects on blood glucose, liver triglyceride, and total cholesterol levels in rats (Méndez-Yáñez et al. 2022).

Table 12.2 Phytochemicals identified in JA (adopted from Oleszek et al. 2019)[a]

S. no.	Part of plants	Metabolite
1	Tubers	Coumarin: Ayapin (6,7-methylenedioxy-coumarin), scopoletin (6-methoxy-7-hydroxycoumarin) Ferulic acids Isomers of caffeoylquinic acid (neo-chlorogenic acid, chlorogenic acid, crypto-chlorogenic acid), isomeric di-caffeoylquinic acids (3,4-*O*-caffeoyl; 3,5-*O*-caffeoyl; 1,3-*O*-dicaffeoyl; 4,5-*O*-caffeoyl esters) Cyclopentanol, hexadecanoic acid, 9-octadecenoic acid, 9-octadecenoic acid, 9-octadadecinoic acid, octadeconic acid, 13-octadecenal and 9-octadeconic acid, uteronoenen-1-ol-3; 2-propen-1-ol; 3-deoxy-D-manneolclactone; heyadecanic acid; 1-pyrrolin,3-ethyl; 9-octadecnoic acid; octadelenoic acid; 13-octa decnal; 1,2-epoxy-1-vinylcyclodecoene and cyclopentadecanone, 2-hydroxy, 2-furan carboxaldehyde; 2-furan carboyaldehyde; dodecane1,1-oxybis; glycine,*n*-methyl-*n*-1-oxadodecyl; hexa-decaneicacid; 9-octadecenoic acid; oleic acid; 9-octadecenal; 9-octadecenal and phthalic acid; disooctyl ester
2	Leaves	Flavone glucosides (kaempferol 3-*O*-glucoside and quercetin 7-*O*-glucoside) Chlorogenic acids (feruloyl quinic acid, caffeoylquinic acids, caffeoylquinic acids) Phenolic compounds (caffeic acid, 3-*O*-caffeoylquinic acid, 3,5-caffeoylquinic acid, 1,5-dicaffeoylquinic acid, 4,5-dicaffeoylquinic acid, 3,5-dicaffeoylquinic acid methyl ether) Chlorogenic acids (two isomers of caffeoylquinic acid, 3-*O*-caffeoylquinic acid, caffeic acid, *p*-coumaroyl-quinic acid, 3,4-dicaffeoyquinic acid, feruloylquinic acid, 3,5-dicaffeoylquinic acid, 1,5-dicaffeoylquinic acid and 4,5-dicaffeoylquinic acid) and other phenolics (isorhamnetin glucoside, caffeoyl glucopyranose, kaempferol glucuronide, kaempferol-3-*O*-glucoside) Sesquiterpene lactones (3-hydroxy-8β-tigloyloxy-1,10-dehydroariglovin; 4,15-isoatripliciolide methyl acrylate; 4,15-iso-atripliciolide tailgate; 4,15-iso-atripliciolide isobutyrate; 4,15-iso-atripliciolide angelate, budlein A methyl acrylate; budlein A tiglate, desacetylovatifolin, 1a-hydroxy pinnatifid in, 1a-acetoxy pinnatifid in; melampolide) and flavones (pedunculin; 5,8-diOH-6,7diMeO-2-(3,4-diMeOPh)-4-benzopyrone)
3	Callus flowers	Lectins Terpenoids: Monoterpenoid (loliolide acetate), kauranediterpenoids (ent-kaur-16-en-19-oic acid, 15-hydroxykaur-16-en-19-oic acid, 17-hydroxykaur-15-en-19-oic acid, 15αtiglinoyloxykaur-16-en-l9-oic acid) and triterpenoids (faradiol 3-*O*-fatty acid ester derivative); coumarins (8-methoxyobliquin, 7-*O*-prenylscopoletin); flavonoids (isoliquiritigenin 2,4,20,40-tetrahydroxychalcone, liquiritigenin, acerosin, quercetin 7-*O*b-glucoside, sulfuretin, sulfurein glycoside) and chromone derivative (7-hydroxychromone)
4	Whole plants	Ent-17-oxokaur-15(16)-en-19-oic acid; ent-17-hydroxykaur-15(16)-en-19-oic acid; ent-15bhydroxykaur-16(17)-en-19-oic acid methyl ester; ent-15-nor-14-oxolabda-8(17); 12E-dien-18-oic acid; 4,15-isoatriplicolide angelate; 4,15-isoatriplicolide methylacrylate; (+)-pinoresinol; (−)-loliolide; vanillin

[a] This table was adopted directly from an Open Access Material distributed under the terms of the Creative Commons Attribution 4.0 International License (Oleszek et al. 2019)

12.6.4.5 Superfood

Due to its impressive nutritional qualities, JA is currently being explored as a superfood with promising applications in various gastronomic preparations, such as flour, chips, noodles, and pastries.

12.6.5 JA: Increasing Quality of Foods for Cooking

JA finds extensive applications in the food industry (Slapkauskaite et al. 2016) conducted a study where they fermented tubers using lactic acid bacteria to enhance the nutritional quality of dairy products. Helianthus tuberosus was preferred over other plants tested due to its superior organoleptic qualities. Moreover, JA powder has been successfully incorporated into glass noodles, a popular Asian dietary item. This addition improved the organoleptic parameters, such as hardness, cohesiveness, springiness, and gumminess, while also boosting the fiber and sugar content, making the noodles more nutritious (Méndez-Yáñez et al. 2022). JA is prepared as a pickle in China, producing a smooth, sweet, aromatic, and crisp product. Additionally, in the baking industry, enriching bread with inulin extracted from JA (at a low concentration of 2.5 g/100 g flour) led to an increase in crumb hardness and chewiness (Méndez-Yáñez et al. 2022). Alternatively, incorporating 5% JA powder in bread formulation yielded products with excellent organoleptic qualities, an extended shelf life, and elevated nutrient content. JA powder has been examined for pastry products to promote healthier options. Like its effect on bread, the powder induced changes in appearance, organoleptic qualities, and nutritional value. Notably, a substantial amount of JA powder reduced the calorie content in cakes, butter biscuits, and honey biscuits (Gedrovica and Karklina 2012). Sensorial studies involving cookies, muffins, cracker bread, and mashed products with JA powder generally received positive feedback from female consumers (Méndez-Yáñez et al. 2022).

12.6.6 JA Leaves Bioactive Compounds

JA leaves have a long history of traditional use as a folk medicine to treat bone fractures, skin wounds, swelling, and pain. These leaves contain valuable bioactive compounds with medicinal effects, including antifungal, antioxidant, and anticancer activities (Hai-wei et al. 2007; Han et al. 2010; Chen et al. 2013). Several studies have isolated and tested bioactive compounds from JA leaves for their antifungal properties. (Hai-wei et al. 2007). Extracted the leaves using different solvents and tested their effects on various fungi. They found that extracts from organic solvents, particularly ethyl acetate, showed the highest inhibitory activity against fungi like *Rhizoctonia solani*, *Gibberellazeae*, *Alternaria solani*, and *Botrytis cinerea* (Han et al. 2010). They conducted research focusing on the inhibitory ability of acetone extracts from Jerusalem artichoke leaves against pepper gray mold and found that the acetone extracts exhibited strong inhibitory effects. In a more (Chen et al. 2013), six phenolic acids were isolated from Jerusalem artichoke leaves using *n*-butanol as a solvent. Among these phenolic acids, caffeic acid, 3,4-caffeoylquinic acid, and 1,5-dicaffeoylquinic acid were identified as responsible for inhibiting Gibberellazeae. These studies suggest that Jerusalem artichoke leaves hold promise as a potential natural source of fungicides due to their antifungal properties.

12.7 Challenges and Future Perspectives

JA is a versatile and promising crop that has gained significant attention for its potential valorization in various fields. Despite its numerous advantages, several challenges must be addressed to exploit its potential fully. One major challenge lies in the cultivation and agronomic practices of Jerusalem artichoke. The crop requires careful management due to its rapid growth, aggressive nature, and tendency to become invasive. Additionally, optimizing cultivation techniques to enhance tuber yield and quality remains a priority to ensure its economic viability. Another significant challenge is the efficient conversion of JA biomass into valuable products. The crop is primarily valued for its tubers, a rich source of inulin—a polysaccharide with various industrial applications. However, other parts of the plant, such as stems and leaves, contain potentially valuable compounds like lignocellulose, proteins, and polyphenols, which are yet to be fully exploited. Developing efficient and cost-effective extraction methods for these compounds is essential to maximize the overall valorization of JA. Furthermore, the variability in inulin composition among different cultivars and environmental conditions poses a challenge for industrial applications. Standardization of inulin extraction processes and developing improved cultivars with consistent inulin profiles are necessary to meet the demands of various industries, including food, pharmaceuticals, and biofuels. Environmental concerns also arise in the valorization of JA. While it is considered a sustainable crop due to its low input requirements and potential for carbon sequestration, there are concerns about its invasive nature in some regions. Ensuring responsible cultivation and addressing potential ecological impacts are vital to maintaining a balanced approach to its valorization.

The future perspectives of JA valorization are promising. Its potential as a renewable resource for bioenergy production, particularly through biogas and bioethanol generation, is gaining interest. Additionally, its role in sustainable agriculture as a rotational crop, improving soil health and reducing nitrogen leaching, enhances its value in modern farming practices. Jerusalem artichoke can contribute to the growing demand for functional foods and nutraceuticals, given the health benefits of inulin consumption, such as improved gut health and blood sugar regulation. Its role in the pharmaceutical industry for producing prebiotics, insulin, and other biologically active compounds presents exciting opportunities.

12.8 Conclusion

In conclusion, the valorization of JA and its crop residues through the implementation of green technologies offers a sustainable and environmentally friendly approach to unlocking the full potential of this versatile feedstock. By adopting innovative and eco-conscious processes, such as enzymatic hydrolysis, fermentation, and biotransformation, a wide range of valuable products can be derived from Jerusalem's artichoke crop residues. These green technologies enable the production of various sought-after commodities, including inulinases, high fructose syrup, fructooligosaccharides, biofuels, organic acids, single-cell oil, single-cell proteins, 2,3-butanediol, sorbitol, mannitol, and more, using Jerusalem artichoke feedstocks. This presents lucrative economic opportunities, reduces waste, and promotes a circular economy. Furthermore, different approaches, such as separate hydrolysis and fermentation (SHF), simultaneous saccharification and fermentation (SSF), and consolidated bioprocessing (CBP), have been explored to utilize Jerusalem artichoke for product synthesis effectively. Each approach brings its advantages and challenges, necessitating careful consideration and optimization for successful implementation at an industrial scale. To fully capitalize on the valorization potential of Jerusalem artichoke crop residues, it is imperative to address the technical, economic, and operational hurdles associated with scaling up these green technologies. Collaborative efforts among researchers, industry stakeholders, and policymakers are crucial in fostering innovation, overcoming barriers, and establishing cost-effective and sustainable processes. By harnessing the power of green technologies, Jerusalem's artichoke can be transformed into valuable commodities, thereby reducing waste generation, promoting the utilization of renewable resources, and contributing to a more sustainable future. The continuous exploration and development of green valorization methods for Jerusalem artichoke crop residues hold significant promise in meeting the growing demand for eco-friendly and resource-efficient solutions across various industries.

References

Bae J-H, Kim H-J, Kim M-J, Sung BH, Jeon J-H, Kim H-S et al (2018) Direct fermentation of Jerusalem artichoke tuber powder for production of l-lactic acid and d-lactic acid by metabolically engineered *Kluyveromyces marxianus*. J Biotechnol 266:27–33. https://doi.org/10.1016/j.jbiotec.2017.12.001

Baston O, Constantin OE (2012) Selection of lactic acid bacteria able to ferment inulin hydrolysates. Ann Univ Dunarea de Jos of Galati Fascicle VI Food Technol 36(2):31–40

Chang W-C, Jia H, Aw W, Saito K, Hasegawa S, Kato H (2014) Beneficial effects of soluble dietary Jerusalem artichoke (*Helianthus tuberosus*) in the prevention of the onset of type 2 diabetes and non-alcoholic fatty liver disease in high-fructose diet-fed rats. Br J Nutr 112(5):709–717. https://doi.org/10.1017/S0007114514001421

Chen F, Long X, Yu M, Liu Z, Liu L, Shao H (2013) Phenolics and antifungal activities analysis in industrial crop Jerusalem artichoke (*Helianthus tuberosus* L.) leaves. Ind Crop Prod 47:339–345. https://doi.org/10.1016/j.indcrop.2013.03.027

Chen L, Xin C, Deng P, Ren J, Liang H, Bai F (2010) Butanol production from hydrolysate of Jerusalem artichoke juice by *Clostridium acetobutylicum* L7. Sheng wu gong cheng xue bao (Chin J Biotechnol) 26(7):991–996

Cheng Y, Zhou W, Gao C, Lan K, Gao Y, Wu Q (2009) Biodiesel production from Jerusalem artichoke (*Helianthus tuberosus* L.) tuber by heterotrophic microalgae *Chlorella protothecoides*. J Chem Technol Biotechnol 84(5):777–781. https://doi.org/10.1002/jctb.2111

Choi H-Y, Ryu H-K, Park K-M, Lee EG, Lee H, Kim S-W, Choi E-S (2012) Direct lactic acid fermentation of Jerusalem artichoke tuber extract using *Lactobacillus paracasei* without acidic or enzymatic inulin hydrolysis. Bioresour Technol 114:745–747. https://doi.org/10.1016/j.biortech.2012.03.075

Cui W, Wang Q, Zhang F, Zhang S-C, Chi Z-M, Madzak C (2011) Direct conversion of inulin into single cell protein by the engineered *Yarrowia lipolytica* carrying inulinase gene. Process Biochem 46(7):1442–1448. https://doi.org/10.1016/j.procbio.2011.03.017

Duvnjak Z, Turcotte G, Duan Z (1991) Production of sorbitol and ethanol from Jerusalem artichokes by *Saccharomyces cerevisiae* ATCC 36859. Appl Microbiol Biotechnol 35:711–715. https://doi.org/10.1007/BF00169882

Dziekońska-Kubczak U, Berłowska J, Dziugan P, Patelski P, Balcerek M, Pielech-Przybylska K, Robak K (2019) Two-stage pretreatment to improve saccharification of oat straw and Jerusalem artichoke biomass. Energies 12(9):1715. https://doi.org/10.3390/en12091715

Gao J, Yuan W, Kong L, Xiang R, Zhong S (2015) Efficient ethanol production from inulin by two-stage aerate strategy. Biomass Bioenergy 80:10–16. https://doi.org/10.1016/j.biombioe.2015.04.013

Gao L, Chi Z, Sheng J, Ni X, Wang L (2007) Single-cell protein production from Jerusalem artichoke extract by a recently isolated marine yeast *Cryptococcus aureus* G7a and its nutritive analysis. Appl Microbiol Biotechnol 77:825–832. https://doi.org/10.1007/s00253-007-1210-7

Ge X-Y, Qian H, Zhang W-G (2010) Enhancement of L-lactic acid production in *Lactobacillus casei* from Jerusalem artichoke tubers by kinetic optimization and citrate metabolism. J Microbiol Biotechnol 20(1):101–109. https://doi.org/10.4014/jmb.0905.05032

Ge X-Y, Zhang W-G (2005) A shortcut to the production of high ethanol concentration from Jerusalem artichoke tubers. Food Technol Biotechnol 43(3):241–246. Retrieved from https://hrcak.srce.hr/110539

Gedrovica I, Karklina D (2012) Influence of Jerusalem artichoke powder on the nutritional value of pastry products. Int J Nutr Food Eng 6(7):524–527. https://doi.org/10.5281/zenodo.1055273

Gunnarsson IB, Svensson S-E, Johansson E, Karakashev D, Angelidaki I (2014) Potential of Jerusalem artichoke (*Helianthus tuberosus* L.) as a biorefinery crop. Ind Crop Prod 56:231–240. https://doi.org/10.1016/j.indcrop.2014.03.010

Hai-wei L, Zhao-Pu L, Ling L, Geng-mao Z (2007) Studies on the antifungal activities and chemical components of extracts from *Helianthus tuberosus* leaves. Nat Prod Res Dev 19(3):405

Han R, Wang L, Zhong Q, Sun K, Li Y (2010) Study on antifungal activity of the extract from the leaves of *Helianthus tuberosus*. Mod Agric Sci Technol 5:123

Huang J, Cai J, Wang J, Zhu X, Huang L, Yang S-T, Xu Z (2011) Efficient production of butyric acid from Jerusalem artichoke by immobilized *Clostridium tyrobutyricum* in a fibrous-bed bioreactor. Bioresour Technol 102(4):3923–3926. https://doi.org/10.1016/j.biortech.2010.11.112

Iraporda C, Rubel IA, Manrique GD, Abraham AG (2019) Influence of inulin rich carbohydrates from Jerusalem artichoke (*Helianthus tuberosus* L.) tubers on probiotic properties of Lactobacillus strains. LWT 101:738–746

Johansson E, Prade T, Angelidaki I, Svensson S-E, Newson WR, Gunnarsson IB, Persson Hovmalm H (2015) Economically viable components from Jerusalem artichoke (*Helianthus tuberosus* L.) in a biorefinery concept. Int J Mol Sci 16(4):8997–9016. https://doi.org/10.3390/ijms16048997

Kang Y-M, Lee K-Y, An H-J (2018) Inhibitory effects of *Helianthus tuberosus* ethanol extract on *Dermatophagoides farina* body-induced atopic dermatitis mouse model and human keratinocytes. Nutrients 10(11):1657. https://doi.org/10.3390/nu10111657

Kays SJ, Nottingham SF (2007) Biology and chemistry of Jerusalem artichoke: *Helianthus tuberosus* L. CRC Press

Khatun MM, Li Y-H, Liu C-G, Zhao X-Q, Bai F-W (2015) Fed-batch saccharification and ethanol fermentation of Jerusalem artichoke stalks by an inulinase producing *Saccharomyces cerevisiae* MK01. RSC Adv 5(129):107112–107118. https://doi.org/10.1039/C5RA23901J

Khatun MM, Liu C-G, Zhao X-Q, Yuan W-J, Bai F-W (2017) Consolidated ethanol production from Jerusalem artichoke tubers at elevated temperature by *Saccharomyces cerevisiae* engineered with inulinase expression through cell surface display. J Ind Microbiol Biotechnol 44(2):295–301. https://doi.org/10.1007/s10295-016-1881-0

Kim DM, Kim HS (1992) Continuous production of gluconic acid and sorbitol from Jerusalem artichoke and glucose using an oxidoreductase of *Zymomonas mobilis* and inulinase. Biotechnol Bioeng 39(3):336–342. https://doi.org/10.1002/bit.260390312

Kim J, Lee Y, Han S, Kim S, Hwang I, Kim D (2019) Antioxidant activity of *Helianthus tuberosus* L. flower in *Caenorhabditis elegans*. Korean J Pharmacogn 50(2):96–101

Kim S, Kim CH (2014) Evaluation of whole Jerusalem artichoke (*Helianthus tuberosus* L.) for consolidated bioprocessing ethanol production. Renew Energy 65:83–91. https://doi.org/10.1016/j.renene.2013.07.025

Kim S, Park JM, Kim CH (2013) Ethanol production using whole plant biomass of Jerusalem artichoke by *Kluyveromyces marxianus* CBS1555. Appl Biochem Biotechnol 169:1531–1545. https://doi.org/10.1007/s12010-013-0094-5

Kim Y-S, Lee S-J, Hwang J-W, Kim E-H, Park P-J, Jeon B-T (2011) Antioxidant activity and protective effects of extracts from *Helianthus tuberosus* L. leaves on t-BHP induced oxidative stress in chang cells. J Korean Soc Food Sci Nutr 40(11):1525–1531. https://doi.org/10.3746/jkfn.2011.40.11.1525

Kleessen B, Schwarz S, Boehm A, Fuhrmann H, Richter A, Henle T, Krueger M (2007) Jerusalem artichoke and chicory inulin in bakery products affect faecal microbiota of healthy volunteers. Br J Nutr 98(3):540–549. https://doi.org/10.1017/S0007114507730751

Li D, Dai J-Y, Xiu Z-L (2010) A novel strategy for integrated utilization of Jerusalem artichoke stalk and tuber for production of 2, 3-butanediol by *Klebsiella pneumoniae*. Bioresour Technol 101(21):8342–8347. https://doi.org/10.1016/j.biortech.2010.06.041

Li K, Qin J-C, Liu C-G, Bai F-W (2016) Optimization of pretreatment, enzymatic hydrolysis and fermentation for more efficient ethanol production by Jerusalem artichoke stalk. Bioresour Technol 221:188–194. https://doi.org/10.1016/j.biortech.2016.09.021

Li L, Li L, Wang Y, Du Y, Qin S (2013) Biorefinery products from the inulin-containing crop Jerusalem artichoke. Biotechnol Lett 35:471–477. https://doi.org/10.1007/s10529-012-1104-3

Liang Z-X, Li L, Li S, Cai Y-H, Yang S-T, Wang J-F (2012) Enhanced propionic acid production from Jerusalem artichoke hydrolysate by immobilized *Propionibacterium acidipropionici* in a fibrous-bed bioreactor. Bioprocess Biosyst Eng 35:915–921. https://doi.org/10.1007/s00449-011-0676-y

Lingyun W, Jianhua W, Xiaodong Z, Da T, Yalin Y, Chenggang C et al (2007) Studies on the extracting technical conditions of inulin from Jerusalem artichoke tubers. J Food Eng 79(3):1087–1093. https://doi.org/10.1016/j.jfoodeng.2006.03.028

Long X-H, Shao H-B, Liu L, Liu L-P, Liu Z-P (2016) Jerusalem artichoke: a sustainable biomass feedstock for biorefinery. Renew Sust Energ Rev 54:1382–1388. https://doi.org/10.1016/j.rser.2015.10.063

Marchal R, Blanchet D, Vandecasteele J (1985) Industrial optimization of acetone-butanol fermentation: a study of the utilization of Jerusalem artichokes. Appl Microbiol Biotechnol 23:92–98

Margaritis A, Bajpai P (1982a) Continuous ethanol production from Jerusalem artichoke tubers. I. Use of free cells of *Kluyveromyces marxianus*. Biotechnol Bioeng 24(7):1473–1482. https://doi.org/10.1002/bit.260240702

Margaritis A, Bajpai P (1982b) Continuous ethanol production from Jerusalem artichoke tubers. II. Use of immobilized cells of *Kluyveromyces marxianus*. Biotechnol Bioeng 24(7): 1483–1493. https://doi.org/10.1002/bit.260240702

Matías J, González J, Cabanillas J, Royano L (2013) Influence of NPK fertilisation and harvest date on agronomic performance of Jerusalem artichoke crop in the Guadiana Basin (southwestern Spain). Ind Crop Prod 48:191–197. https://doi.org/10.1016/j.indcrop.2013.04.010

Méndez-Yáñez A, Ramos P, Morales-Quintana L (2022) Human health benefits through daily consumption of Jerusalem artichoke (*Helianthus tuberosus* L.) tubers. Horticulturae 8(7):620. https://doi.org/10.3390/horticulturae8070620

Nabeshima EH, Moro TM, Campelo PH, Sant'Ana AS, Clerici MTP (2020) Tubers and roots as a source of prebiotic fibers. In: Advances in food and nutrition research. Elsevier, pp 267–293. https://doi.org/10.1016/bs.afnr.2020.06.005

Nguyen DTC, Nguyen TT, Le HT, Nguyen TTT, Bach LG, Nguyen TD et al (2021) The sunflower plant family for bioenergy, environmental remediation, nanotechnology, medicine, food and agriculture: a review. Environ Chem Lett 19:3701–3726. https://doi.org/10.1007/s10311-021-01266-z

Oleszek M, Kowalska I, Oleszek W (2019) Phytochemicals in bioenergy crops. Phytochem Rev 18: 893–927. https://doi.org/10.1007/s11101-019-09639-7

Paixao SM, Alves L, Pacheco R, Silva CM (2018) Evaluation of Jerusalem artichoke as a sustainable energy crop to bioethanol: energy and CO_2eq emissions modeling for an industrial scenario. Energy 150:468–481. https://doi.org/10.1016/j.energy.2018.02.145

Puttha R, Jogloy S, Suriharn B, Wangsomnuk PP, Kesmala T, Patanothai A (2013) Variations in morphological and agronomic traits among Jerusalem artichoke (*Helianthus tuberosus* L.) accessions. Genet Resour Crop Evol 60:731–746. https://doi.org/10.1007/s10722-012-9870-2

Qiu Y, Lei P, Zhang Y, Sha Y, Zhan Y, Xu Z et al (2018) Recent advances in bio-based multi-products of agricultural Jerusalem artichoke resources. Biotechnol Biofuels 11(1):1–15. https://doi.org/10.1186/s13068-018-1152-6

Qiu Y, Sha Y, Zhang Y, Xu Z, Li S, Lei P et al (2017) Development of Jerusalem artichoke resource for efficient one-step fermentation of poly-(γ-glutamic acid) using a novel strain *Bacillus amyloliquefaciens* NX-2S. Bioresour Technol 239:197–203. https://doi.org/10.1016/j.biortech.2017.05.005

Saha BC (2006) Production of mannitol from inulin by simultaneous enzymatic saccharification and fermentation with *Lactobacillus intermedius* NRRL B-3693. Enzym Microb Technol 39(5): 991–995. https://doi.org/10.1016/j.enzmictec.2006.02.001

Sarchami T, Rehmann L (2014) Optimizing enzymatic hydrolysis of inulin from Jerusalem artichoke tubers for fermentative butanol production. Biomass Bioenergy 69:175–182. https://doi.org/10.1016/j.biombioe.2014.07.018

Sawicka B, Skiba D, PszczóÅ P, Aslan I, Sharifi J, Krochmal-Marczak B (2020) Jerusalem artichoke (*Helianthus tuberosus* L.) as a medicinal plant and its natural products. Cell Mol Biol 66(4):160–177. https://doi.org/10.14715/cmb/2020.66.4.20

Shi Z, Wei P, Zhu X, Cai J, Huang L, Xu Z (2012) Efficient production of l-lactic acid from hydrolysate of Jerusalem artichoke with immobilized cells of *Lactococcus lactis* in fibrous bed bioreactors. Enzym Microb Technol 51(5):263–268. https://doi.org/10.1016/j.enzmictec.2012.07.007

Showkat MM, Falck-Ytter AB, Strætkvern KO (2019) Phenolic acids in Jerusalem artichoke (*Helianthus tuberosus* L.): plant organ dependent antioxidant activity and optimized extraction from leaves. Molecules 24(18):3296. https://doi.org/10.3390/molecules24183296

Singh R, Singh R, Kennedy J (2017) Immobilization of yeast inulinase on chitosan beads for the hydrolysis of inulin in a batch system. Int J Biol Macromol 95:87–93. https://doi.org/10.1016/j.ijbiomac.2016.11.030

Singh R, Singh T, Hassan M, Larroche C (2022) Biofuels from inulin-rich feedstocks: a comprehensive review. Bioresour Technol 346:126606. https://doi.org/10.1016/j.biortech.2021.126606

Slapkauskaite J, Sekmokiene D, Kabasinskiene A, Bartkiene E, Juodeikiene G, Sarkinas A. (2016). Influence of lactic acid bacteria–fermented *Helianthus tuberosus* L. and *Lupinus luteus* on quality of milk products. CyTA J Food 14(3):482–488. https://doi.org/10.1080/19476337.2015.1137355

Slimestad R, Seljaasen R, Meijer K, Skar SL (2010) Norwegian-grown Jerusalem artichoke (*Helianthus tuberosus* L.): morphology and content of sugars and fructo-oligosaccharides in stems and tubers. J Sci Food Agric 90(6):956–964. https://doi.org/10.1002/jsfa.3903

Sun L-H, Wang X-D, Dai J-Y, Xiu Z-L (2009) Microbial production of 2, 3-butanediol from Jerusalem artichoke tubers by *Klebsiella pneumoniae*. Appl Microbiol Biotechnol 82:847–852. https://doi.org/10.1007/s00253-008-1823-5

Sung M, Seo YH, Han S, Han J-I (2014) Biodiesel production from yeast *Cryptococcus* sp. using Jerusalem artichoke. Bioresour Technol 155:77–83. https://doi.org/10.1016/j.biortech.2013.12.024

Swanton C, Clements D, Moore M, Cavers P (1992) The biology of Canadian weeds. 101. *Helianthus tuberosus* L. Can J Plant Sci 72(4):1367–1382. https://doi.org/10.4141/cjps92-169

Szambelan K, Nowak J, Jelen H (2005) The composition of Jerusalem artichoke (*Helianthus tuberosus* L.) spirits obtained from fermentation with bacteria and yeasts. Eng Life Sci 5(1): 68–71. https://doi.org/10.1002/elsc.200400052

Tesio F, Weston LA, Ferrero A (2011) Allelochemicals identified from Jerusalem artichoke (*Helianthus tuberosus* L.) residues and their potential inhibitory activity in the field and laboratory. Sci Hortic 129(3):361–368. https://doi.org/10.1016/j.scienta.2011.04.003

Wang D, Li F-L, Wang S-A (2016) Engineering a natural *Saccharomyces cerevisiae* strain for ethanol production from inulin by consolidated bioprocessing. Biotechnol Biofuels 9(1):1–11. https://doi.org/10.1186/s13068-016-0511-4

Wang L-F, Wang Z-P, Liu X-Y, Chi Z-M (2013a) Citric acid production from extract of Jerusalem artichoke tubers by the genetically engineered yeast *Yarrowia lipolytica* Strain 30 and purification of citric acid. Bioprocess Biosyst Eng 36:1759–1766. https://doi.org/10.1007/s00449-013-0951-1

Wang L, Xue Z, Zhao B, Yu B, Xu P, Ma Y (2013b) Jerusalem artichoke powder: a useful material in producing high-optical-purity L-lactate using an efficient sugar-utilizing thermophilic *Bacillus coagulans* strain. Bioresour Technol 130:174–180. https://doi.org/10.1016/j.biortech.2012.11.144

Wang Y-Z, Zou S-M, He M-L, Wang C-H (2015) Bioethanol production from the dry powder of Jerusalem artichoke tubers by recombinant *Saccharomyces cerevisiae* in simultaneous saccharification and fermentation. J Ind Microbiol Biotechnol 42(4):543–551. https://doi.org/10.1007/s10295-014-1572-7

Xia J, Xu J, Liu X, Xu J, Wang X, Li X (2017) Economic co-production of poly (malic acid) and pullulan from Jerusalem artichoke tuber by *Aureobasidium pullulans* HA-4D. BMC Biotechnol 17:1–10. https://doi.org/10.1186/s12896-017-0340-y

Xu Q, Zang Y, Zhou J, Liu P, Li X, Yong Q, Ouyang J (2016) Highly efficient production of D-lactic acid from chicory-derived inulin by *Lactobacillus bulgaricus*. Bioprocess Biosyst Eng 39:1749–1757. https://doi.org/10.1007/s00449-016-1650-5

Xue C, Zhang X, Wang J, Xiao M, Chen L, Bai F (2017) The advanced strategy for enhancing biobutanol production and high-efficient product recovery with reduced wastewater generation. Biotechnol Biofuels 10:1–11. https://doi.org/10.1186/s13068-017-0836-7

Yang L, He QS, Corscadden K, Udenigwe CC (2015) The prospects of Jerusalem artichoke in functional food ingredients and bioenergy production. Biotechnol Rep 5:77–88. https://doi.org/10.1016/j.btre.2014.12.004

Yi H, Zhang L, Hua C, Sun K, Zhang L (2010) Extraction and enzymatic hydrolysis of inulin from Jerusalem artichoke and their effects on textural and sensorial characteristics of yogurt. Food Bioprocess Technol 3:315–319

Zhao C-H, Chi Z, Zhang F, Guo F-J, Li M, Song W-B, Chi Z-M (2011) Direct conversion of inulin and extract of tubers of Jerusalem artichoke into single cell oil by co-cultures of *Rhodotorula mucilaginosa* TJY15a and immobilized inulinase-producing yeast cells. Bioresour Technol 102(10):6128–6133. https://doi.org/10.1016/j.biortech.2011.02.077

Zhao C-H, Cui W, Liu X-Y, Chi Z-M, Madzak C (2010a) Expression of inulinase gene in the oleaginous yeast *Yarrowia lipolytica* and single cell oil production from inulin-containing materials. Metab Eng 12(6):510–517. https://doi.org/10.1016/j.ymben.2010.09.001

Zhao C-H, Zhang T, Li M, Chi Z-M (2010b) Single cell oil production from hydrolysates of inulin and extract of tubers of Jerusalem artichoke by *Rhodotorula mucilaginosa* TJY15a. Process Biochem 45(7):1121–1126. https://doi.org/10.1016/j.procbio.2010.04.002

Zhao X, Wu S, Hu C, Wang Q, Hua Y, Zhao ZK (2010c) Lipid production from Jerusalem artichoke by *Rhodosporidium toruloides* Y4. J Ind Microbiol Biotechnol 37(6):581–585. https://doi.org/10.1007/s10295-010-0704-y

Onion Solid Waste as a Potential Source of Functional Food Ingredients

13

Meenakshi, Yogesh Kumar, Suvarna Bhoj, C. Nickhil, Narashans Alok Sagar, Deepak Kumar, Ranjna Sirohi, and Ayon Tarafdar

Abstract

Onion is consumed globally in several forms and has been recognized as a source of minerals, essential oils, and proteins. It is mostly used in food preparations to enhance the organoleptic quality of cooked foods. While processing onions, a large amount of waste, mostly in the form of peels, fleshy scales, roots, bulbs, etc., are left behind, which are rich sources of bioactive compounds that can be extracted through suitable technologies for the development of value-added products. Onion waste can also be fermented for the production of bioethanol

Meenakshi
Division of Livestock Products Technology, ICAR-Indian Veterinary Research Institute, Izatnagar, Uttar Pradesh, India

Y. Kumar
Department of Food Technology, Faculty of Science and Humanities, SRM University, Sonipat, Haryana, India

S. Bhoj · A. Tarafdar (✉)
Livestock Production and Management Section, ICAR-Indian Veterinary Research Institute, Izatnagar, Uttar Pradesh, India
e-mail: ayon.tarafdar@icar.gov.in

C. Nickhil
Department of Food Engineering and Technology, Tezpur University, Tezpur, Assam, India

N. A. Sagar
Department of Biotechnology, University Centre for Research and Development, Chandigarh University, Mohali, Punjab, India

D. Kumar
DS Group, DS Headquarters, Noida, Uttar Pradesh, India

R. Sirohi
Rajasthan Agricultural Research Institute, SKN Agricultural University, Durgapura, Jaipur, Rajasthan, India

R. C. Ray (ed.), *Roots, Tubers, and Bulb Crop Wastes: Management by Biorefinery Approaches*, https://doi.org/10.1007/978-981-99-8266-0_13

and, therefore, holds great value to the industrial bioprocessing sector. This chapter discusses various innovative, contemporary valorization approaches (chemical, thermal, enzymatic, and biotechnological, etc.) for the extraction of functional components from onion waste and the production of value-added products from them for sustainable growth through the circular economy mode.

Keywords

Flavonoids · Biofuel · Extraction · Onion waste · Value-addition · Bioprocessing

Abbreviations

ACSOs	*S*-alk(en)yl-L-cysteine sulfoxides
AOAC	Association of Official Agricultural Chemists
DF	Dietary fibers
DPPH	2,2-Diphenyl-1-picrylhydrazyl
DW	Dry weight
FOS	Fructooligosaccharide
FRAP	Ferric reducing antioxidant power
FTIR	Fourier transform infrared spectroscopy
GAE	Gallic acid equivalent
GC-MS	Gas chromatography-mass spectrometry
HPLC	High-pressure liquid chromatography
IDF	Insoluble dietary fiber
MAE	Microwave assisted extraction
OJR	Onion juice residues
OS	Onion scales
OSP	Onion scale powder
OWs	Onion wastes
QE	Quercetin equivalents
SDF	Soluble dietary fiber
SPE	Solid phase extraction
SSCF	Saccharification and co-fermentation
SSTF	Saccharification and two-step fermentation
SWE	Supercritical water extraction
TAC	Total anthocyanin content
TFC	Total flavonoid content
TPC	Total phenolic content
UAE	Ultrasound-assisted extraction

13.1 Introduction

Onion (*Allium cepa* L.) is regarded as an indispensable horticulture commodity in various civilizations across the world. It is highly valued as a vegetable and a therapeutic herb that originated in Central Asia and is cultivated globally second to tomato. The major share of global onion production is held by China, India, USA, Egypt, Turkey, Pakistan, Sudan, Bangladesh, Iran, and Russia with China as a key producer followed by India (Kumar et al. 2022). The major onion producing nations are illustrated in Fig. 13.1. The unique odor and flavor of onion are attributed to the presence of a sulfur compound, allyl-propyl disulfides. It is frequently referred to as the "Queen of the kitchen" because of its extensive use in numerous cuisines (Griffiths et al. 2002). Onion is often red, yellow, and white in color with distinct flavors and pungency (Khandagale and Gawande 2019). It is considered a potential source of proteins, sugars, cellulose, minerals, essential oils, and dietary flavonoids with anti-diabetic, anti-thrombotic, anti-inflammatory, anti-spasmodic, anti-allergic and neuro-protective biomedical properties (Bisen and Emerald 2016).

The worldwide usage of onion as a food article in both domestic and commercial setups has raised concerns over the large quantities of onion wastes (OWs), creating sustainability issues. In 2025, an upsurge of 88.6–92.8% in OWs is anticipated globally rendering the valorization strategies of vital importance (Segundo et al. 2022). European Union alone gives rise to nearly 0.6 metric tons of OWs annually mainly produced from onion processing industries leaving 21.4% of OWs unutilized (Kim et al. 2019; Kumar et al. 2022). It can be characterized as industrial waste generated by food processing, household waste, and post-harvest or supply chain waste. Processed onion in the form of dehydrated onion, canned onion, and onion

Fig. 13.1 Top five onion producing nations of the world. (Source: https://www.atlasbig.com/en-in/countries-by-onion-production)

pickle is highly appreciated in Germany, Japan, Netherlands, Russia, Spain, and the UK. Onions are dried to reduce the moisture content by up to 7% or less to further inhibit microbial growth during storage. This generates an enormous quantity of waste in the form of peeled skins, the two outer fleshy scales, roots, and the bulb's top and bottom, including diseased, damaged, or deformed bulbs. It accounts for nearly 38% of fresh onion weight discarded as waste in the processing industry (Sharma et al. 2016; Sagar et al. 2022a) predominant in Spain, the Netherlands, and the UK. Domestic OWs mainly include skin peels, top and bottom parts, fleshy scales, and spoiled bulbs (Bello et al. 2013). A South Korean family reportedly produces 5.45% of the food waste, while a Brazilian food stall generates 0.04–0.26 kg/OWs per day (Adelodun et al. 2021; Brancoli et al. 2022). Inferior logistics/transit facilities, post harvesting of onion crops cause around 10% wastage during food supply because of pathological and biological, and physical damage (Vojvodic Cebin et al. 2020). Uruguay accounts for high post-harvest wastage, i.e., 21.7%, followed by Egypt (19%) and Tanzania (16%), while China, Ethiopia, and Pakistan face damage of about 10–12%, 10.7%, and 9% post-harvest onion, respectively (Kitinoja and Kader 2015). India and the Philippines experienced, up to 35–40% (Anbukkarasi et al. 2013) and 31.49% of huge losses, respectively (Calica and Dulay 2018).

Its characteristic pungent odor, high sulfur content, and rapid growth of phytopathogenic agent, *Sclerotium cepivorum* disqualifies it as fodder and manure leaving landfill and incineration as the only option, posing a serious economic and environmental threat (Sharma et al. 2016; Sagar and Pareek 2020a). However, OWs comprise of high-value bioactive compounds such as flavonoids: flavanols, anthocyanins, etc., as antioxidants, fibers, minerals, and *S*-alk(en)yl-L-cysteine sulfoxides (ACSOs), which can be managed to produce various bio-products, bio enzymes, polyphenols, pigments, and bio-energy (Kumar et al. 2022). This chapter discusses various innovative, contemporary valorization approaches (chemical, thermal, enzymatic, and biotechnological, etc.) for the extraction of functional components from onion waste and the production of value-added products from them for sustainable growth through the circular economic mode.

13.2 Composition of Onion Wastes

After the processing of the onion, waste is produced from its roots, outer fleshy scales, top, and skin (Benítez et al. 2011). OWs have a remarkably different chemical composition. For instance, it contains a good amount of minerals (selenium, Zn, Ca, Mg, and K), dietary fiber, ash, and protein. In the outer skin of the onion, calcium, flavoring compounds, carbohydrates, and polyphenols are present in high proportion, while crude fiber, ash, and protein are found in a lower concentration. Outer fleshy scales contain high amounts of ACSOs, fructans, flavonols, and dietary fiber content. Furthermore, Zn, Mg, Fe, and Mn are found in the top and bottom parts of the onion (Ifesan 2017; Sharma et al. 2016; Benítez et al. 2012). Dietary fiber has a

Table 13.1 Proximate composition of onion waste

Components	Exaction method	Outer scales	Top-bottom	Peel
Crude protein	AOAC method	2–9%	8–15%	5–6%
Crude fiber	AOAC method	26–27%	15–16%	4–9%
Crude fat	AOAC method	15–16%	15–16%	0.3–0.9%
Ash content	AOAC method	3–5%	11–12%	7–13%
Moisture content	AOAC method	8–9%	9–10%	6–11%
Carbohydrates	AOAC method	66–68%	54–55%	65–75%
Non-structural carbohydrates				
Sucrose, %	HPLC method	37–39	25–27	–
Glucose, %	HPLC method	207–214	37–41	–
Fructose, %	HPLC method	192–199	47–55	–
Total fructans, %	Colorimetric method	24–29	7–9	–

Source: Sagar et al. (2021), Bello et al. (2013), Oguezi et al. (2019), Benítez et al. (2011)

high proportion of soluble dietary fiber (SDF): insoluble dietary fiber (IDF) and flavonols show high antioxidant properties (Benítez et al. 2011).

Two main categories of carbohydrates are found in the onion waste, i.e., cell wall polysaccharides and non-structural carbohydrates. Approximately 65% of the dry matter in onions is made up of non-structural carbohydrates such as fructooligosaccharides (FOS), sucrose, fructose, and glucose (Griffiths et al. 2002; Sharma et al. 2016). The nutritional and calorific value of onions is augmented by the sugars (sucrose, fructose, and glucose). FOS are functional components that do not possess any nutritional value rather they are prebiotics and a type of dietary fiber. The proximate composition of onion waste is given in Table 13.1.

Another category of carbohydrate is cell wall polysaccharides. Monomeric sugars are examined to identify the cell wall polysaccharides after the hydrolysis of onion waste. It reveals that glucose (27%) and uronic acids or galacturonic acid (21%) are the dominant components on a dry matter basis (Vojvodic et al. 2016).

In OWs, a high percentage of galacturonic acid indicates the presence of a high amount of soluble fiber. However, the presence of pectin in OWs has structural features that make them intractable for the applicable fiber analysis. Although pectin present in OWs is not properly dissolved in dilute acids and water, it is mostly chelator-soluble. Due to this, insoluble fiber is predominated in OWs. Increased lignin content in the outer portions of the bulb is associated with high insoluble fiber content. Waste with fleshy scales has a dry matter base of around 59% (Jaime et al. 2002; Vojvodic et al. 2016).

High sources of bioactive substances and phytochemicals are found in OWs. Compared to edible flesh, onion peel contains a higher concentration of phytochemicals. There are numerous organic components present in it, including sulfur. The odd smell of onion and several positive health benefits are also caused by these substances. Interaction between thiol and these sulfur molecules is used to determine their larvicidal, anticancer, antifungal, and allergenic effects on the biological system (Sagar et al. 2022b).

Table 2 Sulfur and bioactive components of onion waste

	Skin	Outer scales	Top–bottom
Total sulphur and ACSOs content			
Total sulfur (μmol/g DW)	15–16	98–102	139–146
Total ACSOs (μmol/g DW)	4–5	28–30	22–23
Total S-ACSOs vs total S (%)	29–30	29–30	15–16
PECSO (mg/g DW)	0.4–0.5	3–4	2–3
MCSO (mg/g DW)	0.3–0.4	1.2–1.6	1–2
Total ACSOs (mg/g DW)	0.7–0.8	5–6	3–4
The total content of phenol, flavonoids and flavonols, and antioxidant capacity			
Antioxidant activity (μmol Fe^{2+}/g DM)	224–230	104–106	154–158
Total flavonoids (mg QE/g DM)	41–168	18–21	24–27
Total phenolics (mg GAE/g DM)	51–54	18–21	28–33
Total flavonols (mg/g DM)	7–8	18–21	13–17

Source: Benítez et al. (2011), Prakash et al. (2007), Roldán-Marín et al. (2010)

They are derived from aliin, also known as ACSOs, a non-volatile precursor sulfuric chemical. The three most important ACSOs found in onions are trans-(+)-*S*-(propen-1-yl)-L-cysteine sulfoxide (izoaliin), (+)-*S*-propyl-L-cysteine sulfoxide (propiin), and (+)-*S*-methyl-L-cysteine sulfoxide (metiin) (Table 13.2).

Up to 80% of onion ACSOs are represented by Metin. The most prevalent non-protein amino acids are *S*-alk(en)yl cysteines and associated sulfoxides. When the cellular tissue is ruptured, the odorless *S*-alk(en)yl cysteine sulfoxides undergo enzymatic disintegration to provide the distinctive flavor to onion (Brewster and Rabinowitch 2020; Griffiths et al. 2002; Benítez et al. 2012).

In fruits and vegetables, onions are the most significant dietary sources of flavonoids. The amount of flavonoids in onion peels is substantially larger than in the vegetable's edible section. Anthocyanidins, flavonols, flavanonols, flavanones, leucoanthocyanidins (flavan-3,4-diols), and catechins (flavan-3-ols) are the basic subgroups of flavonoids (Velíšek 2002). There are three main flavonoids: quercetin diglucoside, quercetin 4′-glucoside, and quercetin aglycone (Table 13.3). In some situations, isorhamnetin monoglycoside or kaempferol monoglycoside may also be present (Pareek et al. 2017). Quercetin and its glucoside derivatives (quercetin-3,4′-diglucoside and quercetin-4′-glucoside) predominates among onion's flavonoids, accounting for up to 80% of the vegetable's total flavonoids (Benítez et al. 2012). The brown and yellow color of onion peel is due to the presence of quercetin. The amount of aglycone quercetin increases toward the onion skin while the quercetin glucosides decline toward the outer scales. The maximum concentration of aglycone quercetin is present in the outer peel. The deglucosilation of quercetin glucosides—possibly in the presence of sunlight—leads to an increase in aglycone concentration (Higashio et al. 2004). Protocatechuic acid is produced when quercetin is further oxidized by peroxidase. It possesses antifungal properties (Takahama and Hirota 2000). This implies that in the dried onion skin, an enzymatic process produces anti-infection defenses derived from quercetin. Quercetin exhibits

Table 13.3 Structure and composition of flavonoids in onion waste

Flavonoids	Structure	Concentration	Reference
Quercetin	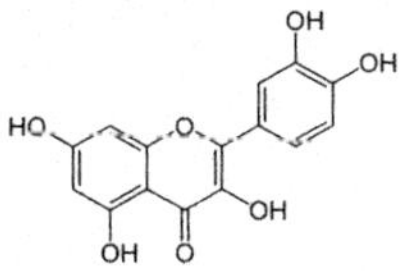	2800–12,000 (mg/kg dw)	Sagar et al. (2020)
Quercetin 3′-glucoside	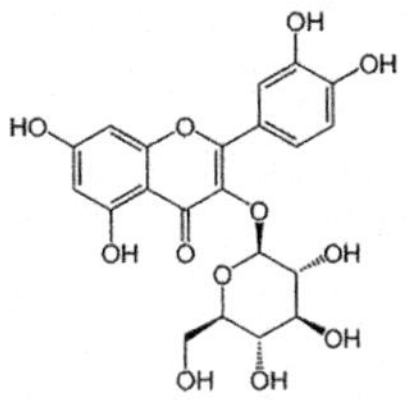	300–1400 (mg/kg dw)	Sagar et al. (2020)
Quercetin 4′-glucoside	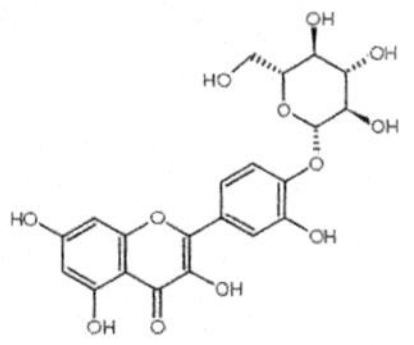	0.32–1.03 (mg/g dw)	Benítez et al. (2011)
Quercetin 3,4′-diglucoside	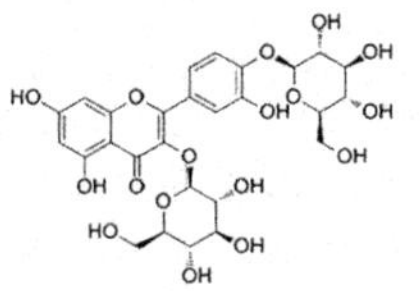	0.34–9.49 (mg/g dw)	Benítez et al. (2011)
Isorhamnetin 3,4′-diglucoside	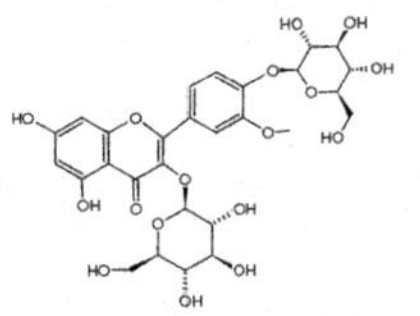	0.19–0.57 (mg/g dw)	Benítez et al. (2011)
Isorhamnetin 4′-glucoside	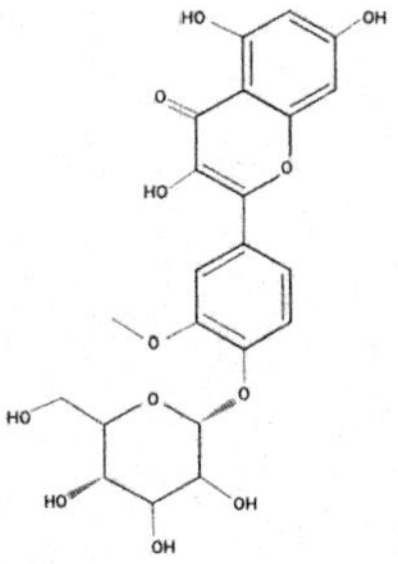	0.32–1.03 (mg/g dw)	Benítez et al. (2011)
Kaempferol	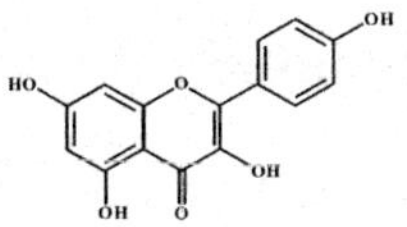	400–700 (mg/kg dw)	Sagar et al. (2020)

(continued)

Table 13.3 (continued)

Flavonoids	Structure	Concentration	Reference
Kaempferol-3-*O*-glucoside	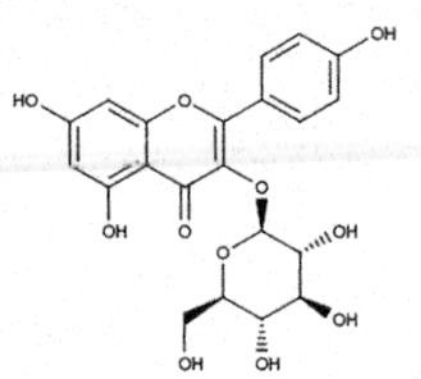	30–130 (mg/g dw)	Burri et al. (2017)
Luteolin	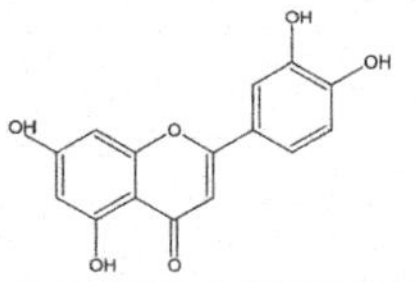	300–1700 (mg/kg dw)	Sagar et al. (2020)

antiviral, anticancer, anti-allergic, antihistamine, anti-inflammatory, antibacterial, and antioxidant properties (Sagar and Pareek 2021; Sagar et al. 2022b).

Protecting onions from the damaging effects of highly reactive oxygen species and radicals of oxygen is an important function of phenolic compounds that are known antioxidants. The phenolic structure of the phenols includes an aromatic ring with a hydroxyl substituent that is primarily responsible for the scavenging capacity of phenols. In addition, total phenolics and antioxidant activity have a favorable and extremely significant correlation (Sagar et al. 2020). Moreover, onion waste is rich in several phenolic compounds, such as *p*-coumaric acid, vanillic acid, tannic acid, ferulic acid, and gallic acid (Sagar et al. 2022a).

13.2.1 Methods to Analyze Onion Waste Composition

For analysis of OWs, a uniform sample is prepared following several steps. The uniformity of the sample improves the interaction between the solvent and the sample by increasing the surface area. Prewashing is the first step: 0.5% chlorinated water was used to wash OWs to remove adherent debris, impurities, and dust, after which it was washed again with distilled water to remove all impurities. Using a porous tray for spreading wash waste for 10 min, then freeze it for 24 h at 40 °C. After prewashing, drying was carried out using a lyophilizer at room temperature of 50 °C at 0.039 mbar pressure for 48 h to retain bioactive compounds. Then, the powder was formed by grinding the freeze-dried skin in a grinder and stored at room temperature (28 ± 2 °C) in airtight plastic containers (Sagar et al. 2021).

13.2.1.1 Analysis of Proximate Composition

The prepared powdered form can be used as a test sample for proximate analysis. The Association of Official Agricultural Chemists (AOAC) (2003) standard methods of analysis were used to determine the proximate contents of the dried onion waste.

The test sample was dried in an oven at 105 °C to a constant weight to determine the moisture content. For determining ash content, light grey ash was produced after the ignition of a test sample at 550 °C in a muffle furnace. By using the conventional macro-Kjeldahl method, crude protein was evaluated. Digestion, distillation, and titration are the three main steps of the macro-Kjeldahl method. The ether extraction method is used for the determination of lipid content. The apparatus used in ether extraction is Soxhlet, in which petroleum ether is used as a solvent, and the following difference formula is used to evaluate carbohydrate content:

$$\text{Percentage of carbohydrates} = 100\% - (\%\text{Ash} + \%\text{Total fat} + \%\text{Moisture} + \%\text{Protein})$$

13.2.1.2 Quantification of Bioactive Compounds

For quantification of the bioactive compounds present in OWs, several chromatographic and spectroscopic techniques have been formulated over the years. The procedure for the extraction and quantification of these bioactive compounds has been described in this section.

Extraction of Bioactive Compounds

A crucial step in the quantification of bioactive compounds is extraction since it is necessary to extract the chemical components for further separation and characterization. Extracts from plant samples must be prepared in a way that prevents the loss, distortion, or destruction of active constituents (Sooch et al. 2022). In many cases, the choice of solvent system depends on the particular nature of the bioactive compound. It is possible to extract bioactive compounds from natural products using various solvent systems. Ethyl-acetate or ethanol, and methanol are polar solvents used for the hydrophilic biomolecules. A 1:1 mixture of dichloromethane/methanol and dichloromethane is also used for lipophilic extraction. For the removal of chlorophyll, hexane extraction is utilized (Cosa et al. 2006). The applicability of extraction techniques should always be considered. Since the target compounds can vary from nonpolar to polar and are thermally labile. Extraction methods include Soxhlet extraction, sonification, heating under reflux, and others (Fabricant and Farnsworth 2001).

There was a significant difference in the amount of total phenols and total flavonoids extracted from onion peels based on the extraction method (Table 13.4). Total flavonoids (183.95 mg quercetin equivalents (QE)/g extract) and phenolics (372.50 mg gallic acid equivalents (GAE)/g extract) were concentrated highest when extracted in ethanol. Following the subcritical water extraction method at 110 °C and hot water extraction. Moreover, the lowest concentration of total flavonoids (27.10 mg QE/g extract) and total phenolics (56.68 mg GAE/g extract) was observed by subcritical water extraction (SWE)-165 °C. As the solvent concentration increased (60–80%) in both continuous mode (130 and 750 W) and pulsed mode (750 W), more phenolics in onion skin were detected in the ultrasound-assisted extraction (UAE). The highest values were noted at

Table 13.4 Different methods for extraction of bioactive compounds

Extraction method	Solvents	Conditions	Mechanism of action	Quantified bioactive compounds	References
Microwave-assisted extraction (MAE)	Ethanol (10–100%)	Time: 10–20 s Power: 700 W	Microwaves resulted in rupturing the cell wall due to expansion as a result of microwave heating, thus liberating quercetin and its glycoside from onion skin	Quercetin and its glycoside	Jin et al. (2011)
Supercritical water extraction (SWE)	Distilled water	Temperature: 110 or 165 °C Time: 15 min Pressure: <500 psi	Higher temperatures are maintained, with high pressure to maintain the liquid state of water. This extraction also assists in lowering the polarity near to ethanol or methanol, thus helping in better extraction of bioactive	Total phenolics and flavonoids	Lee et al. (2011)
Hot water	Distilled water	Temperature: 80 °C Time: 3 h	–	Total phenolics and flavonoids	Lee et al. (2014)
Conventional method	20%, 40%, 60%, and 80% ethanol	Temperature: 25 °C Time: 30, 60, 120, and 240 min		Total phenolic, flavonoids, and anthocyanins content	Viera et al. (2017)
Ultrasound-assisted extraction (UAE)	70% (v/v) ethanol solution	Temperature: 34.02–58.40 °C Time of extraction: 5, 10, 15, 20, and 25 min Solid to	UAE uses ultrasounds that cause plant tissue disruption by acoustic cavitation, thus,	Anthocyanins, flavonoids, and tannins	Oancea et al. (2020)

(continued)

Table 13.4 (continued)

Extraction method	Solvents	Conditions	Mechanism of action	Quantified bioactive compounds	References
		solvent ratio: 30/1 and 20/1 ratio Sonication: 150 W power and 40 kHz frequency, and 70% ultrasonic amplitude	combining DES with UAE enhanced extraction of bioactive compound from onion peel		
Supercritical fluid extraction	Co-solvent ethanol	Temperature: 55 °C Time: 2.5–4 h Condition: CO_2 at 400 bar pressure	Compounds separation: Two separation cells (S40: 200 bar/40 °C and S45: 50 bar/23 °C)	Phenolic acids, quercetin, oleoresin	Constantin et al. (2020)

622.87 mg GAE/g dw in the continuous mode and 117.50 mg GAE/g dw in the pulsed mode. Bioactive compounds from onion waste are quantified using different extraction methods (Viera et al. 2017, 2021; Lee et al. 2014)

Quantification Methods

It remains a challenge to identify and characterize bioactive compounds because extract samples contain a mixture of different types of bioactive compounds or phytochemicals of different polarities. For obtaining pure bioactive compounds, a variety of separation techniques, such as GC-MS, High-Performance Liquid Chromatography (HPLC), and Colorimetric methods [UV(Ultraviolet)Spectroscopy], are commonly practiced (Table 13.5).

Quantification Method Based on HPLC

An HPLC system, a photodiode array detector, an auto-sampler, and a solvent delivery system are used to quantify flavonoids. To separate flavonoids, a reverse phase column is recommended. The data is quantified using a standard curve. Two mobile phases are used: (A) trifluoroacetic acid in water 0.1% (v/v) and (B) methanol. The sample injection volume may differ from 20 to 30 μL. The flow rate can be kept at 1–2 mL/min (Sagar et al. 2020).

Quantification Method Based on Gas Chromatography-Mass Spectroscopy (GC-MS)

A GC-MS is used to evaluate volatile compounds, each equipped with a capillary column. Helium is used as the carrier gas at a constant flow rate of 1 mL/min during the chromatographic separation of a 2 μL sample. During the program, the

Table 13.5 Methods for quantification of bioactive compounds

Bioactive compounds	Quantification method	Reagents	Standards	Wavelength for detection	Quantification value	References
Flavonoids	HPLC	Mobile phase A (trifluoroacetic acid in water 0.1% (v/v)): B (methanol)	Calibration curve as a standard	365 nm	mg/kg dry weight (dw)	Sharma et al. (2015)
Total phenolic contents	Colorimetric method	Folin-Ciocalteu's reagent, sodium carbonate, distilled water	Gallic acid as standard	760 nm	GAE/g dw	Sun et al. (2007)
Total flavonoid content	Colorimetric method	Methanol, aluminium chloride, potassium acetate, distilled water	Quercetin as standard	Absorption measured at 410 nm	(µg Q/g DW)	Benítez et al. (2011)
Organosulfur compounds	GC-MS	Helium gas, commercial standard	Internal standard—response factor equal to the 2-octonal		mg/kg DW	Liguori et al. (2017)
Phenolic acids	HPLC	Mobile phase A—a mixture of methanol/water 5:95, V/V, mobile Phase B—mixture methanol/water 40:60, V/V; H_3PO_4, acetic acid		Detection at $\lambda = 250$ nm	mg/kg DW	Lachman et al. (2003)

temperatures are as follows: 3 min at 40 °C, first ramp at 2 °C/min to 150 °C and second ramp at 4 °C/min to 220 °C. After holding the GC injector temperature in the range of 40 °C for 3 min, the temperature in the column is increased to 150 °C at 2 °C/min, then 220 °C at 4 °C/min (hold for 10 min). The transfer line and detector are usually kept at 250 °C. Volatile compounds were identified by comparing their spectra, and their retention indices with literature references and injecting commercial standards.

Quantification Method Based on UV Spectrophotometer

Folin-Ciocalteu colorimetric method is used to measure the total phenolic content (TPC) in onion waste. The test tubes were filled with extracts (20 mL) of onion waste, 0.5 mL of Folin-Ciocalteu's reagent, and 5 mL of distilled water. After 10 min of incubation, each test tube received 0.5 mL of sodium carbonate (2%) and was again kept for 45 min. Using a UV-Vis Spectrophotometer, the absorbance was recorded at 760 nm (UV-2600, Shimazdu, Kyoto, Japan). The concentration was measured using gallic acid as a standard (Sagar and Pareek 2020b).

A colorimetric technique was used to quantify the total flavonoid content (TFC). A test tube containing 0.5 mL of OW extract is obtained from a sample. Each test tube contained 1.5 mL of methanol (80%). Additionally, 0.1 mL of potassium acetate (1 M) and 0.1 mL of aluminum chloride (10%) were added to each test tube. At room temperature (28 ± 2 °C), 2.8 mL of distilled water was added and rested for 30 min. The absorbance was determined using a UV-Vis Spectrophotometer at 410 nm. The results were represented as mg quercetin/g dw because quercetin is used as a standard (Sagar et al. 2020).

13.3 Onion Waste Processing Technologies and Their Products

Several processing techniques are being implemented to harness industrially important products/compounds from OWs. Some of the most notable technologies have been detailed here.

13.3.1 Solvent Extraction

Phenolic content of 372.5 mg GAE/g was obtained from the onion peel ethanol extract, which may vary according to the extraction method (Lee et al. 2014). Also, Lee et al. (2014) reported quercetin (6.8 mg/100 mg DW) and isoquercetin (8.1 mg/100 mg DW) in onion peel extract, and antioxidant capacity was reported 11.5–89.7% of the DPPH (2,2-diphenyl-1-picrylhydrazyl) radical scavenging activity. These bioactive compounds induce various therapeutic activities (antiadipogenic and antiobesity) (Lee et al. 2011). Also, 3.7 mg/g quercetin, 1.4 mg/g protocatechuic acid, 63–75 mg GAE/g phenols, and 23–26 mg QE/g quercetin were extracted from the onion skin waste using supercritical carbon dioxide/co-solvent extraction method which can be used to develop the nutraceuticals (Campone et al. 2018).

Gokulakrishnan et al. (2022) reported the recovery of the flavonoid compounds from the plant biomass using advanced extraction techniques, such as microwave assisted extraction (MAE), SWE, and UAE. Further, various methods were used for the separation and purification of the flavonoids using various solid-phase extraction (SPE) materials like C-18 silica nanomaterials and molecularly imprinted polymers (MIPS) and molecularly imprinted membranes (MIMS). Similarly, Prokopov et al. (2022) extracted the flavonoid content from the OWs using conventional solvent extraction methods. The methanolic extraction at 53 °C, 66 min, and 61% of methanol provides the highest flavonoid content. Therefore, methanolic extract from OWs can be used as an additive for the development of functional food.

13.3.2 Supercritical Fluid Extraction

Onion skin waste was used to extract the quercetin, which acts as a flavonoid compound, using a combination of supercritical fluid extraction and pretreatment by intense pulsed light (Kim et al. 2019). Kim et al. (2019) reported the maximum concentration of quercetin of 17.32 mg/g onion skin at the supercritical water extraction condition of 145 °C for 15 min with intense pulsed light at 1200 V for 60 s whereas the quercetin concentration of 15.19 mg/g onion skin was observed using supercritical water extraction alone at 145 °C for 15 min.

Dried rotten onion was utilized to obtain good quality onion oleoresin using a supercritical fluid extraction technique (Devani et al. 2020). Devani et al. (2020) optimized the concentration of oleoresin by supercritical extraction method using a central composite rotatable design. The optimized conditions obtained for supercritical fluid extraction of onion oleoresin were 80 °C temperature, 4000 bar pressure, 0.53 nm particle size, and 60 min dynamic time. Oleoresin extract at the optimized condition had a yield of 1.012% oleoresin, 31 g sulfur content per kg of oleoresin, and 10.41 μmol pyruvate per g fresh weight of onion.

The outer dry layer of the onion '*Ramat di Montoro*' is utilized to produce functional ingredients such as flavonoids, which can be used in the formulation of nutraceuticals using the supercritical fluid extraction technique. Campone et al. (2018) optimized the pressure (100 bar) temperature (40 °C) and composition of solvents (85% ethanol) using a response surface Box Behnken design. A total of 19 phenolic compounds were obtained from outer dry skin, which includes protocatechuic acid (140.6 mg/100 mg dry matter), 3(2H)-benzofuranone, quercetin-7,4-diglycoside, and quercetin derivatives. Apart from the extraction of useful products from onion waste using supercritical fluid extraction, the supercritical fluid extraction technique is used for the identification of pesticides in green onions. The 25% (v/v) methanol/supercritical carbon dioxide mixture at 80 °C and 25 MPa were used to identify the neonicatinoid pesticides (acetamiprid (2 mg/kg), dinotefuran (6 mg/kg), imidacloprid (3 mg/kg), thiamethaxan (2 mg/kg)) (Nakamura et al. 2020).

13.3.3 Biological Extraction

Brown onion skins are a suitable waste as carriers for the immobilization of enzymes. Immobilized enzymes are entrapped or localized on a solid carrier to retain their catalytic activity so that they can be used repeatedly and continuously. Additionally, immobilization improves their catalytic and operation properties. The affinity between enzymes and solid carriers depends on the molecular forces that divide the immobilization of lipases. Normally, commercially available carriers are used for enzyme mobilization, which is expensive, so there is a need to find a cheaper carrier for immobilization, such as onion waste. Therefore, brown onion skin wastes could result in cellulose and hemicellulose-based carriers after treatment. Brown onion skin contains about 60% insoluble fiber in dry matter, with the major polysaccharides (insoluble fiber-cellulose), 2–5% protein, 1% lipid, and 5.3% polyphenols (Benítez et al. 2011; Choi et al. 2015). A multistep extraction system was required for enzyme immobilization, non-polar (hexane) and polar solvent and finally, alkali wash was utilized to extract the cellulose and hemicellulose rich substrate from brown onion skin. Therefore, commercially available forms of pure cellulose and its modified derivatives, water-insoluble lignocellulosic material from brown onion skin wastes can be used as a cost-effective carrier for lipase immobilization. Immobilization of lipase was achieved by adsorption to pretreated and processed waste or by covalent binding to the chemically modified carrier (brown onion waste-chemically modified). The most commonly used lipases in the immobilized form are *Thermocyces longinosus* (TLL) and *Candida antartica* lipase B (CALB) (Budžaki et al. 2022).

13.3.4 Assisted Extraction Techniques

Freeze-dried onion skin cultivar (dark red, red, pink, and white) was utilized to obtain the flavonoids extract (4570.82–14,318.05 mg/kg DW), which acts as an antimicrobial agent using sonication-assisted extraction (1:25 w/v (sample: methanol), 50% amplitude and 20 kHz frequency, 10 min). The obtained flavonoid extract from onion skin showed antimicrobial activity against six pathogenic bacteria. The dark red, red, and pink cultivars of onion skin showed antimicrobial activity against *Bacillus cereus* (10 mm of inhibition zone), *Streptococcus agalactiae* (3 mm of inhibition zone), *Staphylococcus aureus* (3 mm of inhibition zone), *Klebsiella pneumoniae* (3 mm of inhibition zone), *Salmonella typhimurium* (3 mm of inhibition zone) whereas white cultivar of onion skin showed antimicrobial activity against *B. cereus*, *P. aeruginosa* and *S. typhimurium* (Sagar and Pareek 2020a).

Onion solid waste was utilized to obtain polyphenol and pigment enriched extracts using ultrasound assisted solid-liquid extraction (Katsampa et al. 2015). The total polyphenol (83.09 mg GAE/g and total pigments (2.09 mg CyE/g) was obtained at 80 °C under sonication (140 W, 37 kHz, 35 W/L) (Katsampa et al. 2015). Similarly, Jang et al. (2013) used ultrasound assisted extraction to obtain quercetin from onion solid wastes. The maximum concentration of quercetin (11.08 mg/g dry weight) of onion solid waste was obtained using 59% of ethanol and at a temperature of 49 °C. Whereas (Pal and Jadeja 2019) used an MAE technique to produce bioactive phenolic compounds from onion peel waste. The optimum conditions were determined using Box-Behnken design (100 W microwave power, 15.03 min irradiation time, and 54.97 mL/g liquid to solid ratio to obtain total phenolic content of 80.45 mg GAE/g DW and ferric reducing antioxidant power (636.18 μmol/AAE DW).

13.3.5 Hydrolysis and Fermentation

Onion waste contains sugars and proteins, which can be hydrolyzed and fermented using suitable enzymes/microbes for the production of industrially important products. This section describes some studies utilizing hydrolysis and fermentation techniques on onion waste.

13.3.5.1 Hydrolysis

Onion juice residue (OJR) was utilized for the production of D-tagatose and bioethanol (Kim et al. 2017). The enzymatic hydrolysis of OJR was carried out by using 10% (w/v) of dry onion, cellulase, pectinase, 0.1% yeast extract, 0.2% (w/v) peptone, and 0.05 M citrate buffer (pH 4.8) at 45 °C for 48 h. The further fermentation process of hydrolysate with 0.5 g dry yeast (*Saccharomyces cerevisiae*, KCTC 7906) was performed at 30 °C for 48 h, which yielded 90.0% of fermentable sugar and 74.6% of ethanol (1.56 g of bioethanol were produced from 10 g of OJR). Onion skin was utilized for the isolation of cellulose nanocrystals using acid hydrolysis (45%, 55%, and 65% H_2SO_4) (Rhim et al. 2015), further nanocrystals were evaluated through the preparation of agar/cellulose nanocrystals nanocomposite films. Out of three acid hydrolysis concentrations, it was suggested that 45% of H_2SO_4 at 60 °C for 3 h was the optimum condition for the extraction of cellulose nanocrystals from onion skin cellulose fiber.

13.3.5.2 Fermentation

Lignocellulosic biomass biorefinery is another solution for the utilization of onion waste from the agri-food industry (Culaba et al. 2022). The utilization of lignocellulosic biomass in a biorefinery involves different processes such as feedstock preparation, pretreatment, saccharification, and fermentation. Ganguly et al. (2021) explored the utilization of onion peel waste for bioethanol production (0.953 g/g) and wastewater treatment using physical, chemical, and biological treatment techniques. Also, the cellulose and lignin obtained from onion peel waste were utilized for the removal of crystal violet dye (96% removal). Bioactive compounds present in onion peel can be extracted for the preparation of therapeutic and nutraceuticals, which prevent various illnesses such as obesity and cardiovascular diseases.

Poe et al. (2020) analyzed the acetone (3.8 g/100 g), butanol (5.6 g/100 g) and ethanol (0.5 g/100 g) and total acetone, butanol, ethanol ratio (9.9 g/100 g) from red onion skin wastes using fermentation method, waste was inoculated with Clostridium culture at 35 °C for 72 h. The acetone–butanol–ethanol fermentation process has the potential to recover energy and value-added chemicals. Similarly, Kim et al. (2017) reported rare sugar, i.e., D-tagatose (0.99 g) from onion juice residue and 1.5 g of bioethanol through a bioconversion and fermentation process, respectively.

13.3.6 Composting/Vermicomposting

Vermicomposting is an efficient technology to treat organic waste such as OWs, which was characterized to assess its potential use in agriculture (Ray 2022). OWs from the food industry were mixed with the low dump and vermicomposted for 154 days, with a moisture of 85% and a temperature of 33 °C. The final vermicompost was a mature material (C/N = 13) with considerable carbon content (15.7%) (Pellejero et al. 2020). Similarly, Morales-Corts et al. (2014) prepared green and pruning waste compost, which was prepared from nine growth substrates: rosemary, ley land cypress, lettuce, onion, petunia, and pansy. Table 13.6 shows the different strategies for onion waste valorization.

13.4 Application of OWs-Based Bioproducts

Recent innovations in the food business have brought attention to the search for onion waste that can be utilized as a source of functional ingredients and environmentally friendly cuisine. Onion skin, a byproduct of onions, is a valuable resource for the development of byproducts with added value (Schieber et al. 2001). Onion skin is abundant in beneficial bioactive components, including flavonoids (Fig. 13.2). The flavonoids, such as polyphenols, are recognized for their antioxidant properties (Martino and Guyer 2004). Quercetin is a flavonoid with anti-inflammatory (Kleemann et al. 2011), anticancer (Dajas 2012), antiviral (Kleemann et al. 2011), and antioxidant (Kleemann et al. 2011; Dajas 2012) properties (Boots et al. 2008). Onion peel contains more than 77 times more quercetin than the edible portion of onions. The findings about the reuse of items obtained from onion trash will assist in the development of waste reduction methods that are environmentally friendly, cost-effective, and efficient (Kim et al. 2019; Downes et al. 2010).

Consequently, discovering a greener and more effective catalyst for synthesizing bisenols is of great interest. OW management constitutes a widespread environmental problem. OWs can be used as food-grade components, which may be added to meals to improve them with bioactive compounds. For this aim, onion waste should be treated and stabilized appropriately. In conclusion, industrial processing significantly affects the bioactive content, producing products with diverse functional applications (Elsebaie and Essa 2018). In addition, pasteurization proved to be the optimal method for producing safe products rich in dietary fiber and fructans, whereas sterilization may be utilized to make alkenyl cysteine sulphoxide-rich products. The biorefining of agricultural waste has the prospect of enhancing

Table 13.6 Strategies for onion waste valorization

Value added products	Strategies	References
Immobilization of enzymes	Solvent extraction	Budžaki et al. (2022)
Lignocellulosic biomass biorefinery	Saccharification, fermentation	Culaba et al. (2022)
Bioethanol production	Chemical, biological	Ganguly et al. (2021)
Flavonoids	Solvent extraction Microwave-assisted extraction Ultrasound-assisted extraction	Lee et al. (2014) Gokulakrishnan et al. (2022)
Ethanol, biogas	Thermochemical conversion pathways	Vigneshwar et al. (2022)
Sugar	KOH extraction	Blue et al. (2023)
Biopolymer	Chemical	Bolaji et al. (2021)
Acetone, butanol	Fermentation	Poe et al. (2020)
D-Tagatose	Fermentation	Kim et al. (2017)

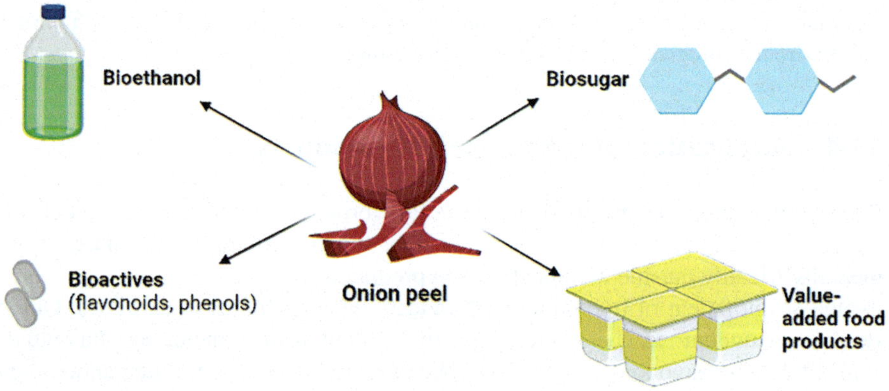

Fig. 13.2 Value-added products from onion peels

environmental sustainability on a global scale. The viability of producing acetic acid from carbohydrate feedstock (onion waste, as an environmentally preferable substitute to feedstock derived from natural gas. In addition, OWs can be employed as an alternative to manufactured food supplements in response to the growing demand for food enhancements and health-related substances. Recently, natural compound research has grown in prominence (Khiari and Makris 2012). OW residues contain significant phytochemicals that can be utilized in the pharmaceutical, culinary, and cosmetics sectors. Numerous researchers exert tremendous effort to transform agricultural waste into a high-value resource (shown in Fig. 13.2) as its exponential growth becomes an ever-increasing challenge.

13.4.1 Onion Skin Peel Paste and Powder

The European Union produces thousands of tonnes of onion peel waste annually. The flavonoids from the peel and its aqueous extracts could be employed as an antioxidant in meat-based products. Many researchers studied the antioxidant activities of onion skin water extracts on cooked pig patties. The following were evaluated: antioxidant activity, total polyphenols, thiobarbituric acid reactive compounds, and sensory evaluation. Antioxidant activity and total polyphenol concentration were considerably greater in burger patties containing additional antioxidants. After 5 days of storage at 5 °C, samples with onion extracts had considerably lower thiobarbituric acid reactive compounds levels than controls. Using liquid chromatography–mass spectrometry with electrospray ionization in negative mode, two major phenolic compounds were discovered in onion skin extracts: quercetin and quercetin monoglucoside. Bedrníček et al. (2019) showed that onion skin water extracts have the potential to be utilized as a source of naturally occurring antioxidants in meat products that exhibit high levels of antioxidant activity.

Onion skin contains the highest concentration of flavonoids, with quercetin and its glycosides being the most abundant antioxidants and free radical scavengers (Albishi et al. 2013). Previous studies have shown that ethanol extracts from onion skins can reduce the oxidation of lipids in meat and stop the growth of microbes that cause spoiling (Alahakoon et al. 2013; Bedrníček et al. 2020).

Several extraction methods use organic solvents, in addition to some newer "green extraction techniques" such as pressurized liquid extraction, supercritical CO_2 extraction, and microwave-assisted extraction (Campone et al. 2018; Bedrníček et al. 2020). To develop a barrier technology to enhance the extraction efficiency of quercetin from onion waste skin by integrating environmentally friendly subcritical water extraction with physical pre-treatment using intense pulsed light. This was done to maximize the recovery of quercetin from the onion waste skin. The highest concentration of quercetin, which was 17.32 mg/g onion skin, was produced by intense pulsed light at 1200 V for 60 s after the onion had been heated to 145 °C for 15 min. The concentration of quercetin was only 15.19 mg/g onion skin when subcritical water extraction was used alone after the onion had been heated to 145 °C for 15 min.

Scanning electron and fluorescence microscopy found that intense pulsed light was responsible for loosening the interstitial onion walls. Intense pulsed light as a pretreatment holds great potential for enhancing quercetin extraction through sub-critical water extraction (Kim et al. 2019).

Segundo et al. (2022) examined the colossal environmental concerns of organic waste due to the huge global population growth. Using waste as a fuel source for energy generation is made possible by microbiological fuel cells. OWs were employed to generate maximum current and voltage peaks of 4.45 mA and 0.99 V, respectively. FTIR diffractogram spectra of onion fragments reveal a decrease in the strong peaks between the first and final diffractogram spectra. After the monitoring, the molecular approach identified *Pseudomonas aeruginosa*,

Acinetobacter bereziniae, *Stenotrophomonas maltophilia*, and *Yarrowia lipolytica* adhering to the anode electrode.

As an efficient catalytic system, the production of bisenol in the water extract of burned ash of onion peel waste was studied by Chia et al. (2018). The outstanding yield of bisenols (62–94%) was achieved by employing different benzaldehyde and 4-hydroxycoumarin catalyzed by the ash-water extract. In addition, all pure yields were achieved through precipitation (Chia et al. 2018).

Many researchers conducted the study to investigate processed onion waste by evaluating the impacts of processing and heat stabilization on bioactive components in onion waste. The byproducts, such as paste, bagasse, and juice, were obtained during processing. The process of pasteurization and sterilization increased the ratio of soluble fibers to insoluble fibers in paste and bagasse. However, the total dietary fiber concentration did not change in the products that were pasteurized, but it did slightly decrease in the products that were sterilized. The loss of fructans that occurred as a consequence of subjecting juice to heat treatments was significantly higher after sterilization (59%) than after pasteurization (39%). However, sterilizing yielded superior alkenyl cysteine sulphoxide content than pasteurization for byproducts. In conclusion, industrial processing significantly affects the bioactive content, producing products with diverse functional applications. In addition, pasteurization proved to be the optimal method for producing safe products rich in dietary fiber and fructans, whereas sterilization may be utilized to make alkenyl cysteine sulphoxide content-rich products (Benítez et al. 2011, 2013).

Extracts made from onion peel waste were found to impact the quality of meat patties positively. At a dosage of 10%, they showed excellent protection against fat oxidation and significantly increased meat patties' antioxidant activity and total polyphenol content. It is primarily due to the presence of quercetin and quercetin monoglucoside that extracts prepared from onion skin waste have such a high level of antioxidant activity. According to the findings, the water extracts from onion peel, which are often regarded as a waste product, have the potential to be an efficient source of antioxidants to enhance cooked pork patties. Onion peels are considered a waste product. There are just a few papers that discuss the possibility of using onion skin as a source of antioxidants for meat products; nevertheless, ethanol or subcritical water was utilized as an extraction solvent in these investigations (Bedrníček et al. 2020). The easy extraction process may have applications in the food business; however, further investigation is required.

The viability of producing acetic acid from carbohydrate feedstock (onion waste) as an environmentally preferable alternative to feedstock derived from natural gas. In particular, onion waste is an excellent feedstock for biorefining due to its high carbohydrate content (69.7%). The simultaneous saccharification and two-step fermentation (SSTF) method produces more acetic acid from onion waste in 5 days than the simultaneous saccharification and co-fermentation (SSCF) method. The SSTF produced 19.3 g/L of acetic acid and had the maximum conversion yield (90.50%) from onion waste (6% substrate loading, w/v). These results indicated that acetic acid might be generated efficiently and sustainably from onion waste using the SSTF method (Kim et al. 2019).

Several investigations demonstrated that valorized onion byproducts are significant natural antioxidant sources. Onion wastes also contain dietary fiber, FOS, flavonoids, and *S*-alk(en)yl-L-cysteine sulfoxide. All of these substances have been shown to have significant health advantages. OWs have been found to contain various chemicals, including total phenolics, flavonoids, flavonols, and antioxidants. The overall amount of phenolic content, flavonoids, flavonols, and the antioxidant activity of whole onions and the byproducts they produce. The onion contains two important flavonoids, flavonols and anthocyanins, both of which are subgroups of the flavonoids (Roldán-Marín et al. 2009). In producing yellow and brown chemicals that can take on various forms, flavonols, which include quercetin and its derivatives, play a vital role and a substantial portion. Anthocyanins are responsible for giving other species a reddish-violet color. Reports indicate that the principal flavonols found in fresh onion are quercetin 4′-glucoside and quercetin 3,4′--diglucoside, whereas onion peels include quercetin 3,4′-diglucoside. The amount of quercetin and the glucosides it contains can vary in fresh onions depending on the circumstances of storage and the length of time they are kept (Mogren et al. 2007). It has been found that the ratio of quercetin to its glucosides significantly increases after the product has been stored for a period of time (Subramoniam 2016). The biological processes that lead to the deterioration of onions and the effects those processes have on quercetin in the time after storage. It is possible that the onion-growing sector and the food processing industry could benefit from the extraction of quercetin from spoiled onions that have been stored for a long time. Therefore, it is possible to find a solution to the disposal problem. One of the important compounds that can be found in onions is called quercetin. It can protect against human cardiovascular diseases and possesses antioxidant and free radical-scavenging characteristics (Bonaccorsi et al. 2008). In addition to this, there is evidence that quercetin possesses antiviral, anti-inflammatory, and anticancer activities (Bonaccorsi et al. 2008). It can be absorbed quickly but then eliminated from the body gradually. The amounts of flavonoid found in the edible portions of Allium plants range from 0.03 to >1 g/kg. However, the flavonoid content of onion skins ranges from 2 to 10 g/kg, which is a significant increase from the levels found in the edible component of the plant (Kim et al. 2005).

Due to the processing-induced loss of bioactive components, onion juice has fewer cell walls than other onion byproducts, such as paste and bagasse. This is in contrast to the skin and top–bottom portions of the onion, which contain the highest concentration of bioactive compounds. It would be fascinating if onion processors segregated the different sections of the onion that they got during the processing, as different parts of the onion contain distinct value-added components. Because of this, we will be able to use all onion byproducts effectively. To acquire various products from onions, such as frozen, chopped, freeze-dried, or ready-to-eat meals, it is necessary to separate the top and bottom layers of the onion bulbs, as well as the outer brown skin and two fleshy scales. Waste onion taken from both the top and bottom is a combination of the inner, outer, and skin layers of the onion, and as a result, it contains a higher overall concentration of essential fiber (Bello et al. 2013). The fiber in the diet is critical to maintaining human health. It prevents the over absorption of water, which can contribute to constipation, as well as the creation of firm stools, which can also cause constipation. Furthermore, fiber lowers the amount of cholesterol found in the body, which in turn lowers the risk of developing cardiovascular disease.

The top–bottom layer and the brown skin both have the highest concentrations of dietary fiber. As a result of the fact that top–bottom waste was made up of inner, outer, and skin sections, higher levels of total dietary fiber might be attributed to the top-high contribution of skin. Nonstructural carbohydrates (such as glucose, fructose, sucrose, and FOS) can make up as much as or even more than 65% of the product's dry weight (DW). Ketose, nystose, and fructofuranosylnystose are the three primary sugars that are found in onion bulbs (Jaime et al. 2002). It has been established that the two outer fleshy leaves are the best sources for the recovery of fructans and FOS. These leaves are located on the outermost layer of the plant. The inner scales of the onion have the largest quantities of nonstructural carbs, total fructans, and FOS. The nonstructural carbohydrate profile of the inner scales is comparable to that of the onion. Between 61% and 87% of the total fructans in onion wastes are contributed by FOS. In addition to this, it was shown that increasing the fructan concentration led to an increased amount of polymerization. This has provided an overview of extracting nutritional fiber from onion waste products. Composition of the onion oil and its industrial byproducts in terms of minerals and fatty acids. Most onion waste is made up of the plant's tops and roots, which are the areas where nutrients are absorbed, leading to a high mineral concentration. The mineral richness of the soil in which the onions were produced directly influenced the distribution of minerals found in onion waste.

This investigation assessed the ability of onion juice residue to produce valuable bio sugar or bioethanol. To demonstrate another application for onion juice residue, it was bio-processed and fermented using 10 g of onion juice residue to produce 1.56 g of bioethanol. These results demonstrate that this can be employed in an industrial context to produce sugars and that onion juice residue can be converted to D-tagatose and bioethanol. The generation of important rare sugars and bioethanol from onion juice residue was efficiently used as the waste product (Choi et al. 2015).

Santos and Martins (2022) investigated the effects of bio-based solvents (ethanol, water, and acetic acid) on the extraction of phenolic (Total Phenolic Content, TPC), anthocyanins (Total Anthocyanin Content, TAC), and flavonoids (Total Flavonoid Content, TFC) from purple onion peel. Additionally, they investigated the effects of storage temperature on the kinetics of anthocyanin degradation and the color stability of the obtained phenolic-rich extract. In the extract of purple onion prepared with 60% ethanol (E60E), the amounts of TPC, TAC, and TFC were roughly 3.45, 2.50, and 4.45 times greater than when using the other solvents. Compared to other solvents, E60E showed much higher antioxidant and antibacterial activity levels. The half-life of E60E was calculated to be 285 days at a temperature of 25 °C, 48.81 days at 38 °C, and 21.80 days at 4 °C. An endothermic and non-spontaneous reaction with an activation energy of 54.65 kJ mol was responsible for the degradation of anthocyanins in E60E. This activation energy is strongly correlated with the phenolic molecule's temperature sensitivity.

The production of onion bulbs results in the creation of onion skin as a byproduct. Recent studies have shown a significant concentration of bioavailable and bioaccessible compounds. As a result, it has the potential to be utilized in the production of unique culinary products. Powdered onion skin (OS) was used in this experiment in place of semolina in pasta at three different weights per 100 g: 2.5, 5, and 7.5 g. Compared to a control sample, the effects of enriched pasta on the chemical composition, antioxidant potential, technological features, and sensory qualities of the pasta were investigated and analyzed. The addition of onion skin led to a statistically significant improvement in nutritional qualities, as shown by an increase in total phenolic compounds, flavonoid content, dietary fiber, and antioxidant activity. Fortification with onion skin led to an improvement in nutritional features [FRAP (Ferric Reducing Antioxidant Power)] and DPPH (2,2-diphenyl-1-picrylhydrazyl)]. The cooking loss increased with higher levels of onion skin incorporation; nonetheless, all pasta samples stayed below the permitted threshold of 8 g of cooking loss per 100 g of pasta. Compared to the sample that served as the control, adding onion skin reduced the ideal cooking time, the water solubility index, and increased redness. According to sensory analysis, the pasta in which this plant component made up 2.5% of the flour had the greatest "overall quality" value. According to research findings, the food sector might use powder made from onion skins as an alternate method to produce nutritionally enhanced pasta (Michalak-Majewska et al. 2020).

Utilizing onion skin as a food ingredient would be helpful, as onion skin is a valuable byproduct with excellent bioactive qualities. Thus, samples of chocolate truffles containing 5% and 10% onion skin powder (OSP) and peanut butter containing 3% and 6% OSP were made. It explored how the OSP affected sensory qualities. As OSP content grew, chocolate truffles' flavor and overall acceptance declined. The addition of OSP diminished the internal hue and greasiness of chocolate samples. There was no difference in bitterness between the control and the 5% OSP containing chocolate truffles. The flavor and sweetness of the chocolate truffles were unaffected by the addition of OSP. The odor, color, greasiness, and consistency of peanut butter did not change when OSP was added. Up to 3%, OSP did not influence the peanut butter's sensory qualities; however, 6% OSP decreased the peanut butter's flavor, sweetness, and overall acceptability compared to the control (Kirkin et al. 2021).

13.5 Conclusions and Perspectives

Onion waste such as peeled skins, outer fleshy scales, and roots are produced in very large quantities worldwide. There is a need to utilize these onion byproducts produced in households and the industrial sector by employing suitable processing technologies and developing their beneficial application. This chapter summarized different bioproduct utilization interventions and various compounds that can be extracted from onion waste for a sustainable approach. Various aspects have been considered while selecting suitable technologies for processing onion byproducts, such as the type of material and qualitative and quantitative characteristics of the onion waste. Elaborated studies were done on onion skin and peel demonstrated that it can be used to develop functional foods as the onion skin is rich in antioxidants and a variety of bioactive compounds. Research has been conducted in this context by testing different technologies and developing sustainable alternate technologies. Further research and scaleup studies on the valorization of onion waste as a whole are required to increase their relevance for industry applications. It will provide additional benefits like circular bioeconomy.

References

Adelodun B, Kim SH, Choi KS (2021) Assessment of food waste generation and composition among Korean households using novel sampling and statistical approaches. Waste Manag 122: 71–80

Alahakoon AU, Bae YS, Kim HJ, Jung S, Jayasena DD, Yong HI et al (2013) The effect of citrus and onion peel extracts, calcium lactate, and phosvitin on microbial quality of seasoned chicken breast meat. Korean J Agric Sci 40(2):131–137

Albishi T, John JA, Al-Khalifa AS, Shahidi F (2013) Antioxidative phenolic constituents of skins of onion varieties and their activities. J Funct Foods 5(3):1191–1203

Anbukkarasi V, Paramaguru P, Pugalendhi L, Ragupathi N, Jeyakumar P (2013) Studies on pre and postharvest treatments for extending the shelf life in onion: a review. Agric Res Commun Centre 34(4):256–268

AOAC (2003) Official methods of analysis of the association of official's analytical chemists, 17th edn. Association of Official Analytical Chemists, Arington, VA

Bedrníček J, Jirotková D, Kadlec J, Laknerová I, Vrchotová N, Tříska J et al (2020) Thermal stability and bioavailability of bioactive compounds after baking of bread enriched with different onion by-products. Food Chem 319:126562

Bedrníček J, Kadlec J, Laknerová I, Mráz J, Samková E, Petrášková E, Hasoňová L, Vácha F, Kron V, Smetana P (2020) Onion peel powder as an antioxidant-rich material for sausages prepared from mechanically separated fish meat. Antioxidants 9(10): 974

Bello MO, Olabanji IO, Abdul-Hammed M, Okunade TD (2013) Characterization of domestic onion wastes and bulb (*Allium cepa* L.): fatty acids and metal contents. Int Food Res J 20(5): 2153

Benítez V, Mollá E, Martín-Cabrejas MA, Aguilera Y, López-Andréu FJ, Cools K et al (2011) Characterization of industrial onion wastes (*Allium cepa* L.): dietary fibre and bioactive compounds. Plant Foods Hum Nutr 66(1):48–57

Benítez V, Mollá E, Martín-Cabrejas MA, Aguilera Y, López-Andréu FJ, Esteban RM (2012) Onion (*Allium cepa* L.) by-products as source of dietary fiber: physicochemical properties and effect on serum lipid levels in high-fat fed rats. Eur Food Res Technol 234(4):617–625

Benítez V, Mollá E, Martín-Cabrejas MA, Aguilera Y, López-Andréu FJ, Terry LA, Esteban RM (2013) The impact of pasteurisation and sterilisation on bioactive compounds of onion by-products. Food Bioprocess Technol 6(8):1979–1989

Bisen PS, Emerald M (2016) Nutritional and therapeutic potential of garlic and onion (Allium sp.). Curr Nutr Food Sci 12(3):190–199

Blue D, Fortela DL, Holmes W, LeBoeuf S, Subramaniam R, Hernandez R, Zappi ME, Revellame ED (2023) Alkali pretreatment of industrial mixed vegetable waste for fermentable sugar production. Biomass Conv Bioref 13:5367–5377

Bolaji I, Nejad B, Billham M, Mehta N, Smyth B, Cunningham E (2021) Multi-criteria decision analysis of agri-food waste as a feedstock for biopolymer production. Resour Conserv Recycl 172:105671

Bonaccorsi P, Caristi C, Gargiulli C, Leuzzi U (2008) Flavonol glucosides in allium species: a comparative study by means of HPLC–DAD–ESI-MS–MS. Food Chem 107(4):1668–1673

Boots AW, Haenen GR, Bast A (2008) Health effects of quercetin: from antioxidant to nutraceutical. Eur J Pharmacol 585(2–3):325–337

Brancoli P, Makishi F, Lima PG, Rousta K (2022) Compositional analysis of street market food waste in Brazil. Sustainability 14(12):7014

Brewster JL, Rabinowitch HD (2020) Onions and allied crops, Biochemistry, food science, and minor crops, vol III. CRC Press, pp 230–251

Budžaki S, Velić N, Ostojčić M, Stjepanović M, Rajs BB, Šereš Z et al (2022) Waste management in the agri-food industry: the conversion of eggshells, spent coffee grounds, and brown onion skins into carriers for lipase immobilization. Foods 11(3):409

Burri SC, Ekholm A, Håkansson Å, Tornberg E, Rumpunen K (2017) Antioxidant capacity and major phenol compounds of horticultural plant materials not usually used. J Funct Foods 38: 119–127

Calica GB, Dulay MMN (2018) Assessment of the postharvest systems and losses of shallots in Ilocos, Philippines. Asian J Postharvest Mech 1(1):81

Campone L, Celano R, Piccinelli AL, Pagano I, Carabetta S, Di Sanzo R et al (2018) Response surface methodology to optimize supercritical carbon dioxide/co-solvent extraction of brown onion skin by-product as source of nutraceutical compounds. Food Chem 269:495–502

Chia PW, Lim BS, Yong FSJ, Poh SC, Kan SY (2018) An efficient synthesis of bisenols in water extract of waste onion peel ash. Environ Chem Lett 16(4):1493–1499

Choi IS, Cho EJ, Moon JH, Bae HJ (2015) Onion skin waste as a valorization resource for the by-products quercetin and biosugar. Food Chem 188:537–542

Constantin OE, Milea AS, Bolea C, Mihalcea L, Enachi E, Copolovici D, Copolovici L, Munteanu F, Bahrim G, Râpeanu G (2020) Onion (*Allium cepa* L.) peel extracts characterization by conventional and modern methods. Int J Food Eng 17(6):485–493

Cosa P, Vlietinck AJ, Berghe DV, Maes L (2006) Anti-infective potential of natural products: how to develop a stronger in vitro 'proof-of-concept'. J Ethnopharmacol 106(3):290–302

Culaba AB, Mayol AP, San Juan JLG, Vinoya CL, Concepcion RS II, Bandala AA et al (2022) Smart sustainable biorefineries for lignocellulosic biomass. Bioresour Technol 344:126215

Dajas F (2012) Life or death: neuroprotective and anticancer effects of quercetin. J Ethnopharmacol 143(2):383–396

Devani BM, Jani BL, Balani PC, Akbari SH (2020) Optimization of supercritical CO_2 extraction process for oleoresin from rotten onion waste. Food Bioprod Process 119:287–295

Downes K, Chope GA, Terry LA (2010) Postharvest application of ethylene and 1-methylcyclopropene either before or after curing affects onion (*Allium cepa* L.) bulb quality during long term cold storage. Postharvest Biol Technol 55(1):36–44

Elsebaie EM, Essa RY (2018) Microencapsulation of red onion peel polyphenols fractions by freeze drying technicality and its application in cake. J Food Proc Preserv 42(7):e13654

Fabricant DS, Farnsworth NR (2001) The value of plants used in traditional medicine for drug discovery. Environ Health Perspect 109(suppl 1):69–75

Ganguly P, Khan A, Das P, Bhowal A (2021) Cellulose from lignocellulose kitchen waste and its application for energy and environment: bioethanol production and dye removal. Indian Chem Eng 63(2):161–171

Gokulakrishnan SA, Arthanareeswaran G, Gnanasekaran G, László Z, Veréb G, Kertész S, Taweepreda W (2022) Advanced extraction and separation approaches for the recovery of dietary flavonoids from plant biomass: a review. Biomass Convers Biorefin:1–23

Griffiths G, Trueman L, Crowther T, Thomas B, Smith B (2002) Onions—a global benefit to health. Phytother Res 16(7):603–615

Higashio H, Hirokane H, Sato F, Tokuda S, Uragami A (2004) Effect of UV irradiation after the harvest on the content of flavonoid in vegetables. Acta Hortic 682:1007–1012

Ifesan BO (2017) Chemical composition of onion peel (*Allium cepa*) and its ability to serve as a preservative in cooked beef. Int J Sci Res Methodol 7(4):1–10

Jaime L, Mollá E, Fernández A, Martín-Cabrejas MA, López-Andréu FJ, Esteban RM (2002) Structural carbohydrate differences and potential source of dietary fiber of onion (Allium cepa L.) tissues. J Agric Food Chem 50(1):122–128

Jang M, Asnin L, Nile SH, Keum YS, Kim HY, Park SW (2013) Ultrasound-assisted extraction of quercetin from onion solid wastes. Int J Food Sci Technol 48(2):246–252

Jin EY, Lim S, Kim SO, Park Y-S, Jang JK, Chung M-S, Park H, Shim K-S, Choi YJ (2011) Optimization of various extraction methods for quercetin from onion skin using response surface methodology. Food Sci Biotechnol 20:1727–1733

Katsampa P, Valsamedou E, Grigorakis S, Makris DP (2015) A green ultrasound-assisted extraction process for the recovery of antioxidant polyphenols and pigments from onion solid wastes using Box–Behnken experimental design and kinetics. Ind Crop Prod 77:535–543

Khandagale K, Gawande S (2019) Genetics of bulb colour variation and flavonoids in onion. J Hortic Sci Biotechnol 94(4):522–532

Khiari Z, Makris DPJ (2012) Food Sci Technol 49(4):489–494

Kim MJ, Kim CY, Park I (2005) Prevention of enzymatic browning of pear by onion extract. Food Chem 89(2):181–184

Kim SW, Ko MJ, Chung MS (2019) Extraction of the flavonol quercetin from onion waste by combined treatment with intense pulsed light and subcritical water extraction. J Clean Prod 231: 1192–1199

Kim HM, Song Y, Wi S G, Bae HJ (2017) Production of D-tagatose and bioethanol from onion waste by an intergrating bioprocess. J Biotechnol 260: 84–90

Kirkin C, Çinar M, Derindere B, Erikman K, Onural N (2021) Sensory properties of chocolate truffles and peanut butter as affected by onion skin powder addition. J Tourism Gastron Stud 9(3):1547–1553

Kitinoja L, Kader AA (2015) Measuring postharvest losses of fresh fruits and vegetables in developing countries. PEF White Paper 15:26

Kleemann R, Verschuren L, Morrison M, Zadelaar S, van Erk MJ, Wielinga PY, Kooistra T (2011) Anti-inflammatory, anti-proliferative and anti-atherosclerotic effects of quercetin in human in vitro and in vivo models. Atherosclerosis 218(1):44–52

Kumar M, Barbhai MD, Hasan M, Punia S, Dhumal S, Rais N et al (2022) Onion (*Allium cepa* L.) peels: a review on bioactive compounds and biomedical activities. Biomed Pharmacother 146: 112498

Lachman J, Pronek D, Hejtmánková A, Dudjak J, Pivec V, Faitová K (2003) Total polyphenol and main flavonoid antioxidants in different onion (*Allium cepa* L.) varieties. Hortic Sci 30(4): 142–147

Lee KA, Kim KT, Nah SY, Chung MS, Cho SW, Paik H-D (2011) Antimicrobial and antioxidative effects of onion peel extracted by the subcritical water. Food Sci Biotechnol 20:543–548

Lee KA, Kim KT, Kim HJ, Chung M-S, Chang P-S, Park H, Paik H-D (2014) Antioxidant activities of onion (*Allium cepa* L.) peel extracts produced by ethanol, hot water, and subcritical water extraction. Food Sci Biotechnol 23(2):615–621

Liguori L, Califano R, Albanese D, Raimo F, Crescitelli A, Di Matteo M (2017) Chemical composition and antioxidant properties of five white onion (*Allium cepa* L.) landraces. J Food Qual 9:6873651

Martino KG, Guyer D (2004) Supercritical fluid extraction of quercetin from onion skins. J Food Process Eng 27(1):17–28

Michalak-Majewska M, Teterycz D, Muszyński S, Radzki W, Sykut-Domańska E (2020) Influence of onion skin powder on nutritional and quality attributes of wheat pasta. PLoS One 15(1): e0227942

Mogren LM, Olsson ME, Gertsson UE (2007) Quercetin content in stored onions (*Allium cepa* L.): effects of storage conditions, cultivar, lifting time and nitrogen fertiliser level. J Sci Food Agric 87(8):1595–1602

Morales-Corts MR, Gómez-Sánchez MÁ, Pérez-Sánchez R (2014) Evaluation of green/pruning wastes compost and vermicompost, slumgum compost and their mixes as growing media for horticultural production. Sci Hortic 172:155–160

Nakamura K, Otake T, Hanari N (2020) Evaluation of supercritical fluid extraction for the determination of neonicotinoid pesticides in green onion. J Environ Sci Health B 55(7):604–612

Oancea S, Radu M, Olosutean H (2020) Development of ultrasonic extracts with strong antioxidant properties from red onion wastes. Roman Biotechnol Lett 25(2):1320–1132

Oguezi V, Kelle H, Ngbede E, Ajiwe V (2019) Nutritional, antinutritional and food value of onion skin waste. J Chem Soc Nigeria 44(4)

Pal CBT, Jadeja GC (2019) Microwave-assisted deep eutectic solvent extraction of phenolic antioxidants from onion (*Allium cepa* L.) peel: A Box–Behnken design approach for optimization. J Food Sci Technol 56: 4211–4223

Pareek S, Sagar NA, Sharma S, Kumar V (2017) Onion (*Allium cepa* L.). In: Fruit and vegetable phytochemicals: chemistry and human health, vol 2. Wiley, pp 1145–1162

Pellejero G, Rodriguez K, Ashchkar G, Vela E, García-Delgado C, Jiménez-Ballesta R (2020) Onion waste recycling by vermicomposting: nutrients recovery and agronomical assessment. Int J Environ Sci Technol 17(6):3289–3296

Poe NE, Yu D, Jin Q, Ponder MA, Stewart AC, Ogejo JA et al (2020) Compositional variability of food wastes and its effects on acetone-butanol-ethanol fermentation. Waste Manag 107:150–158

Prakash D, Singh BN, Upadhyay G (2007) Antioxidant and free radical scavenging activities of phenols from onion (*Allium cepa*). Food Chem 102(4):1389–1393

Prokopov T, Nikolova M, Tasheva S, Taneva D (2022) Optimization of methanol extraction parameters for total flavonoids from onion processing waste. In: BIO web of conferences, vol 45. EDP Sciences

Ray RC (ed) (2022) Fruits and vegetable wastes: valorization into bioproducts and platform chemicals. Springer, Singapore, 448p

Rhim JW, Reddy JP, Luo X (2015) Isolation of cellulose nanocrystals from onion skin and their utilization for the preparation of agar-based bio-nanocomposites films. Cellulose 22(1):407–420

Roldán-Marín E, Sánchez-Moreno C, Lloría R, de Ancos B, Cano MP (2009) Onion high-pressure processing: flavonol content and antioxidant activity. LWT Food Sci Technol 42(4):835–841

Roldán-Marín E, Jensen RI, Krath BN, Kristensen M, Poulsen M, Roldán-Marín E, Jensen RI, Krath BN, Kristensen M, Poulsen M, Cano MP et al (2010) An onion byproduct affects plasma lipids in healthy rats. J Agric Food Chem 58(9):5308–5314

Sagar NA, Pareek S (2020a) Antimicrobial assessment of polyphenolic extracts from onion (*Allium cepa* L.) skin of fifteen cultivars by sonication-assisted extraction method. Heliyon 6(11): e05478

Sagar NA, Pareek S (2020b) Dough rheology, antioxidants, textural, physicochemical characteristics, and sensory quality of pizza base enriched with onion (Allium cepa L.) skin powder. Sci Rep 10(1):1–11

Sagar NA, Pareek S (2021) Fortification of multigrain flour with onion skin powder as a natural preservative: effect on quality and shelf life of the bread. Food Biosci 41:100992

Sagar NA, Pareek S, Gonzalez-Aguilar GA (2020) Quantification of flavonoids, total phenols and antioxidant properties of onion skin: a comparative study of fifteen Indian cultivars. J Food Sci Technol 57(7):2423–2432

Sagar NA, Khar A, Tarafdar A, Pareek S (2021) Physicochemical and thermal characteristics of onion skin from fifteen Indian cultivars for possible food applications. J Food Qual 2021:1

Sagar NA, Kumar Y, Singh R, Nikhil C, Kumar D, Sharma P et al (2022a) Onion waste based-biorefinery for sustainable generation of value-added products. Bioresour Technol 362:127870

Sagar NA, Pareek S, Benkeblia N, Xiao J (2022b) Onion (*Allium cepa* L.) bioactives: chemistry, pharmacotherapeutic functions, and industrial applications. Food Front:1–30. https://doi.org/10.1002/fft2.135

Santos LG, Martins VG (2022) Recovery of phenolic compounds from purple onion peel using bio-based solvents: thermal degradation kinetics and color stability of anthocyanins. J Food Proc Preserv e17161

Schieber A, Stintzing FC, Carle R (2001) By-products of plant food processing as a source of functional compounds—recent developments. Trends Food Sci Technol 12(11):401–413

Segundo RF, De La Cruz-Noriega M, Milly Otiniano N, Benites SM, Esparza M, Nazario-Naveda R (2022) Use of onion waste as fuel for the generation of bioelectricity. Molecules 27(3):625

Sharma K, Ko EY, Assefa AD, Ha S, Nile SH, Lee ET, Park SW (2015) Temperature-dependent studies on the total phenolics, flavonoids, antioxidant activities, and sugar content in six onion varieties. J Food Drug Anal 23(2):243–252

Sharma K, Mahato N, Nile SH, Lee ET, Lee YR (2016) Economical and environmentally-friendly approaches for the usage of onion (*Allium cepa* L.) waste. Food Funct 7(8):3354–3369

Sooch BS, Mann MK, Sharma P, Ray RC (2022) Phenolics and other bioactive compounds from vegetables and fruit wastes: extraction methods and their possible utilizations. In: Ray RC (ed) Fruits and vegetable wastes: valorization to bioproducts and platform chemicals. Springer, pp 81–114

Subramoniam A (2016) Anti-diabetes mellitus plants: active principles, mechanisms of action and sustainable utilization. CRC Press

Sun T, Powers JR, Tang J (2007) Evaluation of the antioxidant activity of asparagus, broccoli and their juices. Food Chem 105(1):101–106

Takahama U, Hirota S (2000) Deglucosidation of quercetin glucosides to the aglycone and formation of antifungal agents by peroxidase-dependent oxidation of quercetin on browning of onion scales. Plant Cell Physiol 41(9):1021–1029

Velíšek J (2002) Chemie potravin 3. OSSIS, Tábor. 368 p. ISBN: 80-902391-5-3

Viera VB, Piovesan N, Rodrigues JB, Mello RDO, Prestes RC, Santos RCVD, Vaucher RDA, Hautrive TP, Kubota EH (2017) Extraction of phenolic compounds and evaluation of the antioxidant and antimicrobial capacity of red onion skin (*Allium cepa* L.). Int Food Res J 24(3):990–999

Viera BV, Piovesan N, Mello RDO, Barin JS, Fogaça ADO, Bizzi CA et al (2021) Ultrasonic-assisted extraction of phenolic compounds with evaluation of red onion skin (*Allium cepa* L.) antioxidant capacity. J Culin Sci Technol:1–17. https://doi.org/10.1080/15428052.2021.1910095

Vojvodic Cebin A, Šeremet D, Mandura A, Martinić A, Komes D (2020) Onion solid waste as a potential source of functional food ingredients. Eng Power Bull Croatian Acad Eng 15(3):7–13

Vojvodic A, Komes D, Vovk I, Belšcak-Cvitanovic A, Bušic A (2016) Compositional evaluation of selected agro-industrial wastes as valuable sources for the recovery of complex carbohydrates. Food Res Int 89:565–573

Vigneshwar SS, Swetha A, Gopinath KP, Goutham R, Pal R, Arun J, SundarRajan P, Bhatnagar A, Chi NTL, Pugazhendhi A (2022) Bioprocessing of biowaste derived from food supply chain side-streams for extraction of value added bioproducts through biorefinery approach. Food Chem Toxicol 165:113184

Biovalorization of Garlic Waste to Produce High Value-Added Products

14

Fatma Kallel and Semia Ellouz Chaabouni

Abstract

Due to increasing environmental concerns and strict laws about solid biological wastes from industries, it becomes critical to create strategies to reuse these wastes by transforming them for usage in other processes/industries. Information in the literature about the use of garlic waste as a bioresource is scarce. This chapter summarizes the physicochemical and biological characterization of garlic by-products and provides the usage of garlic waste under specific topics: feed alimentation, soil amendment, bioenergy, adsorbent for pollutants, and value-added products.

Keywords

Garlic waste · Biovalorization · High value-added biomolecules · Bioenergy recovery · Feed alimentation · Soil amendment

Abbreviations

AFM Atomic force microscope
CNC Cellulose nanocrystals

F. Kallel (✉)
Laboratory for Plant Improvement and Valorization of Agroresources, National School of Engineering of Sfax (ENIS), Sfax, Tunisia

S. E. Chaabouni
Laboratory for Plant Improvement and Valorization of Agroresources, National School of Engineering of Sfax (ENIS), Sfax, Tunisia

Common Service Unit of Bioreactor Coupled with an Ultrafilter, National School of Engineering, Sfax University, Sfax, Tunisia

R. C. Ray (ed.), *Roots, Tubers, and Bulb Crop Wastes: Management by Biorefinery Approaches*, https://doi.org/10.1007/978-981-99-8266-0_14

DPPH	α, α-diphenyl-β-picrylhydrazyl
FTIR	Fourier transforms infrared spectroscopy
GC-MS	Gas chromatography-mass spectrometry
NMR	Nuclear magnetic resonance
SEM	Scanning electron microscope
TEM	Transmission electron microscope
XOS	Xylooligosaccharides
XRD	X-Ray diffraction

14.1 Introduction

Industries need to pay quick attention to serious issues related to waste treatment before disposal since these issues are reducing their profit margins. These concerns highlight how crucial it is for all businesses to recover, recycle, and upgrade processing waste. The food processing sector currently creates significant amounts of waste, residues, and by-products and has a massive annual revenue of billions of dollars (Machado et al. 2014; Ray 2022). These readily available resources must be correctly acknowledged as the raw materials for producing new goods with significant economic value or goods with additional benefits (Laufenberg et al. 2003).

The worldwide production of garlic by-products is about 3.7 million tons. Garlic waste material holds great strength and can be utilized effectively to develop as a good biological resource (Fig. 14.1) (Kallel and Ellouz Chaabouni 2017). The majority of garlic waste is currently disposed of in landfills or used as soil amendments causing negative environmental impacts and plant phytotoxicity (Lawson 1998). Therefore, they need to be disposed of in environmentally friendly and cost-effective ways.

This chapter focuses on all works related to the properties of garlic waste (stem, peel, straw, husk). This chapter provides up-to-date development on the application of garlic possessing waste in different domains such as feed alimentation, soil amendment, and production of biogas and bio-oil. This chapter also describes the use of these wastes as an adsorbent for pollutants and a source of high value-added products.

14.2 Abundance and Production of Garlic

Garlic has been consumed either as a substantial part of the diet or as a flavoring on every continent for over 5000 years. It is cultivated extensively all over the world. China is by far the world's largest producer and exporter of garlic. More recently, sales to specialty processors for medicinal or health food use have increased. It is grown extensively throughout the world as both a vegetable crop and as an herb or a medicinal plant. Garlic (*Allium sativum* L.) belongs to the family *Alliaceae* and is the second most widely cultivated *Allium* species next to onion (Ammarellou et al. 2022). Garlic yield grew gradually from 15.4 tons/ha in 2007 to 18.1 tons/ha in 2016. Global garlic consumption grew per year, amounting to 26.5 million tons in 2016, and will grow further to 31.1 million tons by 2030 (Rauf et al. 2022).

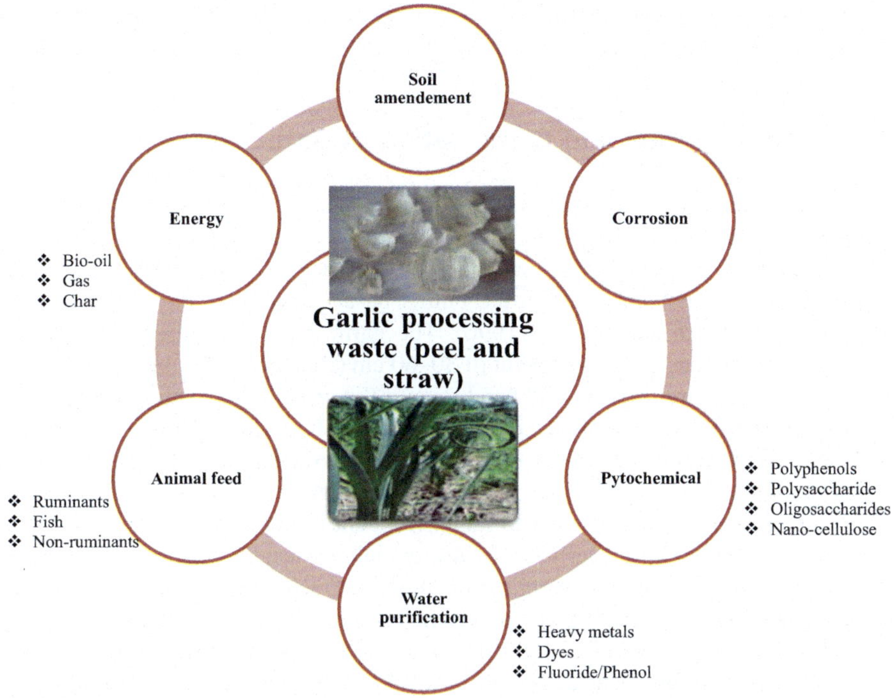

Fig. 14.1 The perspective of garlic processing wastes for the production of high-value-added products

China is the leading country in garlic production, providing about 74% of this crop (Etana 2018). As per data available from the Food and Agriculture Organization Corporate Statistical Database, garlic production is the highest in China (11093.5 kt), followed by India (500 kt), the Republic of Korea (445 kt), Russia (230 kt), and the USA (236.96 kt). Egypt's production rate is much lower (162.077 kt) even though it is one of the leading countries in garlic production (Gustafsson et al. 2013). Garlic has always been a favorite food in Egypt and throughout the Mediterranean region. Ibrahim et al. (2018) have proposed these areas as the secondary center of diversity for *A. sativum.*

14.3 Harvesting and Storage of Garlic

Hand harvesting of garlic is a common practice in many parts of the world (El Mashad et al. 2019). Fresh garlic can be stored for up to 3 months in a standard warehouse, 6 months in cold storage, or 1 year in a controlled atmosphere storage (Sinha et al. 2010). Although most garlic are stored in the form of bulbs, processed, pre-peeled, and chopped garlic products are also stored. Hong and Kim (2001) showed that storing chopped garlic under high hydrostatic pressure combined with citric and/or ascorbic acids retarded its browning which resulted from nonenzymatic autoxidation of phenolic compounds. These storage conditions did not inhibit the production of a green pigment in dormancy-terminated bulbs.

14.4 Morphology and Composition of Garlic Waste

Garlic has been used throughout recorded history for both culinary and medicinal purposes. Garlic is classified into two subspecies: hard neck (stiff neck) and soft neck. Hard neck garlics, also known as *ophio* or top-set garlic, are defined by woody center stalks that extend down to the bulb's basal plate. They develop scapes or flower stalks (Koch and Lawson 1996). The garlic bulb can only be reproduced vegetatively and can grow to a height of 30–60 cm (Bourlière et al. 1962). When the leaf tips begin to dry, darken, and bend toward the ground, harvesting begins.

The proximate compositions of a bulb of fresh garlic (g/100 g) are water (62–68 g) and carbohydrates (26–30 g), whereas proteins are only found in very small amounts (1.5–2.1 g). Given the high dry matter content of the bulbs, protein concentrations of 6.2–6.6 g are also rather frequent in different cultivars, while fat contents are around 0.5%. Garlic bulbs also contain noticeable fractions of vitamins, β-carotene, thiamin, riboflavin, niacin, pantothenic acid, pyridoxine, folacin, and ascorbic acid. Besides, fresh garlic also contains many bioactive constituents which include chlorogenic acid, diallyl disulfide, ferulic acid, kaempferol, linalool, oleanolic acid, coumaric acid, phytic acid, allyl cysteine, saponins, stigmasterol, etc. (Takagi 1990; Lawson 1998; Iciek et al. 2009; Brewster and Rabinowitch 2020). The distinctive aroma and bioactive qualities of dry bulbs of garlic are ascribed to their abundant supply of volatile chemicals (Table 14.1). There are ajoenes, as well

Table 14.1 Biological functions of the most representative bioactive constituents from garlic

Name	Biological potential	References
Alliin	Antioxidant Antimicrobial	Rahman (2007), Martins et al. (2016)
Allicin	Anticancer antimicrobial Anti-inflammatory Cardioprotective Immunomodulatory	Borlinghaus et al. (2014), Rahman (2007), Kopeć et al. (2013)
Ajoene	Anticancer Antimicrobial Antioxidant Cardioprotective	Capasso (2013), Harris et al. (2001), Rahman (2007), Yoshida et al. (1987)
Allyl sulfide	Anticancer Antimicrobial Antioxidant Antithrombotic	Khanum et al. (2004), Kopeć et al. (2013), Rahman (2007), Rose et al. (2005)
1,2-vinyldithium	Antimicrobial Antioxidant Antithrombotic	Higuchi et al. (2003)

as several sulfur-containing compounds such as alliin, allicin, 1,2-vinyldithiin, allixin, and *S*-allyl-cysteine. There are several studies demonstrating that these volatile compounds can reduce many diseases such as anticarcinogenic, antidiabetic, anti-inflammatory, antimicrobial, antioxidant, cardioprotective, and immunomodulatory activities (Table 14.1) (Yoshida et al. 1987; Harris et al. 2001; Higuchi et al. 2003; Rose et al. 2005; Capasso 2013; Borlinghaus et al. 2014; Martins et al. 2016). Important biological properties of most representative bioactive constituents from garlic are mentioned in Table 14.1.

14.5 Processing and Wastes Production

Garlic is often consumed without doing any pre-processing operation. Lately, it has found uses in its dried form, as a component in precooked dishes and rapid convenience foods such as sauces, gravies, and soups, increasing the demand for dried garlic. To meet this demand, garlic should be processed swiftly and efficiently to keep its quality. Bulb breaking, peeling, dehydration, grinding, and other unit activities are all part of the garlic processing process. Garlic peeling is a critical key unit action prior to any subsequent processing activity. During garlic peeling, the inedible thin membrane skin is removed from the segments (Manjunatha et al. 2014).

In fact, a rise in worldwide output from 1998 to 2003 led to increases in yield and crop area of 13% and 18%, respectively (Cardelle et al. 2010). The increase in production results in a massive waste generation [25–30% solid waste], consisting of peel and straw. Garlic bulbs generate about 760 g of cloves and 240 g of outer and inner peel per kg (Qureshi et al. 1983). Over 25% of the total industrial production of garlic is usually discarded and is characterized as an industry by-product. This enormous quantity of garlic waste has been burned without further use or used as a biosorbent in the chemical industry. Thus, garlic harvesting and processing wastes including the husk, stem, peels, straw, and unwanted bulbs could hold great strength to be used as a good biological resource and environment-friendly alternative for residues (Kallel and Ellouz Chaabouni 2017).

Proximate compositions of the garlic wastes are presented in Fig. 14.2. They might be a good source of carbohydrates and ash. Bampidis et al. (2005) and Kallel et al. (2014b) showed that garlic peel, composed of four to five layers separated into multiple (5–20) sublayers, had a low lipid content and a high protein content. These findings were consistent with previous research on various agro-industrial wastes (Madhava Naidu et al. 2011; Vriesmann et al. 2011). Garlic skins contain 7.8% protein, 9.9% ash, 27% pectin, 11.4% combined rhamnose, and 5.6% galactose (Abdel-Fattah and Edrees 1972; Alexander and Sulebele 1973; Chitsaz et al. 2018). Kallel et al. (2014a, b) presented the chemical composition of garlic husk. As can be seen from Fig. 14.2, the skin contains high concentrations of dietary fiber, calcium, and potassium. Garlic husk also contains noticeable fractions of lignin. Garlic straw, like all lignocellulose, is mainly built by the three major plant polymers cellulose, hemicellulose, and lignin (Kallel et al. 2015a, b). The chemical composition of garlic straw mainly consists of 41% cellulose, 18% hemicelluloses, and 6.3% lignin. The

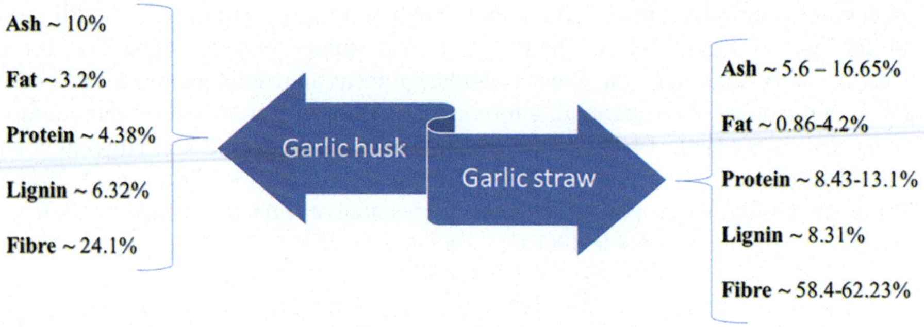

Fig. 14.2 Proximate composition of garlic by-products

protein and lipid contents of garlic straw were 4.38% and 2.66%, respectively. The higher content of extractives and ash in garlic straw, 1.2% and 10.08%, respectively, may be caused by the higher content of epidermal cells which contain a higher amount of silica. The ash content was higher than the ash content in garlic (Za and Ae 2016).

14.6 Structural Characterization of Garlic Waste

The proximate compositions, morphology, and other characteristics are discussed.

14.6.1 Fourier Transforms Infrared Spectroscopy (FTIR) Analysis

FTIR spectroscopy, which is a useful method for determining functional groups, has been used for the determination of the molecular conformation and the reactivity of the surface functional groups of garlic by-products (Reddy et al. 2009). The infrared spectra and the related functional groups of crude fiber of garlic straw are given in Fig. 14.3a. It shows several sorption peaks in the 4000–500 cm^{-1} a wave number range, suggesting the complex nature of the garlic waste (Kallel et al. 2014a). The FTIR analysis revealed wide bands between 3400 and 3200 cm^{-1}, which was due to the O–H stretching vibration of the hydrogen-bonded hydroxyl group of cellulose and lignin (Sun et al. 2005). The peak around 1630 cm^{-1} was associated with the C-O stretching which might be caused by the presence of water. It would be more probably attributed to the aromatic C=C stretch of aromatic rings of lignin (Mohebby 2008). Intense peaks at 1734 cm^{-1} and 1251 cm^{-1} were attributed to the acetyl uronic ester groups of the hemicelluloses, the ester linkage of the carboxylic group of the ferulic and *p*-coumaric acids of lignin or hemicelluloses (Sun and Sun 2002).

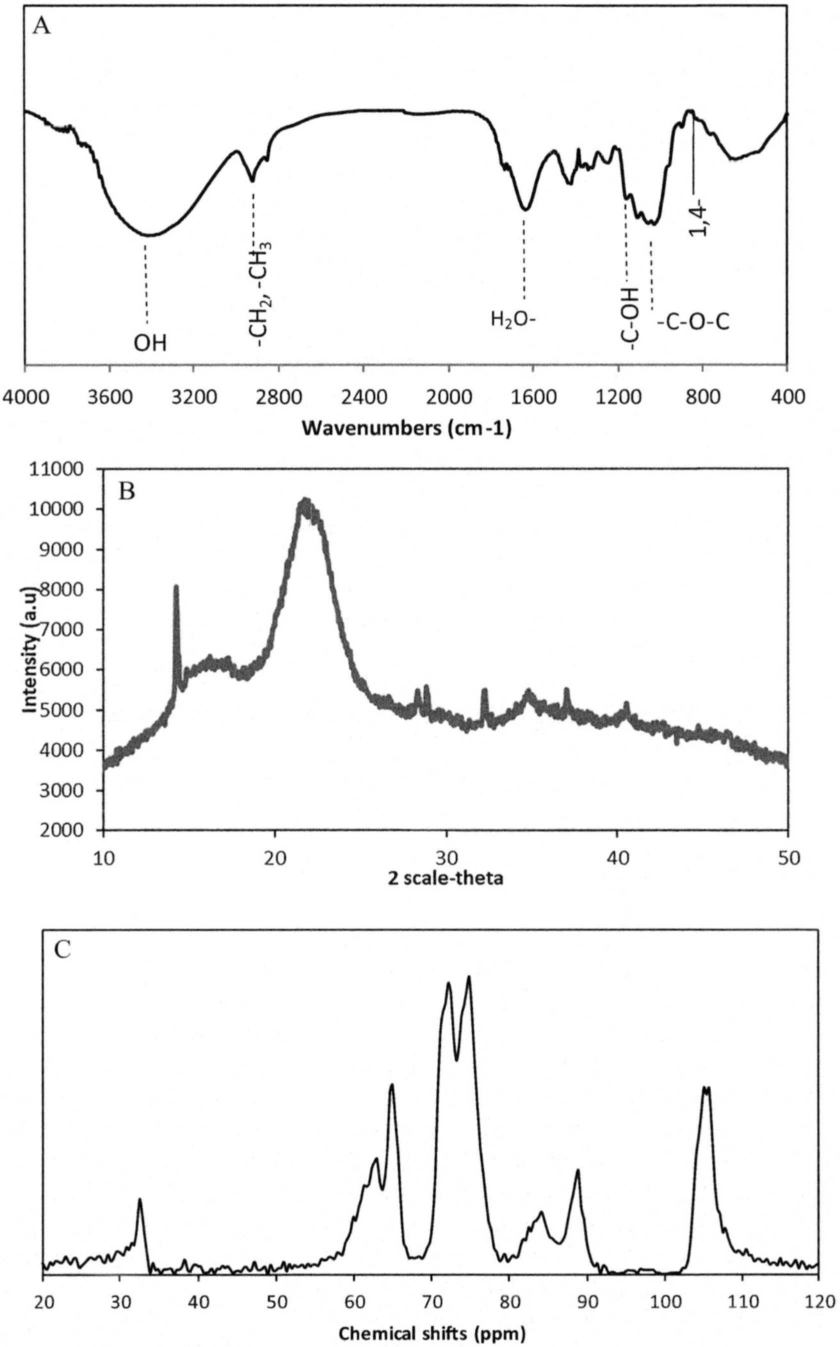

Fig. 14.3 (**a**) FTIR spectra, (**b**) X-ray diffraction, and (**c**) ^{13}C MAS NMR spectra characteristics of garlic straw (Kallel et al. 2016a)

14.6.2 X-Ray Diffraction (XRD) Analysis

Figure 14.3b shows the diffraction patterns obtained for the crude fiber of garlic straw. The diffractogram exhibited sharp peaks around 16° and 22.1° 2θ angles, attributed to the diffraction planes of (101) and (002), respectively, which commonly represent typical cellulose I crystal form (Jonoobi et al. 2015). The crystallinity index of garlic straw was obtained by the peak height method and its value was 37.4% (Kallel et al. 2016a). The crystallinity index of garlic straw was higher than the crystallinity index of garlic skin (Reddy and Rhim 2014).

14.6.3 Solid-State ^{13}C Nuclear Magnetic Resonance (NMR)

The ^{13}C NMR spectra of crude fiber from garlic straw are presented in Fig. 14.3c. The six carbon atoms that are assigned to the cellulose molecule dominate the spectrum. Starting on the up-field side of the spectrum, the two peaks at 63.4 and 65.4 ppm were assigned to C-6 of cellulose. The cluster of resonances from 72.5 to 74.8 ppm represents C-2, C-3, and C-5 carbons of cellulose. The next two peaks, at 82.3–83.7 and 88.8 ppm, were related to C-4 and finally, a peak at 104.7 ppm was attributed to C-1 of cellulose (Kallel et al. 2016a).

14.7 Valuables Applications of Garlic Processing Waste

The following section deals with the various applications of garlic by-products.

14.7.1 Animal Feed

Garlic waste may be a potential source of roughage for ruminant animals, fish, and non-ruminants. They contain a moderate amount of protein (9.3–13.0%) and a high concentration of non-dietary fiber (NDF) (48–59%) (Bidura 1999; Chu et al. 2003; Bampidis et al. 2005; Thanikachalam et al. 2010; Kamruzzaman et al. 2011, 2014; Lee et al. 2017; Liu et al. 2019; Liao et al. 2022).

Feeding garlic stem and leaf silage to sheep resulted in greater N and energy utilization without detrimental effects on ruminal fermentation. Garlic leaf has positive impacts on N metabolism in sheep. They inhibit protein and amino acid degradation, hyper-ammonia-producing bacteria, and methane production in ruminants. Besides, garlic stem components have a positive role in total volatile fatty acid production and reducing feed scarcity in the future for ruminants (Kamruzzaman et al. 2011, 2014).

Chu et al. (2003) have reported that garlic straw can be used as supplemental roughage. Beef quality and eye muscle area were higher in the garlic straw silage-fed *Hanwoo* steers than in *Hanwoo* steers-fed rice straw. Bampidis et al. (2005) showed that garlic husks can be incorporated into concentrate mixtures for growing lambs with no adverse effects on performance or carcass characteristics in isonitrogenous and iso (net energy) energetic diets.

Thanikachalam et al. (2010) have demonstrated that the addition of powdered garlic peels enhances the serum protein, albumin, and globulin in African catfish [*Clarias gariepinus*] fingerlings compared to the control group. The results also demonstrate that low survival rates in control groups and significantly higher survival rates were recorded in all the garlic peel-fed groups after challenging with *Aeromonas hydrophila*.

Bidura (1999) showed that the utilization of garlic straw in diets increased egg yolk percentage and decreased egg cholesterol. Besides, water consumption, body weight gain, and feed efficiencies were increased significantly with increasing levels of garlic leaf meal in diets. Liu et al. (2019) showed that garlic waste tended to increase the relative weight of immune organs, total protein, and serum IgG and IgA in the growing rabbit. Therefore, garlic straw also plays a role in improving the immune function of growing rabbits. Lee et al. (2017) showed that garlic stalks incubated in situ in the rumen of two Hanwoo steers had lower NDF, higher ADF, and greater effective degradabilities of DM and NDF. Liao et al. (2022) showed that dietary supplementation of different levels of garlic straw powder can improve the meat quality and antioxidant capacity of yellow-feathered broilers without affecting growth performance and intestinal mucosal morphology.

14.7.2 Soil Amendment

The use of garlic stalks as a source of animal feed would help to minimize the problem of feed inadequacy and potentially reduce the feed costs and the problems associated with the disposal of garlic waste. Many literature data have confirmed the ability of *Allium sativum* residues to reduce the damage to vegetable crops by nematode and other soil-borne diseases (Mallek et al. 2007; Lechner and Monaldi 2011; Han et al. 2012, 2013; Gong et al. 2013). Mallek et al. (2007) reported the effects of *Allium* spp. residues are concentration-dependent and these by-products inhibited the germination of weeds. Gong et al. (2013) also showed that garlic straw can be used as a natural biofumigant. The effect of garlic straw on the incidence of *Meloidogyne incognita* of tomato indicated that the immobility and mortality of *M. incognita* juveniles were significantly increased with increasing water extract concentrations of raw garlic straw, and hatchability was significantly inhibited by application of the extract. Compared to raw garlic straw, fermenting garlic straw increased the mortality of *M. incognita* rate. Raw garlic straw can reduce *M. incognita* infestation of tomato roots with the increasing volume shown in pot trials. However, the higher concentration of garlic straw allelopathy compound could inhibit the tomato plant growth. The inhibition of *M. incognita* could be due to the content of sulfur-containing compounds in garlic straw.

Besides, Han et al. (2012) showed that garlic stalk significantly improves the quality of soil by changing nutrient dynamics and soil enzyme activity, compared to soil without garlic stalk. Also, garlic stalk decomposition significantly improves the growth of lettuce and promotes physiological indices and activities of plant enzymes in a lettuce leaf Han et al. (2013). These findings provide evidence that the decomposed garlic straw may contain some allelochemicals and cause allelopathy by releasing the allelochemicals into the environment.

Lechner and Monaldi (2011) showed that garlic waste was found to be a suitable medium for growing mushrooms (*Pleurotus ostreatus*) watered with olive-mill wastewater.

Recently, the application of raw garlic straw to soil has promoted the growth and development of eggplant and regulated the antioxidant plant defense system, which ensured better plant survival under stress (Ghani et al. 2019).

14.7.3 Garlic Waste Used as a Pollutant Absorbent

Garlic wastes can be used directly or after processing as biosorbents and bioremediation agents (Hameed and Ahmad 2009; Huang et al. 2011; Zhou et al. 2012; Chowdhury et al. 2012; Negi et al. 2012; Liang et al. 2013; Liu et al. 2014; Asfaram et al. 2014; Parlayıcı and Pehlivan 2015; Muthamilselvi et al. 2016; Kallel et al. 2016b, c; Zhao et al. 2017, 2022; Abu-Shahba et al. 2022; Raaj et al. 2022).

The maximum adsorption capacity for metal ions was described in Fig. 14.4. (Liang et al. 2013) studied the removal of Pb^{2+}, Cu^{2+}, and Ni^{2+} from contaminated water using garlic peels. The potential of garlic peels mercerized garlic peels to remove Pb^{2+} from the aqueous solution was evaluated by Liu et al. (2014). They reported that the adsorption capacity was found to be 109 mg/g. In another report, garlic wastes have been used as an unconventional bio-adsorbent of Pb^{2+}, Sn^{2+}, Fe^{2+}, Hg^{2+}, As^{3+}, and Cd^{2+} at different adsorbent/metal ion ratios. The adsorption of heavy metals using dry garlic stems was also investigated using the batch adsorption technique by Zhou et al. (2012). The results showed adsorption capacity was achieved at 8.52 mg/g for Pb^{2+} and 20.90 mg/g for Cu^{2+}. Removal of Cu^{2+} from aqueous solution by garlic skin was also studied by Chowdhury et al. (2012). The adsorption capacity of garlic straw for Cr (VI) was found to be 103.09 mg/g. The efficiency and performance of garlic peel adsorbent for the removal of direct red 12B from wastewater was also investigated (Asfaram et al. 2014). The maximum direct red sorption capacity was found to be 37.96 mg/g. The potential of garlic peels to remove methylene blue and crystal violet from aqueous solutions was evaluated by Hameed and Ahmad (2009) and Raaj et al. (2022). The phenol adsorption onto garlic peels was also conducted by Muthamilselvi et al. (2016). The maximum adsorption capacity was calculated as 14.49 mg/g. Also, garlic peels exhibited quite well up-taking behavior for fluoride anion adsorption (Huang et al. 2011). The maximum adsorption capacity is evaluated to be 20.9 mg/g of garlic peels. In another report, garlic straw was used for the removal of two hazardous textile dyes such as direct red 80 and methylene blue with maximum capacities of 107.53 mg/g and 256.41 mg/g, respectively (Kallel et al. 2016b, c).

Zhao et al. (2017) investigated garlic peels as a sorbent medium to remove quinolone antibiotics from excretion from humans and animals. Results showed

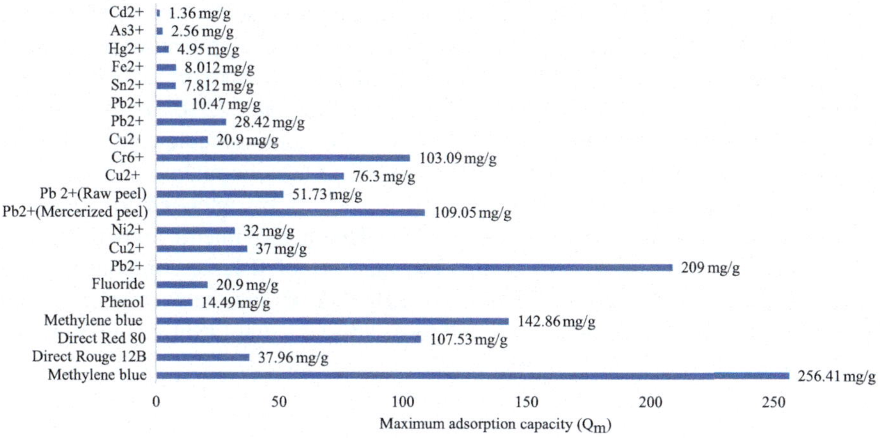

Fig. 14.4 Maximum adsorption capacities of garlic waste for pollutant removal

that pretreated garlic peels with nitric acids had improved adsorption capacity than that of the untreated peels.

Abu-Shahba et al. (2022) showed that indigenous garlic peel or mercerized garlic peel significantly increased the nonenzymatic and enzymatic antioxidants in leaves and roots under cadmium stress. Lettuce plants exposed to garlic peel improved growth attributes, chlorophyll, and carotenoid levels. Garlic peel extract possesses sulfhydryl, which plays an essential role in the sequestration and separation process of heavy metals detoxification.

The study of adsorption capacities for enrofloxacin by concentrated sulfuric acid carbonized garlic peel biochar and pyrolysis-derived garlic peel biochar provided a new sustainable perspective for the design of efficient biochars and explored the interfacial interaction mechanism of antibiotics removal on biochars (Zhao et al. 2022).

14.7.4 Bio-energy

Garlic waste has been used to produce energy via biochemical and thermochemical processes. Garlic peel and stem are characterized by a C/N ratio of 35.58 and 21.7, respectively (Park et al. 2012). Garlic possessing waste materials is a potentially important energy source. The carbon, hydrogen, oxygen, and nitrogen contents are 14.42%, 9.26%, 74.95%, and 1.37%, respectively. Besides, the calorific value of garlic stem was around 4388 Kcal/kg. Similar calorific value was found in other lignocellulosic materials such as pepper stem (4636 Kcal/kg), soya bean husk (4170 Kcal/kg), and cotton husk (4400 Kcal/kg) (Murali et al. 2007; Dhyani and Bhaskar 2018).

Park et al. (2012) reported the pyrolysis characteristics of garlic stems to examine their prospective potential for the production of chemicals and bio-oil. Maximum yields of bio-oil, synthesis gas, and biochar were approximately 39.6%, 32.3%, and 28.1%, respectively, at the optimum temperature of 500 °C. Pyrolysis of garlic stalks was concluded and the maximum bio-oil production was found at 550 °C and increased significantly with the KOH catalyst (Singh et al. 2020). The pyrolysis of this garlic peel waste was accomplished to obtain various valuable solid and liquid products such as acetic acid as usual; methanol and acetone mainly analyzed with TGA/DTA and GS/MS. The less oxidized molecules are intended for regarding energy capacity (Caglar and Aydinli 2018). Thermogravimetric experiments of garlic husk pyrolysis were carried out. The majority of volatile components were found to thermally degrade in the range of 210–370 °C and the average activation energy from Starink methods was found to be 154.32 kJ/mol (Singh and Sawarkar 2020; Singh et al. 2020).

14.7.5 Dietary Fiber, Reducing Sugars and Cellulose

Various vegetables and fruit wastes such as garlic skin, peel, and straw have been used to extract polysaccharides, oligosaccharides, and sugar alcohols (Kallel et al. 2014a, 2015a, b; Kotsanopoulos et al. 2022). Kallel et al. (2014a) have isolated a novel polysaccharide from garlic straw. Hot water polysaccharide (HWP) had high antioxidant activities and moderate antimicrobial activity. The inoculation of HWP in minced beef meat showed inhibitions of lipid oxidation of aerobic storage and also improved the meat color stability.

Xylooligosaccharides (XOS) obtained by enzymatic breakdown of the hemicellulosic fraction from garlic straw showed important antioxidant activity and as a special nutrient for lactic bacteria such as *Bifidobacterium adolescentis* and *Lactobacillus acidophilus*. Furthermore, XOS had high antimicrobial activity against *Klebsiella pneumoniae, Enterococcus faecalis, Bacillus thuringiensis,* and *Pseudomonas aeruginosa* (Kallel et al. 2015a, b).

Reddy and Rhim (2014) and Kallel et al. (2016a) have studied the scanning electron microscopy (SEM) of garlic skin and garlic straw, respectively. The crude fiber of garlic skin was composed of bundles of fibers with a smooth surface, which was mainly due to the presence of wax and other extractives in the fiber (Fig. 14.5a). The morphology and structure of the external aspect of garlic straw show that garlic straw possesses high specific surface as presented in Fig. 14.5b. Figure does not show individualized cellulose fibers because of the presence of lignin and hemicelluloses that act as cementing components around the fiber bundles. Cellulose microfibers and nanocrystals were isolated from garlic waste by alkali treatment and acid hydrolysis and were characterized by transmission electron microscopy. The obtained cellulose nanocrystals (CNCs) obtained from garlic skin exhibited a spherical shape with a size of 58–96 nm (Fig. 14.5c). The isolated CNCs from garlic stalks appear as black, spherical spots in the micrograph. It can also be observed that the CNCs are highly aggregated, with some particles overlapping each other (Agustin et al. 2013). Besides, CNCs obtained from garlic straws are spherical and have an average diameter of 35 nm (Fig. 14.5d). They presented an aspect ratio of

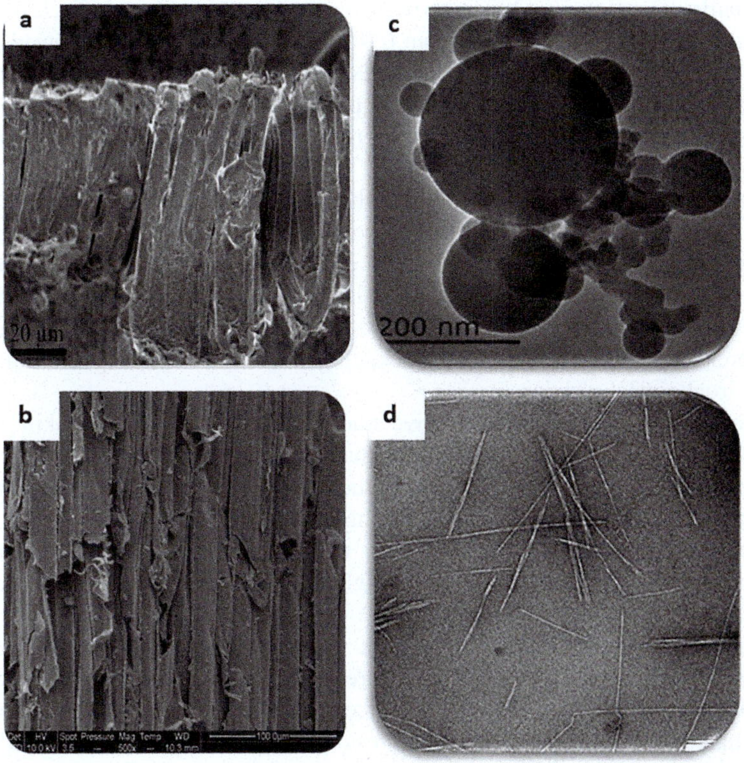

Fig. 14.5 Scanning electron microscopy and transmission electron microscopy micrographs of garlic waste and obtained cellulose nanocrystals. (**a**) SEM micrographs of garlic peel, (**b**) SEM micrographs of garlic straw, (**c**) CNC isolated from garlic peel, (**d**) CNC isolated from garlic straw

approximately 80 (Kallel et al. 2016a). The crystallinity index value of CNCs extracted from garlic straw was 68.8% which was higher than the crystallinity index of CNCs extracted from garlic skin (Kallel et al. 2016a, b, c; Reddy and Rhim 2014).

The use of biodegradable polymer or biopolymer for composite materials has been of great interest in the past decades due to escalating environmental problems posed by non-biodegradable and petroleum-based polymers. The material obtained from garlic by-products has been utilized to make which in turn can be reinforced with natural fibers to make green composites (Agustin et al. 2013; Selvamani et al. 2016; Lee et al. 2021; Chaudhary et al. 2021; Luo et al. 2022; Hernández-Varela et al. 2022; Salim et al. 2022; Geng et al. 2022).

Agustin et al. (2013) have reported reinforcing starch-based biocomposite films using CNCs from garlic stalks. The results of this study showed homogeneous dispersion of CNC in the starch matrix, improvement in tensile strength, and CNC-reinforced films had lower moisture uptake than nonreinforced films.

Recently, Ding et al. (2023) showed a novel active chitosan packaging film containing garlic leaf extract and stem CNC was prepared. Results showed that the addition of CNC to the chitosan film increased its tensile strength, thermal stability, and water/oxygen barrier. Garlic leaf improves the antioxidant and antibacterial activities of the chitosan film. Then, coating black garlic with electrostatic spraying delays its water loss and purification.

Chaudhary et al. (2021) have reported the preparation of a sustainable multifunctional food packaging composite film containing waste garlic peel extract and chitosan. The results showed that the film exhibited antimicrobial, antioxidant, and non-cytotoxic activities. Films showed also improved thermal stability after the addition of garlic peel extract. The results suggest that films can find widespread commercial applications like food packaging materials, replacing the commonly used petrochemical plastics.

Hernández-Varela et al. (2022) have studied the addition of garlic skin particles and the changes in physicochemical, mechanical, and microstructural properties of films formulated with potato starch and gellan gum. It was shown that concentrations at 2% of garlic skin particles in the size range 212–38 μm provide films with a good surface and color, as well as enhancing their physicochemical, microstructural, mechanical, and thermal properties through structural reinforcement of the films' polymeric network.

Geng et al. (2022) have shown that combined garlic residues and acrylic monomers to prepare sprayable garlic residue-acrylate liquid mulch films via emulsion polymerization effectively reduced soil water evaporation and increased soil temperature. The optimized garlic residue-acrylate mulch film reached a biodegradation rate of 74.6% and increased the germination rate of pakchoi seeds. The experimental results presented herein demonstrate that the full utilization of garlic residues can improve the performance of sprayable kitchen waste-based mulch films for agricultural applications.

Salim et al. (2022) showed that films incorporating both cellulose nanocrystals and garlic extract from garlic skin hold more promise for active food packaging applications due to a combination of enhanced physical characteristics and antibacterial activity.

Luo et al. (2022) have reported that the development of a novel shape-stable phase-change material for thermal energy storage was based on activated garlic peel, derived from widespread and cheap garlic peels.

Lee et al. (2021) and Selvamani et al. (2016) showed the importance of garlic peel as a carbon framework for the sodium ion and sulfur encapsulating in Li-S batteries. The porous carbon materials synthesized using biomass waste garlic peels exhibited excellent rate capability as well as steady-state cycling performance.

14.7.6 Biological Properties of Garlic Waste Extracts

The use of plant extracts as food additives contributes to the reduction of synthetic additives, meeting the demand for natural foods. Garlic peel or husk extracts can be used as a source of phytonutrients. Ichikawa et al. (2003) suggested the presence of phenylpropanoids (*N*-trans-coumaroyloctopamine, *N*-trans-feruloyl octopamine, diacylglycerol-ferulic acid ether, and coumaric and ferulic acids) in garlic skins extract (Table 14.2). The extraction of polyphenol compounds could be considered as an alternative of great interest for the change in the value of food processing by-products (Thach 2022). These phenolic compounds could be related to a high antioxidant potential and bacteriostatic and bactericidal effects (dos Santos et al. 2022).

14.7.7 Antibacterial and Antioxidant

Kallel et al. (2014b) observed that the highest extraction yield was obtained in water and the high content of phenolic compounds was shown by the extract obtained using 50% methanol/water as a solution. The antioxidant and antimicrobial activities could be attributed to containing high amounts of caffeic, *p*-coumaric, ferulic, and di-ferulic acids.

Besides, Ifesan et al. (2014) demonstrated that garlic peel ethanol extract exhibited strong antioxidant activities and reduced the bacterial population (*Streptococcus aureus*, *Escherichia coli*, *Bacillus subtilis*, *B. cereus,* and *Proteus vulgaris)* inoculated into the cooked beef compared to control. Dos Santos et al. (2022) observed that the hydroalcoholic extracts of garlic peel had a higher content of bioactive compounds and exhibited antioxidant and antimicrobial activities. These extracts did not show any toxic effect on *Artemia salina* and adult *zebrafish.*

Recently, Carreón-Delgado et al. (2023) and Noda et al. (2019) have also shown that garlic peel and husk were found to be a good source of phytochemicals and exhibited antioxidant and antimicrobial activities. These properties were significantly enhanced by pre-processing the garlic peel by either solid-state fermentation and roasting pretreatments or using two hydrothermal methods, steam explosion, and microwave irradiation methods.

In addition, according to Chhouk et al. (2017), the antioxidant activity of extracts from garlic husk using carbon dioxide expanded ethanol method could be attributed to the containing of five major phenolic compounds such as garlic acid, 4-hydroxybenzoic acid, caffeic acid, p-coumaric acid, and trans-ferulic acid (Table 14.2).

The aqueous extract of garlic straw application inhibited the proliferation rate of some fungi such as *Fusarium oxysporum* f. sp. *niveum, F. oxysporum* f. sp. *cucumerinum,* and *Phytophthora capsici* (Wang et al. 2019). The inhibitory action could be attributed to the containing of dimethyl 2-methoxyhexane-1,6-dioate, dibutyl phthalate, 4-hydroxy-3-methoxy benzoic acid, and 4-hydroxy-benzoic acid, and the phenol like compounds such as mequinol (Table 14.2) (Han et al. 2012, 2016).

Table 14.2 Phytochemicals identified in garlic peel extract

Extract	Identification methods	Phenolic compounds	Reference
50/50 (v/v) methanol-water extract	LC-MS	Caffeic acid *p*-coumaric Ferulic acid di-ferulic acid	Kallel et al. (2014c)
Ethanol extract	LC-MS and NMR	*N*-trans-coumaroyloctopamine, *N*-trans-feruloyl octopamine, diacylglycerol-ferulic acid ether *p*-coumaric acid Ferulic acid	Ichikawa et al. (2003)
Aqueous extract	GS-MS	Dimethyl 2-methoxyhexane-1,6-dioate, Dibutyl phthalate, 4-hydroxy-3-methoxy benzoic acid, 4-hydroxy-benzoic acid Phenol (mequinol)	Han et al. (2012)
80% (v/v) ethanol extract	GS-MS	di-2-propenyl trisulfide or allyl trisulfide, Thioacetic acid *S*-(tetrahydro-pyran-3-yl) ester, Disulfide, Methyl 2-propenyl, diallyl sulfide, 5-(3,4-dimethoxyphenyl)-2,3-dihydro-2-thioxo-4-thiazolecarboxaldehyde, 2,2-dimethyl-4-ethynyl-tetrahydrothiopyran-4-ol, 2-amino-4-(p-aminophenyl)-thiazole 4-methyl-6-phenyl-pyrimidine; d-glycero-l-gluco-heptose; Glycerin; Catechol; Butanoic acid, 1-methylethyl ester; Hexanoic acid 9,12-octadecadienoic acid 5-ethyl-1,3-dioxane-5-methanol Octadecenoic acid 2-methoxy-4-vinylphenol Lactose 2-hydroxy-gamma-butyrolactone 4-((1E)-3-hydroxy-1-propenyl)-2-methoxyphenol 2,3-dihydro-3,5-dihydroxy-6-methyl-4H-pyran-4-one 5-hydroxymethylfurfural 3-deoxy-D-mannoic lactone 2,3,5,6-tetrafluoroanisole Eugenol	Teixeira et al. (2023)

14.7.8 Antifungal

Garlic straw extracts were used in the Integrated Pest Management of apple orchards. The extracts have the potential as an inexpensive tool against *Tetranychus urticae* and *T. viennensis* proliferation (Geng et al. 2014, 2022).

Carreón-Delgado et al. (2023) showed that garlic peel extracts obtained with 60% methanol under steam cooking pretreatment presented acceptable inhibition percentages for in vitro and in vivo assays against *Colletotrichum gloeosporioides* and *Botrytis cinerea*, respectively. Moreover, the application of garlic extracts as a preventive treatment presented a significant reduction in fungal growth and a reduction of rot disease in papaya and strawberry fruits.

Besides, garlic peel extract was effective against several phytopathogenic fungi. It significantly reduced the lesion size in apples caused by subepidermal inoculation with *Colletotrichum acutatum* spores. The antifungal activity might account for the presence of a considerable amount of phytochemicals, such as di-2-propenyl trisulfide or allyl trisulfide, thioacetic acid S-(tetrahydro-2H-pyran-3-yl) ester, disulfide, methyl 2-propenyl, diallyl sulfide, and 2-amino-4-(*p*-aminophenyl)-thiazole and other phytochemicals such as 4-methyl-6-phenyl-pyrimidine; d-glycero-l-gluco-heptose; catechol; butanoic acid, 3-oxo-, 1-methyl ethyl ester; hexanoic acid and its esters; 9,12-octadecadienoic acid and its esters; 5-ethyl-1,3-dioxane-5-methanol; octadecenoic acid; 2-methoxy-4-vinylphenol; lactose; 2-hydroxy-gamma-butyrolactone; 4-((1E)-3-hydroxy-1-propenyl)-2-methoxy phenol (Table 14.2) (Teixeira et al. 2023).

According to Naqvi et al. (2020), garlic peel extracts exhibited strong positive inhibition against *Colletotrichum falcatum* while medium inhibition against *Fusarium moniliforme* and *Rhizoctonia solani*. However, extracts of the decomposed garlic wastes gave inhibition against several fungal phytopathogens mainly; *R. solani*, *F. moniliforme*; *Pythium* sp., and *Phytophthora* sp. (Yaseen 2020).

14.7.9 Antidiabetic

Lolok et al. (2019) showed that the oral administration of the ethanol extract of garlic peel (800 mg/kg body weight) was able to increase tolerance to glucose thus effective in the decrease of blood glucose levels in normal rats compared to the group receiving metformin and glibenclamide combination therapy. The antidiabetic activity was probably due to the presence of significant bioactive components such as organoselenium and tellurium, kaemferol-3-O-b-D-glukopirosa, Iso-ikamnetin-3-O-b-D-glukopirosa, L-arginine, beta-carotene, thiamine, riboflavin, pantothenic acid (B5), alicin and alliin, vitamin B6, folate (Vit.B9), vitamin C, calcium, iron, magnesium, phosphorus, potassium, sodium, and zinc.

14.8 Conclusion and Future Perspectives

The processing of garlic, which is an important crop, produces a significant amount of waste including peels, straw, and skin. These wastes can cause negative environmental problems if they are not disposed of in environmentally friendly and cost-effective ways. They are rich in nutrients, soluble and insoluble dietary fibers, antioxidants, antibacterial and antifungal compounds. They can be used to produce sugars, cellulose, microfibers, and nanocrystals. Besides, garlic wastes can be used directly or after processing as bioadsorbents for wastewater treatment. Pyrolysis processes were also applied to produce bioenergy and bio-oil from garlic waste. Future investigations on the bioactivity, bioavailability, and toxicology of phytochemicals from garlic waste along with their stable interactions with other food ingredients should be carried out carefully both in vitro and in vivo studies.

References

Abdel-Fattah AF, Edrees M (1972) A study on the composition of garlic skins and the structural features of the isolated pectic acid. J Sci Food Agric 23(7):871–877. https://doi.org/10.1002/jsfa.2740230708

Abu-Shahba MS, Mansour MM, Mohamed HI, Sofy MR (2022) Effect of biosorptive removal of cadmium ions from hydroponic solution containing indigenous garlic peel and mercerized garlic peel on lettuce productivity. Sci Hortic 293:110727. https://doi.org/10.1016/j.scienta.2021.110727

Agustin MB, Ahmmad B, De Leon ERP, Buenaobra JL, Salazar JR, Hirose F (2013) Starch-based biocomposite films reinforced with cellulose nanocrystals from garlic stalks. Polym Compos 34(8):1325–1332. https://doi.org/10.1002/pc.22546

Alexander MM, Sulebele GA (1973) Pectic substances in onion and garlic skins. J Sci Food Agric 24(5):611–615. https://doi.org/10.1002/jsfa.2740240514

Ammarellou A, Yousefi AR, Heydari M, Uberti D, Mastinu A (2022) Biochemical and botanical aspects of *Allium sativum* L. Sowing BioTech (Basel) 11(2):16. https://doi.org/10.3390/biotech11020016

Asfaram A, Fathi MR, Khodadoust S, Naraki M (2014) Removal of direct red 12B by garlic peel as a cheap adsorbent: kinetics, thermodynamic and equilibrium isotherms study of removal. Spectrochim Acta A Mol Biomol Spectrosc 127:415–421. https://doi.org/10.1016/j.saa.2014.02.092

Bampidis VA, Christodoulou V, Christaki E, Florou-Paneri P, Spais AB (2005) Effect of dietary garlic bulb and garlic husk supplementation on performance and carcass characteristics of growing lambs. Anim Feed Sci Technol 121(3):273–283. https://doi.org/10.1016/j.anifeedsci.2005.02.003

Bidura IG (1999) The effect of garlic (*Allium sativum*) leaf meal in diets on performance of growing duck. Majalah Ilmiah Peternakan (Indonesia)

Borlinghaus J, Albrecht F, Gruhlke MCH, Nwachukwu ID, Slusarenko AJ (2014) Allicin: chemistry and biological properties. Molecules 19(8):12591–12618. https://doi.org/10.3390/molecules190812591

Bourlière F, Quezel P, et Santa S (1962) Nouvelle flore de l'Algérie et des régions désertiques méridionales. Tome I, 1962. Editions du Centre National de la Recherche Scientifique Paris Revue d'Écologie (La Terre et La Vie) 16(4):459–459

Brewster JL, Rabinowitch HD (2020) Onions and allied crops: volume III. biochemistry, food science, and minor crops. CRC Press, Boca Raton, FL

Caglar A, Aydinli B (2018) The pyrolysis of industrial alliaceous plant wastes: illustration of process and characterization of products. Energy Explor Exploit 36:014459871875955. https://doi.org/10.1177/0144598718759559

Capasso A (2013) Antioxidant action and therapeutic efficacy of *Allium sativum* L. Molecules 18(1):690–700. https://doi.org/10.3390/molecules18010690

Cardelle A, Soria A, Corzo-Martínez M, Villamiel M (2010) A comprehensive survey of garlic functionality. pp 1–60

Carreón-Delgado DF, Hernández-Montesinos IY, Rivera-Hernández KN, del Sugeyrol Villa-Ramírez M, Ochoa-Velasco CE, Ramírez-López C (2023) Evaluation of pretreatments and extraction conditions on the antifungal and antioxidant effects of garlic (*Allium sativum*) peel extracts. Plan Theory 12(1):217. https://doi.org/10.3390/plants12010217

Chaudhary BU, Lingayat S, Banerjee AN, Kale RD (2021) Development of multifunctional food packaging films based on waste garlic peel extract and chitosan. Int J Biol Macromol 192:479–490. https://doi.org/10.1016/j.ijbiomac.2021.10.031

Chhouk K, Uemori C, Wahyudiono, Kanda H, Goto M (2017) Extraction of phenolic compounds and antioxidant activity from garlic husk using carbon dioxide expanded ethanol. Chem Eng Process Process Intensif 117:113–119. https://doi.org/10.1016/j.cep.2017.03.023

Chitsaz H, Oraji H, Keramat Amirkolaei A, Akrami R (2018) Effect of garlic peel on haematological, biochemical and digestive enzyme activity in beluga juvenile (*Huso huso*). Sustain Aquac Health Manag J 4(1):13–28. https://doi.org/10.29252/ijaah.4.1.13

Chowdhury A, Bhowal A, Datta S (2012) Equilibrium, thermodynamic and kinetic studies for removal of copper (II) from aqueous solution by onion and garlic skin. Water 4:37–51

Chu GM, Lee HJ, Park JS, Cho HW, Ahn BH (2003) Effect of garlic stalk silage on performance and carcass characteristics of Hanwoo steers. J Anim Sci Technol 45:1007–1018. https://doi.org/10.5187/JAST.2003.45.6.1007

Dhyani V, Bhaskar T (2018) A comprehensive review on the pyrolysis of lignocellulosic biomass. Renew Energy 129:695–716. https://doi.org/10.1016/j.renene.2017.04.035

Ding Y, Jiang Y, Zhong Y, Wang D, Deng Y, Meng F, Li Y, Zhang M, Zhang C (2023) Preparation of garlic stem cellulose nanocrystal/leaf extract/chitosan film for black garlic preservation by electrostatic spraying. Int J Biol Macromol 225:518–525. https://doi.org/10.1016/j.ijbiomac.2022.11.108

dos Santos PCM, da Silva LMR, Magalhaes FEA, Cunha FET, Ferreira MJG, de Figueiredo EAT (2022) Garlic (*Allium sativum* L.) peel extracts: from industrial by-product to food additive. Appl Food Res 2(2):100186. https://doi.org/10.1016/j.afres.2022.100186

El Mashad HM, Zhang R, Pan Z (2019) Chapter 11: Onion and garlic. In: Pan Z, Zhang R, Zicari S (eds) Integrated processing technologies for food and agricultural by-products. Academic Press, pp 273–296

Etana MB (2018) Review on the agronomic management practices of garlic (*Allium sativum* L.). J Biol Agric Healthcare 8:1–7

Geng S, Chen H, Zhang J, Tu H (2014) Bioactivity of garlic-straw extracts against the spider mites, *Tetranychus urticae* and *T viennensis1*. J Agric Urban Entomol 30(1):38–48. https://doi.org/10.3954/1523-5475-30.0.38

Geng H, Wang D, Li B, Ma J, Li W (2022) Improvement on pest resistance of sprayable mulch films by fully utilizing garlic residues. J Clean Prod 364:132596. https://doi.org/10.1016/j.jclepro.2022.132596

Ghani MI, Ali A, Atif MJ, Ali M, Amin B, Anees M, Cheng Z (2019) Soil amendment with raw garlic stalk: a novel strategy to stimulate growth and the antioxidative defense system in monocropped eggplant in the north of China. Agronomy 9(2):89. https://doi.org/10.3390/agronomy9020089

Gong B, Bloszies S, Li X, Wei M, Yang F, Shi Q, Wang X (2013) Efficacy of garlic straw application against root-knot nematodes on tomato. Sci Hortic 161:49–57. https://doi.org/10.1016/j.scienta.2013.06.027

Gustafsson J, Cederberg C, Sonesson U, Emanuelsson A (2013) The methodology of the FAO study: global food losses and food waste-extent, causes and prevention-FAO, 2011

Hameed BH, Ahmad AA (2009) Batch adsorption of methylene blue from aqueous solution by garlic peel, an agricultural waste biomass. J Hazard Mater 164(2):870–875. https://doi.org/10.1016/j.jhazmat.2008.08.084

Han X, Cheng Z, Meng H (2012) Soil properties, nutrient dynamics, and soil enzyme activities associated with garlic stalk decomposition under various conditions. PLoS One 7(11):e50868. https://doi.org/10.1371/journal.pone.0050868

Han X, Cheng Z, Meng H, Yang X, Ahmad I (2013) Allelopathic effect of decomposed garlic (*Allium Sativum* L.) stalk on lettuce (*L. sativa* var. *crispa* L.). Pak J Bot 45:225–233

Han X, Cheng Z, Meng H (2016) Growth of wheat and lettuce and enzyme activities of soils under garlic stalk decomposition for different durations: function and utilization of garlic stalk decomposition for different durations. J Sci Food Agric 97(9):2727–2735. https://doi.org/10.1002/jsfa.8099

Harris JC, Cottrell SL, Plummer S, Lloyd D (2001) Antimicrobial properties of *Allium sativum* (garlic). Appl Microbiol Biotechnol 57(3):282–286. https://doi.org/10.1007/s002530100722

Hernández-Varela JD, Chanona-Pérez JJ, Resendis-Hernández P, Gonzalez Victoriano L, Méndez-Méndez JV, Cárdenas-Pérez S, Calderón Benavides HA (2022) Development and characterization of biopolymers films mechanically reinforced with garlic skin waste for fabrication of compostable dishes. Food Hydrocoll 124:107252. https://doi.org/10.1016/j.foodhyd.2021.107252

Higuchi O, Tateshita K, Nishimura H (2003) Antioxidative activity of sulfur-containing compounds in *Allium* species for human low-density lipoprotein (LDL) oxidation in vitro. J Agric Food Chem 51(24):7208–7214. https://doi.org/10.1021/jf034294u

Hong S-I, Kim D (2001) Storage quality of chopped garlic as influenced by organic acids and high-pressure treatment. J Sci Food Agric 81:397–403. https://doi.org/10.1002/1097-0010(200103)81:4<397::AID-JSFA831>3.0.CO;2-R

Huang K, Shao J, Zhu H, Inoue K (2011) Removal of fluoride from aqueous solution onto Zr-loaded garlic peel (Zr-GP) particles. J Cent S Univ Technol 18(5):1448–1453. https://doi.org/10.1007/s11771-011-0860-x

Ibrahim M, Shafiullah, Ilyas M, Shah FA, Khan A, Rukh S, Haq IU (2018) Comparison of different garlic (*Allium sativum*) varieties for yield and yield components grown at agriculture research station, Buner. Int J Environ Sci Nat Res 13(5):1–5. https://doi.org/10.19080/IJESNR.2018.13.555873

Ichikawa M, Ryu K, Yoshida J, Ide N, Kodera Y, Sasaoka T, Rosen RT (2003) Identification of six phenylpropanoids from garlic skin as major antioxidants. J Agric Food Chem 51(25):7313–7317. https://doi.org/10.1021/jf034791a

Iciek M, Kwiecień I, Włodek L (2009) Biological properties of garlic and garlic-derived organosulfur compounds. Environ Mol Mutagen 50(3):247–265. https://doi.org/10.1002/em.20474

Ifesan BOT, Fadipe EA, Ifesan BT (2014) Investigation of antioxidant and antimicrobial properties of garlic peel extract (*Allium sativum*) and its use as natural food additive in cooked beef. J Sci Res Rep 3(5):711–721. https://doi.org/10.9734/JSRR/2014/5726

Jonoobi M, Oladi R, Davoudpour Y, Oksman K, Dufresne A, Hamzeh Y, Davoodi R (2015) Different preparation methods and properties of nanostructured cellulose from various natural resources and residues: a review. Cellulose 22(2):935

Kallel F, Ellouz Chaabouni S (2017) Perspective of garlic processing wastes as low-cost substrates for production of high-added value products: a review. Environ Prog Sustain Energy 36(6):1765–1777. https://doi.org/10.1002/ep.12649

Kallel F, Driss D, Bouaziz F, Belghith L, Zouari-Ellouzi S, Chaari F, Haddar A, Chaabouni SE, Ghorbel R (2014a) Polysaccharide from garlic straw: extraction, structural data, biological properties and application to beef meat preservation. RSC Adv 5(9):6728–6741. https://doi.org/10.1039/C4RA11045E

Kallel F, Driss D, Chaari F, Belghith L, Bouaziz F, Ghorbel R, Chaabouni SE (2014b) Garlic (*Allium sativum* L.) husk waste as a potential source of phenolic compounds: influence of extracting solvents on its antimicrobial and antioxidant properties. Ind Crop Prod 62:34–41. https://doi.org/10.1016/j.indcrop.2014.07.047

Kallel F, Driss D, Bouaziz F, Neifer M, Ghorbel R, Ellouz Chaabouni S (2015a) Production of xylooligosaccharides from garlic straw xylan by purified xylanase from *Bacillus mojavensis* UEB-FK and their in vitro evaluation as prebiotics. Food Bioprod Process 94:536–546. https://doi.org/10.1016/j.fbp.2014.07.012

Kallel F, Driss D, Chaabouni SE, Ghorbel R (2015b) Biological activities of xylooligosaccharides generated from garlic straw xylan by purified xylanase from *Bacillus mojavensis* UEB-FK. Appl Biochem Biotechnol 175(2):950–964. https://doi.org/10.1007/s12010-014-1308-1

Kallel F, Bettaieb F, Khiari R, García A, Bras J, Chaabouni Ellouz S (2016a) Isolation and structural characterization of cellulose nanocrystals extracted from garlic straw residues. Ind Crop Prod 87:287–296. https://doi.org/10.1016/j.indcrop.2016.04.060

Kallel F, Bouaziz F, Chaari F, Belghith L, Ghorbel R, Chaabouni SE (2016b) Interactive effect of garlic straw on the sorption and desorption of direct red 80 from aqueous solution. Process Saf Environ Prot 102:30–43. https://doi.org/10.1016/j.psep.2016.02.012

Kallel F, Chaari F, Bouaziz F, Bettaieb F, Ghorbel R, Chaabouni SE (2016c) Sorption and desorption characteristics for the removal of a toxic dye, methylene blue from aqueous solution by a low cost agricultural by-product. J Mol Liq 219:279–288. https://doi.org/10.1016/j.molliq.2016.03.024

Kamruzzaman M, Torita A, Sako Y, Al-Mamun M, Sano H (2011) Effects of feeding garlic stem and leaf silage on rates of plasma leucine turnover, whole body protein synthesis and degradation in sheep. Small Rumin Res 99(1):37–43. https://doi.org/10.1016/j.smallrumres.2011.03.052

Kamruzzaman M, Liang X, Sekiguchi N, Sano H (2014) Effect of feeding garlic leaf on microbial nitrogen supply, kinetics of plasma phenylalanine, tyrosine and protein synthesis in sheep. Anim Sci J 85(5):542–548. https://doi.org/10.1111/asj.12190

Khanum F, Anilakumar KR, Viswanathan KR (2004) Anticarcinogenic properties of garlic: a review. Crit Rev Food Sci Nutr 44(6):479–488. https://doi.org/10.1080/10408690490886700

Koch HP, Lawson LD (1996) Garlic: the science and therapeutic application of *Allium sativum* L. and related species. Williams & Wilkins

Kopec A, Pia E, Leszczyn T, Elzbieta S (2013) Healthy properties of garlic. Curr Nutr Food Sci 9(1):59–64

Kotsanopoulos K, Behera S, Ray R (2022) Microbial production of polysaccharides, oligosaccharides, and sugar alcohols from vegetables and fruit wastes. In: Ray RC (ed) Fruits and vegetable wastes: valorization to bioproducts and platform chemicals. Springer Nature, pp 343–364

Laufenberg G, Kunz B, Nystroem M (2003) Transformation of vegetable waste into value added products:: (A) the upgrading concept; (B) practical implementations. Bioresour Technol 87(2): 167–198. https://doi.org/10.1016/S0960-8524(02)00167-0

Lawson LD (1998) Garlic: a review of its medicinal effects and indicated active compounds. In: Phytomedicines of Europe. American Chemical Society, pp 176–209

Lechner BE, Monaldi S (2011) Utilization of garlic and maize wastes supplemented with olive mill waste water for Pleurotus ostreatus cultivation

Lee YH, Kim YI, Oh YK, Ahmadi F, Kwak WS (2017) Yield survey and nutritional evaluation of garlic stalk for ruminant feed. J Anim Sci Technol 59(1):22. https://doi.org/10.1186/s40781-017-0147-3

Lee S-Y, Choi Y, Kim J-K, Lee S-J, Bae JS, Jeong ED (2021) Biomass-garlic-peel-derived porous carbon framework as a sulfur host for lithium-sulfur batteries. J Ind Eng Chem 94:272–281. https://doi.org/10.1016/j.jiec.2020.10.046

Liang S, Guo X, Tian Q (2013) Adsorption of Pb2+, Cu2+ and Ni2+ from aqueous solutions by novel garlic peel adsorbent. Desalin Water Treat 51(37–39):7166–7171. https://doi.org/10.1080/19443994.2013.769919

Liao S, Liao L, Huang P, Wang Y, Zhu S, Wang X, Lv T, Li Y, Fan Z, Liu T, Lin Q (2022) Effects of different levels of garlic straw powder on growth performance, meat quality, antioxidant and intestinal mucosal morphology of yellow-feathered broilers. Front Physiol 13:902995

Liu W, Liu Y, Tao Y, Yu Y, Jiang H, Lian H (2014) Comparative study of adsorption of Pb(II) on native garlic peel and mercerized garlic peel. Environ Sci Pollut Res Int 21(3):2054–2063. https://doi.org/10.1007/s11356-013-2112-0

Liu L, Zhao X, Liu Y, Zhao H, Li F (2019) Dietary addition of garlic straw improved the intestinal barrier in rabbits1. J Anim Sci 97(10):4248–4255. https://doi.org/10.1093/jas/skz277

Lolok N, Mashar HM, Annah I, Saleh A, Yuliastri WO, Isrul M (2019) Antidiabetic effect of the combination of garlic peel extract (*Allium sativum*) and onion peel (*Allium cepa*) in rats with oral-glucose tolerance method. Res J Pharm Technol 12(5):2153–2156. https://doi.org/10.5958/0974-360X.2019.00357.3

Luo Y, Zhang F, Li C, Cai J (2022) Biomass-based shape-stable phase change materials supported by garlic peel-derived porous carbon for thermal energy storage. J Energy Storage 46:103929. https://doi.org/10.1016/j.est.2021.103929

Machado S, Grosso JP, Nouws HPA, Albergaria JT, Delerue-Matos C (2014) Utilization of food industry wastes for the production of zero-valent iron nanoparticles. Sci Total Environ 496:233–240. https://doi.org/10.1016/j.scitotenv.2014.07.058

Madhava Naidu M, Shyamala BN, Pura Naik J, Sulochanamma G, Srinivas P (2011) Chemical composition and antioxidant activity of the husk and endosperm of fenugreek seeds. LWT Food Sci Technol 44(2):451–456. https://doi.org/10.1016/j.lwt.2010.08.013

Mallek SB, Prather TS, Stapleton JJ (2007) Interaction effects of *Allium* spp. residues, concentrations and soil temperature on seed germination of four weedy plant species. Appl Soil Ecol 37(3):233–239. https://doi.org/10.1016/j.apsoil.2007.07.003

Manjunatha M, Samuel DVK, Anurag RK, Gaikwad N (2014) Development and performance evaluation of a garlic peeler. J Food Sci Technol 51(11):3083–3093. https://doi.org/10.1007/s13197-012-0879-5

Martins N, Petropoulos S, Ferreira ICFR (2016) Chemical composition and bioactive compounds of garlic (*Allium sativum* L.) as affected by pre- and post-harvest conditions: a review. Food Chem 211:41–50. https://doi.org/10.1016/j.foodchem.2016.05.029

Mohebby B (2008) Application of ATR infrared spectroscopy in wood acetylation. J Agric Sci Technol 10:253–259

Murali S, Shrivastava R, Saxena M (2007) Quantification of agricultural residues for energy generation—a case study. J Inst Public Health Eng 3:27–31

Muthamilselvi P, Karthikeyan R, Kumar BSM (2016) Adsorption of phenol onto garlic peel: optimization, kinetics, isotherm, and thermodynamic studies. Desalin Water Treat 57(5): 2089–2103. https://doi.org/10.1080/19443994.2014.979237

Naqvi SAZ, Irfan A, Zahoor AF, Zafar M, Maria A, Chand AJ, Ashfaq S (2020) Determination of antimicrobial and antioxidant potential of agro-waste peels. An Acad Bras Cienc 92(2): e20181103. https://doi.org/10.1590/0001-3765202020181103

Negi R, Satpathy G, Tyagi Y, Gupta R (2012) Biosorption of heavy metals by utilising onion and garlic wastes. Int J Environ Pollut 49:179–196. https://doi.org/10.1504/IJEP.2012.050898

Noda Y, Asada C, Sasaki C, Nakamura Y (2019) Effects of hydrothermal methods such as steam explosion and microwave irradiation on extraction of water soluble antioxidant materials from garlic husk. Waste Biomass Valorizat 10:3397–3402. https://doi.org/10.1007/s12649-018-0353-3

Park Y-K, Yoo ML, Lee HW, Park SH, Jung S-C, Park S-S, Kim S-C (2012) Effects of operation conditions on pyrolysis characteristics of agricultural residues. Renew Energy 42:125–130. https://doi.org/10.1016/j.renene.2011.08.050

Parlayıcı Ş, Pehlivan E (2015) Natural biosorbents (garlic stem and horse chestnut shell) for removal of chromium(VI) from aqueous solutions. Environ Monit Assess 187(12):763. https://doi.org/10.1007/s10661-015-4984-6

Qureshi AA, Din ZZ, Abuirmeileh N, Burger WC, Ahmad Y, Elson CE (1983) Suppression of avian hepatic lipid metabolism by solvent extracts of garlic: impact on serum lipids. J Nutr 113(9):1746–1755. https://doi.org/10.1093/jn/113.9.1746

Raaj E, Bhuvaneshwari K, Lakshmipathy R, Devi VV, Rico IL, Barik D (2022) Garlic Peel surface modification and fixed-bed column investigations towards crystal violet dye. Adsorpt Sci Technol 2022:1–9. https://doi.org/10.1155/2022/6904842

Rahman MS (2007) Allicin and other functional active components in garlic: health benefits and bioavailability. Int J Food Prop 10(2):245–268. https://doi.org/10.1080/10942910601113327

Rauf A, Abu-Izneid T, Thiruvengadam M, Imran M, Olatunde A, Shariati MA, Bawazeer S, Naz S, Shirooie S, Sanches-Silva A, Farooq U, Kazhybayeva G (2022) Garlic (*Allium sativum* L.): its chemistry, nutritional composition, toxicity, and anticancer properties. Curr Top Med Chem 22(11):957–972

Ray RC (ed) (2022) Fruits and vegetable wastes: valorization to bioproducts and platform chemicals. Springer Nature, Singapore, 448p

Reddy J, Rhim J-W (2014) Isolation and characterization of cellulose nanocrystals from garlic skin. Mater Lett 129:20–23. https://doi.org/10.1016/j.matlet.2014.05.019

Reddy KO, Maheswari CU, Reddy DJP, Rajulu AV (2009) Thermal properties of Napier grass fibers. Mater Lett 27(63):2390–2392. https://doi.org/10.1016/j.matlet.2009.08.035

Rose P, Whiteman M, Moore PK, Zhu YZ (2005) Bioactive S-alk(en)yl cysteine sulfoxide metabolites in the genus *Allium*: the chemistry of potential therapeutic agents. Nat Prod Rep 22(3):351–368. https://doi.org/10.1039/b417639c

Salim MH, Kassab Z, Abdellaoui Y, García-Cruz A, Soumare A, Ablouh E, El Achaby M (2022) Exploration of multifunctional properties of garlic skin derived cellulose nanocrystals and extracts incorporated chitosan biocomposite films for active packaging application. Int J Biol Macromol 210:639–653. https://doi.org/10.1016/j.ijbiomac.2022.04.220

Selvamani V, Ravikumar R, Suryanarayanan V, Velayutham D, Gopukumar S (2016) Garlic peel derived high capacity hierarchical N-doped porous carbon anode for sodium/lithium ion cell. Electrochim Acta 190:337–345. https://doi.org/10.1016/j.electacta.2016.01.006

Singh S, Sawarkar A (2020) Thermal decomposition aspects and kinetics of pyrolysis of garlic stalk. Energy Sourc Part A Rec Utilizat Environ Effects:1–11. https://doi.org/10.1080/15567036.2020.1716891

Singh RK, Patil T, Sawarkar AN (2020) Pyrolysis of garlic husk biomass: physico-chemical characterization, thermodynamic and kinetic analyses. Bioresour Technol Rep 12:100558. https://doi.org/10.1016/j.biteb.2020.100558

Sinha NK, Hui YH, Evranuz EO, Siddiq M, Ahmed J (2010) Handbook of vegetables and vegetable Processing. John Wiley, 788p

Sun RC, Sun XF (2002) Fractional and structural characterization of hemicelluloses isolated by alkali and alkaline peroxide from barley straw. Carbohydr Polym 49:415–423. https://doi.org/10.1016/S0144-8617(01)00349-6

Sun XF, Xu F, Sun RC, Fowler P, Baird MS (2005) Characteristics of degraded cellulose obtained from steam-exploded wheat straw. Carbohydr Res 340(1):97–106. https://doi.org/10.1016/j.carres.2004.10.022

Takagi H (1990) Garlic *Allium sativum* L. In: Onions and allied crops. CRC Press, Boca Raton, FL

Teixeira A, Sánchez-Hernández E, Noversa J, Cunha A, Cortez I, Marques G, Martín-Ramos P, Oliveira R (2023) Antifungal activity of plant waste extracts against phytopathogenic fungi: *Allium sativum* peels extract as a promising product targeting the fungal plasma membrane and cell wall. Horticulturae 9(2):136. https://doi.org/10.3390/horticulturae9020136

Thach NA (2022) Investigation of the effects of extraction temperature and time on bioactive compounds content from garlic (*Allium sativum* L.) husk. Front Sustain Food Syst 6:1004281

Thanikachalam K, Marimuthu K, Xavier R (2010) Effect of garlic peel on growth, hematological parameters and disease resistance against *Aeromonas hydrophila* in African catfish *Clarias gariepinus*(Bloch) fingerlings. Asian Pac J Trop Med 3:614–618. https://doi.org/10.1016/S1995-7645(10)60149-6

Vriesmann LC, de Mello Castanho Amboni RD, de Oliveira Petkowicz CL (2011) Cacao pod husks (*Theobroma cacao* L.): composition and hot-water-soluble pectin. Ind Crops Prod 34(1): 1173–1181. https://doi.org/10.1016/j.indcrop.2011.04.004

Wang Y, Wei K, Han X, Zhao D, Zheng Y, Chao J, Gou J, Kong F, Zhang C-S (2019) The antifungal effect of garlic essential oil on *Phytophthora nicotianae* and the inhibitory component involved. Biomol Ther 9(10):632. https://doi.org/10.3390/biom9100632

Yaseen R (2020) Antifungal potential of extracts produced from decomposed agricultural wastes by *Bosea thiooxidans*. Novel Res Microbiol J 4(3):790–807. https://doi.org/10.21608/nrmj.2020.95323

Yoshida S, Kasuga S, Hayashi N, Ushiroguchi T, Matsuura H, Nakagawa S (1987) Antifungal activity of ajoene derived from garlic. Appl Environ Microbiol 53(3):615–617

Za M, Ae E (2016) Proximate analysis of garlic (*Allium sativum*) paste treated with ascorbic and citric acids. J Food Process Technol 7(550):2. https://doi.org/10.4172/2157-7110.1000550

Zhao Y, Li W, Liu J, Huang K, Wu C, Shao H, Chen H, Liu X (2017) Modification of garlic peel by nitric acid and its application as a novel adsorbent for solid-phase extraction of quinolone antibiotics. Chem Eng J 326:745–755. https://doi.org/10.1016/j.cej.2017.05.139

Zhao Y, Liu X, Li W, Huang K, Shao H, Qu C, Liu J (2022) One-step synthesis of garlic peel derived biochar by concentrated sulfuric acid: enhanced adsorption capacities for Enrofloxacin and interfacial interaction mechanisms. Chemosphere 290:133263. https://doi.org/10.1016/j.chemosphere.2021.133263

Zhou N, Zhou Z, Qin Y, Zeng CJ, Huang ZQ (2012) Adsorption of two heavy metal ions on dry garlic stem. Adv Mater Res 396–398:2443–2446. https://doi.org/10.4028/www.scientific.net/AMR.396-398.2443

15 Life Cycle Assessment of Valorization of Root and Tuber Crop Wastes for Bio-commodities and Biofuels: Cassava as a Case Study

Noé Aguilar-Rivera

Abstract

The worldwide agribusiness from sugarcane, beet, cereals, and non-conventional crops as sources of carbohydrates (sugar and starch) are facing important sustainability, competitiveness, and profitability issues. Cassava (*Manihot esculenta* Crantz) is a multipurpose crop that represents important benefits for human health, animal feed, and economic well-being in subsistence culture. In different regions of the world, the whole plant, root, and by-products (peels, stems, straw, bagasse, leaves, and wastewater) have different uses ranging from agro-industrial use as a starch source, animal feed or medicinal, ornamental, religious, soil regenerator and with potential to produce bioethanol, biofertilizers, bioplastics, biofuels, and numerous bio commodities in a biorefinery without compete with the food security and creates jobs for unskilled labor force. However, it is necessary to use novel analytical tools to evaluate socioeconomic and environmental sustainability for each project based on cassava wastes. Therefore, the life cycle analysis (LCA) is a toolbox that will allow the stakeholder to make decisions for the success of productive projects. This work carried out an analysis of the methodological framework of the LCA and the circular economy (CE) for application to the biorefinery of cassava wastes and by-products, as well as the evaluation of challenges and constraints. Therefore, the efficient use of cassava as raw material can help to minimize waste of biological origin for disposal and the extraction, mining, and petrochemical of mineral and fossil resources and improve the sustainability of the value chain, biodiversity, and economic well-being.

N. Aguilar-Rivera (✉)
Facultad de Ciencias Biológicas y Agropecuarias, Universidad Veracruzana, Córdoba, Veracruz, Mexico
e-mail: naguilar@uv.mx

R. C. Ray (ed.), *Roots, Tubers, and Bulb Crop Wastes: Management by Biorefinery Approaches*, https://doi.org/10.1007/978-981-99-8266-0_15

Keywords

Cassava · Social-ecological system · Sustainability · Environmental impact

Abbreviations

AOP	Areas of protection
CBS	Cassava bagasse
CE	Circular economy
CE-LCA	Circular economy life cycle assessment
CPs	Cassava peels
CSF	Cassava starch facilities
GHG	Greenhouse gases
GWP	Global warming potential
LCA	Life cycle assessment
LCC	Life cycle costing
LCSA	Life cycle sustainability assessment
R&T	Roots and tubers
SES	Social-ecological system
S-LCA	Social life cycle assessment
SSF	Simultaneous saccharification and fermentation
WPTSSF	Without pre-hydrolysis time simultaneous saccharification and fermentation

15.1 Introduction

Root and tuber crops (R&T) are underutilized worldwide despite their nutritional and medicinal characteristics. Such crops are fundamental for survival, security, and food sovereignty in tribal, indigenous, and marginal areas by providing a diversified diet of carbohydrates and macro and micronutrients. Rural and tribal populations regularly use some roots, tubers, and bananas as plantations, backyard, or collection food sources. These plants are the fundamental basis of nutrients, culture, and tradition in tribal food habits. Nowadays, 90% of the calories from the diet are supplied by a small number of crops. However, the effects of urbanization, expansion of planting crops, climate change, and dwindling water supplies on agricultural output threaten R&T availability. Processing and value addition in R&T-based products nutritiously made from natural indigenous or rural ingredients to address food and nutritional security and health is a sustainable solution (Arutselvan et al. 2023). Therefore, the main technological solution to meet the growing needs of humanity is to increase the surface of the crops to produce food, feed, biofuel, biochemical, and new biological raw materials. Hence, innovations, transfer, and adoption of emerging and highly efficient technologies are required for sustainable crop production for the next generations.

Sustainable bioprocess with multi-variable situations involving many scenarios must be developed to multiply the environmental and socio-economic welfare with the development of new products. Exploring innovative applications for agro-waste by-products and biomass generated from several industries providing new "greener" technology of carbon neutral and low usage of chemicals to create strategies of human capital development and job opportunities, especially in rural areas (Kassim et al. 2020; Igbokwe et al. 2022).

15.2 Waste Biorefinery

The issues of agribusiness, agroindustry, and biorefinery sustainability have been widely investigated during the last decades by researchers from disciplinary multi, inter, and transdisciplinary approaches and numerous stakeholders and decision-makers. Nonetheless, productivity increase is generally based on actions derived from conventional management systems with high consumption of agrochemicals, machinery, and fossil fuels instead of incorporating agroecological practices.

Agro-industrial residues are marginally being used as animal feed and solid fuel for primary energy generation in rural areas, but the greatest proportion of waste is confined in the agricultural fields without having any sustainable use. In this sense, the search for transformation options and new value chains is a significant environmental and socioeconomic issue as productivity is threatened by the deterioration of natural resources in each agricultural cycle (Prasad et al. 2020).

According to Shirkhan et al. (2022) production of agricultural goods (fruits, vegetables, cereals, etc.), food security and sovereignty, generation, characterization and use of waste, conservation of the environment, and its relationship with human healthcare are issues to consider for crop. It is urgent to revalue these materials instead of considering them as residues to create new technological routes for sustainable use.

Nevertheless, it is urgently required to establish smart decisions based on transverse knowledge to replace highly polluting traditional agricultural practices with green techniques to guarantee higher agricultural productivity, food security, and recovery of agricultural waste for sustainable use in biorefineries as raw materials, revaluing these by-products.

Biorefineries aim to produce green bioproducts from diverse waste and by-products from agribusiness. However, new developments lack technological maturity and are subject to constraints due to the lack of sustainability or circular economy (CE) bases. Therefore, it is necessary to consider the application of frameworks such as life cycle assessment (LCA) to ensure the socioeconomic and environmental sustainability of the waste projects.

Mengistu and Panizzolo (2023) proposed indicators to improve industrial sustainability performance considering the triple-bottom-line approach.

1. *Economic*: Profit, revenue, material cost, labor cost, energy cost, operating cost, maintenance cost, production cost, packaging cost, inventory cost, R&D expenditure, product quality, lead time, and on-time delivery.
2. *Environmental*: Water consumption, energy consumption, material consumption, renewable energy use, renewable material use, recycled water use, recycled material use, hazardous material use, electricity consumption, energy efficiency, energy intensity, packaging material consumption, land use, GHG (Green House Gases) emissions, air emissions, ozone-depleting substances, wastewater discharge, hazardous waste, solid waste, recyclable waste, waste generated.
3. *Social*: Employee turnover, employee satisfaction, occupational health and safety, training and development, fair salary, work-related injuries, working hours, lost working days, number of employees, customer satisfaction, customer complaints, employment/job opportunity, corruption.

15.3 Life Cycle Assessment for Waste Biorefinery

Life Cycle Assessment (LCA) is a toolbox for evaluating and quantifying potential environmental burdens and socioeconomic impact of inputs, outputs, products, and processes throughout end-of-life disposition named cradle-to-grave (extraction of resources to disposal of wastes). It is an indispensable complement to CE approaches.

Blanco et al. (2023) identified three key aspects to be considered in the comparison of LCA appraisal in biorefineries: cradle-to-gate scope (mostly intermediate synthetic chemicals derived from mining or petrochemical), prospective analysis (emerging technologies under development), and numerous processes (biorefineries with several inputs and valuable outputs).

The main components are Life Cycle Assessment (LCA), Life Cycle Costing (LCC), Social Life Cycle Assessment (S-LCA), Life Cycle Sustainability Assessment (LCSA), and Circular Economy Life Cycle Assessment' (CE-LCA) is a novel and integrated tool to analyze the environmental performance in a CE framework (Larsen et al. 2022).

Therefore, LCA is the framework widely used to assess the environmental performance of crop systems at various scales, regions, productive systems, value chains, etc. Consequently, LCA tools have been applied in several geographical regions with different socioeconomic and agroecological constraints to their environmental performance to transition to sustainability (Gulisano et al. 2018).

LCA consists of four different and iterative steps defined by the ISO standards (ISO 14040, 14044):

Step 1: Goal definition and scoping. Identify the LCA's goal, the products, services, or activities of the evaluation, and determine the boundaries of the system.

Step 2: Life-cycle inventory. Calculating the energy sources and raw material inputs from nature and economy and environmental emissions releases associated with each life cycle phase (output).

Step 3: Impact analysis. Assess the impacts on human health, renewable and non-renewable natural resources, and ecosystems.

Step 4: Interpret and report results. Evaluate options to reduce fossil fuels, chemicals, as inputs, or environmental impacts at each stage of the product life-cycle expressed in kg CO_2 equivalents (Blanco et al. 2023; Ramos Huarachi et al. 2023; Saravanan et al. 2023).

Nonetheless, the different results obtained derived from the application of different LCA tools have challenged the credibility of the assessments, policy development, and progress toward or compliance with several goals such as GHG mitigation, productivity, and welfare. In agro-industrial systems and the use of renewable and natural resources, LCA tools often highlight how agricultural production, agribusiness, and biomass biorefineries are the weak points in the life cycle despite environmental impacts. This issue tends to vary according to agro-ecological characteristics, productivity, socioeconomic profile of crop growers and farming systems, for example, greenhouse vs. the open field, organic vs. conventional, and cultivated under different agro-climatic conditions and soil characteristics, properties and land tenure besides the generation and use of wastes, residues, and by-products from crops fields (Timpanaro et al. 2021).

LCA tools involve a comprehensive assessment of the factors that impact humans or pressure on ecosystems, characterizing these in the impact assessment stage using a 'midpoint' describing environmental 'problems.' Some of the commonly related impact categories of midpoint methods are climate change, ozone layer depletion, human toxicity, ecotoxicity, eutrophication, acidification, resource depletion, and photochemical ozone formation. These are, in some cases, linked to 'endpoint' indicators which describe environmental 'damage', severity or consequences of midpoint impacts categories in the areas of protection at the endpoint level, which are areas of protection (AOP): human health, natural environment, and natural resources or scarcity. These endpoint indicators describe and aggregate the 'damage' to aspects of social concern caused by a wide range of environmental problems from product systems (Cavalett et al. 2013; Hardaker et al. 2022).

The two evaluations are complementary because the midpoint has a stronger relation to the environmental flows and relatively low uncertainty, while the endpoint is related to environmental flows (Huijbregts et al. 2016).

Basset-Mens et al. (2021) concluded that the critical challenges to improve the quality and usefulness of LCA tools and carbon footprint in agro-industrial and agribusiness are:

- Lack of operational methods to capture the diversity of farming systems in field sampling procedures.
- Lack of consensual approaches to deal with agriculture multifunctionality (including various issues related to allocation among rotational crops, within multi-cropping systems, etc.).
- Lack of universally valid direct and indirect field emission models for all agriculturally relevant emissions under contrasted pedoclimatic conditions.

- Lack of suitable terrestrial ecotoxicity models.
- Lack of suitable models to account for agricultural impacts on soil quality, including biodiversity and salinization.

The above is particularly more critical in developing economies because occur three main constraints:

- A great diversity of production systems with little reliable data.
- Highly specific natural contexts with little data, knowledge, and tools for informing the inventory and impact assessment phases (especially for tropical systems).
- Stakeholders' varying awareness and capacities in relation to the environment and environmental assessment.

15.4 Cassava (*Manihot esculenta* Crantz)

Starch is the main biomaterial produced from cassava. Cassava starch is used for manufacturing several products, mainly in the food, pulp, and paper industries, and bioplastics obtained in cassava starch facilities (CSF) because of its high capacity to form films and ethanol E1. The cassava starch stepwise process can be divided into four main stages: (1) preparation (reception, weighing, and unloading of the roots, peeling, and washing); (2) rasping/pulping/grating; (3) recovery (starch sedimentation, washing, dewatering, drying); and (4) finishing (milling and packaging).

In relation to the above, starch-based bioethanol production consists of three stages: (1) starch liquefaction, (2) enzymatic saccharification, and (3) the fermentation process. Aruwajoye et al. (2020) evaluated simultaneous saccharification and fermentation(SSF) processes from cassava peel wastes (periderm and cortex) with a previous step of prehydrolysis (PTSSF) and without prehydrolysis (WPTSSF). The evaluation of the kinetics of yeast growth and bioethanol conversion and preliminary scale-up assessment obtained the highest bioethanol concentration with the optimized WPTSSF process.

According to Trakulvichean et al. (2019), to produce 1000 kg of starch requires approximately 4400 kg of cassava roots, 10.9 m^3 of water, 207.8 kWh of electricity, 1898 MJ of heat for drying the starch, 0.9 kg of chemicals, 93.1 m^3 of biogas necessary for heat and electricity generation. Wastes from the cassava starch production process typically consist of 600 kg of rhizomes, 170 kg of sand, 100 kg of cassava peels, 2500 kg of wet pulp, 8.4 m^3 of wastewater, and 4.0 kg of low-grade starch.

The by-product of the starch industry from cassava is called "thippi." The leaves and immature roots and stems are sources of organic matter and nutrients for manures such as vermicompost to improve soil fertility. However, these are very often burned or left in the field for decomposition (Nedunchezhiyan et al. 2011; Chithra and Sreekumar 2020).

The research of Thuppahige et al. (2023) evaluated the potential for the production of starch from cassava waste (peel and straw). The result showed that cassava peel is suitable to obtain starch over cassava bagasse with similar properties; therefore, the researchers concluded that there is a high potential of replacing starch from roots with cassava peel and bagasse as non-food-based starch for several applications.

Moreover, Fronza et al. (2023) determined that the starch extracted from cassava peels presents high purity and low values of ash, lipids, and protein contents. The starch has an amylopectin-to-amylose ratio of 5:1, a value within the limits considered adequate for applications in food matrices. The solubility and swelling power values were within the range of those necessary for applications in food products as thickeners and water retainers. The starch viscosity is suitable for products that need to flow through pipes in the food industry while providing the desired texture to the final product. They concluded that cassava peel starch presented great potential to be applied in the food industry, mainly in developing edible films, contributing to the reduction of environmental impacts from its inappropriate disposal of cassava wastes.

The review of Andrade et al. (2022) concluded that the environmental sustainability challenges for cassava lead to uncertainties due to energy balance and potential environmental and social impact as a complex social-ecological system. Therefore, data-generated LCA tools can help reduce bottlenecks found in the productive stages, making production more competitive to improve the economic well-being of regions of high marginalization. When applying an LCA to the cassava value chain, several possible impacts must be evaluated.

Besides, S-LCA is a useful tool to assess the social and socio-economic effects as part of the whole supply chains (production, processing, wholesale, retail, and consumer) (Desiderio et al. 2022; Prasara and Gheewala 2021) (Fig. 15.1).

Usubharatana and Phungrassami (2015) calculated the carbon footprint of cassava starch products. The results indicated that the cultivation and harvesting procedure was the major source of emitting GHGs, around 40–59% of the total. Moreover, the linear economy model "extraction–production–use" or 'take-make-use and dispose' that leads to environmental problems has been gradually replaced by the concept of CE as an alternative to linear production processes, which aims to maximize resource efficiency and waste reduction (Davis et al. 2023). Therefore, CE is a regenerative system in which resource input and waste, emission, and energy leakage are minimized by slowing, closing, and narrowing material and energy loops (van Stijn et al. 2021) (Fig. 15.2).

CE is a novel framework that adds up to sustainable development. The circularity is considered to have great potential to save primary mineral and fossil resources, preserve ecosystems, and minimize global warming. Moreover, the concept is expected to have social benefits, such as generating jobs, wealth, and welfare (Luthin et al. 2023).

Besides, the main existing LCA cassava studies are focused on the LCA of starch production or bioethanol from starch with a focus on the upstream processes, in-field emissions, and a specific starch plant, specific logistics, and production to provide

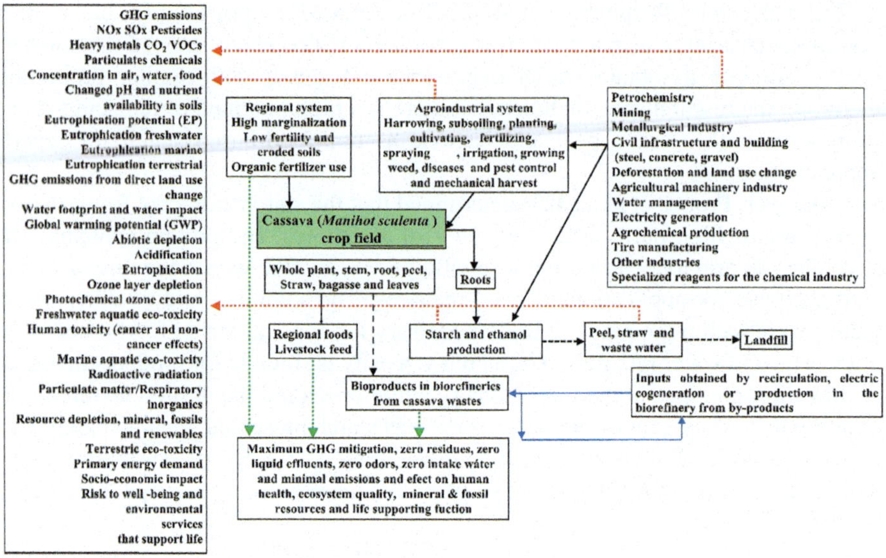

Fig. 15.1 Life cycle assessment for conventional, traditional production of cassava and waste biorefinery

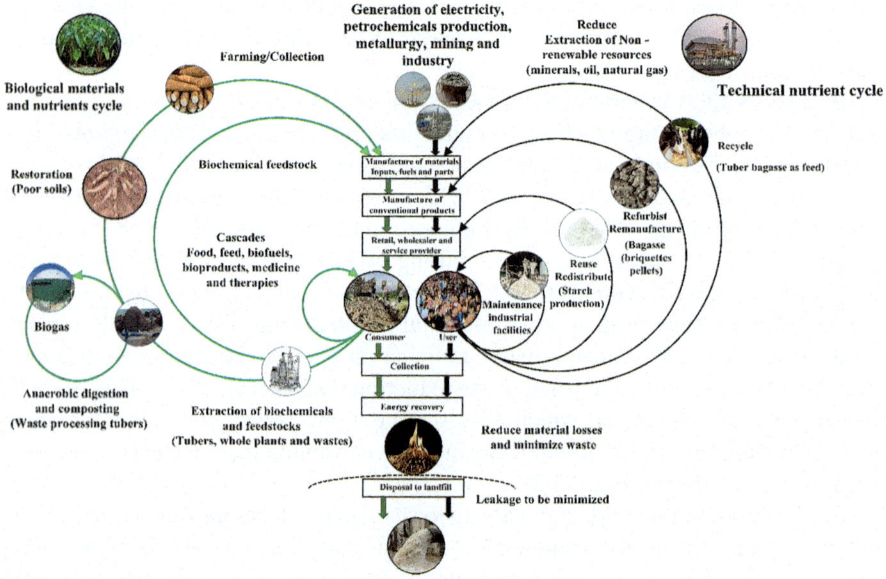

Fig. 15.2 Model of the circular economy for cassava value chain ecosphere

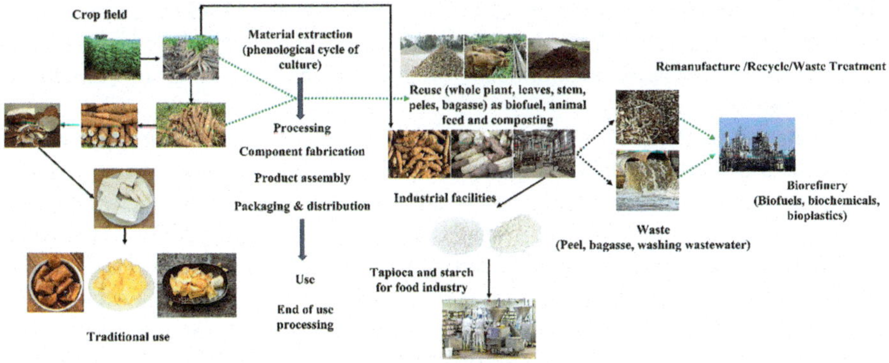

Fig. 15.3 Cassava value chain for conventional, traditional production of cassava and waste biorefinery

local stakeholders with an accurate environmental assessment (Padi and Chimphango 2021; Matheus et al. 2022; Andrade et al. 2022).

Therefore, biorefining, zero-waste technologies, and frameworks of sustainability assessment are essential components for achieving a sustainable economy by converting organic waste into profit. Waste conversion into bioproducts must be carried out with minimal external resources, such as freshwater, fossil fuel, electricity, and land use. The adoption of the framework of CE to valorize wastes and by-products from the cassava starch can lead to sustainable development, higher economic profit, and more efficient resource usage as biogas, bioethanol, fermented animal feed, and mushroom growth media through waste minimization, as well as environmental benefits identifying drivers and barriers (technology, economic, regulatory, and social responsibility) to the CE concept implementation for the cassava production (Lerdlattaporn et al. 2021) (Fig. 15.3).

15.5 Bioethanol from Cassava

Bioethanol from starchy crops, such as cassava, holds a remarkable potential because of its high production per hectare and comparatively less demand for use as food and fodder. The review by Thatoi et al. (2016) focuses on bioethanol production from cassava, sweet potato, potato, yam, aroids, sugar beet, etc., fermentation techniques, and microorganisms used in the fermentation process along with its prospects. The authors stand out as a priority in the biotechnological tools to convert biomass starchy into ethanol.

Bioethanol production from cassava root starch began in the 1970s, and stakeholders are currently doing numerous research on cassava by-products such as peel, bagasse, stem, rhizome, and leaves to determine potential uses in bioproducts (Paul et al. 2022) process. According to the above, the cassava-based process has a lower bioethanol yield than other raw materials (175 L ethanol/tonne).

Nevertheless, bioethanol from cassava plants gave the highest return on investment from this starchy material. This return on investment showed that the cassava plant is a more technically and economically viable starchy feedstock for bioethanol production.

For the production of bioethanol from cassava, the global warming potential is due to the transportation of large quantities of roots required for the process, tractor operations, and nitrogenous and potassium fertilizer application, but with severe environmental problems and long-term sustainability as acidification potential eutrophication potential. In this regard, cassava production contributes 30.15% of CO_2. Hence, chemical fertilizer application was identified as the main bottleneck. Therefore, as mentioned above, the application of liquid and solid organic inputs (compost, vermicompost, bokashi, etc.) derived from cassava plant residue left in the field and other wastes can reduce CO_2 emission by 30.15% for cassava production with a reduction in annual operating costs (Sanni et al. 2022).

Wang et al. (2013) concluded that eutrophication is the most significant environmental impact indicator, followed by acidification, global warming, human toxicity, and photochemical oxidation. They proposed the total elimination of fossil fuels for steam and electricity generation, replacing a cogeneration system, and the promotion of best management practices in the cultivation of cassava.

In this regard, Lyu et al. (2020) showed that the ethanol produced by integral plant (cassava whole plant and by-products) could improve the comprehensive utilization of straw, reduce land occupation, increase the energy benefits, and reduce environmental emissions because whole plant cassava is the most energy-saving, renewable and environmentally friendly bioethanol feedstock, followed by cassava root and finally cassava straw.

15.6 Cassava Bioproducts

Cassava contains more than 70% water. The bioproducts derived from cassava make it especially fitting as low-cost and high-calorific food for many poor people. It produces solid residues (peels, stems, and leaves) and wastewater, which is characterized by a high chemical oxygen demand, mineral nutrients (phosphorus, potassium, calcium, magnesium, sulfur, iron, zinc, manganese, copper, and nitrate) (de Oliveira Schmidt et al. 2023), and cyanogenic glucosides. The processing of one-ton roots can generate between 250 and 600 L of wastewater. Their potential is as a substrate for microalgae biomass production, biodiesel production as well and other value-added products. Solid wastes are 18% of the whole plant. In this regard, peels pose the greatest nuisance to the environment. It can be as high as 20–35% of the total weight of processed root. Leaves, stem, bagasse, and rhizome may account for 10%, 35%, 15%, and 10%, respectively, of the weight of the fresh tubers and need to be managed too (Ubalua 2007; Ekop et al. 2019; Oghenejoboh et al. 2021).

The solid and liquid waste from cassava processing is organic and acidic, which leads to various global concerns—primarily global warming and biodiversity loss. However, the valorization of cassava wastes and conversion into value-added products such as compost, biochar, and biofuels such as biogas, bioethanol, biohydrogen, and biobutanol (second-generation bioenergy production) associated with environmental management directs toward sustainable development.

Besides nanomaterials, nanofibrils and nanocrystals are produced by cassava starch and cellulose to the removal of contaminants from the wastewater and carriers for drug delivery in the biomedical field (Nizzy and Kannan 2022).

The value-adding from cassava has several applications in medical, industrial, biochemical, platform chemicals, biopolymers, or biofuels (Kamsuwan et al. 2023). Notwithstanding cassava's importance in food and bioenergy, its cultivation is faced with a myriad of challenges. The review of Fathima et al. (2023) concluded the importance of improving cassava's resilience to challenges affecting its products for both food and bioenergy as alternative feedstock such as stems, leaves, and peels. These strategies would ensure farmers can venture into cassava for bioenergy without jeopardizing its status as an important food crop.

Cassava waste presented 70% of starch, with a high potential for bioproducts. The advantage of investing in cassava's waste valorization is that starch by-products can be chemically easier to transform than lignocellulosic ones. Another advantage is the large quantity of starch waste that is adhered to the lignocellulosic peel. Therefore, residues are a source of high-added-value products such as fermentable sugars, 5-hydroxymethylfurfural, and dextrin. Therefore, investing in the hydrolysis of cassava's total residue can be more profitable than hydrolyzing only the peels to obtain fermentable sugars or dextrin due to the more fragile glycosidic bonds of amylose and amylopectin (Martins et al. 2023).

Cruz et al. (2021) explored the anaerobic digestion for biogas generation, process intensification, scale scale-up strategies, and challenges. Besides, the authors concluded that it is necessary for research institutes, universities, researchers, and government to work together to create public incentives for its utilization to diversify the energy matrix.

Cassava peel being is a by-product of the storage roots, which will have a degree of preservation under environmental conditions because cassava roots are susceptible to post-harvest physiological deterioration. However, this by-product is usually disposed of as waste. There are currently a lot of research reports that have been driven to determine the nutritional potential and benefits in several regions and cassava cultivars.

The composition of this by-product is 11–12% moisture content, 58–62% volatile matter, 28–30% fixed carbon, and 0.2–0.4% for ash content, respectively. Cassava peels are a raw material to produce activated carbon due to having commendable physiochemical and ultimate characteristics for applications in water industries, beverages, breweries, and pharmaceuticals, according to the research of Kayiwa et al. (2021) considering process technology, economics, and environmental issues at each stage.

Aisien and Aisien (2020) carried out a proximal analysis of cassava peel, concluding that 8–15% of the total is a dry matter of the root, the composition consists of 20–31% hemicelluloses, 16–42% cellulose, and 6–8% lignin; also contain 81.9–93.9% organic matter and 4.1–6.5% crude protein. Their mineral content of cassava peel is 48.7% C; 1% N; 1.1% K; 1.6% P; 0.16% NO_3; 0.15% Na; 0.9% Ca; 125 mg/kg Zn; 15 mg/kg Cu; 180 mg/kg Mn; 16.7 mg/kg Pb; 48.7 C/N; and 52.6% ash.

The review of Tonukari et al. (2023) highlights the biochemical composition of cassava peels, which are 1–4 mm thick and account for 10–12% of the total dry matter of the root, the effects of fermentation on its nutritional and anti-nutrient constituents, as well as several bioproducts obtainable from its biotransformation as sugars and confectionaries, bioethanol, animal feeds, industrial enzymes, biogas, bioethanol, organic acids (including citric acid), tetracycline, alternative fiber sources for pulp and paper making and treatment of wastewater. The chemical composition of cassava peels is composed of organic carbon 48.7 (%), total nitrogen 1.0 (%), C/N ratio 48.7, K 1.1 (%), P 1.6 (%), NO_3 0.16 (%), Zn 125 (mg/kg), Cu 15 (mg/kg), Mn 180 (mg/kg), pH 6.4, Na 0.15 (%), Ca 0.9 (%), Pb 16.7 (mg/kg), and Ash 52.6 (%).

The use of cassava leaves and peel in food and feed applications is constrained due to their high levels of antinutritional components, such as cyanogenic glycosides (linamarin and lotaustralin, cyanohydrin, and free cyanide) (Mukhtar et al. 2023). Cassava leaves, a biomass resource available in many countries, are largely underutilized because high fibrous content and toxic chemicals. For these reasons, Olude et al. (2023) proposed their utilization as cassava leaf meal (CLM) in aquafeed exploring strategies that will ensure higher utilization of cassava leaves to the extent of complete substitution of defatted soybean meal in fish diet.

Miranda et al. (2023) developed suitable cassava leaves extract-based films evaluating the influence of extraction parameters (solid-liquid ratio, temperature, and pH) on the amount of protein, carbohydrates, and total solids on thickness, color, mechanical properties, morphology, surface properties, and water vapor permeability.

15.7 Prospective for Cassava Sustainability

Currently, the cassava plant is cultivated at a lower cost, labor, and inputs as chemical fertilizer. Numerous bioproducts can be developed using cassava roots or whole plants because the main chemical components are starch (about 20% by w/w), fiber, cellulose, and hemicellulose. The applications of the cassava plant are various as food, feed, textile, pharmaceutical, detergent, alcohols, organic acids, glucose, and bio-plastic industries. Moreover, the waste generated during the production of cassava-based products, including peels, wastewater, bagasse, pulp, etc., can be converted into value via bioprocessing with novel technologies and strategies for sustainable management (Fernando et al. 2022) (Table 15.1).

It is noteworthy to mention that Fernando et al. (2022) review the development of cassava plant biorefinery to add more value to the whole plant root, leaves, and stem, waste management processes, and finally, the problems and challenges associated with cassava-based biorefinery system:

Table 15.1 Technological development for sustainable management

Technological development	Reference
Evaluation of the potential for producing bioelectricity, biogas, biochar, and manures from cassava peel and stem	Aduba et al. (2023)
Economic and environmental performances of cassava waste bioenergy (CWB) in Africa with SimaPro® based gate-to-gate life cycle analysis for sustainability enhancement in cassava industries	Padi et al. (2022)
The use of cassava waste pulps (CWP) to produce biofuel biohydrogen (bio-H_2) with a yield of 99.89% bio-H_2 component of the total biogas produced through dark fermentation. One kilogram of fresh CWP can produce 12.39 ± 0.8 L bio-H_2	Balbin et al. (2022)
Optimization of cassava waste composting for the application of agriculture	He et al. (2023)
Bio-oil production via fast pyrolysis of cassava rhizome yield ranged between 58% and 60%, char represented 17–19%, and gas 21–24%	Rueangsan et al. (2023)
Production of torrefied briquettes as a thermochemical process to increase the calorific value from cassava rhizome to a relaxed density of 1270 kg/m^3	Granado et al. (2023)
Technological trends on anaerobic treatment of wastewater from cassava processing toward a green and low-carbon economy	dos Santos et al. (2022)
Evaluation of starch, fibers, polymers, and composites obtained from cassava and their potential industrial applications	Abotbina et al. (2022)

1. Cassava plant
2. Morphology of cassava plant
3. Harvesting and production capacities of cassava
4. Cassava as a food source
5. Cassava products for animal feeding (cassava chips, pellets, and meals for pigs, cattle, and poultry)
6. Cassava starch in the food industry
7. Industrial uses of cassava (adhesive production, paper industry, plywood industry, textile industry, pharmaceutical industry, detergent industries, board production from cassava stalks, and bio-plastics production
8. Waste management process (peels, fibrous by-products from crushing and sieving as bagasse or pulp, starch residues after starch settling, and wastewater effluents are generated during the starch and ethanol)
9. Cassava plant-based biorefinery

Notwithstanding, tradition, population, and culture determined the production of cassava. Besides, climate change affects cassava production due to increased erosion menace, scarcity of irrigation water, reduced soil fertility, increased disease of crops, and reduced profitability of cassava growers.

Osuji et al. (2023) mentioned that the main adaptation strategies for the sustainability of cassava crops are:

1. Planting improved cassava varieties
2. Insurance
3. Planting different crops

4. Livelihood diversification
5. Soil and water conservation techniques
6. Adjusting planting and harvesting dates
7. Irrigation
8. Reliance on climate information and forecasts
9. Collaboration with extension workers/agents
10. Adjusting planting and harvesting dates
11. Efficient and effective use of pesticides
12. Increased land access
13. Erosion control measures

Therefore, the use of tools derived from Agriculture 4.0, such as digitalization and agroecological zoning, will be useful for reducing the environmental footprint and maximizing cassava productivity.

Additionally, de Jesus et al. (2022) recommended eco-innovation in agro-industry as cassava must be carried out in eight lines of action: (1) environmental, (2) social, (3) economic, (4) human resource training, (5) institutional development, (6) introduction of innovation, (7) unwanted occurrences, and (8) characteristics of environmental management.

15.8 Conclusions

This work analyzed the applications, trends, challenges, and the constraints for possible future starch from cassava (whole plant, root, leaves, stem, and wastewater) as a promising field in green chemistry without competing with other crops for food or feed. Besides, Cassava is an underutilized crop with promising economic value. Therefore, several strategies related to the sustainability and valorization of integral cassava as a raw material, innovative production processes, and natural and environmental impacts must be discussed in future projects.

The cassava's potential for food, feed, biofuels, and bioproducts requires a synergy of conventional and emergent technologies such as agroecological zoning, LCA, and improvements in the institutional and policy environment because cassava is produced at the tropical agro-climatic environment in great abundance without agrochemicals.

Cassava waste from whole plants is a good source of fiber for animal feed, pulp, and paper industries and as biodegradable packaging. In addition, cassava peel can also be processed to produce bioethanol and bioenergy. Therefore, the efficient utilization of cassava plants and by-product processing can help reduce waste and improve the sustainability of cassava production.

References

Abotbina W, Sapuan SM, Ilyas RA, Sultan MTH, Alkbir MFM, Sulaiman S, Bayraktar E (2022) Recent developments in cassava (*Manihot esculenta*) based biocomposites and their potential industrial applications: a comprehensive review. Materials 15(19):6992. https://doi.org/10.3390/ma15196992

Aduba CC, Ndukwe JK, Onyejiaka CK, Onyeiwu SC, Moneke AN (2023) Integrated valorization of cassava wastes for production of bioelectricity, biogas and biofertilizer. Waste Biomass Valoriz:1–17. https://doi.org/10.1007/s12649-023-02126-3

Aisien FA, Aisien ET (2020) Biogas from cassava peels waste. Detritus 10:100–108. https://doi.org/10.31025/2611-4135/2020.13910

Andrade LRS, Felisardo RJA, Cruz IA, Bilal M, Iqbal HM, Mulla SI, Ferreira LFR (2022) Integrated biorefinery and life cycle assessment of cassava processing residue–from production to sustainable evaluation. Plants 11(24):3577. https://doi.org/10.3390/plants11243577

Arutselvan R, Raja K, Pati K, Chauhan VBS, Nedunchezhiyan M (2023) Tuber crops and their potential in food and nutritional security. In: Pandemics and innovative food systems. Taylor & Francis Group, pp 122–136. https://www.taylorfrancis.com/chapters/edit/10.1201/9781003191223-7/tuber-crops-potential-food-nutritional-security-arutselvan-raja-kalidas-pati-vbs-chauhan-nedunchezhiyan

Aruwajoye GS, Sewsynker-Sukai Y, Kana EG (2020) Valorisation of cassava peels through simultaneous saccharification and ethanol production: effect of prehydrolysis time, kinetic assessment and preliminary scale up. Fuel 278:118351. https://doi.org/10.1016/j.fuel.2020.118351

Balbin AJFT, Gonzales HA, Bornillo LAM, Tagupa JHA, Jaro RH, Ido A, Arazo RO (2022) High biohydrogen yield from fresh cassava waste pulps via gas dissolution in a dark fermentative environment. Clean Waste Syst 3:100039. https://doi.org/10.1016/j.clwas.2022.100039

Basset-Mens C, Avadí A, Bessou C, Acosta-Alba I, Biard Y, Payen S (2021) Life cycle assessment of agre-food systems: an operational guide dedicated to developing and emerging economies. Editions Quae, p 210. https://library.oapen.org/handle/20.500.12657/57571

Blanco J, Martín S, Linares M, Iglesias J, Moreno J (2023) Life cycle assessment applied to bio-based platform molecules: critical review of methodological practices. J Clean Prod 414:137513. https://doi.org/10.1016/j.jclepro.2023.137513

Cavalett O, Chagas MF, Seabra JE, Bonomi A (2013) Comparative LCA of ethanol versus gasoline in Brazil using different LCIA methods. Int J Life Cycle Assess 18:647–658. https://doi.org/10.1007/s11367-012-0465-0

Chithra S, Sreekumar J (2020) Mineralization of Thippi (cassava starch factory solid residue) compost under incubation. J Root Crops 46(1):33–40. http://journal.isrc.in/index.php/jrc/article/view/571

Cruz IA, Andrade LRS, Bharagava RN, Nadda AK, Bilal M, Figueiredo RT, Ferreira LFR (2021) Valorization of cassava residues for biogas production in Brazil based on the circular economy: an updated and comprehensive review. Clean Eng Technol 4:100196. https://doi.org/10.1016/j.clet.2021.100196

Davis MM, Vallejo A, Criollo P, Domenech T (2023) The 'Human Sphere' and the figure of 8 as the enabler of circular economy in developing countries: a case study. In: Urban and transit planning: city planning: urbanization and circular development. Springer International Publishing, Cham, pp 207–218. https://doi.org/10.1007/978-3-031-20995-6_19

Desiderio E, García-Herrero L, Hall D, Segrè A, Vittuari M (2022) Social sustainability tools and indicators for the food supply chain: a systematic literature review. Sustainable Prod Consumption 30:527–540. https://doi.org/10.1016/j.spc.2021.12.015

de Jesus MAS, Aguiar Dutra ARD, Cirani CBS, Jesus KRE, Neto RCS, Guerra JBA (2022) Eco-innovation assessment of biodigesters technology: an application in cassava processing industries in the south of Brazil, Parana state. Clean Technol Environ Policy:1–18. https://doi.org/10.1007/s10098-021-02232-7

de Oliveira Schmidt VK, de Vasconscelos GMD, Vicente R, de Souza CJ, Della-Flora IK, Degang L, de Andrade CJ (2023) Cassava wastewater valorization for the production of biosurfactants: surfactin, rhamnolipids, and mannosileritritol lipids. World J Microbiol Biotechnol 39(2):65. https://doi.org/10.1007/s11274-022-03510-2

dos Santos AL, Castro ALS, Salomon KR, de Souza TS, Vich DV (2022) Global research trends on anaerobic digestion and biogas production from cassava wastewater: a bibliometric analysis. J Chem Technol Biotechnol 97(6):1379–1389. https://doi.org/10.1002/jctb.6976

Ekop IE, Simonyan KJ, Evwierhoma ET (2019) Utilization of cassava wastes for value added products: an overview. Int J Sci Eng Sci 3(1):31–39. http://ijses.com/wp-content/uploads/2019/01/171-IJSES-V3N1.pdf

Fathima AA, Sanitha M, Tripathi L, Muiruri S (2023) Cassava (*Manihot esculenta*) dual use for food and bioenergy: a review. Food Energy Secur 12(1):e380. https://doi.org/10.1002/fes3.380

Fernando NML, Amaraweera APSM, Gunawardane OHP, Wanninayaka WMDB, Manipura A, Manamperi WAR, Kulathunga KMAK (2022) Sustainable biorefinery approach for cassava. Engineer 55(2):71–88. https://doi.org/10.4038/engineer.v55i2.7510

Fronza P, Costa ALR, Franca AS, de Oliveira LS (2023) Extraction and characterization of starch from cassava peels. Starch-Stärke 75(3–4):2100245. https://doi.org/10.1002/star.202100245

Granado MPP, Gadelha AMT, Rodrigues DS, Antonio GC, De Conti AC (2023) Effect of torrefaction on the properties of briquettes produced from agricultural waste. Bioresour Technol Rep 21:101340. https://doi.org/10.1016/j.biteb.2023.101340

Gulisano G, Strano A, De Luca AI, Falcone G, Iofrida N, Stillitano T (2018) Evaluating the environmental, economic, and social sustainability of agro-food systems through life cycle approaches. In: Sustainable food systems from agriculture to industry. Academic Press, pp 123–152. https://doi.org/10.1016/B978-0-12-811935-8.00004-4

Hardaker A, Styles D, Williams P, Chadwic D, Dandy N (2022) A framework for integrating ecosystem services as endpoint impacts in life cycle assessment. J Clean Prod 370:133450. https://doi.org/10.1016/j.jclepro.2022.133450

He X, Cong R, Gao W, Duan X, Gao Y, Li H, Luo J (2023) Optimization of composting methods for efficient use of cassava waste, using microbial degradation. Environ Sci Pollut Res 30(17): 51288–51302. https://doi.org/10.1007/s11356-023-25818-8

Huijbregts MA, Steinmann ZJ, Elshout PM, Stam G, Verones F, Vieira MDM, van Zelm R. (2016). ReCiPe 2016: a harmonized life cycle impact assessment method at midpoint and endpoint level report I: characterization. https://rivm.openrepository.com/handle/10029/620793

Igbokwe VC, Ezugworie FN, Onwosi CO, Aliyu GO, Obi CJ (2022) Biochemical biorefinery: a low-cost and non-waste concept for promoting sustainable circular bioeconomy. J Environ Manag 305:114333. https://doi.org/10.1016/j.jenvman.2021.114333

Kamsuwan C, Tanprasert S, Nimmanterdwong P, Piemjaiswang R, Saencharee K, Pumchumpol T, Chalermsinsuwan B (2023) Process simulation of cassava cellulosic ethanol production with low to medium-pressure steam heat integration. Energy Rep 9:11–17. https://doi.org/10.1016/j.egyr.2023.04.246

Kassim MA, Meng TK, Serri NA, Yusoff SB, Shahrin NAM, Seng KY, Keong LC (2020) Sustainable biorefinery concept for industrial bioprocessing. In: Biorefinery production technologies for chemicals and energy. Wiley, pp 15–53. https://doi.org/10.1002/9781119593065.ch2

Kayiwa R, Kasedde H, Lubwama M, Kirabira JB (2021) The potential for commercial scale production and application of activated carbon from cassava peels in Africa: a review. Bioresour Technol Rep 15:100772. https://doi.org/10.1016/j.biteb.2021.100772

Larsen VG, Tollin N, Sattrup PA, Birkved M, Holmboe T (2022) What are the challenges in assessing circular economy for the built environment? A literature review on integrating LCA, LCC and S-LCA in life cycle sustainability assessment, LCSA. J Build Eng 50:104203. https://doi.org/10.1016/j.jobe.2022.104203

Lerdlattaporn R, Phalakornkule C, Trakulvichean S, Songkasiri W (2021) Implementing circular economy concept by converting cassava pulp and wastewater to biogas for sustainable

production in starch industry. Sustain Environ Res 31(1):20. https://doi.org/10.1186/s42834-021-00093-9
Luthin A, Traverso M, Crawford RH (2023) Assessing the social life cycle impacts of circular economy. J Clean Prod 386:135725. https://doi.org/10.1016/j.jclepro.2022.135725
Lyu H, Zhang J, Zhai Z, Feng Y, Geng Z (2020) Life cycle assessment for bioethanol production from whole plant cassava by integrated process. J Clean Prod 269:121902. https://doi.org/10.1016/j.jclepro.2020.121902
Martins F, Abreu MCC, Diniz PAF, Marques GN, Oliveira MM, Mendonça CDJS, Cavalcante KSB (2023) Hydrolysis of cassava's (*Manihot esculenta* Crantz) waste polysaccharides using the natural zeolite stilbite-Ca as catalyst. Bioresour Technol Rep 21:101384. https://doi.org/10.1016/j.biteb.2023.101384
Matheus JRV, de Farias PM, Satoriva JM, de Andrade CJ, Fai AEC (2022) Cassava starch films for food packaging: trends over the last decade and future research. Int J Biol Macromol 225:658. https://doi.org/10.1016/j.ijbiomac.2022.11.129
Mengistu AT, Panizzolo R (2023) Analysis of indicators used for measuring industrial sustainability: a systematic review. Environ Dev Sustain 25(3):1979–2005. https://doi.org/10.1007/s10668-021-02053-0
Miranda CG, Speranza P, Sato ACK (2023) Evaluation of cassava leaves extract as a material to produce biopolymer-based films. Food Hydrocoll 108944:108944. https://doi.org/10.1016/j.foodhyd.2023.108944
Mukhtar A, Latif S, Barati Z, Müller J (2023) Valorization of cassava by-products: cyanide content and quality characteristics of leaves and peel. Appl Sci 13(10):6340. https://doi.org/10.3390/app13106340
Nedunchezhiyan M, Jata SK, Byj G, Veena SS (2011) Effect of tuber crop wastes/byproducts on nutritional and microbial composition of vermicomposts and duration of the vermicomposting process. J Bot 2011. https://downloads.hindawi.com/archive/2011/801703.pdf
Nizzy AM, Kannan S (2022) A review on the conversion of cassava wastes into value-added products towards a sustainable environment. Environ Sci Pollut Res 29(46):69223–69240. https://doi.org/10.1007/s11356-022-22500-3
Oghenejoboh KM, Orugba HO, Oghenejoboh UM, Agarry E (2021) Value added cassava waste management and environmental sustainability in Nigeria: a review. Environ Challenges 4:100127. https://doi.org/10.1016/j.envc.2021.100127
Olude OO, Sahu NP, Sardar P, Nuzaiba PM (2023) Utilization of valorized cassava leaf meal as an alternative feedstuff to defatted soybean meal in feed for rohu, labeorohita fingerlings. Bioresour Technol Rep 22:101400. https://doi.org/10.1016/j.biteb.2023.101400
Osuji EE, Igberi CO, Ehirim NC (2023) Climate change impacts and adaptation strategies of cassava farmers in Ebonyi state, Nigeria. J Agric Extension 27(1):35–48. https://doi.org/10.4314/jae.v27i1.4
Padi RK, Chimphango A (2021) Comparative sustainability assessments for integrated cassava starch wastes biorefineries. J Clean Prod 290:125171. https://doi.org/10.1016/j.jclepro.2020.125171
Padi RK, Chimphango A, Roskilly AP (2022) Economic and environmental analysis of waste-based bioenergy integration into industrial cassava starch processes in Africa. Sust Prod Consum 31:67–81. https://doi.org/10.1016/j.spc.2022.02.002
Paul NE, Sudeshkumar M, Duraimurugan P, Jayaseelan V (2022) Synthesis and characterization of cardanol oil and cassava tuber peel biochar toughened epoxy composite coating for structural application. Biomass Convers Biorefin 13:7301. https://doi.org/10.1007/s13399-022-03128-2
Prasad S, Singh A, Korres NE, Rathore D, Sevda S, Pant D (2020) Sustainable utilization of crop residues for energy generation: a life cycle assessment (LCA) perspective. Bioresour Technol 303:122964. https://doi.org/10.1016/j.biortech.2020.122964
Prasara-A J, Gheewala SH (2021) An assessment of social sustainability of sugarcane and cassava cultivation in Thailand. Sustainable Prod Consumption 27:372–382. https://doi.org/10.1016/j.spc.2020.11.009

Ramos Huarachi DA, Hluszko C, Ulloa MIC, Moretti V, Ramos Quispe JA, Puglieri FN, Francisco ACD (2023) Life cycle thinking for a circular bioeconomy: current development, challenges, and future perspectives. Sustainability 15(11):8543. https://doi.org/10.3390/su15118543

Rueangsan K, Heman A, Kraisoda P, Tasarod H, Duanguppama K, Trisupakitti S, Morris J (2023) Bio-oil production via fast pyrolysis of cassava residues combined with ethanol and volcanic rock in a free-fall reactor. Cogent Eng 10(1):2156054. https://doi.org/10.1080/23311916.2022.2156054

Sanni A, Olawale A, Sani YM, Kheawhom S (2022) Sustainability analysis of bioethanol production from grain and tuber starchy feedstocks. Sci Rep 12(1):20971. https://doi.org/10.1038/s41598-022-24854-7

Saravanan A, Karishma S, Kumar PS, Rangasamy G (2023) A review on regeneration of biowaste into bio-products and bioenergy: life cycle assessment and circular economy. Fuel 338:127221. https://doi.org/10.1016/j.fuel.2022.127221

Shirkhan F, Mostafidi M, Zahedi MT, Ziarat P, Vambo V, Vambol S (2022) Green technologies and environmental management: a new understanding and approach to the use of agricultural waste. Lett Appl NanoBioSci 11:3065–3075. https://nanobioletters.com/wp-content/uploads/2021/06/22846808111.30653075.pdf

Thatoi H, Das PK, Mohapatra S, Swain MR (2016) Bioethanol production from tuber crops using fermentation technology: a review. Int J Sustain Energy 35(5):443–468. https://doi.org/10.1080/14786451.2014.918616

Thuppahige VTW, Moghaddam L, Welsh ZG, Wang T, Xiao HW, Karim A (2023) Extraction and characterisation of starch from cassava (Manihot esculenta) agro-industrial wastes. LWT 182:114787. https://doi.org/10.1016/j.lwt.2023.114787

Timpanaro G, Branca F, Cammarata M, Falcone G, Scuderi A (2021) Life cycle assessment to highlight the environmental burdens of early potato production. Agronomy 11(5):879. https://doi.org/10.3390/agronomy11050879

Tonukari N, Anigboro AA, Avwioroko OJ, Egbune EO, Ezedo T, Ajoh AI, Aganbi E (2023) Biochemical properties and biotechnological applications of cassava peels. Biotechnol Mol Biol Rev 14(1):1–8. https://doi.org/10.5897/BMBR2023.0292

Trakulvichean S, Chaiprasert P, Otmakhova J, Songkasiri W (2019) Integrated economic and environmental assessment of biogas and bioethanol production from cassava cellulosic waste. Waste Biomass Valoriz 10:691–700. https://doi.org/10.1007/s12649-017-0076-x

Ubalua AO (2007) Cassava wastes: treatment options and value addition alternatives. Afr J Biotechnol 6(18):2065–2073

Usubharatana P, Phungrassami H (2015) Carbon footprint of cassava starch production in North-Eastern Thailand. Procedia CIRP 29:462–467. https://doi.org/10.1016/j.procir.2015.02.031

van Stijn A, Eberhardt LM, Jansen BW, Meijer A (2021) A circular economy life cycle assessment (CE-LCA) model for building components. Resour Conserv Recycl 174:105683. https://doi.org/10.1016/j.resconrec.2021.105683

Wang M, Shi Y, Xia X, Li D, Chen Q (2013) Life-cycle energy efficiency and environmental impacts of bioethanol production from sweet potato. Bioresour Technol 133:285–292. https://doi.org/10.1016/j.biortech.2013.01.067

Prospective for Biorefineries Development from Roots, Tubers, and Bulb Crop Wastes and By-Products: Value Addition and Circular Economy

16

Ramesh C. Ray, Noé Aguilar-Rivera, and B. B. Sooch

Abstract

Roots, tubers, and bulb crops are primarily used for human consumption and, to a lesser extent in the food, starch, pharmaceutical, and animal feed industrial sectors. During the processing of these crops, a lot of waste in the forms of peels, skin, pulp, and bagasse is generated. For sustainable industrial processes focused on waste valorization, critical aspects such as (1) the technical feasibility of the process at the industrial level, (2) techno-economic analysis of the process and products, (3) life-cycle analysis of benefits: cost factors and (4) socio-economic impacts and well-being need to be considered. Most studies reviewed on the roots and tuber crop waste valorizations are limited to lab-scale studies and only a few at pilot-scale levels. Further, research and technical know-how are necessary to assist up-scale performance, feedstock security, and economic and environmental aspects of root and tuber crops' waste valorization processes.

Keywords

Bagasse · Biorefinery · Circular economy · Bulbs · Roots and tubers · RTBCWs · Techno-economic assessment · Valorization

R. C. Ray (✉)
ICAR-Central Tuber Crops Research Institute, Regional Center, Bhubaneswar, India

N. Aguilar-Rivera
Facultad de Ciencias Biológicas y Agropecuarias, Universidad Veracruzana, Córdoba, Veracruz, Mexico

B. B. Sooch
Enzyme Biotechnology Laboratory, Department of Biotechnology, Punjabi University, Patiala, India

R. C. Ray (ed.), *Roots, Tubers, and Bulb Crop Wastes: Management by Biorefinery Approaches*, https://doi.org/10.1007/978-981-99-8266-0_16

Abbreviation

RTBCWs Roots, tubers, bulb crop wastes

16.1 Introduction

The main roots and tuber crops include potato, sweet potato, cassava, beetroot, carrot and turnip, yams and aroids, and Jerusalem artichoke and bulb crops such as onion and garlic, Roots, tubers, and bulb crops are primarily used for human consumption as energy source and recently as functional foods (Ashogbon 2021), to a lesser extent in the food, starch, pharmaceutical, and animal feed industrial sectors. Tuber crops or subterranean crops are considered as plants among the oldest on Earth, they are also the crops of the future (Sunitha et al. 2023). During the processing of these crops, a lot of waste, either dry or wet, in the forms of peels, skin, pulp, straw, bagasse, and wastewater is generated. There are several options to valorize these wastes. Agricultural waste biorefineries can significantly reduce environmental burden and provide sustainable resources to synthesize value-added biochemicals, biomaterials, and biofuels (Panda et al. 2018). This will significantly contribute to the transition from a linear-based economy to a more circular economy. A variety of bio-based materials, chemicals, enzymes, organic acids, and biofuels can be produced from roots, tuber, and bulb crop wastes (RTBCWs) by integrating various technologies such as pyrolysis, anaerobic digestion, gasification, aerobic composting, and fermentation to gain phytonutrients, energy, and material recovery (Chaps. 1–15 in this book). Bio-waste biorefinery contributes as a driving force to cope with the challenges of resource scarcity, food waste disposal, climate change, and increased demand (Roy et al. 2023).

For sustainable industrial processes focused on waste valorization, critical aspects such as (1) the technical feasibility of the process at the industrial level, (2) techno-economic analysis of the process and products, and (3) life-cycle analysis of benefits: cost factors need to be considered; and (4) socio-economic impacts and well-being. In this chapter, we analyze all these aspects concerning RTBCWs based on the reviews drawn from the foregoing chapters in this book.

16.2 Waste-Based Value-Added Products from Bulb, Roots, and Tuber Crop Wastes

Roots and tubers are crops as a food source of carbohydrates and energy widely harvested (Figs. 16.1, 16.2, and 16.3).

Therefore, management and valorization of roots and tuber wastes should be carried out with a multi, inter and transdisciplinary research area as strategic crops, mainly in the poorest regions of the world, to achieve food safety and sovereignty,

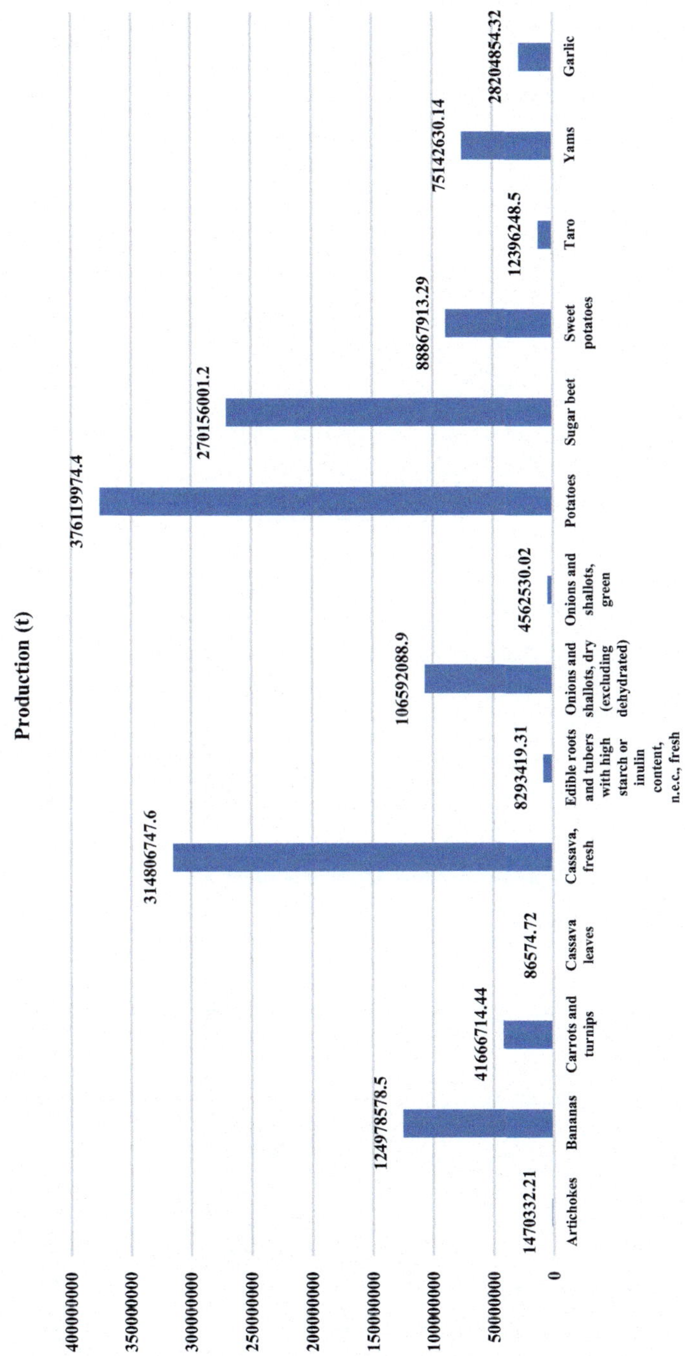

Fig. 16.1 Worldwide production (t) of root and tubers. (Data from FAOSTAT 2023, https://www.fao.org/faostat/en/#data/QCL)

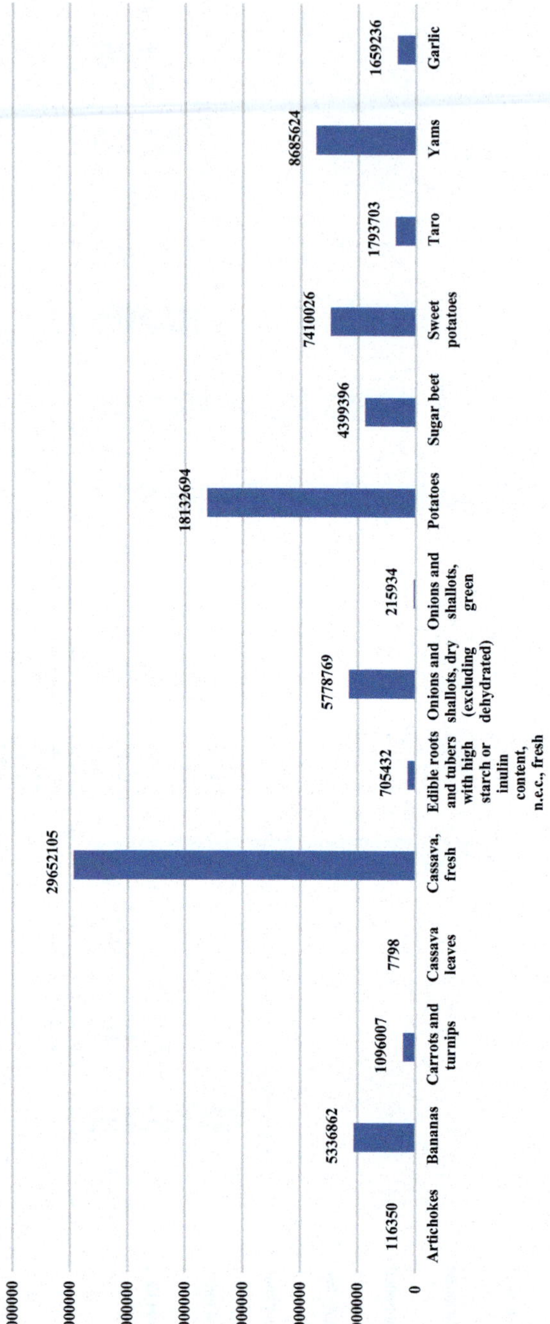

Fig. 16.2 Worldwide harvested area (ha) of roots and tubers. (Data from FAOSTAT 2023, https://www.fao.org/faostat/en/#data/QCL)

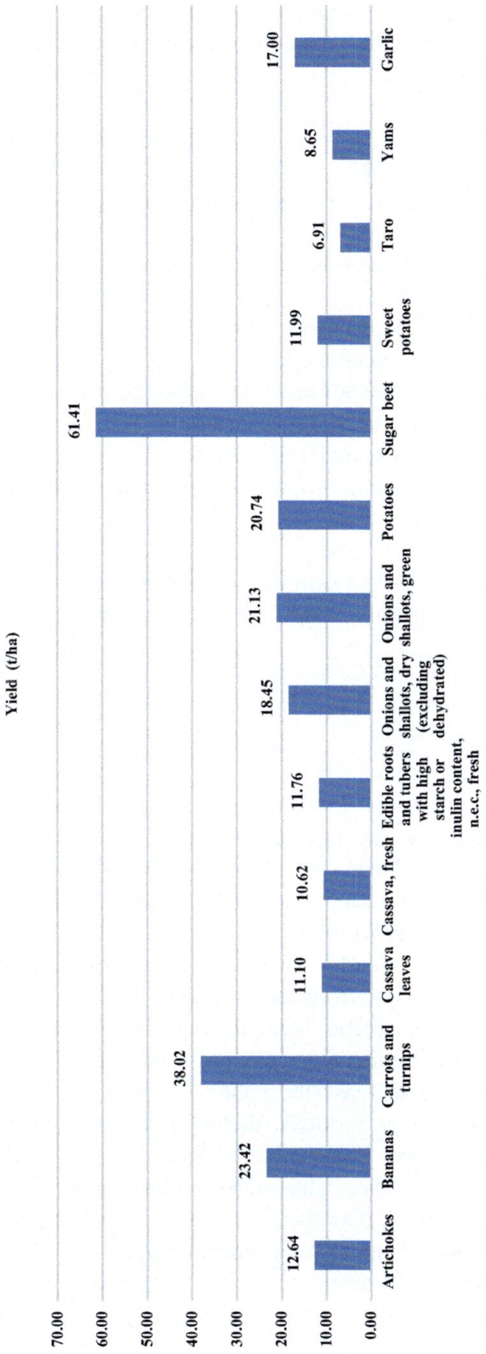

Fig. 16.3 Worldwide productivity (t ha) of roots and tubers. (Data from FAOSTAT 2023, https://www.fao.org/faostat/en/#data/QCL)

access to new renewable raw materials and obtaining bioproducts with sustainable practices and with the principles of circular bioeconomy.

Nevertheless, roots and tubers as subterranean species, according to Ali and Bhattacharjee (2023) are:

1. Crops that have known native uses in specific localized areas with scientific or ethnobotanical evidence of their nutritional value.
2. Adaptation to agroecological niches or marginal areas and representation by ecotypes/landraces.
3. These crops must be grown less frequently as compared to traditional staple crops.
4. Adaptation and cultivation are based on indigenous knowledge and practices.
5. Rare depiction in the ex-situ collection.
6. Grain supply networks are either underdeveloped or non-existent.
7. They have received scant attention from scientists, extension agents, farmers, policymakers, technologists, and consumers.
8. These crops might be extremely nutritive and/or possess therapeutic potential with other multiple uses.

This section briefly provides a sum of the valorization pathways for wastes and bagasse generated from the harvesting and processing of the main bulb, root, and tuber crops.

16.2.1 Potato

The potato (*Solanum tuberosum* L.) is one of the world's most widely grown and consumed food crops, and its production has significantly increased in recent years. Its industrial processing produces several waste products that negatively impact the environment; potato peel stands out among them, accounting for up to 10% of all the potato waste obtained after processing (Pathak et al. 2018; Abdelraof et al. 2019). Potato peel is a valuable source of bioactive substances that can be processed to be converted into value-added bioproducts. It is one of the essential substrates for industrial aspects such as cosmetic, medicinal, bioenergy, etc., to make economically profitable and environmentally safe industries. The integrated biorefinery model for potato peel can increase the value of this waste by producing several value-added products such as biofuels, dietary fiber, biofertilizers, antioxidants, organic acids, livestock feed, bioadsorbents, biopolymers, biogas, biocomposite, energy storage devices, and other value-added products (Pathak et al. 2018). There are three chapters in this book (Chaps. 2–4) that exclusively deal with valorization options of potato peel waste in a circular bioeconomy.

The valorization options of potato peel wastes are illustrated in Fig. 16.4.

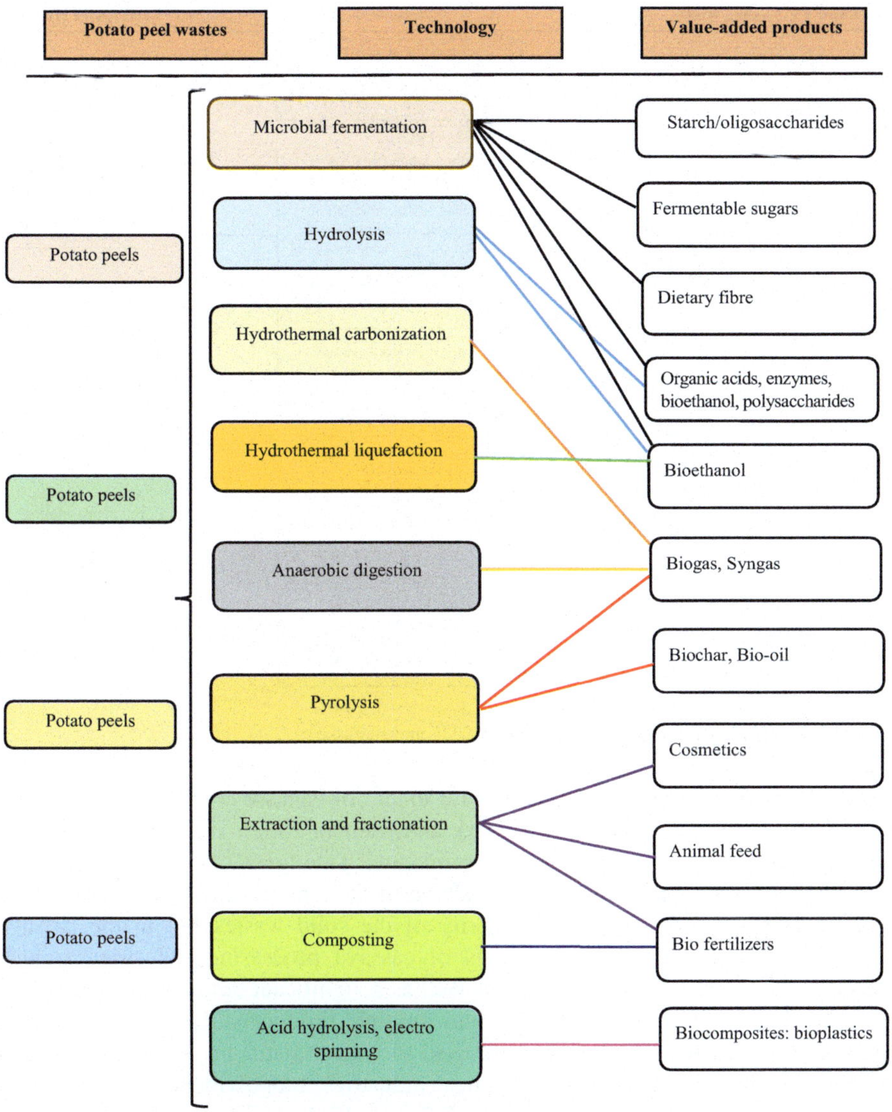

Fig. 16.4 Valorization options for potato peel wastes

16.2.2 Cassava

Cassava (*Manihot esculenta* Crantz) is a major staple food crop in tropical and sub-tropical countries. It is the third most significant food carbohydrate source after rice and maize in the tropics and the fifth most common starchy crop produced worldwide. The cassava-based industry produces large amounts of cassava bagasse (cassava pulp) every day, which ferments uncontrollably and emits foul odors.

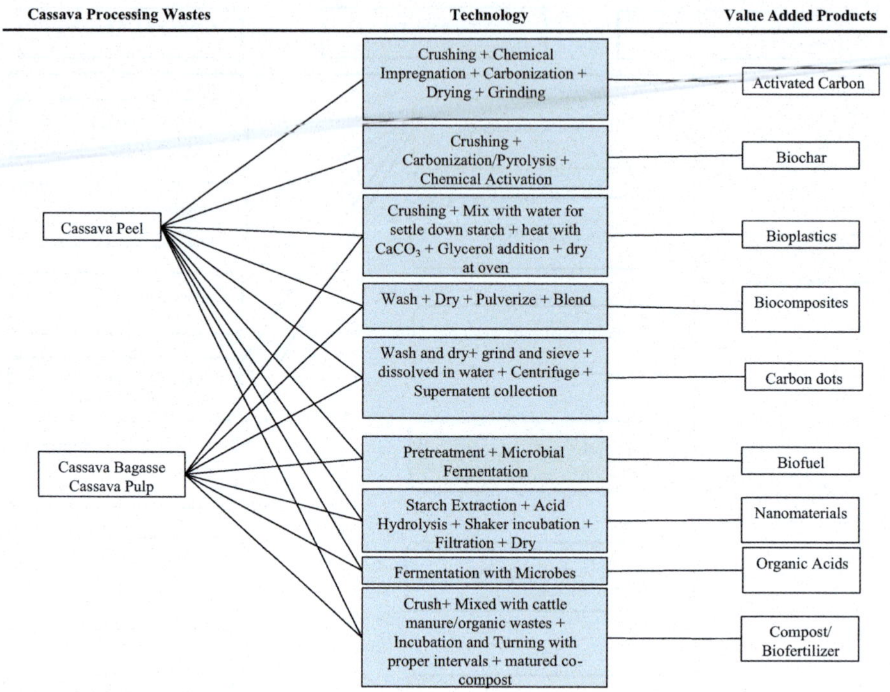

Fig. 16.5 Valorization options of cassava wastes and bagasse

About 0.38 tonnes of cassava peel and 1.4 tonnes of bagasse are produced for every tonne of cassava starch produced (Nizzy and Kannan 2022). However, a relatively small quantity of peels is used to feed ruminants. Two types of waste are produced by cassava processing industries: solid waste and liquid waste. Cassava stalks, leaves, and straw make up the majority of the solid wastes left in the cassava cultivating field, and the by-products of cassava processing industries include bagasse, bark, bran, peel, and fiber, as well as a significant amount of spent water. Cassava bagasse has been valorized into multiple products such as bioadsorbents and activated carbon biochar that are used in water treatment, antimicrobials and nanomaterials used in the pharmaceutical industries, biopolymers, biocomposites, bioplastics, and organic acids (Ray and Swain 2011; Nizzy and Kannan 2022), vermicompost and biofertilizer in organic farming, bioethanol, biobutanol, biohydrogen, biomethane, and volatile fatty acids in the biofuel industry, etc. (Pandey et al. 2000).

The three chapters in this book (Chaps. 5, 6, and 15) deal with the valorization options of cassava bagasse and wastes in biorefineries. The valorization options are illustrated in Fig. 16.5.

16.2.3 Sweet Potato

Sweet potato (*Ipomoea batatas* L.) is the seventh most important food crop in the world (Ray and Tomlins 2010). Sweet potato outranks most carbohydrate foods in vitamins, minerals, dietary fiber, and protein content. The processing of this root tuber into starch, alcoholic beverages, ethanol, flour, purees, food colorants, and other value-added products (Ray and Tomlins 2010; El Sheikha and Ray 2017) generates by-products that currently have little utilization and go to waste into landfills and water bodies (Akoetey et al. 2017). Leaves, vines, and tuber peels contain polyphenols and carotenoids that could be used as antioxidants and food colorants. Sweet potato wastes generated during starch processing are rich sources of many functional bioactive compounds such as polyphenols, vitamins, fibers, enzymes, bio-oils, etc., which can be recovered through several extraction methods (Akoetey et al. 2017; Vithu et al. 2019). These valuable materials have been valorized into value-added products especially meant for many health-promoting, pharmaceuticals, and food applications (Ray and Tomlins 2010). Chapter 7 in this book exclusively deals with the valorization processes of sweet potato wastes and bagasse.

The valorization options of sweet potato wastes and bagasse are given in Fig. 16.6.

16.2.4 Yams and Aroids

Yams are starchy root vegetables belonging to the *Dioscorea* genus and Dioscoreaceae family. The four common edible species are *D. rotundata*, *D. alata*, *D. esculenta*, and *D. hispida*. They are widely grown in tropical and subtropical regions of Africa, Asia, Latin America, and the Pacific Islands. The aroids are an important group of indigenous tuber crops, grown widely for their leaves, petioles, stolons, corms, and cormels primarily include elephant foot yams [*Amorphophallus paeoniifolius* (Dennst.) Nicolson], taro (*Colocasia esculenta* var. *esculenta* and *C. esculenta* var. *antiqurom*), and tannia (*Xanthosoma sagittifolium* (L.) Schott). These crops generate a huge amount of peel waste, which otherwise valorized land in landfills and composting. However, there are several options for valorizing the yams and aroid wastes into compost and vermicompost, bioethanol, biogas, energy generation, bioremediation, production of enzymes and organic acids, single-cell protein, animal feed, and food additives (Akponah and Akpomie 2011; Makinde and Odokuma 2015; Amenaghawon et al. 2023) (Chap. 8 in this book).

The various valorization options for yams and aroids are given in Fig. 16.7.

16.2.5 Sugar Beet and Beetroot

Sugar beet comes from the same family as beetroots. However, it is genetically and nutritionally different. Sugar beets are white, and manufacturers tend to use them for

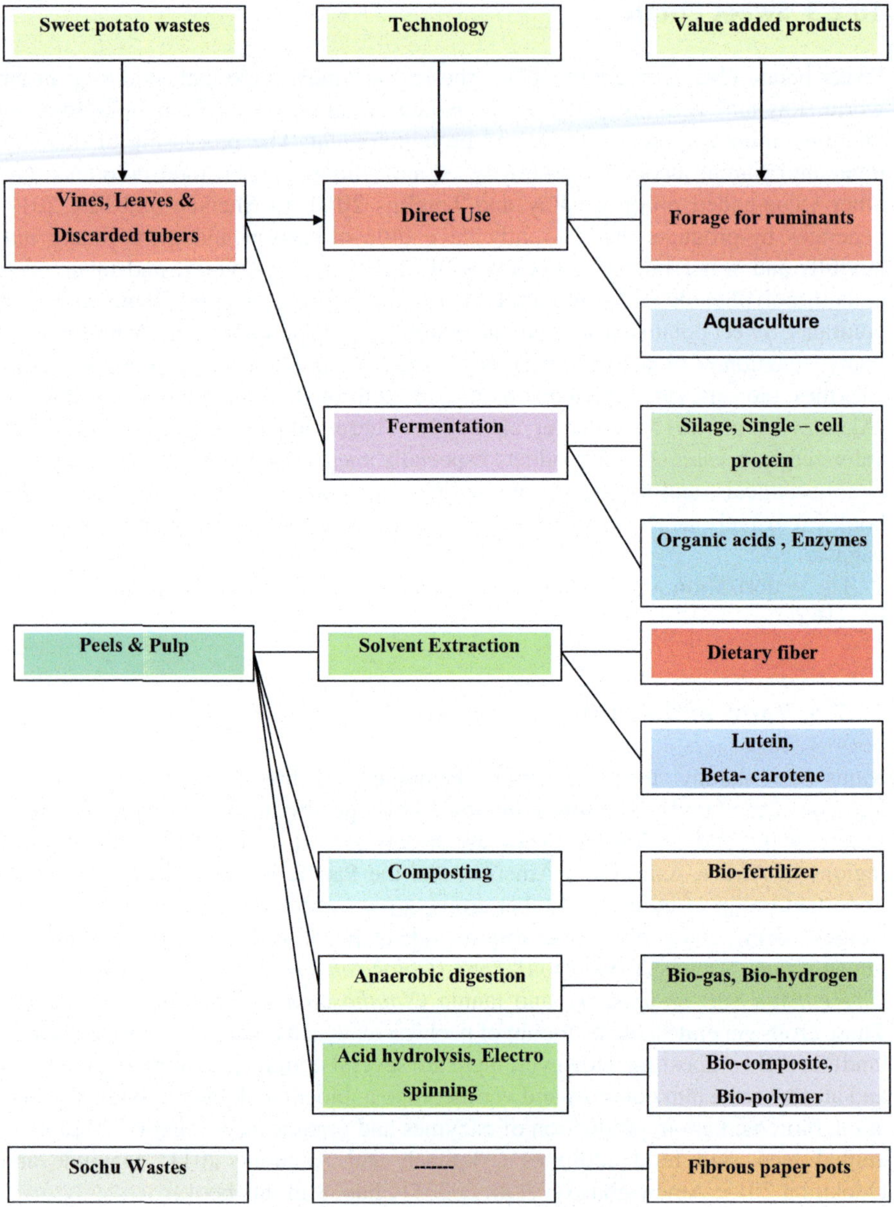

Fig. 16.6 Valorization options of sweet potato wastes and bagasse

extracting sugar and sweetening processed foods. It is not possible to extract sugar from beetroot, which is mostly red or gold.

Sugar beet (*Beta vulgaris* ssp. *vulgaris*) is a major crop used for large-scale sugar production. During sugar beet processing to sugar, two types of wastes are produced,

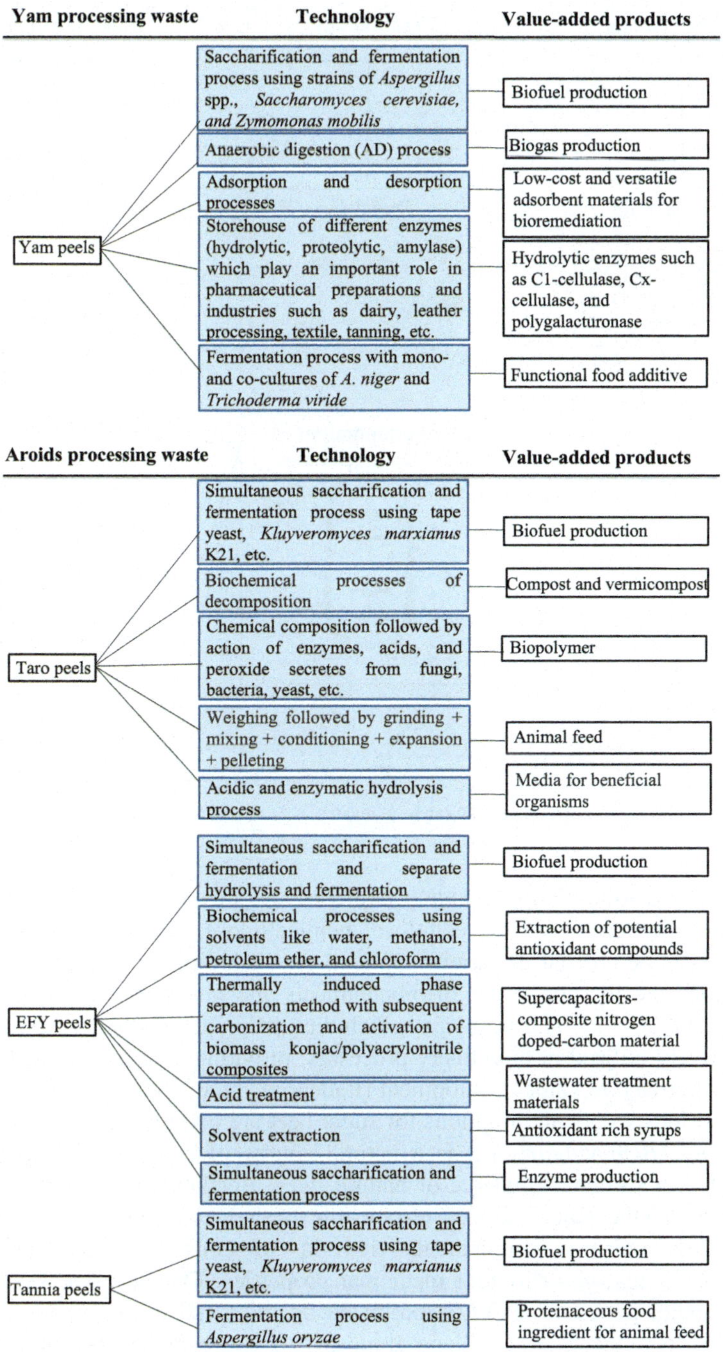

Fig. 16.7 Valorization options of yams and aroid wastes

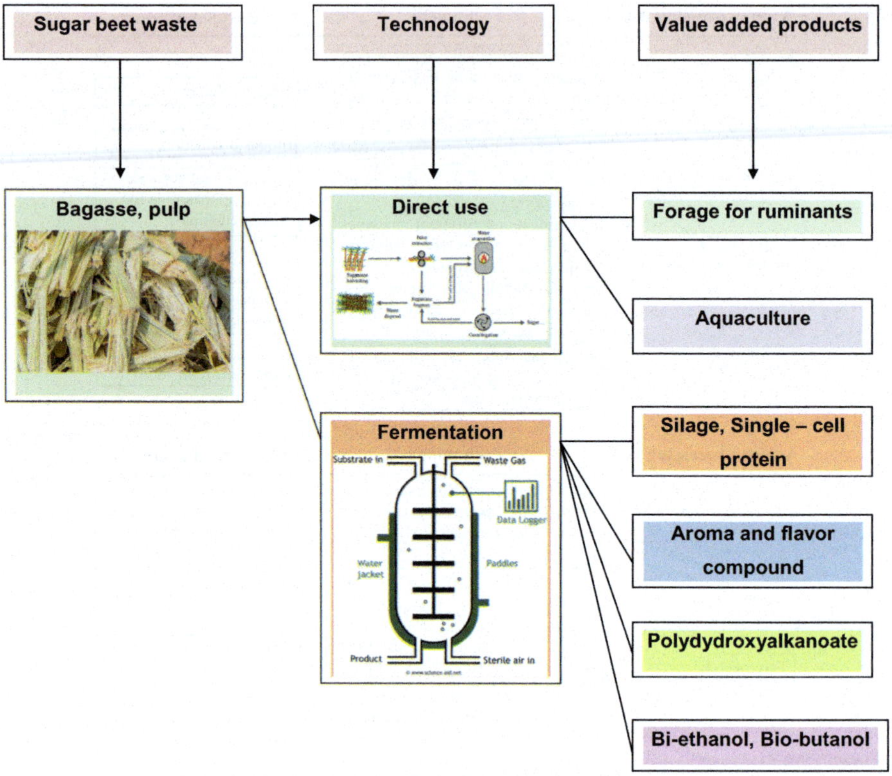

Fig. 16.8 Valorization options for sugar beet wastes

i.e., liquid and solid waste, including molasses and sugar beet pulp, which contain valuable compounds such as cellulose, hemicellulose, and pectin. These waste materials can be utilized in various applications, including animal feed, bioethanol production, and the generation of biochemicals and pharmaceutical intermediates (Cosyn et al. 2011). Sugar beet waste, including pulp and residual components, can be effectively utilized in biorefinery processes, providing a long-term and comprehensive approach to value development (Ladenović et al. 2016).

The various valorization options for sugar beet are given in Fig. 16.8.

Beetroot (*Beta vulgaris* L.) is a natural source of color and holds immense potential as a sustainable source of natural dyes. Betalains, present in abundance in beetroot, are the major pigments responsible for their color. In addition, beetroot is an excellent dietary supplement, rich in minerals, vitamins, and many phytoconstituents with different medicinal properties. The beetroot waste, which includes pulp, peel, and stalks, is produced by industrial juice, jams, drinks, and sugar beet manufacturers (Chhikara et al. 2019). The valorization of this beetroot waste for the extraction of natural colors can contribute to its valuable use in addition to the reduction of this waste.

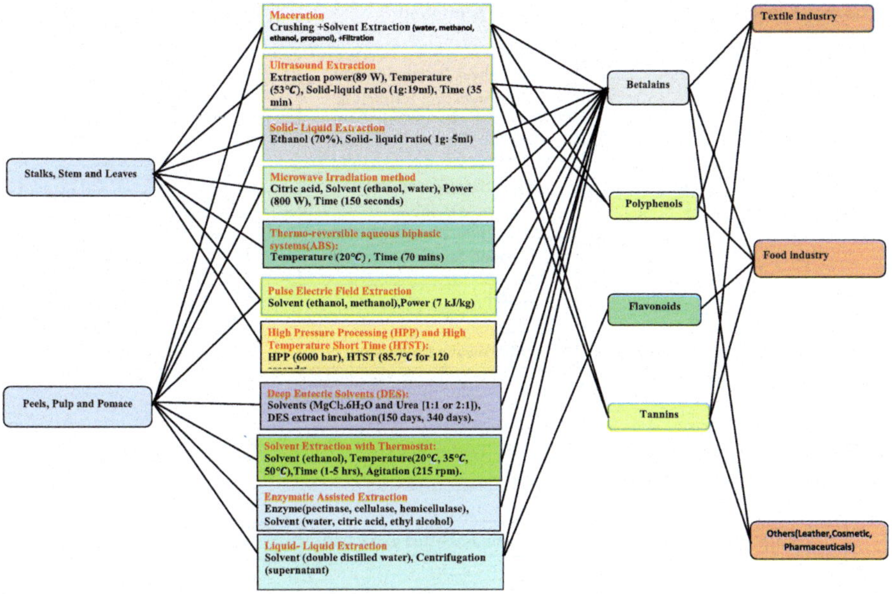

Fig. 16.9 Valorization options of beetroot wastes for extraction of natural dyes

The various valorization options for beetroot wastes for extraction of natural dyes and applications in food industries are given in Fig. 16.9.

16.2.6 Carrots and Turnips

Carrots (*Daucus carota* var. *sativus*) and turnips ((*Brassica rapa* L. ssp. *rapa*) are the two important root vegetables in the world. The underused wastes (seed, peel, rind, and pomace) of these crops have the potential to be excellent sources of beneficial bioactive substances, such as functional nutrients, amylopectin, phytochemicals, vitamins, enzymes, dietary fibers, and oils (Agcam et al. 2021). Fruit and vegetable wastes are among the most common agro-industrial wastes (Ray 2022). The discarded biomass from carrots and turnips can be used to create biofuels such as bioethanol, biomethane, biohydrogen, and biobutanol (Chap. 9 in this book).

Carrot and turnip waste can also be used as raw material to recover bioactive substances such as antioxidants, enzymes, or antibiotics, as well as to create industrially useful organic acids such as acetic acid, lactic acid, and citric acid (Coman et al. 2020).

The valorization options of carrot and turnip wastes are given in Fig. 16.10.

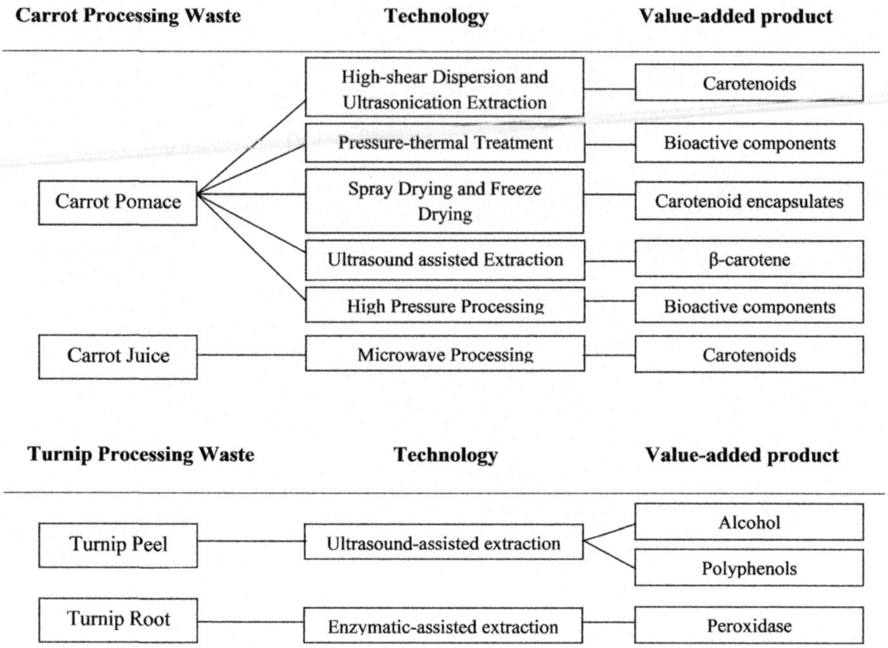

Fig. 16.10 Valorization options of carrots and turnip wastes

16.2.7 Jerusalem Artichoke

The Jerusalem artichoke (*Helianthus tuberosus* L.) is a plant species native to North America (Yang et al. 2015). The valorization of industrial by-products from Jerusalem artichoke using environmentally friendly green technologies has emerged as a significant research area in recent years. Green technologies, including enzymatic hydrolysis, fermentation, extraction, and biorefinery processes, offer sustainable and efficient pathways to convert these by-products into valuable materials. Enzymatic hydrolysis of residual biomass can yield sugars, serving as feedstock for producing biofuels, organic acids, and other biochemicals. Microbial strains can transform the extracted compounds into products such as single-cell oil, single-cell proteins, and 2,3-butanediol through fermentation. Additionally, extraction techniques enable the recovery of bioactive compounds with applications in the food, pharmaceutical, and cosmetic industries.

The valorization options of Jerusalem artichoke tuber wastes are given in Fig. 16.11.

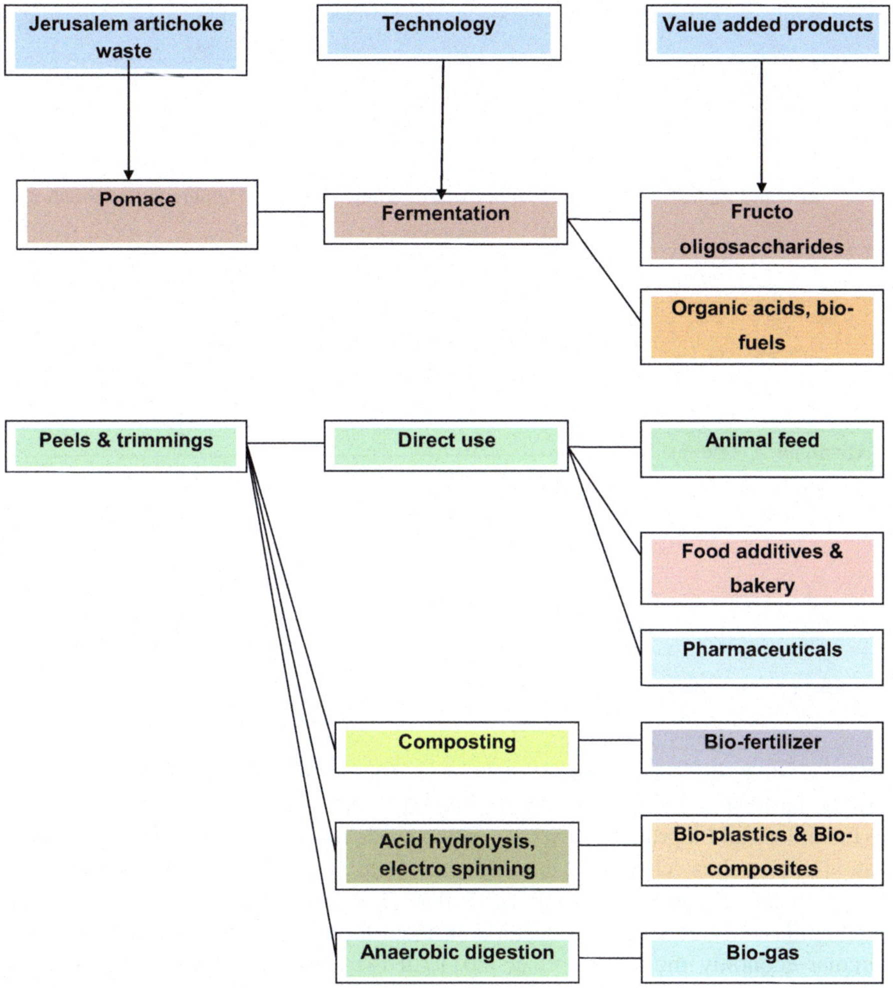

Fig. 16.11 Valorization options of Jerusalem artichoke tuber wastes

16.2.8 Onion

Onion (*Allium cepa* L.) is a highly valued vegetable and a therapeutic herb that originated in Central Asia and is cultivated globally second to tomato. While processing onions, a large amount of waste, mostly in the form of peels, fleshy scales, roots, bulbs, etc., are left behind, which are rich sources of bioactive compounds (Benítez et al. 2011, 2012) which can be extracted through suitable technologies for the development of value-added products (Cosa et al. 2006). The onion waste has certain problems in terms of disposal: (1) they cannot be used as fodder due to specific or strong aroma and toxicity, (2) because of high sulfur-containing compounds and high susceptibility for the growth of phytopathogen, limited for composting, and (3) due to high moisture content, the combustion of

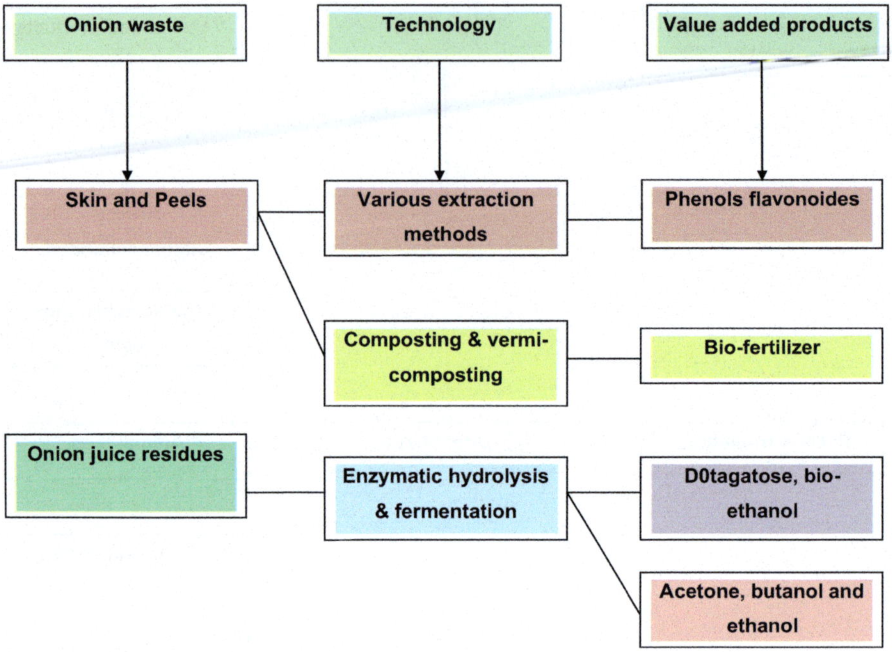

Fig. 16.12 Valorization options for onion wastes

onion waste is quite expensive (Salem et al. 2023). Nevertheless, onion waste can also be fermented for the production of bioethanol and, therefore, holds great value to the industrial bioprocessing sector. The various innovative, contemporary valorization approaches (chemical, thermal, enzymatic, and biotechnological, etc.) for the extraction of functional components from onion waste (Jin et al. 2011) and the production of value-added products from them for sustainable growth through the circular economy mode (Lee et al. 2011, 2014; Choi et al. 2015) (Chap. 13 in this book).

The valorization options of onion wastes are given in Fig. 16.12.

16.2.9 Garlic

Garlic (*Allium sativum* L.) belongs to the family *Alliaceae* and is the second most widely cultivated *Allium* species. Next to onion (Rubatzky and Yamaguchi 1997). Garlic bulbs yield approximately 760 g/kg cloves and 240 g/kg outer and inner husks (Qureshi et al. 1983). Thus, garlic by-products, which comprise skins, outer leaves, tops, and bottoms, could hold great strength to be used as a good biological resource, and their valorization might constitute an efficient, profitable, and environment-friendly alternative for residues (Makris et al. 2007). The physicochemical and biological characterization of garlic by-products, such as straw and peels, provides various applications in different domains and unveils their potential for industrial

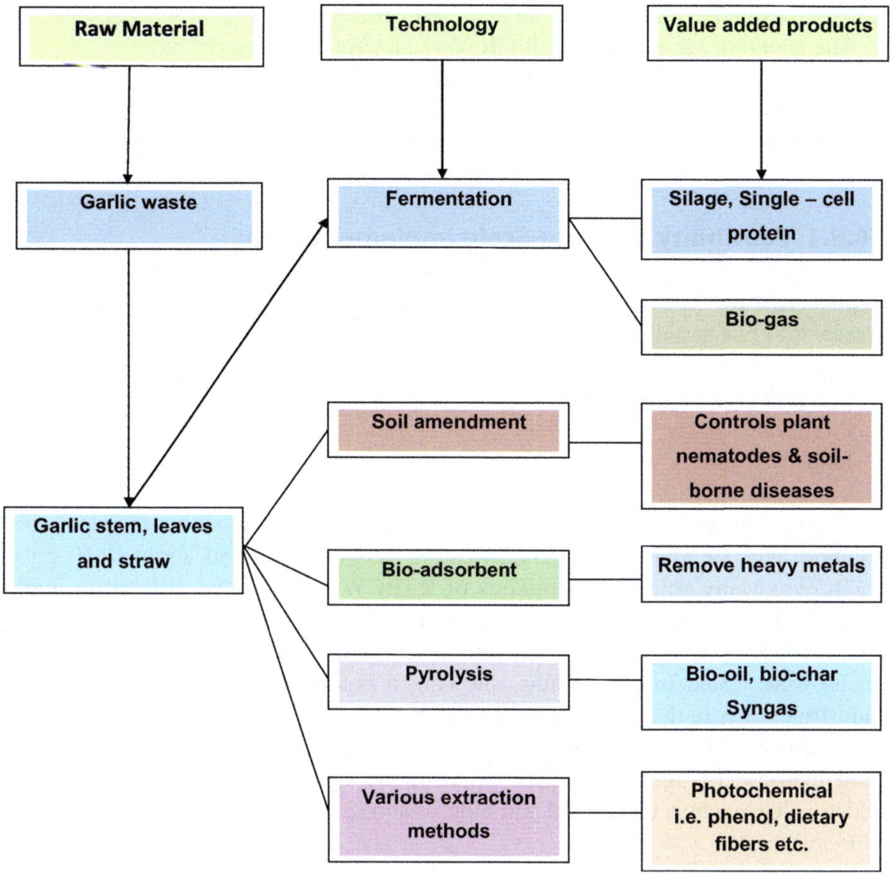

Fig. 16.13 Valorization options for garlic wastes

applications such as feed alimentation, soil amendment, and bioenergy (Kallel et al. 2015). These by-products could be exploited for further industrial purposes such as the extraction of bioactive components and extraction of functional ingredients, the production of phenolic compounds and biofuel, as well as their applications as bioadsorbents for wastewater treatment (Chap. 14 in this book).

The valorization options of garlic wastes are given in Fig. 16.13.

16.3 Techno-economic Assessment

As illustrated in the previous sections, several alternatives exist for the valorization of RTBCWs and bagasse. To explore the implementations in biorefineries, relevant techno-economic aspects need to be considered (Panda et al. 2018; Caldeira et al. 2020), including

- The feasibility of implementing at an industrial scale
- The economic assessment of RTBCWs valorization pathways
- The crop waste supply chain management and logistics
- Consumers acceptability
- The market price of waste-valorized products versus conventional products

16.3.1 Feasibility of Large-Scale Implementations

In general, three types of equipment are used to study the valorization of agricultural wastes into bio-based products.

- Equipment at the laboratory scale
- Pilot scale
- Industrial scale

Most of the studies reviewed tested the bulb, roots, and tuber waste valorization solution at a laboratory scale, i.e., treating between 1 g and about 5–10 kg of RTBCWs. Many valorization options of RTBCWs were found difficult to implement at a larger scale and could not be upgraded to a pilot scale except composting and vermicomposting. Other valorization solutions (i.e., cassava bagasse, potato peels) were tested in pilot plants, showing a potential to be implemented at an industrial scale. In the case of cassava bagasse, the diverse studies aimed at seeking a viable use for cassava bagasse can be ranked into three large areas: (1) use in the food industry, (2) biotechnological applications to generate products with greater added value, such as bioethanol and biogas, and (3) use as a reinforcement material in biomaterials.

16.3.2 Economic Assessment of RTBCWs Valorization Pathways

Very few articles have performed an economic assessment of RTBCWs valorization scenarios using economic indicators such as the return on the investment and the payback period. Only a few studies applied to Life Cycle Costing to identify the benefits of some valorized products. To perform a comprehensive economic assessment of RTBCWs valorization pathways, it is imperative to know whether such wastes are generated in sufficient quantities, whether seasonal or round–the–year availability, proximity to the potential location of the biorefinery, energy cost, etc.

16.3.3 RTBCW Supply Chain Management and Logistics

Logistic issues of RTBCWs valorization have a relevant impact on the profitability of the conversion of some substrates.

Although relatively few studies have adapted Life Cycle Analysis (LCA) when investigating new RTBCWs valorization processes, the studies reviewed demonstrate that the assessment of the environmental performance of novel valorization technologies employing LCA can prove extremely useful to identify potential trade-offs in the design phase and ensure environmentally beneficial choices are made.

16.3.4 Consumers' Acceptability

Consumers' acceptability is very important, especially for the products to be used as food supplements, cosmetics, and pharmaceutical uses. Rigorous guidelines and respective government protocols are being followed before launching the product in the markets with proper notifications. The ultimate factor lies in the consumers' acceptability of the products.

16.3.5 The Market Price of Waste-Valorized Products Versus Conventional Products

The market price of waste-valorized products should be competitive and may be lower than conventional products, especially when used in food and cosmetics.

16.4 4. Limitations and Challenges

Chapters 1–15 showcase several potential options for the valorization of RTBCWs, illustrating their high potential to contribute to the development of a circular bioeconomy. Most of the processes reported in these studies analyzed are developed at a low technology readiness level and, therefore, lacking information on their performance at the industrial level. Other challenges are the availability of year-round feedstock, the putative nature of the feedstock, logistic issues, product cost, and marketing.

It is generally assumed that the valorization of agricultural wastes (including RTBCWs) is expected to bring economic and environmental benefits as the majority of them are land filled with costs associated with and in which organic carbon decomposes, contributing to greenhouse gas emissions and global warming.

According to Fertahi et al. (2023), the complex socioeconomic and political issues of biomass applications, along with their challenges, opportunities, and future research perspectives, must be evaluated with various sustainability frames to promote concrete technologies transferable into an industrial level. Novel research is required on the environmental assessment of root and tuber crops, especially on wastes to help achieve their full potential as a source of bioproducts in different agroecological conditions and effective utilization of local inputs such as compost, manure, regional seeds, labor force, etc. to achieve more sustainable productivity.

Notwithstanding, the greatest challenge for the valorization of these crops is the transfer of technology to stakeholders after a complex process of research, development, and innovation into the value chain with principles of circular bioeconomy (Table 16.1).

Table 16.1 Trends in the valorization of root and tubers waste

Root and tuber	Trend in research	Reference
Potato	Techno-economic analysis for managing potato peel is still scarce. Therefore, the integration of different routes and multiple processes for potato peel value, such as extraction, anaerobic digestion, and composting, are some technological options to close the carbon cycle.	Almeida et al. (2023)
Sweet potato	The sustainable production of sweet potatoes as specialty crops with the participation of stakeholders, such as growers, researchers, and policymakers, to address economic, environmental, and social challenges related to the root can assist in mitigating hunger and malnutrition while creating opportunities, possibilities and sustainable value chain in a biorefinery	Tedesco et al. (2023)
Cassava	Emerging crops from developing countries, such as cassava could be optimized as high-value genetic resources to promote agro-ecological sustainability by improving crop resilience, reducing the use of chemicals, and minimizing environmental impact	Bacha and Iqbal (2023)
Yams and aroids	For the production and valorization of yam, aroids, and various roots and tubers, the latest techniques, such as meristem culture, in vitro conservation, cryopreservation, molecular marker-based fingerprinting, production of transgenics, etc. can be exploited for its non-food productive diversification	Nair (2023)
Beetroot	Currently, sustainable food must be organic, local, and seasonal food. According to this food tendency, beetroots, like other roots and tubers, are relatively cheap, cultivation is not complicated, can be produced locally worldwide, and have long shelf life without deterioration through the year, and their by-products are high in phytochemicals as carotenoids, phenolic acids, flavonoids, and highly bioactive water-soluble nitrogenous pigments known as betalains	Paulauskienė et al. (2023)
Carrots and turnips	They are by-products from the juice and beverage industries. The strategies to add value to chemical complexity from these wastes involve the extraction of specific components with metabolomics tools to support the identification of biorefinery strategies	Granucci et al. (2023)
Jerusalem artichoke	JA is a tuber with great potential in the pharmaceutical and gastronomic areas. Nevertheless, cultural barriers and lack of knowledge are the principal problems to use as a superfood. The nutritional compounds present health benefits and are a very low-cost food in poor regions of the world, while in more privileged regions, it is now consumed as gourmet food.	Méndez-Yáñez et al. (2022)
Onion	The demand is increasing due to medicinal properties and the food industries for onion-based products such as peeled onion, onion paste, onion flakes, onion oils, etc., with the generation of solid waste (skins, stalks, roots, petiole,	Salem et al. (2023)

(continued)

Table 16.1 (continued)

Root and tuber	Trend in research	Reference
	flowers, and damaged bulbs) with potential as a source of fiber, polyphenols, flavonoids, antioxidants, and pigments.	
Garlic	The use of garlic waste as skin, leaf, stalk, and damage bulbs to the diets of ruminants is suitable because their chemical composition helps growth and health, reducing methane emissions and improving ruminant product quality.	Ding et al. (2023)

16.5 Conclusion and Future Perspectives

Nowadays, the overdependence on a few dominant starchy crops such as rice, corn, barley, oat, wheat sorghum, and oleaginous such as soy, canola, rapeseed, sunflower, safflower, etc. is insufficient to meet the global challenges for food, feed, fiber, biofuels, and bioproducts without considering other underutilized crop species such as bulbs, roots and tubers and their wastes. Although several valorization options exist, most of them are laboratory studies, and studies analyzing their feasibilities at the industrial level are somewhat missing. Hence, relevant aspects are to be explored to assist up-scale performance for the techno-economic assessment of the biorefinery, such as feedstock availability and logistics, production cost, the market value of the product, brand values, economic and environmental aspects, and above all consumers' acceptability.

Acknowledgments The contribution from Drs. Gargi Ghosal, Deep J. Kalita, and Nizzy Albert Mariathankam are sincerely acknowledged.

References

Abdelraof M, Hasanin MS, El-Saied H (2019) Ecofriendly green conversion of potato peel wastes to high productivity bacterial cellulose. Carbohydr Polym 211:75–83. https://doi.org/10.1016/j.carbpol.2019.01.095

Agcam E, Akyildiz A, Kamat S, Balasubramaniam VM (2021) Bioactive compounds extraction from the black carrot pomace with the assistance of high-pressure processing: an optimization study. Waste Biomass Valoriz 12:5959–5977

Akoetey W, Britain MM, Morawicki RO (2017) Potential use of by-products from cultivation and processing of sweet potatoes. Ciência Rural Santa Maria 47(5):1–8

Akponah E, Akpomie OO (2011) Analysis of the suitability of yam, potato and cassava root peels for bioethanol production using *Saccharomyces cerevisiae*. Int Res J Microbiol 2(10):393–398. Available: http://www.interesjournals.org/IRJM

Ali A, Bhattacharjee B (2023) Nutrition security, constraints, and agro-diversification strategies of neglected and underutilized crops to fight global hidden hunger. Front Nutr 10:1144439. https://doi.org/10.3389/fnut.2023.1144439

Almeida PV, Gando-Ferreira LM, Quina MJ (2023) Biorefinery perspective for industrial potato peel management: technology readiness level and economic assessment. J Environ Chem Eng 11:110049. https://doi.org/10.1016/j.jece.2023.110049

Amenaghawon AN, Omoruyi BO, Kenneth I, Okedi MO, Esenogho GO, Oyefolu PK, Muojama OE, Otuya IC, Eshiemogie SO, Okoh R, Anyalewechi CL (2023) Biotechnological conversion of yam peels for enhanced citric acid production: data-driven machine learning modeling and global sensitivity analysis of the impact of metabolic stimulants. Ind Crop Prod 191:116022

Ashogbon AO (2021) Roots and tubers: functionality, health benefits, and applications. In: Handbook of cereals, pulses, roots, and tubers. CRC Press, pp 531–550. https://doi.org/10.1201/9781003155508

Bacha SAS, Iqbal B (2023) Advancing agroecological sustainability through emerging genetic approaches in crop improvement for plants. Funct Integr Genomics 23(2):145. https://doi.org/10.1007/s10142-023-01074-4

Benítez V, Mollá E, Martín-Cabrejas MA, Aguilera Y, López-Andréu FJ, Cools K et al (2011) Characterization of industrial onion wastes (*Allium cepa* L.): dietary fibre and bioactive compounds. Plant Foods Hum Nutr 66(1):48–57

Benítez V, Mollá E, Martín-Cabrejas MA, Aguilera Y, López-Andréu FJ, Esteban RM (2012) Onion (*Allium cepa* L.) by-products as source of dietary fiber: physicochemical properties and effect on serum lipid levels in high-fat fed rats. Eur Food Res Technol 234(4):617–625

Caldeira C, Vlysidisa A, Fiorea G, Laurentiisa VD, Vignalib G, Salaa S (2020) Sustainability of food waste biorefinery: a review on valorization pathways, techno-economic constraints, and environmental assessment. Bioresour Technol 312:123575

Chhikara N, Kushwaha K, Sharma P et al (2019) Bioactive compounds of beetroot and utilization in food processing industry: a critical review. Food Chem 272:192–200. https://doi.org/10.1016/J.FOODCHEM.2018.08.022

Choi IS, Cho EJ, Moon JH, Bae HJ (2015) Onion skin waste as a valorization resource for the by-products quercetin and biosugar. Food Chem 188:537–542

Coman V, Teleky BE, Mitrea L, Martau GA, Szabo K, Calinoiu LF, Vodnar DC (2020) Bioactive potential of fruit and vegetable wastes. Adv Food Nutr Res 91:157–225

Cosa P, Vlietinck AJ, Berghe DV, Maes L (2006) Anti-infective potential of natural products: how to develop a stronger in vitro 'proof-of-concept'. J Ethnopharmacol 106(3):290–302

Cosyn S, Woude KVD, Sauvenier X, Evrard J-N (2011) Sugar beet: a complement to sugar cane for sugar and ethanol production in tropical and subtropical areas. Int Sugar J 113:120–123

Ding H, Ao C, Zhang X (2023) Potential use of garlic products in ruminant feeding: a review. Anim Nutr 14:343. https://doi.org/10.1016/j.aninu.2023.04.011

El Sheikha AF, Ray RC (2017) Potential impacts of bio-processing of sweet potato: review. Crit Rev Food Sci Nutr 57 93:455–471

FAOSTAT (2023) Crops and livestock products. https://www.fao.org/faostat/en/#data/QCL

Fertahi S, Elalami D, Tayibi S, Taarji N, Lyamlouli K, Bargaz A, Barakat A (2023) The current status and challenges of biomass biorefineries in Africa: a critical review and future perspectives for bioeconomy development. Sci Total Environ 870:162001. https://doi.org/10.1016/j.scitotenv.2023.162001

Granucci N, Harris PJ, Villas-Boas SG (2023) Chemical compositions of fruit and vegetable pomaces from the beverage industries. Waste Biomass Valoriz:1–16. https://doi.org/10.1007/s12649-023-02095-7

Jin EY, Lim S, Kim SO, Park Y-S, Jang JK, Chung M-S, Park H, Shim K-S, Choi YJ (2011) Optimization of various extraction methods for quercetin from onion skin using response surface methodology. Food Sci Biotechnol 20:1727–1733

Kallel F, Driss D, Bouaziz F, Belghith L, Zouari-Ellouzi S, Chaari F, Haddar A, Ellouz Chaabouni S, Ghorbel R (2015) Polysaccharide from garlic straw: extraction, structural data, biological properties and application to beef meat preservation. RSC Adv 5:6728

Ladenović DD, Djukić-Vuković AP, Kocić-Tanackov SD, Pejin JD, Mojović LV (2016) Lactic acid production on a combined distillery stillage and sugar beet molasses substrate. J Chem Technol Biotechnol 91:2474–2479

Lee KA, Kim KT, Nah SY, Chung MS, Cho SW, Paik H-D (2011) Antimicrobial and antioxidative effects of onion peel extracted by the subcritical water. Food Sci Biotechnol 20:543–548

Lee KA, Kim KT, Kim HJ, Chung M-S, Chang P-S, Park H, Paik H-D (2014) Antioxidant activities of onion (*Allium cepa* L.) peel extracts produced by ethanol, hot water, and subcritical water extraction. Food Sci Biotechnol 23:615–621

Makinde OA, Odokuma LO (2015) Comparative study of the biogas potential of plantain and yam peels. Brazil J Sci Technol 9:354–359

Makris DP, Boskou G, Andrikopoulos NK (2007) Polyphenolic content and *in vitro* antioxidant characteristics of wine industry and other agri-food solid waste extracts. J Food Compos Anal 20:125–132

Méndez-Yáñez A, Ramos P, Morales-Quintana L (2022) Human health benefits through daily consumption of Jerusalem artichoke (*Helianthus tuberosus* L.) tubers. Horticulturae 8(7):620. https://doi.org/10.3390/horticulturae8070620

Nair KP (2023) Biotechnological potential and interventions in subterranean crops and some thoughts for future research in these crops. In: Global commercial potential of subterranean crops. Springer, Cham. https://doi.org/10.1007/978-3-031-29646-8_9

Nizzy AM, Kannan S (2022) A review on the conversion of cassava wastes into value-added products towards a sustainable environment. Environ Sci Pollut Res 29:69223–69240. https://doi.org/10.1007/s11356-022-22500-3

Panda SK, Ray RC, Mshra SS, Kayitesi K (2018) Microbial processing of fruit and vegetable wastes into potential biocommodities: a review. Crit Rev Biotechnol 38:1–16

Pandey A, Soccol CR, Nigam PS, Soccol VT, Vandenberghe LP, Mohan R (2000) Biotechnological potential of agro-industrial residues. II: cassava bagasse. Bioresour Technol 74:81–87. https://doi.org/10.1016/S0960-8524(99)00143-1

Pathak PD, Mandavgane SA, Puranik NM, Jambhulkar SJ, Kulkarni BD (2018) Valorization of potato peel: a biorefinery approach. Crit Rev Biotechnol 38(2):218–230. https://doi.org/10.1080/07388551.2017.1331337

Paulauskienė A, Šileikienė D, Karklelienė R, Tarasevičienė Ž, Česonienė L (2023) Quality research of the beetroots (*Beta vulgaris* L., ssp. *vulgaris* var. *Conditiva* Alef.) grown in different farming systems applying chemical and holistic research methods. Sustainability 15(9):7102. https://doi.org/10.3390/su15097102

Qureshi AA, Din ZZ, Abuimeileh N, Burger WC, Ahmad Y, Elson CE (1983) Suppression of avian hepatic lipid metabolism by solvent extracts of garlic: impact on serum lipids. J Nutr 113:1746–1755

Ray RC (2022) Fruits and vegetable wastes: valorization into bioproducts and platform chemicals. Springer, Singapore, 448p

Ray RC, Swain MR (2011) Bio-ethanol, bioplastics, and other fermented industrial products from cassava starch and flour. In: Colleen MP (ed) Cassava: farming, uses, and economic impacts. Nova Science Publishers Inc., Hauppauge, New York, pp 1–32

Ray RC, Tomlins KI (eds) (2010) Sweet potato: post-harvest aspects in food, feed, and industry. Nova Science Publishers, Inc., 316pp

Roy P, Mohanty AK, Dick P, Misra M (2023) A review on the challenges and choices for food waste valorization: environmental and economic impacts. ACS Environ Au 3(2):58–75. https://doi.org/10.1021/acsenvironau.2c00050

Rubatzky VE, Yamaguchi M (1997) World vegetable. In: Principles, production and nutritive values, 2nd edn. International Thomson Publishing, New York

Salem MA, Mansour HEA, Mosalam EM, El-Shiekh RA, Ezzat SM, Zayed A (2023) Valorization of by-products derived from onions and potatoes: extraction optimization, metabolic profile, outstanding bioactivities, and industrial applications. Waste Biomass Valoriz:1–36. https://doi.org/10.1007/s11157-022-09642-z

Sunitha S, Akash AU, Sheela MN, Kumar JS (2023) The water footprint of root and tuber crops. Environ Dev Sustain:1–23. https://doi.org/10.1007/s10668-023-02955-1

Tedesco D, de Almeida Moreira BR, Júnior MRB, Maeda M, da Silva RP (2023) Sustainable management of sweet potatoes: a review on practices, strategies, and opportunities in nutrition-sensitive agriculture, energy security, and quality of life. Agric Syst 210:103693. https://doi.org/10.1016/j.agsy.2023.103693

Vithu P, Dash SK, Rayaguru K (2019) Post-harvest processing and utilization of sweet potato: a review. Food Rev Int 35:726. https://doi.org/10.1080/87559129.2019.1600540

Yang L, He QS, Corscadden K, Udenigwe CC (2015) The prospects of Jerusalem artichoke in functional food ingredients and bioenergy production. Biotechnol Rep 5:77–88

GPSR Compliance

The European Union's (EU) General Product Safety Regulation (GPSR) is a set of rules that requires consumer products to be safe and our obligations to ensure this.

If you have any concerns about our products, you can contact us on ProductSafety@springernature.com

In case Publisher is established outside the EU, the EU authorized representative is:

Springer Nature Customer Service Center GmbH
Europaplatz 3
69115 Heidelberg, Germany

The manufacturer's authorised representative in the EU is Springer Nature Customer Service Centre GmbH, Europaplatz 3, 69115 Heidelberg, Germany. If you have any concerns regarding our products, please contact ProductSafety@springernature.com

Printed and bound by CPI Group (UK) Ltd, Croydon, CR0 4YY
01/10/2026
02225900-0014